VERTEBRATE STRUCTURES AND FUNCTIONS

Readings from
SCIENTIFIC AMERICAN

VERTEBRATE STRUCTURES AND FUNCTIONS

With Introductions by
Norman K. Wessells
Stanford University

W. H. Freeman and Company
San Francisco

Library of Congress Cataloging in Publication Data

Wessells, Norman K   comp.
  Vertebrate structures and functions.

  Edition for 1969 published under title:
Vertebrate adaptations.
  Bibliography: p.
  1. Vertebrates. 2. Adaptation (Biology)
I. Scientific American. II. Title.
[DNLM: 1. Adaptation, Physiological—Collected
works. 2. Vertebrates—Collected works. QH546
V567 1973]
QL605.W42 1974   596′.05   73–17004
ISBN 0–7167–0890–6
ISBN 0–7167–0889–2 (pbk.)

All of the SCIENTIFIC AMERICAN articles in
VERTEBRATE STRUCTURES AND FUNCTIONS are
available as separate Offprints. For a complete list of
more than 950 articles now available as Offprints, write
to W. H. Freeman and Company, 660 Market Street,
San Francisco, California 94104.

Printed in the United States of America

10 9 8 7 6 5 4 3 2 1

The unprecedented triumphs of molecular biology since the 1950's tend to make us forget the broad and ancient base on which modern life sciences are built. As we succeed in constructing mechanistic interpretations of life processes, it is sobering to realize how perceptive and accurate were the observations of those phenomena made hundreds of years ago. Though the following quotation was written by Pliny the Elder 1900 years ago, nearly every topic in it is discussed in this book and is still under active investigation.

> The swiftest of all animals, not only those of the sea, is the dolphin: it is swifter than a bird and darts faster than a javelin, and were not its mouth much below its snout, almost in the middle of its belly, not a single fish would escape its speed. But nature's foresight contributes delay, because they cannot seize their prey except by turning over on their backs. . . . They have a habit of sallying out on to the land for an unascertained reason, and they do not die at once after touching earth—in fact they die more quickly if the gullet is closed up. . . . For a voice they have a moan like that of a human being. . . . The dolphin is an animal that is not only friendly to mankind but is also a lover of music, and it can be charmed by . . . the sound of the water-organ. . . . (Pliny, *Natural History*, IX, vii, viii.)

Today we believe the swiftness is due in part to loose skin that can form "standing waves" of wrinkles and so permit water to flow smoothly over the dolphin's surface without creating turbulence and drag. The sallies of dolphins or whales onto beaches that one occasionally reads of in newspapers probably stem from failure of the echo-locating navigational system to detect certain shallow, sloping sea bottoms. The "voice" is of course an integral part of the echo-locating system and is also used for communication between the remarkably "brainy" cetaceans. And, what else but "nature's forethought" (natural selection in our terms) could explain so well the origin of these adaptations or the equally marvelous ones described throughout this collection of articles from *Scientific American*?

The purpose of this book is to provide information about the structures and functions of vertebrate animals. Two themes will dominate: first, the way that our bodies "work," whether at the molecular, cellular, or organ level; second, the ways that those molecules, cells, and organs have been modified in the course of evolution to permit the fantastic diversity in bodies and behaviors that is, in a sense, the hallmark of the vertebrates.

The collection is designed to supplement courses in introductory biology and vertebrate biology. The vertebrates have been studied by generations of college students, largely because knowledge of these organisms has been deemed essential for premedical students. Recently, the traditional "comparative anatomy" courses have been influenced by the impact of molecular biology and comparative physiology. Neither students nor faculty are satisfied any longer with a narrow and specialized view of any group of organisms, even our nearest relatives; hence, courses in "vertebrate biology" have become increasingly prevalent, and, I think, the students coming from them have a much more useful understanding of animals and man.

As a supplement, this collection is designed to fill some of the gaps and answer some of the questions that might arise from a general survey of vertebrate biology. Orientation to the field and articles is provided by Introductions to the sections of the book. *Scientific American* has not of course published articles on all the subjects that might interest a student. Nevertheless, the scope is quite broad, and references to Offprints on related subjects are included. In addition, a special list of recent references for each section is found in the bibliography at the end of the book.

I have assumed that teachers and students will be able to consult the *Annual Review of Physiology, Biological Reviews*, or *Physiological Reviews*, for many useful papers that are not cited specifically here. The important book by Hochachka and Somero, *Strategies of Biochemical Adaptation*, and Gordon's second edition, *Animal Physiology*, are, to my mind, the most useful means of amplifying the many intriguing aspects of vertebrate life treated in these *Scientific American* articles.

My thanks go to Charlea Massion, friend and student in my Vertebrate Biology course in 1972, who wrote the poem found at the close of An Essay on Vertebrates. Warm thanks also go to Jeanne Kennedy for another excellent job at preparing the index.

July 1973   Norman K. Wessells

CONTENTS

Note on cross-references: References to articles included in this book are noted by the title of the article and the page on which it begins; references to articles that are available as Offprints, but are not included here, are noted by the article's title and Offprint number; references to articles published by SCIENTIFIC AMERICAN, but which are not available as Offprints, are noted by the title of the article and the month and year of its publication.

VERTEBRATE STRUCTURES AND FUNCTIONS

AN ESSAY
ON VERTEBRATES

Adaptations are specializations in structure and function that permit animals to survive under a given set of environmental conditions. Whether they are physiological or structural, adaptations are variations upon pre-existing themes: a change in a metabolic pathway occurs, causing a new form of nitrogen excretion; or the shape of a pelvic bone alters, and erect, bipedal locomotion becomes possible. Adaptations develop in the course of generations as responses of continually varying, sexually reproducing organisms to an ever changing environment. Nowhere is this process better illustrated than in the remarkable evolution of vertebrates from filter feeder to abstract thinker.

Many adaptations seem so marvelously complex that to view one individually is, at first glance, to strain credulity. But as one studies the evolution of living forms, one acquires a perspective which makes it possible to see how the individual parts of an organism have been altered to yield an animal specialized or "adapted" for some special circumstance. In this introductory essay we will attempt to gain this perspective from a brief, historical survey of the vertebrates. We begin with vertebrate origins.

One of the oldest characteristics unique to those organisms that gave rise to vertebrates was a filter-feeding apparatus with gill slits and cilia. (It is found in some primitive types of organisms alive today—adult tunicates, or sea squirts, amphioxus, and even more primitive, worm-like creatures called *Hemichordates*.) These marine organisms fed by ingesting water, detritus, and suspended food particles through the mouth. The food and detritus were then trapped in mucus, as the water left the body through gill slits located in each lateral wall of the anterior gut (the pharynx). Beating hairlike cilia propelled the stream of water.

Interestingly, the mucus used for trapping the food was secreted by cells that are thought to be a portion of the precursor of the vertebrate endocrine gland, the thyroid. Other cells of the same part of the pharynx apparently had the ability to bind iodine into organic molecules, one of which was the hormone thyroxine. This chemical has widespread effects upon both the developmental processes and the adult metabolic functions of today's fishes. Because animals as large and complex as the earliest vertebrates probably required a substantial set of control machinery, we should not be surprised that parts of the endocrine system evolved early in vertebrate history. In fact, the presence of a pituitary complex even in certain primitive chordate larvae may have been one of the crucial factors that led to the origins of free-swimming adult vertebrates.

Figure 1. A simplified view of the organs in Amphioxus. In an adult animal as many as 90 gill slits may be found on each side of the body, although only a small number is shown here for clarity. The stiff notochord and segmental muscle masses make up the swimming apparatus, and the nerve cord lies directly above the notochord. [After Wischnitzer, S., *Atlas and Dissection Guide for Comparative Anatomy*, Second Edition, W. H. Freeman and Company. Copyright © 1972.]

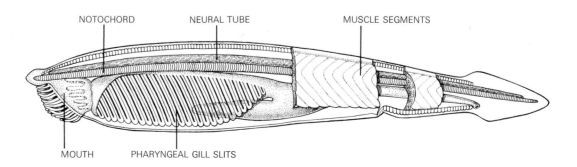

NOTOCHORD   NEURAL TUBE   MUSCLE SEGMENTS

MOUTH   PHARYNGEAL GILL SLITS

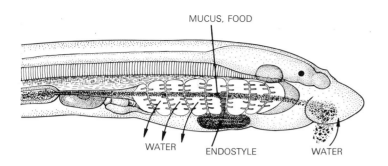

MUCUS, FOOD

WATER ENDOSTYLE WATER

Figure 2. The anterior end of a lamprey ammocoete, showing the feeding apparatus. Water, food, and detritus enter the mouth and proceed posteriorly. The water leaves through the gill slits while the food is trapped in the mucus strand rising from the ventral endostyle.
[After Young, J. Z., *The Life of Vertebrates*, Oxford, 1962.]

A second ancient characteristic of the ancestors of vertebrates was a locomotor complex that differed from that of any other organism. It consisted of segmental masses of muscle arranged down the length of the body, a stiff central rod (the notochord or, in later forms, the vertebral column) that prevented shortening of the body when muscle contraction occurred, and a nervous control system for coordination of the muscle masses. This combination of structures is seen today in larval tunicates, adult amphioxus, and all vertebrates. Even in the most primitive forms there is a hollow nerve tube, located above the notochord, that shows expansion at the anterior end into a brain. The brain had areas of specialization associated with the various senses (seeing, smelling, tasting), and it was presumably the control center that could override the basic reflex activity of the spinal cord of the central nervous system.

The next major vertebrate adaptation is a muscular, rather than ciliary, filter-feeding apparatus. It is found in the earliest vertebrate fossils, as well as in primitive organisms alive today—the larvae of both lampreys and hagfish (cyclostomes). Lamprey larvae (ammocoetes), for example, do not have cilia, but propel the feeding current by expanding and contracting the pharyngeal cavity. The action thus produced is analogous to the regular movements of the gill cover (operculum) on the sides of common higher fish; such as trout or goldfish. The "suction pump" of ammocoetes is driven by muscles in the wall of the pharynx that pull upon the skeletal support rods between the gill slits. A substantial blood supply serves the gill muscles. One might guess that, in the ancient relatives of the cyclostomes, the presence of this blood near a point of rapid water flow over a body surface (the gill slits) could well provide conditions leading to the formation of the first vertebrate respiratory organ, the "gill."

Although the transition from ciliary to muscular filter feeding does not sound very astounding, its consequences were momentous. Feeding in the new way was apparently much more efficient: larger organisms suddenly appear in the fossil record (see Figure 3), and a smaller percentage of the body is devoted to food gathering by filtering (e.g., the pharynx-gill slit complex has ten or fewer pairs of gill slits, as opposed to fifty or more in amphioxus and other ciliary filter feeders). In the Ordovician period (about 450 million years ago), there were a variety of primitive vertebrate fishes, and all of them probably fed by the muscular filter-feeding method or a variation of it. Unfortunately, we have no fossils of the intermediate forms between the ciliary filter feeders and these primitive agnathans (jawless vertebrates). In addition to the muscular filter-feeding apparatus, agnathans possessed many other "vertebrate" characteristics: bone, a skeletal material not found in invertebrates; an internal ear used in sensing orientation in space or body movement; eyes very like those of modern fishes; nasal openings, indicating the organism's ability to smell; lateral-line canals like those of the higher fishes; a light-sensing pineal organ (see page 350); and cranial nerves, some of which led to the gill regions. The circulatory system of these fishes was probably "closed"; that is, it probably consisted of an interconnected set of tubular arteries, capillaries, veins, and heart. Because of the large body size, it seems likely that the vertebrate respiratory pig-

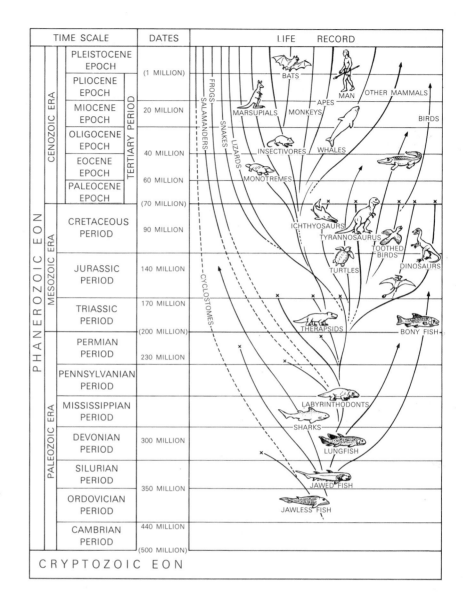

TIME SCALE			DATES	LIFE RECORD

Figure 3. The geological time scale and the appearance of various vertebrate types. The dates in parentheses are estimates, whereas the others are based on analysis of radioactive minerals in the rock strata (see "Radiocarbon Dating" by Edward S. Deevey, Jr., *Scientific American*, February, 1952; Offprint 811). The probable relationship between most of the organisms mentioned in the text is shown. Lines ending in a cross indicate extinction; those ending in an arrow indicate that descendants of that line are alive today. [After Dunbar, Carl O., *Historical Geology*, Second Edition, Wiley, 1960.]

ment, hemoglobin, had to be present. We can only guess about other systems composed of soft parts that were not preserved in the fossil record, but there is no reason not to suppose that endocrine and excretory systems, much like those of modern fish groups, were already present.

Bone deserves special comment as a vertebrate innovation. For many years, it has been assumed that bone arose due to selective pressures favoring development of hard protective armor or skeletal support materials. The British biologist Halstead has pointed out, however, that a more likely original use of bone was simply as a reservoir for phosphate, the crucial ion required for ATP production. Presence of such stored phosphate would have protected our ancient ancestors from the variations in the phosphate cycle in the sea. Summer or autumn phytoplankton "blooms," for instance, can drastically reduce dissolved phosphate. It may have been fortunate happenstance that the storage form of phosphate, the apatites, was hard and so could be employed secondarily as armor or endoskeleton.

The next major advance for the vertebrate stock was also in feeding: it was the transition from filtering food to biting it. Moveable jaws probably arose as modifications of the anterior gill support bones and their musculature. The importance of this step can hardly be exaggerated—for the first time a vertebrate could eat a large invertebrate, or large plant material, or another vetebrate! Immediately, an added complexity of food chains resulted: big fish could eat little fish, or be eaten by bigger fish. No longer were most vertebrates bottom-dwellers, scooping up mud and food. Instead, a remarkable radiation into a variety of early fishes that had jaws (the placoderms) took place. The placoderms and their descendants came to dominate the oceans and the fresh waters of the earth. Their bodies were highly variable in structure, and many new vertebrate organs evolved. Since food was ingested in morsels, a storage and digestion reservoir, the stomach, was necessary. The pursuit of prey and the escape from predators required mechanisms for controlling orientation in the three-dimensional world of water. Thus paired fin systems anterior and posterior to the center of gravity developed, as did changes in body and tail shape. Mechanisms for increasing buoyancy also appeared. As a result of such alterations, an extraordinary diversity of fishes lived in both the marine and fresh waters of the earth during the Devonian period (the "Age of Fishes") which lasted from about 320 to 265 million years ago. Among them were the ancestors of today's sharks, the Elasmobranchs, those of the higher bony fish, the Teleosts, and those of the lines that gave rise to terrestrial vertebrates.

One of the important features of the early history of bony fishes was the development of sacs extending from the anterior gut. Presumably these sacs were vascularized and were one of the respiratory organs of fresh-water fishes. In some salt-water fishes, a sac extending from the gut and similar to a lung in structure became modified into the swim bladder. This is a gas-filled organ that allows bony fishes to remain in neutral buoyancy so that they can hover in the water without expending muscular energy. Other types of fishes had primitive lungs, like those of the first amphibians who made the transition to life on land.

The next major advancements were several that occurred when vertebrates left the water and invaded the land. Adaptations for desiccation control, support of the body, locomotion, and gas exchange had to be evolved. Probably, most attempts at meeting the challenge of life on land failed. Indeed, most of the surviving descendants of the earliest land dwellers—the frogs, toads and salamanders—are even today only marginally emancipated from water.

Frogs and salamanders have moist skins and so are subject to evaporative water loss. Their kidneys are like those of fresh-water fish—they are ideally suited to eliminate large volumes of dilute urine, but hardly adapted for water conservation. Moreover, because the eggs, sperm and embryos have no protection from drying, at the time of reproduction the organisms must return to the water for breeding. But certain useful advances toward control of desiccation can be seen in these amphibians. A new type of endo-dermal bladder, a storage site for urine from which water can be resorbed, is present for the first time among vertebrates. Coincidentally the structure and function of a portion of the pituitary gland changed, causing the release of a new hormone, vasopressin. This hormone affects the permeability of certain tissues, allowing water to enter the body spaces (through the skin, bladder wall, or kidney tubules). Ready availability of vasopressin is assured because of anatomical changes that have (1) affected the localization of the nerve endings so that release of the hormone occurs in a discrete region, the pars nervosa of the posterior pituitary gland; and (2) caused the appearance of a special "portal" blood supply that drains the portion of the pituitary that stores vasopressin. Finally, although most amphibians still reproduce in bodies of fresh water, behavioral adaptations permit a few to remain away from lakes and streams during breeding. The Javanese tree frog, for example, seals its eggs between two leaves; the egg jelly liquefies and the embryos and larvae live in the pool between the leaves until metamorphosis produces a young frog.

Transition to land posed great mechanical problems for the vertebrate body. No longer was the body supported and pressed on all sides by water: instead, it rested on the ground or was held up by paired derivatives of the pectoral and pelvic fins, the legs. The vertebral column, formerly a simple

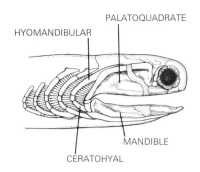

Figure 4. The jaws of an acanthodian placoderm. The upper and lower jaw bones are thought to be modified forms of the gill support bones that are shown posterior to the jaws. [After Romer, A. S., *The Vertebrate Body,* Third Edition, Saunders, 1962.]

strut for preventing compression, now tended to move dorsally so that the bulk of the body mass hung below it. Finally, the points at which weight was transferred from the vertebral column to the legs—the pectoral and pelvic girdles—were highly modified and greatly strengthened. We see all these bones in a transitional condition in fossilized early amphibians or in today's salamanders.

The problem of gas exchange on land was especially serious. Primitive lungs simply don't do the job very well. In frogs and salamanders, the work of the simple saccular lungs is supplemented by respiratory exchange across the moist skin. In frogs, about 35% of the respiratory capillaries are in the skin, while in salamanders, 70% are located there. Incidentally, this is the reason why water loss from the skin still occurs in most of today's Amphibia. In addition, Amphibia lack an efficient mechanism for filling and emptying the lungs. A frog swallows bubbles of air much as a lungfish does; unlike all the terrestrial vertebrates, it cannot expand its chest cavity to draw air into the lungs.

Thus we see that the first land vertebrates and most of their surviving amphibian relatives were only partly successful in adapting to the demands of the terrestrial habitat. It was the reptiles, descendants of the ancient amphibians, that succeeded where their predecessors had failed.

Reptiles have a dry and relatively impermeable skin. Their kidneys excrete a small volume of concentrated urine. Their nitrogen metabolism has been modified so that uric acid, rather than urea, is the main end product. Since a molecule of uric acid contains four nitrogen atoms, instead of the two present in urea, only half as many molecules (i.e., osmotically active particles) are produced for a given amount of protein catabolism. Therefore, only half as much osmotic water need be lost from the body as wastes are carried away. In addition, uric acid precipitates out of solution as its concentration is raised in the special storage organs of the reptilian body, so that even more water is freed for resorption into the body.

In the reptiles, an egg that was enclosed in a protective shell evolved. The female reproductive ducts covered the fertilized zygote with various protective layers that effectively sealed it from the atmosphere. Then, once the eggs had been laid in a terrestrial environment, they would develop if appropriately incubated so that evaporation was reduced. The male acquired a copulatory organ, which prevented the exposure of sperm and eggs to drying conditions, and insured fertilization of the egg far enough within the female reproductive tract so that there would be time for the zygote to be sealed within the protective coverings.

Other reptilian adaptations centered on the respiratory system. Because the surface of the skin, the epidermis, had become thick, dry, and dead, gases could no longer traverse it, and gas exchange necessarily became restricted to the lungs. This was made possible by the increased structural complexity and surface area of the lungs, and by the newly evolved, expandable rib cage, which filled and emptied the lungs by pumping large volumes of air to and from the respiratory surfaces.

These adaptations freed the reptile from dependence on water—except for the need to drink it—at every stage in its life cycle. Other organ systems also changed radically. The limbs and girdles were modified so that the legs were located more directly beneath the body, raising it higher off the ground. Bends and rotations of the bones at the elbow and knee allowed the limbs to move in an anterior-posterior direction, parallel with the long axis of the body, instead of in sweeping arcs to each side as did the primitive amphibian limbs that protruded outward much like fish fins. With the advent of the reptiles, there occurred for the first time great variation in the shapes and the uses of the legs as these organisms diversified and lived in new ways on the land.

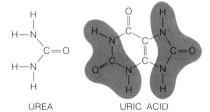

UREA   URIC ACID

Figure 5. Structure of urea and uric acid. Note that two molecules of urea occur in the uric acid. Despite its larger size, uric acid is equivalent to urea as an osmotic particle.

Just as the fishes diverged into an extensive variety of forms after they acquired jaws, so did the reptiles as they overcame the major problems of terrestrial life. During much of the Mesozoic Era (185 to 60 million years ago) they were the dominant forms of animal life on earth.

Nevertheless, reptiles had one significant limitation: their susceptibility to fluctuations in temperature. A terrestrial animal, because it is directly exposed to the atmosphere, is subject to much greater environmental variation than is a marine animal: temperature alters markedly from day to night; seasonal temperature variation in higher latitudes is dramatic; and daily changing weather conditions can produce rapid alterations in the environment. The aquatic vertebrates are not beset by such sudden variations because they are immersed in water, which warms and cools much more slowly than air. For these animals, the only rapid temperature changes would be those encountered in going from one depth of water to another. This basic difference between the two habitats directly affected the adaptations of aquatic and terrestrial vertebrates. A particular lizard so adapted that it could catch insects efficiently on warm summer afternoons might be unable to move rapidly, or function efficiently, in the cool evening or the cold winter. But fishes never having been exposed to the high temperature of the atmosphere, are physiologically adapted to their relatively constant cold surroundings and can thus function year round.

The relative inability to tolerate environmental extremes, then, was one reason why the reptiles' utilization of the terrestrial habitat was limited. Add to it the fact that large regions of the earth were, because of the cold, simply beyond their reach, and one can imagine the selective pressure that must have operated to give rise to the two temperature-regulating groups, the birds and the mammals.

All biochemical reactions are somewhat inefficient in that they generate a small amount of heat. Therefore heat is continually being produced by normal animal metabolism. To conserve this heat, birds and mammals acquired feathers or fur to insulate their bodies. Both of these types of insulators consist of dead epidermal cells that require no blood supply. (The presence of superficial vascular networks in the skin is an important cause of heat loss in all vertebrates.) The feathers or hairs are arranged in overlapping layers that trap and immobilize a layer of insulating air next to the skin. To complement these means for conserving heat, three separate control systems evolved—one for heat production, another for heat dissipation, and another for heat retention. And, interestingly, even though birds and mammals arose at different times from different types of reptiles, there developed in all of them the same basic control features that permit maintenance of a constant high body temperature.

In both of these "warm-blooded" groups, the ability to regulate temperature was dependent upon key changes in the blood vascular system and its control machinery. A high-pressure system with rapid blood flow resulted from such changes as the complete separation of the pulmonary (lung) circulatory pathway from the systemic (body) pathway. Equally vital alterations occurred in the protein (globin) portion of hemoglobin, the respiratory pigment of the blood, so that avian and mammalian hemoglobin could bind and carry oxygen at high temperatures (37 degrees centigrade); the respiratory pigment of fishes or amphibians can carry oxygen only at lower temperatures. The most important alteration in avian and mammalian hemoglobin, however, was its new lower affinity for oxygen; this allows a higher oxygen content in the body tissues, which is essential for the high rates of metabolism required by life at 37 to 40 degrees centigrade.

As a result of the many adaptations that had made temperature control possible, a variety of birds and mammals appeared. Most could be active for longer periods of the day and, in the temperate regions, for greater parts of the year; some could even live in the polar regions of the earth.

Although feathers of primitive birds were important heat insulators, we don't know whether they originated for this purpose or for that of increasing wing surface area to permit short, gliding flights. Certainly, the feathers contributed to the great difference between avian and mammalian locomotion. They provide a light surface area that makes up much of the aerofoils of wings, the control surface of the tail, and the contouring that assures efficient airflow over the body during flight.

Most organ systems of the avian body have come to resemble those of mammals, but one, the reproductive system, is very different. A bird lays a shelled egg like that of the reptiles; this type of reproduction dictates certain procedures for nesting and for feeding of the young by both parents. In the mammals, a new means of embryo incubation developed. The female reproductive tract and the hormonal system that controls its activity have been altered so that the embryo is retained and nourished within the body. Some advantages of this type of incubation are that the female is mobile during the gestation period, and that the mates do not have to remain together in order to take turns at incubating eggs. (A few species of birds are like mammals in that the female incubates the eggs and cares for the nestlings.) For the mammalian embryo, incubation within the mother's body provided a constant, high temperature. A much longer potential period of gestation became possible because of the availability of an essentially unlimited source of food—the mother's blood—for the embryo. In contrast, all the food for a bird's embryonic development must be put into the yolk before the egg shell is sealed off. An important innovation associated with mammalian embryos maturing in a uterus (and in fact with avian ones in the cleidoic egg) is the evolution of various types of "fetal" hemoglobin. This embryonic respiratory pigment invariably has a greater affinity for oxygen than does adult hemoglobin of the same species; in this way, net transfer of oxygen from maternal to fetal blood is assured.

Mammals acquired still another convenience associated with reproduction—the mammary gland and its hormonal control network. This gland provided a ready source of a constant food type (i.e., a species-specific balance of protein, fats, and carbohydrates) and eliminated the necessity of special food gathering to support early life of offspring. We think that mammary glands evolved from sweat glands, structures normally used as a means of lowering elevated body temperature. Thus an organ originally developed for one purpose, temperature regulation, was modified to serve a completely different function.

Other adaptations for life on land affected the animal's ability to gather information from the environment. Changes in the cornea and lens of the eye were particularly important because they enabled the animal living in air to focus light. In fishes the outermost structure of the eye, the cornea, has a refractive index close to that of water (the refractive index is the ratio of the velocity of light in a vacuum to the velocity of light in another medium, such as water or the cornea). Hence the cornea is of little use in bending and focusing light on the retina; consequently, it tends simply to be shaped like the side of the fish so that it offers the least resistance to movement through the water. Fishes do almost all focusing with the lens, by moving it out or in, or occasionally by changing its shape. But in terrestrial vertebrates, the cornea is the critical structure for focusing light because its refractive index (1.376) is so much greater than that of the air (1.00). As C. Ladd Prosser has pointed out, in air the cornea functions as the "coarse" adjustment on a microscope, and the lens as the "fine" adjustment.

Terrestrial adaptations of the ear affected primarily one of the chambers (the lagena) of the internal ear. The lagena has become greatly elongated in most terrestrial vertebrates (in which it is termed the cochlea), and it

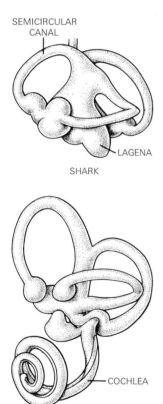

SEMICIRCULAR CANAL

LAGENA

SHARK

COCHLEA

MAMMAL

Figure 6. The internal ears of a shark and of a mammal. Note how the lagena is expanded to form the long, coiled cochlea of the terrestrial vertebrate. The semicircular canals are similar in structure and in function; in these and other vertebrates the canals are part of the apparatus that senses acceleration, deceleration, and relative position of the body in the water.

contains large numbers of sensory hairs that respond to shearing movements of certain membranes. In addition, a "middle" ear developed to amplify and transmit sound from the external ear drum to the membrane system of the internal ear. In fact, these alterations in the ear are among the most complex morphological changes that took place in the transition from aquatic to terrestrial life.

The most striking way in which mammals differ from birds is, of course, in the larger mammalian brain in which the most complex cognitive processes are carried out in the cerebral cortex. For unknown reasons, the most complex types of bird behavior are controlled by an entirely different region of the brain, the hyperstriatum (see Stettner and Matyniak, Offprint 515). In seeking to understand why the mammalian brain, and particularly that of the primates (man and his close relatives), became so large, we must examine many aspects of mammalian biology.

One factor that may have contributed to the expansion of the primate brain is the unique alteration in the reproductive activity of these animals. Virtually all other vertebrates are seasonal breeders: all females mate only when they are in a condition of estrus, or heat, and a condition in which ovulation or release of fertilizable eggs is likely; all females of a given population of a species come into estrus together, and mating occurs at such a time that the young will be born at a season propitious for their survival. In nonhuman primates (and in some other animals), we see an initial alteration in this pattern: females are asynchronous in estrus. They still mate only when ovulation is likely to occur, but as a result of the asynchrony, breeding within a population takes place throughout the year. Only in man do we see the additional step that dissociates mating from ovulation. The nervous and endocrine systems have been modified in such a way that the individual female is a potential breeder all year long. It has been argued, in fact, that this change is the most significant physiological difference between ourselves and the apes. The "competitive" societal organization of the apes could have resulted in large part from the sexual asynchrony of the females. Males repeatedly fight to establish their position in the dominance hierarchy month after month as new females enter estrus and become sexually receptive. Polygamy and harems are the rule in such societies. Mature males tend not to take part in rearing the young, and the oldest, most experienced males are often driven from the group when they are no longer dominant. In man, the dissociation of copulation from reproduction has eliminated at least one condition fostering competition between males—the constant competition for a new sex partner; stable relations between one male and one female can be established, since the female is continually available sexually to the male. It has been proposed that this change established the conditions in which the "cooperative" society of humans could appear. Thus, we can argue that the unique feature of human physiology—the behavior that most distinguishes us from other animals—is sex for pleasure alone, rather than merely for reproduction.

Had such an alteration occurred in the structure of society of prehuman anthropoids, the resultant permanent relationships between males and females, and the cooperative living between old, reproductive, and young non-reproductive animals, could have greatly stimulated development of communication systems, transmittal of experience, and other factors conducive to expanded learning capacity.

Another possible cause contributing to the expansion of the brain might be the evolution of the hand. Indeed, the development of the hand is one of the most intriguing aspects of man's own recent evolutionary history. Several reptiles and mammals (such as *Tyrannosaurus* and the kangaroo) abandoned the usual tetrapod stance for a bipedal one. The forelimbs of these animals became much smaller than those of most other terrestrial vertebrates. The one exception to this tendency has been man. This we take to be a strong piece of indirect evidence that man's primate ancestors lived in the trees. Monkeys and small primates have small, light bodies that they can balance easily above the limbs of trees. As a result, these creatures use four limbs to walk and run on *top* of the branches. It is probable that some of the ancestors of today's pongids developed a new type of locomotion—brachiation—in which they swung by their forelimbs *beneath* the branches. These creatures certainly gave rise to the gibbons of today. It is not clear whether the ancestors of other apes also passed through a brachiating condition or whether they became "knuckle walkers" like today's chimpanzees and gorillas.

Perhaps because of brachiation which put new stresses and strains on the forelimbs and pectoral girdle, and perhaps because of knuckle-walking and semi-erect locomotion on the ground, the forelimbs became accentuated in development. Witness the result in a chimp or a gibbon: the forelimbs are longer than the hindlimbs; the shoulder-girdle bones and muscles are expanded in size (perhaps to support total body weight hanging beneath the branches); and the typical mammalian "paw" has been modified for grasping branches. Several million years ago, when our ancestors gradually left the trees to become bipedal terrestrial dwellers, they probably already possessed large forelimbs and hands, which, because of the ways they were used, tended to be preserved as such, rather than shrinking to relative inconsequence as did those of *Tyrannosaurus*. With these forelimbs and hands, then, man's immediate ancestors acquired their skills in manipulating objects, and it is very possible that this increased dexterity contributed to the expansion of the brain.

So far we have argued that endocrinological and behavioral alterations in sexual activity made development of a cooperative society possible, and that the altered skeleton of hominids permitted bipedal locomotion and use of the hands in new ways. Ecological factors, too, were probably partly responsible for the changes in the societies of prehumans. It is thought that, when early man left the trees to live in the savannah forest-edge habitat, he ate both plant and animal food. Dentition, animal remains, and occasional tools or weapons that have been found support this idea. Some anthropologists have speculated that the practice of hunting large game was the key to cooperation between early hominids. Obvious advantages and consequences of hunting in groups can be imagined. Probably we

Figure 7. Comparison of Tyrannosaurus, kangaroo, and man. Note that Tyrannosaurus and the kangaroo have much smaller forelimbs relative to man.

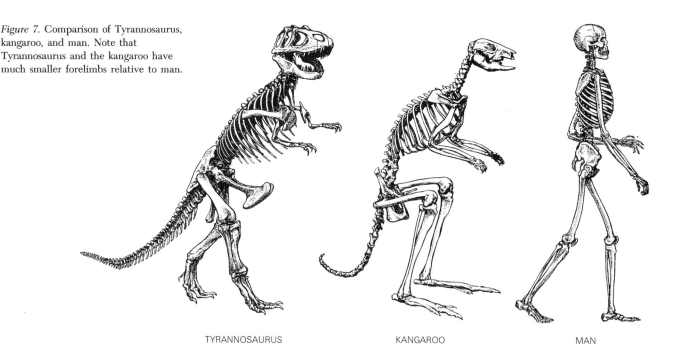

TYRANNOSAURUS   KANGAROO   MAN

shall never know whether this type of cooperative food gathering pre-dated and precipitated the altered sex behavior and family structure.

Perhaps the main conclusion to be drawn from this discussion is that cultural and biological evolution of man are intimately linked—feedback from each affects the other so that both change in time. The focus of change, of course, is the brain and its increase in size and complexity. Though the only data available at the moment are sparse, controversial, and on rodents, there is evidence of altered quantities and types of macromolecules present in the brains of animals subjected to learning experiences or enriched en-vironments. These alterations in individual animals may mirror what occurred during evolution to yield the modern primate brain. In particular, the number of brain associational neurons (intermediate neurons between the sensory input cells and the motor output nerve cells) grew until ulti-mately a new property—self-awareness or abstract thought—appeared. We cannot say, as yet, at what point the number and complexity of inter-actions between neurons became great enough to generate this property, but there is no reason to invoke any nonbiological element to explain it.

Our survey, then, ends with man. We see how an incredibly intricate series of changes, occurring in the course of millions of years, has led to the diversity of vertebrate life and to modern man. In man we can see traces of many of those steps: filter feeding, swimming and its control, feeding mech-anisms, the invasion of land, the control of high body temperature, and finally the cultural-biological interplay that led to thought; all have left their marks. Yet these are merely focal points for further study. We have looked at some questions that are still unanswered and at some speculations that are still controversial. As long as they remain so, our survey is not really at an end, but only just begun.

N.K.W.

SPECULATIONS ON A THEORY BY WESSELLS

"Thus, we can argue that the unique feature of human physiology—the behavior that most distinguishes us from other animals—is sex for pleasure alone, rather than merely for reproduction."

Although in cities we appear
to stand apart from eel and bear,
our heart is chambered like the hawk's;
our parts, connected limb to bone,
are analogs of antelopes';
and eyes and ears, your sweet tongue too,
supply the heads of skink and newt.
Placentas fed the shark and you.

Shiny fish can shift their prism;
the tree toad, switch from green to brown;
the bat and dolphin see by sound.
While geese, by light, detect direction,
by night the owl can hear to hunt.
Young chimps choose grass to capture ants.
And you, love, caught on the rim of sleep,
ask if humans are unique.

Objective, naked inspection shows,
oh not huge heads, odd thumbs or stance,
divide us from the beasts below—
instead: our vibrant body-dance.
You smile, sex for pleasure alone,
without intent, cycle or season;
the fire, caught to cross the pair,
becomes our reason dreaming here.

Charlea Massion

I

STRUCTURAL ADAPTATIONS
OF VERTEBRATE BODIES

*Almost all species except man and monkeys, both the viviparous
and the oviparous, have tails corresponding to the requirements of
their bodies, bare with the hairy species, like boars, small with the
shaggy ones, like bears, very long with the bristly, like horses. With
lizards and snakes when cut off they grow again. The tails of fishes
steer their winding courses after the manner of a rudder, and even
serve to propel them like a sort of oar by being moved to the right and
left. Actual cases of two tails are found in lizards. Oxen's tails have
a very long stem, with a tuft at the end, and in asses it is longer than
in horses, but it is bristly in beasts of burden. A lion's tail is shaggy
at the end, as with oxen and shrewmice, but not so with leopards;
foxes and wolves have a hairy tail, as have sheep, with which it is
longer. Pigs curl the tail, dogs of low breeds keep it between their legs.*

Pliny
NATURAL HISTORY, XI, CXi.

I

STRUCTURAL ADAPTATIONS OF VERTEBRATE BODIES

INTRODUCTION

Figure 1. A diagrammatic representation of a shark swimming. The bends in the body are caused by waves of contraction of the muscle masses that start at the front and proceed toward the rear of the fish. The darkened muscle masses are those that are contracted; an instant later, these would be relaxed and the next most posterior muscle masses would be contracted. The next wave of contraction would start on the opposite side (right side in the drawing) of the animal.

Life in water, life on land, life in the air—each places very different demands upon the vertebrate body. The muscle-bone complex used for support and locomotion and its neuronal control network have evolved into a bewildering variety of types. In this introduction, we will discuss adaptations for each of the environmental habitats and see how the basic body components changed in time.

The elements of vertebrate swimming are illustrated more diagrammatically in the elongate fish, such as sharks, than in the higher bony fish that are described by Sir James Gray in the first article "How Fishes Swim." When sharks are swimming, their long, thin bodies assume the shape of moving sine waves. This shape results from alternate series of contractions of the muscle masses (myotomes) that make up each side of the fish body. In swimming, a wave of muscle contraction starts in the most anterior muscle mass on one side of the body and then proceeds posteriorly, each myotome contracting in turn. Then, before the first wave proceeds too far, another wave starts at the anterior end on the opposite side of the body. The process continues, first on one side, then on the other, and the result is a bending motion like that of sine waves. This action causes each portion of the sides of the fish to be swept back and forth through the water as if it is a moving inclined plane. Forward propulsion results. In higher fishes, the sine-wave motions are much less obvious, and most of the propulsive thrust comes from the specialized tail region where amplitude of the waves is greatest.

The pattern of the chains of contraction that occur in swimming is largely controlled by nerves in the spinal cord. If the spinal cord of a dog-fish shark is severed from the brain, the organism can be made to swim. It will continue to do so for long periods. Some physiologists believe that the whole pattern of coordinated muscle activity is programmed in the network of motor nerves within the spinal cord, and that no sensory input (or exceedingly little) is necessary to coordinate the chains of contractions. Interestingly, the swimming speeds of such "spinal" animals are fairly constant: the spinal cord control system shows little capacity to alter the rate of muscle activity. If small portions of the cord's sensory input are intact, however, appropriate stimuli can evoke violent muscle contraction or faster swimming; such responses do not require presence of a brain. However, the brain (particularly the medulla) is necessary for normal initiation and control of swimming speed. Thus we may view the brain of a normal animal as the control circuit that can start, stop, or modify sets of activity in the motor nerve switchboard of the spinal cord. One might imagine, in fact, that the long neural tube (i.e., the spinal cord) originated in the pre-vertebrates as a motor control system for the segmental muscle masses.

Primitive vertebrate fishes probably propelled themselves as do today's sharks. Like sharks, they possessed asymmetric tails in which the dorsal half was longer and more massive than the ventral half. When such a tail sweeps to and fro through the water, it tends to drive water down and back; the tail, of course, reacts in the opposite direction: it is driven up and forward. The latter force provides propulsion; the former, is a "lift" at the rear end of the body. As a result, the tail tends to rise and the head to sink during swimming. This was the motion of most agnathans, and it was perhaps a useful adaptation for easy movement along the bottom of the sea

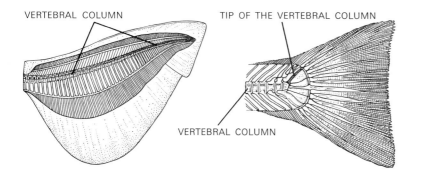

VERTEBRAL COLUMN

TIP OF THE VERTEBRAL COLUMN

VERTEBRAL COLUMN

Figure 2. The caudal "tail" fins of a shark and of a bony fish are seen in these sketches. The asymmetric shape of the shark fin results from an upward bend of the vertebral column and from a large flap of tissue that projects ventrally. The very tip of the bony fish's vertebral column also bends dorsally, perhaps reflecting its evolutionary derivation from an asymmetric tail. Nevertheless, the other bony fin supports are symmetrical so that the whole structure no longer generates a "lift," as does the shark fin.

or fresh water. For free swimming above the bottom, however, turning cartwheels in the water is hardly adaptive! Sharks overcome the difficulty with their broad head, which is flattened ventrally, and a set of wide, flat pectoral fins anterior to the center of gravity. Both flat surfaces tilt up at their anterior ends so they are like planes being pushed through the water; the result is a lift at the front that counteracts that at the rear. By controlling the angle of the anterior lift planes, the fish can direct its movement in the dorsoventral plane. Equivalent fin systems and flattened heads are seen in many types of early agnathan fishes. In addition, in those creatures and in almost all their descendants, there was at least one other pair of fins (the pelvics), located posterior to the center of gravity. The acquisition of the anterior and posterior pairs of fins was important, not only for fishes but also for terrestrial vertebrates, because the paired fins later became the limbs.

Most of the higher bony fish in the world differ from the sharks or ancestral fishes by having symmetrical tails. The reason is, no doubt, the existence of a swim bladder, which is a source of buoyancy. Unlike sharks that sink if they stop moving forward in water, bony fish can add or remove gas from their swim bladder to keep their body in neutral buoyancy with water at a given depth (i.e., they alter the total specific gravity of their body and so keep its mass exactly equal to the mass of the water it displaces). Clearly, any tendency to drive the body up or down, such as lift from an asymmetric tail, would hardly be advantageous to an organism that would otherwise be stable at one depth. So the tail fin became symmetrical and, because the tail lift was eliminated as a result, the paired anterior fins were freed for functions other than counteracting tail lift. Hence, fins came to be used as brakes, tilting rudders, or even wings. One wonders, in fact, what strange and wonderful differences might have taken place in vertebrate evolution if swim bladders and symmetric tails had been present in the earliest vertebrates so that the paired fins need never have developed.

Buoyancy considerations also have led to specialized changes in the shape of certain fishes. The reason is that protein itself is denser than water (1.33 versus 1.026 for sea water). In extreme deep-water fishes (bathypelagics), a common adaptation is a gross reduction in the size of the main body musculature (which is mostly protein in content), as well as in bone. A tiny body often seems to be associated with a relatively huge set of jaws in such weirdly shaped creatures (see Figure 3).

One of the most surprising features of all aquatic vertebrates is the high rate of speed at which some marine mammals can swim. Seemingly accurate measurements indicate speeds of 18 knots for some dolphins. As Gray points out in his article, this speed is thought to be possible because friction is reduced by laminar flow of water next to the body. Laminar flow patterns can form because the whole skin of dolphins is loose and slack. In addition, the epidermis is thought to be perforated with minute tubes.

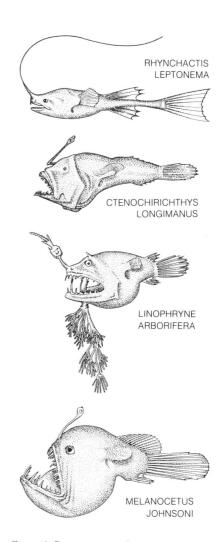

RHYNCHACTIS LEPTONEMA

CTENOCHIRICHTHYS LONGIMANUS

LINOPHRYNE ARBORIFERA

MELANOCETUS JOHNSONI

Figure 3. Representative deep-sea fishes with accentuated jaw and head regions and reduced trunk and tail musculature. The antenna-like appendages extending from the heads of these so-called "angler" fish are apparently used to attract prey. [Courtesy of E. Bertelsen and Dana Reports 39–40, 1951–1953. *Melanocetus* redrawn from Bertelsen, 1951; *Rhynchatis* and *Linophryne* redrawn from Regan, 1926; *Ctenochirichthys* redrawn from Regan and Trewavas, 1932.]

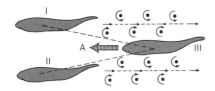

Figure 4. A view from above of three fish in a school. As the tails sweep back and forth in the water, columns of vortices of turbulent water are left in the wake of fishes I and II. Note that directly behind each fish the vortices induce a water velocity opposite to the fishes movement. But, between I and II the vortices create a "street" or channel of flow forward (A). If fish III swims in that street, it receives a boost and so conserves energy. [After D. Weihs, *Nature*, 241, 1973.]

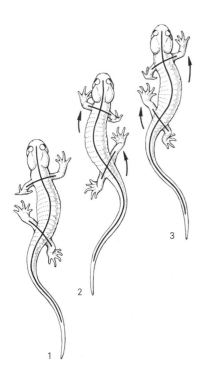

Figure 5. The body movement of a walking salamander. The vertebral column moves back and forth, as does a shark's (see Figure 1). In comparison, when a more advanced terrestrial vertebrate moves, the column remains relatively straight and the limbs move more independently to provide propulsion.
[After Romer, A. S., *The Vertebrate Body*, Third Edition, Saunders, 1962.]

When the fish rapidly accelerates or decelerates, or sustains rapid swimming, numerous standing (stationary) waves, or folds, appear in the skin; that is, unlike a stiff, smooth surface, this flexible skin becomes creased at these times. Apparently the creases in the skin conform to the waves of water flowing past the animal. Consequently, the moving water flows smoothly over the surface and does not break up into eddies and swirls that would produce drag. Without such turbulence, the animal can move forward at a much higher rate than would otherwise be possible.

Still another means of conserving energy during swimming is seen in fishes that commonly move about in schools. The biologist Evelyn Shaw has noted the great regularity of spacing in such schools. Analysis by D. Weihs of Cambridge University suggests that this spacing allows a considerable saving in swimming energy for properly placed trailing fish. An analogous explanation has been offered to explain the V-shaped formation of migratory bird flocks. Rising vortices of air are created by the descending wing beats of birds in front; trailing birds in the V fly in the resultant updrafts and so conserve energy just as schooling fish do. Both these vertebrate types are exploiting a basic physical property of the fluid in which they move. That opportunity is not available to terrestrial creatures who must thrust against the solid earth for locomotion.

The swimming motion of fishes is still seen in the most primitive terrestrial vertebrates alive today, the salamanders (see Figure 5). A salamander uses its limbs for paddling or walking slowly, but, if it is frightened, it tucks them in close to the body and swims or wriggles away by twisting its body in a series of sine-wavelike movements. As was discussed in the introductory essay on vertebrates, it was in the descendants of amphibians, the reptiles, that limbs became the primary source of movement on land.

Life on a solid substrate led to major alterations in the vertebral column and the limbs of vertebrates. In the terrestrial vertebrate, the column has shifted toward the dorsal surface, so that most of the body mass hangs below the chain of bones. Because this places great stresses on individual vertebrae, they tend to slip past each other in the dorsoventral plane. To counteract this tendency, individual vertebrae acquired various flanges of bone (zygapophyses, projections from centra, and others) that brace the column and prevent slippage. Particularly in the ancient reptiles (the therapsids) that subsequently gave rise to mammals, the limbs—which had protruded outward somewhat perpendicular to the body axis—gradually moved in under the body: the hind limb turned forward and in, and the knee bent sharply so that the foot still pointed anteriorly; the forelimb initially bent back at the shoulder, so that a compensatory twist was necessary at the elbow; thus, the forearm rotated forward 180 degrees so that the forefoot also pointed anteriorly. These various bends produced the condition from which mammalian limbs radiated. (See Milton Hildebrand's "How Animals Run.")

Recent studies of so-called "bipedal" locomotion provide interesting perspective on Napier's discussions of the origins of human walking. When a chimp walks in an upright stance, it does so primarily by straightening up the pelvis relative to the femur and leg. The leg is held in the same position as during quadrupedel knuckle walking (see Figure 7). If one observes a chimp in an upright stance from the front, the knee is held slightly lateral to the hip. Hence, the femur is said to be in an abducted position. In comparison, the femur of modern man is directed inward at its lower end, so that our knees are closer together than are the hip joints (i.e., the femur is adducted). *Australopithecines*, in contrast, held their femur in a directly vertical position or in a slightly adducted position, reflecting their closeness to modern man in this important respect.

An interesting sidelight on the upright human stance comes from P. R. Zelazo and his colleagues who have studied "parental" behavior designed to accelerate motor development related to walking among African foraging peoples. Infants are commonly carried in a vertical position in a sling or held vertically in the mother's lap. A sample of such infants shows precocity in sitting, standing, and mature walking compared to non-African controls. Other workers have found that African children brought up in the European manner do not display this precocious development. This sort of phenomenon is another form of the social-biological interplay that may well have important evolutionary consequences. We can only wonder whether such behavior contributed to the origin of human walking in the ancestors of the *Australopithecines* and individual no. 1470 (the fossil discovered by R. E. Leakey; see p. 19).

Whereas most terrestrial vertebrates display marked evolutionary modifications of their limbs, one group in particular, the reptilian snakes, has lost its limbs and depends upon the basic undulatory movements of the vertebral column to generate movement. Carl Gans ("How Snakes Move") points out, however, that even among snakes there are many distinctive ways to get from here to there: side winding, jumping, concertina, and others are variations of the underlying sequential undulations of the somatic muscle masses.

The modifications of the tetrapod limb to make possible an arboreal existence are described by Napier, with emphasis on primates and the early relatives of man. Besides significant alterations in the legs and arms of primates, there occurred a remarkable evolutionary trend in vision. Insectivores—small shrew-like animals—are close the the stem group of all modern mammals and are particularly close to the primates. Insectivores are usually nocturnal and largely dependent on the senses of smell and of high-sensitivity "rod" vision for orientation. (Rods and cones are the two types of light-sensitive cells in the neural retina of the eye). But the higher primates, perhaps as a result of life in the trees, move about principally during the day. Sight is their dominant sense, and they have many low-sensitivity "cone" photoreceptor cells and accentuated color vision.

The sharpness, or acuity, of vision of an eye is affected by such factors as the size of the photosensitive elements, the way information from such

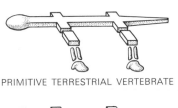

PRIMITIVE TERRESTRIAL VERTEBRATE

MAMMAL

Figure 6. Representations showing the twists that occurred during evolution of the terrestrial vertebrate limbs. The hind limb bent in at the hip; the knee and the ankle bent. The forelimb rotated back at the shoulder; the upper arm bone rotated outward, and, as the elbow bent, the outer bone of the forearm bent forward and, as this occurred, the outer bone (the posterior bone of the primitive terrestrial vertebrate, shown in the top part of the illustration) twisted over the other forearm bone so that, in the mammal, its lower end is on the inner side of the wrist. As a result, the lower surface of the forepaw is on the ground; otherwise it would face upward.
[After Ballard, W. W., *Comparative Anatomy and Embryology.* Copyright © 1964, The Ronald Press Co.]

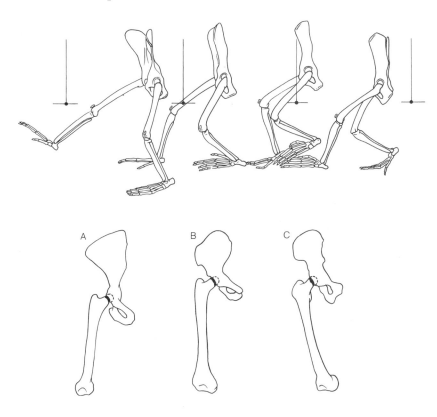

A   B   C

Figure 7. At the top are sketches of a chimpanzee's pelvic girdle and lower limbs during a single stride. Note that, in contrast to man, the legs are held well in advance of the girdle at all phases of the stride; thus, the body is not truly balanced above the limbs and feet. There is also a marked dorso-ventral movement of the pelvis (relative to the line of reference), instead of the much smoother forward motion in man. At the bottom are views from the front of one-half of the pelvic girdle and femur of (A) a chimpanzee, (B) an *Australopithecine robustus,* and (C) modern man. Note how the lower end of the femur has pointed in, toward a "knock-kneed" condition (i.e., an adducted one). This helps to bring the human foot in under the center of gravity both during a stride and when standing still.
[Courtesy of F. A. Jenkins, Jr., and *Science,* vol. 178, pp. 877–879, Figs. 1, 2; 24 November 1972. Copyright 1972 by the American Association for the Advancement of Science.]

elements is routed to the brain, and optical properties of the structures through which the light waves pass (see "Eye and Camera," *Scientific American*, August 1950; Offprint 46). In primates, acuity has been maximized by a layer of yellow pigment, the macula lutea, that developed over the fovea of the retina. The macula lutea filters out blue wavelengths of light, which are the ones that produce the greatest chromatic aberration in optical systems. The fovea was already the region of highest acuity in other mammals, because it is structured in a special way: that is, each cone cell is linked directly to an individual nerve cell that carries impulses toward the brain; elsewhere in the retina each neuron leading to the brain may receive input from many light-sensitive cells, and so such neurons "report" on a larger retinal area, thus yielding less acute vision.

Improved depth perception (i.e., binocular vision) has obvious advantages for animals that run or brachiate through the trees. Anatomical alterations that have permitted better depth perception include reduction in the snout, flattening of the face, and movement of the eyes toward one another so that the fields of vision overlap substantially. A particularly intriguing visual adaptation for depth perception was change in the routes of the optic nerve fibers between the eye and the brain. In most lower vertebrates, optic nerve fibers extend from the eyes across the optic chiasma to the opposite side of the brain. In primates, particularly, this routing has been altered so that many fibers (40 percent of them in man) from the left half of the left eye and the right half of the right eye proceed to the optic chiasma, but then turn outward again to make junctions with brain neurons on the *same* side of the body (fibers from the other halves of each eye continue to cross at the chiasma). The result is that fibers from the left half of each eye go to the left brain, and those from the right halves go to the right brain. This means that the same portions of the visual field are sent to the same side of the brain for comparison. In some still unknown way this yields an advantage in assaying depth relationships. In summary, then, alterations in the chemistry of yellow pigment production, in facial structure, and in nerve circuitry all occurred as adaptations in the visual system of the primates.

The series of primate fossils that has appeared since Raymond Dart, the pioneering anthropologist at the University of Witwatersrand, South Africa, discovered the first remains of the *Australopithecines* in 1924, has led to one surprising conclusion about the evolution of man: appearance of an upright stance and "human" hand preceded development of the large brain. In the early *Australopithecines*, who lived some two million

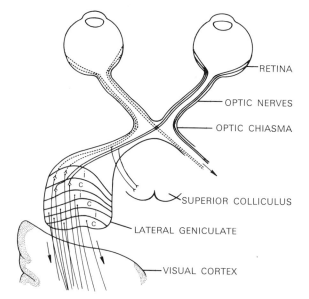

Figure 8. Distribution of optic nerve fibers in a mammal. In a fish, all fibers from each eye would cross the optic chiasma. Note how the lateral geniculate body is subdivided into layers that receive input from the same (I) or opposite (C) retina.
[After Grinnell, A. D.; in M. S. Gordon, ed., *Animal Function: Principles and Adaptations*, Macmillan, 1968.]

RETINA

OPTIC NERVES

OPTIC CHIASMA

SUPERIOR COLLICULUS

LATERAL GENICULATE

VISUAL CORTEX

years ago, the pelvic girdle and leg structure are indicative of a bipedal stance much like our own, and certainly not of the normal stooping stance of the apes. Coupled with this advanced posture and a prehensile hand was a remarkably small brain, which had a volume of about 510 cubic centimeters (about the same as an adult male gorilla). Until 1973, it has been assumed that a series of humanoid fossils with progressively larger cranial capacities reflected the evolutionary line of man. For instance, *Homo habilis*, the next most advanced pre-human who was a contemporary of at least the later *Australopithecines*, had a cranial capacity of 680 cubic centimeters, and the next form, *Homo erectus* (Pithecanthropus) a cranial capacity of 975 cubic centimeters. Thereafter, as the various forms of *Homo sapiens* advanced, the brain volumes increased until the current average size of 1,400 cubic centimeters was reached.

The significance of this seeming sequence is thrown into confusion by Richard E. Leakey's discovery of a fossil skull that is clearly of the genus *Homo*, with a cranial capacity of about 800 cubic centimeters and an age of about 2.8 million years. This creature may have been a contemporary of *Australopithecines* (whose fossils date from 2.9 to 1 million years ago). The new skull, identified for the moment simply as skull no. 1470, lacks the heavy brow ridges characteristic of our pongid relatives and other ancient ancestors. Leg bones from the same geological level as the new skull are even more like those of *Homo sapiens* than are ones of the *Australopithecines*. For example, the thickness of the shaft just beneath the head of the femur is thick in the new find and much thinner in *Australopithecines*; thus, it could probably support more sustained bipedal locomotion. What do these recent finds mean? Probably there were two ancient lines of terrestrial anthropoids living as contemporaries—the *Australopithecines* and the new form. The new skull may represent the ancestral relative of *Homo habilis* and *erectus* and, so, indirectly of *sapiens*.

Brain volume alone is not a reflection of learning capacity, although it provides an excellent approximation in any single type of organism with similar body shapes, surface areas, etc. Perhaps an even better index might be the number of "excess" cortical neurons, that is, neurons that cannot be accounted for by body size or special adaptation (emphasis on one sensory system, for instance, can lead to expansion of the corresponding portion of the brain). The anthropologist Philip Tobias has estimated the number of excess nerve cells for each of the cranial capacities cited above as follows: *Australopithecus*, 3.5 billion; *H. habilis*, 5.3 billion; *H. erectus*, 7.1 billion; and modern man, 8.6 billion. This trend correlates with a series of increasingly complex behavioral patterns: first, use of tools garnered from the environment, then the manufacture of tools, and finally the complex societal relations of man—each an indication of an increase in learning capacity. The early australopithecine and *habilis* fossils demonstrate clearly that an upright stance and a prehensile hand—both discussed in detail by John Napier in "The Antiquity of Human Walking" and "The Evolution of the Hand"—occurred well before substantial expansion of the brain. In fact, as we outlined in the introductory chapter, it seems likely that the brain expanded as a result of the new habits that early man acquired for life in the forest and in the more open grassy savanna: these habits included walking, grasping, making tools, and finally, getting along in a more and more complex society in which the establishment of a more efficient means of communication would be beneficial. It is remarkable indeed that natural selection should have placed such premium upon increased learning capacity, that in only about 70,000 generations the cerebral cortex expanded from that of a primitive primate to that of early *Homo sapiens*.

Having examined some features of the marine and terrestrial vertebrates, let us turn to the major types that fly. From the reptiles arose the pterosaurs

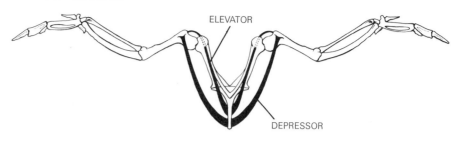

Figure 9. The major bones and muscle-tendon systems of a bird. The depressor muscles originate from the ventral keel. The elevator muscles start at the upper keel and extend over the shoulder joint to insert on the upper surface of the humerus, the proximal wing bone. [After Darling, Lois and Louis, *Bird,* Houghton Mifflin, 1962.]

and the birds; from the mammals, the bats. We assume that the flight of all three groups was originally gliding. Only when the capacity to generate lift and propulsion had increased, did active flight become possible. The gliding stage can be observed today in some of the "flying" squirrels. Since flight places rigid demands on the construction of any animal body, it is not surprising that the vertebrate organization of these three major flying groups responded to those demands in quite similar ways. Light thin bones, elongate forelimbs and hands, reduced hind limbs, and increased wing surface area are features common to all three groups. The wing of the pterosaur was a large, thin sheet of skin stretched behind an enormously elongate fourth digit. The wing of the bat, too, consists of sheets of skin stretched between digits two through five, all of which are greatly lengthened. The bird wing terminates in a fused set of bones—the second, third, and fourth digits—and the great bulk of the wing aerofoil is made up of the flight feathers. The feathered construction has obvious physiological advantages for temperature control since feathers are made of dead, keratinized cells; thus, feathers do not require a supply of the bird's warm blood with the attendant possibility of heat loss. In addition, feathers are more versatile aerodynamic surface than the sheets of skin comprising bat and pterosaur wings. As an example, wing "slots" for reducing drag can be created by twisting individual feathers. Such adjustments are impossible in bat or pterosaur wings. This is one of the reasons that bat flight appears so uneven and "fluttery" in comparison with the level flight of most birds. It is the flight of birds that is so remarkable, and the principles that govern it are discussed in the article "Birds as Flying Machines" by Carl Welty.

The driving force of bird flight comes from the huge pectoralis muscles. Their action is a superb example of engineering design. Since greater aerodynamic stability is gained if the center of gravity of a flying body is at or below the level of the wings, the location of a heavy mass of muscle above and between the wings of a bird would certainly present a problem. Thus the muscles that elevate the wings are actually located below the wings. Both the muscles that depress the wings (the pectoralis majors) and those that raise them (the pectoralis minors, or as some anatomists with a penchant for new names call them, the supracoracoideus muscles) are attached to a flange of bone (the keel) that projects downward from the central chest bone, the sternum. The depressor muscle, as might be expected, simply inserts on the ventral surface of the wing to lower it. The elevator muscle, however, ends at a tendon that proceeds dorsally over the shoulder joint to insert on the upper surface of the proximal wing bone. Thus, by a pulley action the muscles located well below the wings can raise them in flight! Bats and fossil pterosaurs, too, have keels from which the flight-depressor muscles originate—this is a good example of the evolutionary process producing the same functional condition at three different times in vertebrate history.

An important variable in vertebrate skeletal muscle is whether it is "red" or "white" in character. Red muscle is rich in oxygen-storing myoglobin, oxidative enzymes, and frequently lipids, which serve as the energy source. Such muscle has great "staying" power and is used for prolonged activity whether it be flight, swimming, or running. White muscle lacks much myoglobin, contains glycolytic enzymes in quantity, and glycogen for fuel; it is used for rapid, violent escape movements since it fatigues easily. As one might guess, the tiny humming birds that migrate nonstop for over 500 miles across the Gulf of Mexico have pure red pectoral muscles that possess relatively huge lipid deposits prior to the flight.

The manner in which adaptations in one body system influence other systems is demonstrated by the effects of flight on temperature control of birds. Feathers, as effective insulators, present a problem to birds that fly by actively flapping the wings. Sweat glands as a means of counteracting rising temperature have never developed in birds, presumably because water poured onto the body surface would mat the feathers and destroy aerodynamic properties. Instead of sweating, birds pant and employ evaporation from the respiratory system for cooling. Particularly large quantities of heat are generated by the flight muscles of the chest. Much of this heat is dissipated by evaporation from the walls of the anterior air sacs. In the duck, for instance, these sacs are ventilated regularly only during flight. The biologists G. W. Salt and E. Zeuthen have calculated that as a pigeon goes from rest to a maximum speed of 43 miles per hour, its metabolism of flight increases about 26-fold. Ventilation of the respiratory system during flight increases 76-fold. This means that a pigeon in flight loses 37.7 kilocalories of heat per hour, and, on the ground, only 0.5 kilocalories per hour. One other factor that is vital to the operation of this adaptive mechanism is the high speed at which air moves through the ducts of the respiratory system (the trachea, the bronchi, and so on); because of it, there is less opportunity for condensation of water—a process that would retain, rather than dissipate, heat—from the humidified air as it leaves the body.

All vertebrate animals have species-recognition signals of some type or other. Birds—with their wonderful colors and songs—exhibit a variety of interesting signals. We will discuss bird song in Part VIII. The colors of bird feathers are caused by pigments made in the body or by certain structural features of the feathers themselves. Melanin pigments yield browns, blacks, and yellows; fat-soluble lipochromes, yellows, oranges, and reds. No blue pigment is found in birds; rather, blue results from refraction and reflection of blue wavelengths by parts of the feather barbs, while other wavelengths are absorbed by melanin-containing cells beneath the reflecting surface. The exquisite shimmering iridescence of a darting hummingbird is the result of similar highly elaborate feather structures.

Color *patterns* are the result of an interplay between genes, dietary constituents, hormones, and such environmental factors as the length of the day or the temperature. Future melanin-containing cells from the embryo of one species may be transplanted to the embryo of another species and there produce patterns and color characteristic of the donor. This occurs even though the structure of the feathers in which the pigment cells come to reside continues to be typical for the host. Changes in day length seem to be the most critical environmental factor in altering hormone levels of the pituitary glands and the gonads of birds. One result of heightened hormone levels is moulting and the appearance of a breeding plumage with characteristic color patterns. The direct effect of sex hormones upon color pattern is shown easily by injections or by castration; thus, a bright red female parakeet of the genus *Eclectus* turns to the typical male green if the ovaries are removed. In this particular bird, the female hormones normally inhibit development of the male pigmentation. In other birds, male or female hormones may act, not as inhibitors, but as stimulators that elicit developmental capabilities of the pigment cells. Many fascinating questions about the distribution of pigmented or structural color-cell groups remain almost completely unanswered.

Much speculation has surrounded the origin of birds and bats. Donald

Griffin, whose article on bat radar appears in Part VI, has reasoned as follows. Birds are descendants of reptiles, animals that are most active during the day because of their lack of complete temperature regulation. Birds originated some 150 million years ago and diversified substantially before the first bats appeared. Most of them have remained predominantly diurnal, with vision being their main orienting sense. Bats arose from insectivores, the small nocturnal creatures that regulated temperature and gave rise to all other modern mammalian types. Griffin points out that even today most primitive gliding flight—a stage through which early bats must have passed—occurs at night. The reason, of course, is predation during the day by hawks or other large birds. Therefore, one might guess that the bats retained their original habit of flying at night and thus developed their amazing navigation system, based on sound and the sense of hearing. There was, after all, a tremendously large untapped source of food—the nocturnal flying insects—luring the primitive bats into the air! We have, then, a lovely illustration of how and why two types of vertebrate flying machines should arise and be so different in their sensory-navigational complex.

The last article of this section illustrates a different sort of species recognition signal—in this case antlers. Walter Modell, in "Horns and Antlers," describes the way that these distinctive structures form. Horns, as skin derivatives, are permanent offensive or defensive weapons found on both sexes of a species. Antlers on the other hand are bony outgrowths that form each year on males as secondary sexual characteristics. One can only wonder about the unique and fantastic shapes that have evolved for both horns and antlers among different vertebrates.

1

HOW FISHES SWIM

SIR JAMES GRAY

August 1957

In which the speed of small fishes is measured in the laboratory and their power calculated. Similar observations in nature suggest that water may flow over a dolphin completely without turbulence

The submarine and the airplane obviously owe their existence in part to the inspiration of Nature's smaller but not less attractive prototypes —the fish and the bird. It cannot be said that study of the living models has contributed much to the actual design of the machines; indeed, the boot is on the other foot, for it is rather the machines that have helped us to understand how birds fly and fish swim [see "Bird Aerodynamics," by John Storer SCIENTIFIC AMERICAN Offprint 1115]. But engineers may nevertheless have something to learn from intensive study of the locomotion of these animals. Some of their performances are spectacular almost beyond belief, and raise remarkably interesting questions for both the biologist and the engineer. In this article we shall consider the swimming achievements of fishes and whales.

Looking at the propulsive mechanism of a fish, or any other animal, we must note at once a basic difference in mechanical principle between animals and inanimate machines. Nearly all machines apply power by means of wheels or shafts rotating about a fixed axis, normally at a constant speed of rotation. This plan is ruled out for animals because all parts of the body must be connected by blood vessels and nerves: there is no part which can rotate freely about a fixed axis. Debarred from the use of the wheel and axle, animals must employ levers, whipping back and forth, to produce motion. The levers are the bones of its skeleton, hinged together by smooth joints, and the source of power is the muscles, which pull and push the levers by contraction.

The chain of levers comprising a vertebrate's propulsive machine appears in its simplest form in aquatic animals. Each vertebra (lever) is so hinged to its neighbors that it can turn in a single plane. In fishes the backbone whips from side to side (like a snake slithering along the ground), in whales the backbone undulates up and down. A swimming fish drives itself forward by sweeping its tail sidewise; as the tail and caudal fin are bent by the resistance of the water, the forward component of the resultant force propels the fish [see drawing on page 25]. As the tail sweeps in one direction, the front end of the body must tend to swing in the opposite direction, since it is on the opposite side of the hinge, but this movement is usually small—partly because of the high moment of inertia of the front end of the body and partly on account of the resistance which the body offers to the flow of water at right angles to its surface. Thus the head end of the fish acts as a fulcrum for the tail, operating as a flexible lever.

At the moment when the tail fin sweeps across the axis of propulsion, it is traveling rapidly but at a constant speed. During other phases of its motion the speed changes, accelerating as the tail approaches the axis and decelerating after it passes the axis. The whole cycle can be regarded as comparable to that of a variable propeller blade which periodically reverses its direction of rotation and changes pitch as it does so.

How efficient is this propulsion system? Can the oscillating tail of a fish approach in efficiency the steadily running screw propellers that drive a submarine, in terms of the ratio of speed to applied power? To attempt an answer to this question we must first know how fast a fish can swim. Here the biologist finds himself in an embarrassing position, for our information on the subject is far from precise.

As in the case of the flight of birds, the speed of fish is a good deal slower than most people think. When a stationary trout is startled, it appears to move off at an extremely high speed. But the human eye is a very unreliable instrument for judging the rate of this sudden movement. There are, in fact, very few reliable observations concerning the maximum speed of fish of known size and weight. Almost all the data we have are derived from studies of fish under laboratory conditions. These are only small fish, and in addition there is always some question whether the animals are in as good athletic condition as fish in their native environment.

With the assistance of a camera, a number of such measurements have been made by Richard Bainbridge and others in our zoological laboratory at the University of Cambridge. They indicate that ordinarily the maximum speed of a small fresh-water fish is about 10 times the length of the fish's body per second; these speeds are attained only briefly at moments of great stress, when the fish is frightened by a sudden stimulus. A trout eight inches long had a maximum speed of about four miles per hour. The larger the fish, the greater the speed: Bainbridge found, for instance, that a trout one foot long maintained a speed of 6.5 miles per hour for a considerable period [see table on page 28].

It is by no means easy to establish a fair basis of comparison between fish of different species or between different-sized members of the same species. Individual fish—like individual human beings—probably vary in their degree of athletic fitness. Only very extensive observations could distinguish between average and "record-breaking" performances. On general grounds one would expect the speed of a fish to increase

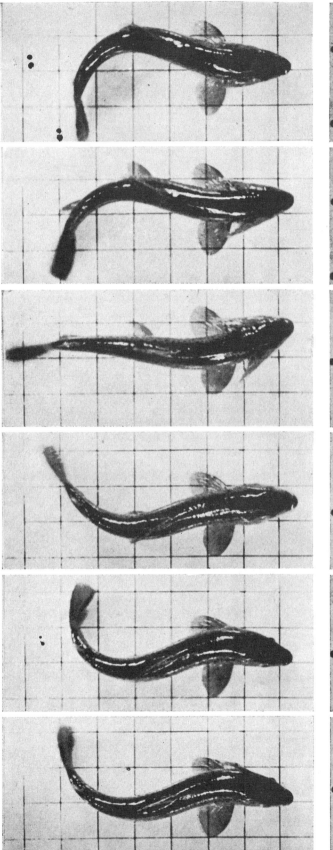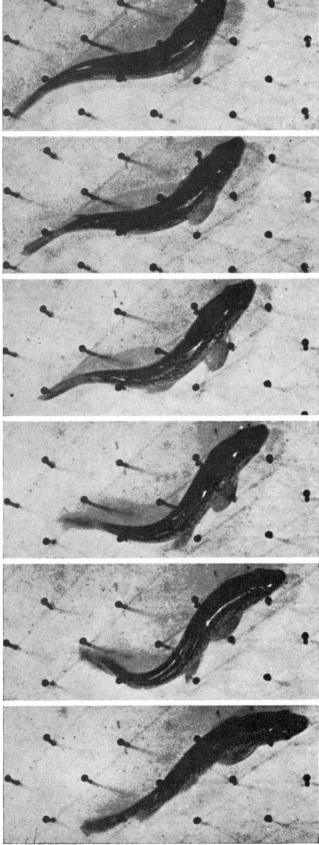

HOW FISH EXERTS FORCE against its medium is illustrated by these two sequences of photographs showing trout out of the water. In the sequence at left the fish has been placed on a board marked with squares; it wriggles but makes no forward progress. In the sequence at right the fish has been placed on a board covered with pegs; its tail pushes against the pegs and moves it across the board.

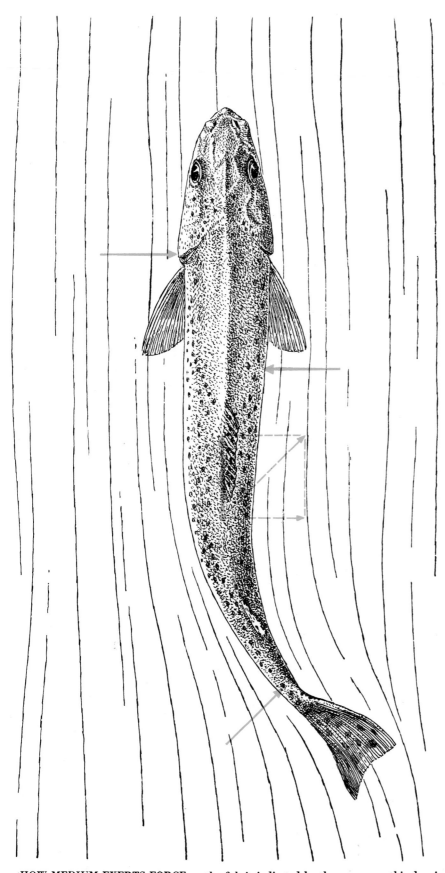

HOW MEDIUM EXERTS FORCE on the fish is indicated by the arrows on this drawing of a trout. As the tail of the fish moves from right to left the water exerts a force upon it (*two diagonal arrows*). The forward component of this force (*heavy vertical arrow*) drives the fish forward. The lateral component (*broken horizontal arrow*) tends to turn the fish to the side. This motion is opposed by the force exerted by the water on sides of the fish.

with size and with the rapidity of the tail beat. Bainbridge's data suggest that there may be a fairly simple relationship between these values: the speed of various sizes of fish belonging to the same species seems to be directly proportional to the length of the body and to the frequency of beat of the tail—so long as the frequency of beat is not too low. In all the species examined the maximum frequency at which the fish can move its tail decreases with increase of length of the body. In the trout the maximum observed frequencies were 24 per second for a 1.5-inch fish and 16 per second for an 11-inch fish—giving maximum speeds of 1.5 and 6.5 miles per hour respectively.

The data collected in Cambridge indicate a very striking feature of fish movement. Evidently the power to execute a sudden spurt is more important to a fish (for escape or for capturing prey) than the maintenance of high speed. Some of these small fish reached their maximum speed within one twentieth of a second from a "standing" start. To accomplish this they must have developed an initial thrust of about four times their own weight.

This brings us to the question of the muscle power a fish must put forth to reach or maintain a given speed. We can calculate the power from the resistance the fish has to overcome as it moves through the water at the speed in question, and the resistance in turn can be estimated by observing how rapidly the fish slows down when it stops its thrust and coasts passively through the water. It was found that for a trout weighing 84 grams the resistance at three miles per hour was approximately 24 grams—roughly one quarter of the weight of the fish. From these figures it was calculated that the fish put out a maximum of about .002 horsepower per pound of body weight in swimming at three miles per hour. This agrees with estimates of the muscle power of fishes which were arrived at in other ways. It seems reasonable to conclude that a small fish can maintain an effective thrust of about one half to one quarter of its body weight for a short time.

As we have noted, in a sudden start the fish may exert a thrust several times greater than this—some four times its body weight. The power required for its "take-off" may be as much as .014 horsepower per pound of total body weight, or .03 per pound of muscle. The fish achieves this extra force by a much more violent maneuver than in ordinary swimming. It turns its head end sharply to one side and with its markedly flexed tail executes a wide and powerful sweep against the water—in short, the fish takes off by arching its back.

This sort of study of fishes' swimming performances may seem at first sight to be of little more than academic inter-est. But in fact it has considerable prac-tical importance. The problem of the salmon industry is a case in point. The seagoing salmon will lay eggs only if it can get back to its native stream. To reach its spawning bed it must journey upriver in the face of swift currents and sometimes hydroelectric dams. In de-signing fish-passes to get them past these obstacles it is important to know pre-cisely what the salmon's swimming ca-pacities are.

Contrary to popular belief, there is little evidence that salmon generally sur-mount falls by leaping over them. Most of the fish almost certainly climb the falls by swimming up a continuous sheet of water. Very likely the objective of their

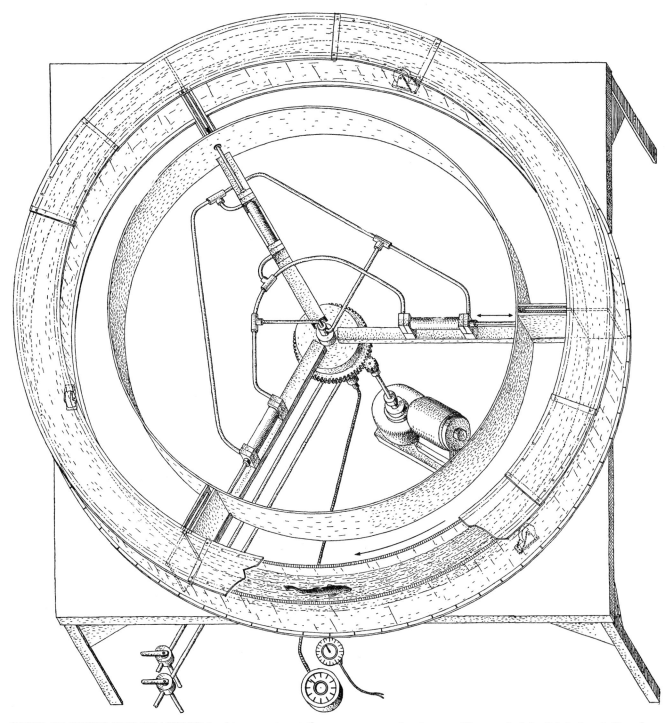

SPEED OF FISHES WAS MEASURED in this apparatus at the University of Cambridge. The fish swims in a circular trough which is rotated by the motor at right center. The speed of the trough is adjusted so that the swimming fish is stationary with respect to the observer. The speed of the fish is then indicated on the speedometer at bottom. Above the speedometer is a clock. When the apparatus is started up, the water is made to move at the same speed as the trough by doors which open to let the fish pass.

leap at the bottom of the fall is to pass through the fast-running water on the surface of the torrent and reach a region of the fall where the velocity of flow can be negotiated without undue difficulty. The brave and prodigious leaps into the air at which spectators marvel may well be badly aimed attempts of the salmon to get into the "solid" water!

A salmon is capable of leaping about six feet up and 12 feet forward in the air; to accomplish this it must leave the water with a velocity of about 14 miles per hour. The swimming speed it can maintain for any appreciable time is probably no more than about eight miles per hour. Accurate measurements of the swimming behavior of salmon in the neighborhood of falls are badly needed—and should be possible to obtain with electronic equipment.

At this point it may be useful to summarize the three main conclusions that have been reached from our study of the small fish. Firstly, a typical fish can exert a very powerful initial thrust when starting from rest, producing an acceleration about four times greater than gravity. Secondly, at times of stress it can exert for a limited period a sustained propulsive thrust equal to about one quarter or one half the weight of its body. Thirdly, the resistance exerted by the water against the surface of the moving fish (*i.e.*, the drag) appears to be of the same order as that exerted upon a flat, rigid plate of similar area and speed. Fourthly, the maximum effective power of a fish's muscles is equivalent to about .002 horsepower per pound of body weight.

Such is the picture drawn from studies of small fishes in tanks. It has its points of interest, and its possible applications to the design of fish-passes, but it poses no particularly intriguing or baffling hydrodynamic problems. Recently, however, the whole matter of the swimming performance of fishes was given a fresh slant by a discovery which led to some very puzzling questions indeed. D. R. Gero, a U. S. aircraft engineer, announced some startling figures for the speed of the barracuda. He found that a four-foot, 20-pound barracuda was capable of a maximum speed of 27 miles per hour! This figure not only established the barracuda's claim to be the world's fastest swimmer but also prompted a new look into the horsepower of aquatic animals.

A more convenient subject for such an examination is the dolphin, whose attributes are somewhat better known than those of the barracuda. (The only essential difference between the propulsive machinery of a fish and that of a dolphin, small relative of the whale, is that the dolphin's tail flaps up and down instead of from side to side.) The dolphin is, of course, a proverbially fast swimmer. More than 20 years ago a dolphin swimming close to the side of a ship was timed at better than 22 miles per hour, and this speed has been confirmed in more recent observations. Now assuming that the drag of the animal's body in the water is comparable to that of a flat plate of comparable area and speed, a six-foot dolphin traveling at 22 miles per hour would require 2.6 horsepower, and its work output would be equivalent to a man—of the same weight as the dolphin—climbing 28,600 feet in one hour! This conclusion is so clearly fantastic that we are forced to look for some error in our assumptions.

Bearing in mind the limitations of animal muscle, it is difficult to endow the dolphin with much more than three tenths of one horsepower of effective output. If this figure is correct, there must be something wrong with the assumption about the drag of the animal's surface in the water: it cannot be more than about one tenth of the assumed value. Yet the resistance could have this low value only if the flow of water were laminar (smooth) over practically the whole of the animal's surface—which an aerodynamic or hydrodynamic engineer must consider altogether unlikely.

The situation is further complicated when we consider the dolphin's larger relatives. The blue whale, largest of all the whales, may weigh some 100 tons. If we suppose that the muscles of a whale are similar to those of a dolphin, a 100-ton whale would develop 448 horsepower. This increase in power over the dolphin is far greater than the increase in surface area (*i.e.*, drag). We should therefore expect the whale to be much faster than the dolphin, yet its top speed appears to be no more than that of the dolphin—about 22 miles per hour. There is another reason to doubt that the whale can put forth anything like 448 horsepower. Physiologists estimate that an exertion beyond about 60 or 70 horsepower would put an intolerable strain on the whale's heart. Now 60 horsepower would not suffice to drive a whale through the water at 20 miles per hour if the flow over its body were turbulent, but it would be sufficient if the flow were laminar.

Thus we reach an impasse. Biologists are extremely unwilling to believe that

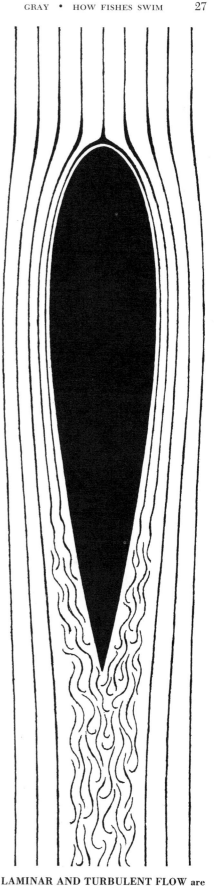

LAMINAR AND TURBULENT FLOW are depicted in this diagram of a streamlined body passing through the water. The smooth lines passing around the body indicate laminar flow; the wavy lines, turbulent flow.

SPECIES	LENGTH (FEET)	MAXIMUM OBSERVED SPEED		RATIO OF MAXIMUM SPEED TO LENGTH
		(FEET PER SECOND)	(MILES PER HOUR)	
TROUT	.656 .957	5.552 10.427	3.8 6.5	8.5 11
DACE	.301 .594 .656	5.229 5.552 8.812	3.6 3.8 5.5	17.8 9 13.5
PIKE	.529 .656	6.850 4.896	4.7 3.3	13 7.5
GOLDFISH	.229 .427	2.301 5.552	1.5 3.8	10.3 13
RUDD	.730	4.240	2.9	6
BARRACUDA	3.937	39.125	27.3	10
DOLPHIN	6.529	32.604	22.4	5
WHALE	90	33	20	33

SPEED OF FISHES IS LISTED in this table. The speed of the first five fishes from the top was measured in the laboratory; that of the barracuda, dolphin and whale in nature. The barracuda is the fastest known swimmer. Whale in the table is the blue whale.

fishes or whales can exert enough power to drive themselves through the water at the recorded speeds against the resistance that would be produced by turbulent flow over their bodies, while engineers are probably equally loath to believe that laminar flow can be maintained over a huge body, even a streamlined body, traveling through the water at 20 miles per hour.

Lacking direct data on these questions, we can only speculate on possible explanations which might resolve the contradiction. One point that seems well worth re-examining is our assumption about the hydrodynamic form of the swimming animal. We assumed that the resistance which the animal (say a dolphin) has to overcome is the same as that of a rigid body of the same size and shape moving forward under a steady propulsive force. But the fact of the matter is that the swimming dolphin is not a rigid body: its tail and flukes are continually moving and bending during each propulsive stroke. It seems reasonable to assume, therefore, that the flow of water over the hind end of the dolphin is not the same as it would be over a rigid structure. In the case of a rigid model towed through the water, much of the resistance is due to slowing down of the water as it flows past the rear end of the model. But the oscillating movement of a swimming animal's tail accelerates water in contact with the tail; this may well reduce or prevent turbulence of flow. There is also another possibility which might be worth investiga-

tion. When a rigid body starts from rest, it takes a little time for turbulence to develop. It is conceivable that in the case of a swimming animal the turbulence never materializes, because the flukes reverse their direction of motion before it has an opportunity to do so.

It would be foolish to urge these speculative suggestions as serious contributions to the problem: they can only be justified insofar as they stimulate engineers to examine the hydrodynamic properties of oscillating bodies. Few, if any, biologists have either the knowledge or the facilities for handling such problems. The questions need to be studied by biologists and engineers working together. Such a cooperative effort could not fail to produce facts of great intrinsic, and possibly of great applied, interest.

DOLPHINS (called porpoises by seamen) were photographed by Jan Hahn as they swam beside the bow of the *Atlantis*, research vessel of the Woods Hole Oceanographic Institution, in the Gulf of Mexico. The speed of these dolphins was about 11 miles per hour.

2

HOW ANIMALS RUN

MILTON HILDEBRAND

May 1960

Many animals, both predators and prey, have evolved the ability to run two or three times faster than a man can. What are the adaptations that make these impressive performances possible?

A man (but not necessarily you or I!) can run 220 yards at the rate of 22.3 miles per hour, and a mile at 15.1 miles per hour. The cheetah, however, can sprint at an estimated 70 miles per hour. And the horse has been known to maintain a speed of 15 miles per hour not just for one mile but for 35 miles.

Other animals are capable of spectacular demonstrations of speed and endurance. Jack rabbits have been clocked at 40 miles per hour. The Mongolian ass is reported to have run 16 miles at the impressive rate of 30 miles per hour. Antelopes apparently enjoy running beside a moving vehicle; they have been reliably timed at 60 miles per hour. The camel has been known to travel 115 miles in 12 hours. Nearly all carnivorous mammals are good runners: the whippet can run 34 miles per hour; the coyote, 43 miles per hour; the red fox, 45 miles per hour. One red fox, running before hounds, covered 150 miles in a day and a half. A fox terrier rewarded with candy turned a treadmill at the rate of 5,000 feet per hour for 17 hours.

I have been attracted by such performances as these to undertake an investigation of how the living running-machine works. The subject has not been thoroughly explored. One study was undertaken by the American photographer Eadweard Muybridge in 1872. Working before the motion-picture camera was invented, Muybridge set up a battery of still cameras to make photographs in rapid sequence. His pictures are still standard references. A. Brazier Howell's work on speed in mammals and Sir James Gray's studies on posture and movement are well known to zoologists. Many investigators have added to our knowledge of the anatomy of running vertebrates, but the analysis of function has for the most part been limited to deductions from skeletons and muscles. The movements of the running animal are so fast and so complex that they cannot be analyzed by the unaided eye.

In my study I have related comparative anatomy to the analysis of motion pictures of animals in action. The method is simple: Successive frames of the motion picture are projected onto tracing paper, where the movements of the parts of the body with respect to one another and to the ground can be analyzed. The main problem is to get pictures from the side of animals running at top speed over open ground. With an electric camera that exposes 200 frames per second I have succeeded in photographing the movements of a cheetah that had been trained by John Hamlet of Ocala, Fla., to chase a paper bag in an enclosure 65 yards long. However, the animal never demonstrated its top speed, but merely loped along at about 35 miles per hour. I have used the same

STRIDE OF A CANTERING HORSE is shown in these photographs from Eadweard Muybridge's *The Horse in Motion*, published in 1878. The sequence runs right to left across the

camera to make pictures of horses running on race tracks, and I am presently collecting motion-picture sequences of other running animals from commercial and private sources.

All cursorial animals (those that can run far, fast and easily) have evolved from good walkers, and in doing so have gained important selective advantages. They are able to forage over wide areas. A pack of African hunting dogs, for example, can range over 1,500 square miles; the American mountain lion works a circuit some 100 miles long; individual arctic foxes have on occasion wandered 800 miles. Cursorial animals can seek new sources of food and water when their usual supplies fail. The camel moves from oasis to oasis, and in years of drought the big-game animals of Africa travel impressive distances. The mobility of cursorial animals enables them to overcome seasonal variations in climate or in food supply. Some herds of caribou migrate 1,600 miles each year. According to their habit, the predators among the cursorial animals exploit superior speed, relay tactics, relentless endurance or surprise to overtake their prey. The prey species are commonly as

swift as their pursuers, but sometimes they have superior endurance or agility.

Speed and endurance are the capacities that characterize all cursorial vertebrates. But one could not make a definitive list of the cursorial species without deciding quite arbitrarily how fast is fast and how far is far. Even then the list would be incomplete, because there are reliable data on speed for only a few animals; in most cases authors quote authors who cite the guesses of laymen. Many cursors are extinct. On the basis of fossils, however, we can surmise that many dinosaurs were excellent runners; that some extinct rhinoceroses, having had long and slender legs, were very fast; and that certain extinct South American grazing animals, having evolved a horse-like form, probably had horselike speed.

In order to run, an animal must overcome the inertia of its body and set it into motion; it must overcome the inertia of its legs with every reversal in the direction of their travel; it must compensate for forces of deceleration, including the action of the ground against its descending feet. A full cycle of motion is called a stride. Speed is the product of length of stride times rate of stride. The giraffe achieves a moderate speed with

a long stride and a slow rate of stride; the wart hog matches this speed with a short stride and a rapid rate. High speed requires that long strides be taken at a rapid rate, and endurance requires that speed be sustained with economy of effort.

Although longer legs take longer strides, speed is not increased simply by the enlargement of the animal. A larger animal is likely to have a lower rate of stride. Natural selection produced fast runners by making their legs long in relation to other parts of the body. In cursorial animals the effective length of the leg—the part that contributes to length of stride—is especially enhanced. The segments of the leg that are away from the body (the foot, shank and forearm) are elongated with respect to the segments close to the body (the thigh and upper arm). In this evolutionary lengthening process the bones equivalent to the human palm and instep have become the most elongated.

Man's foot does not contribute to the length of his leg, except when he rises on his toes. The bear, the opossum, the raccoon and most other vertebrates that walk but seldom run have similar plantigrade ("sole-walking") feet. Carnivo-

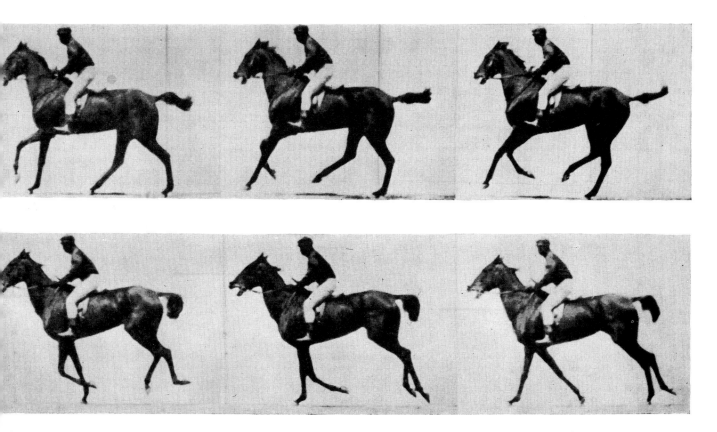

top row and continues across the bottom row. With these and similar photographs Muybridge settled the controversy of whether or not a horse "even at the height of his speed [has] all four of his feet . . . simultaneously free from contact with the ground."

rous mammals, birds, running dinosaurs and some extinct hoofed mammals, on the other hand, stand on what corresponds to the ball of the human foot; these animals have digitigrade ("finger-walking") feet. Other hoofed mammals owe an even further increase in the effective length of their legs to their unguligrade ("hoof-walking") posture, resembling that of a ballet dancer standing on the tips of her toes. Where foot posture and limb proportions have been modified for the cursorial habit, the increased length and slenderness of the leg is striking [*see illustration on page 36*].

The effective length of the front limb of many runners is also increased by the modification of the structure and function of the shoulder. The shoulder joint of amphibians, reptiles and birds is virtually immobilized by the collarbone, which runs from the breast bone to each shoulder blade, and by a second bone,

the coracoid bone. Because mammals do not have a coracoid bone their shoulder blade has some freedom of movement. In the carnivores this freedom is increased by the reduction of the collarbone to a vestige; in the ungulates the collarbone is eliminated. In both carnivores and ungulates the shoulder blade is oriented so that it lies against the side of a narrow but deep chest rather than against the back of a broad but shallow chest, as it does in man. Thus mounted, the shoulder blade pivots from a point about midway in its length, and the shoulder joint at its lower end is free to move forward and backward with the swing of the leg. The exact motion is exceedingly difficult to ascertain in a running animal, but I have found that it adds about 4.5 inches to the stride of the walking cheetah.

The supple spine of the cat and the dog increases the length of stride of these animals still further. The body of such an animal is several inches longer

when the back is extended than when it is flexed. By extending and flexing its back as its legs swing back and forth the animal adds the increase in its body length to its stride. Timing is important in this maneuver. If the animal were to extend its back while its body was in mid-air, its hindquarters would move backward as its forequarters moved forward, with no net addition to the forward motion of the center of mass of its body. In actuality the running animal extends its back only when its hind feet are pushing against the ground. The cheetah executes this maneuver so adeptly that it could run about six miles per hour without any legs.

With the extra rotation of its hip and shoulder girdles and the measuring-worm action of its back, the legs of the running cursor swing through longer arcs, reaching out farther forward and backward and striking and leaving the ground at a more acute angle than they would if the back were rigid. This clear-

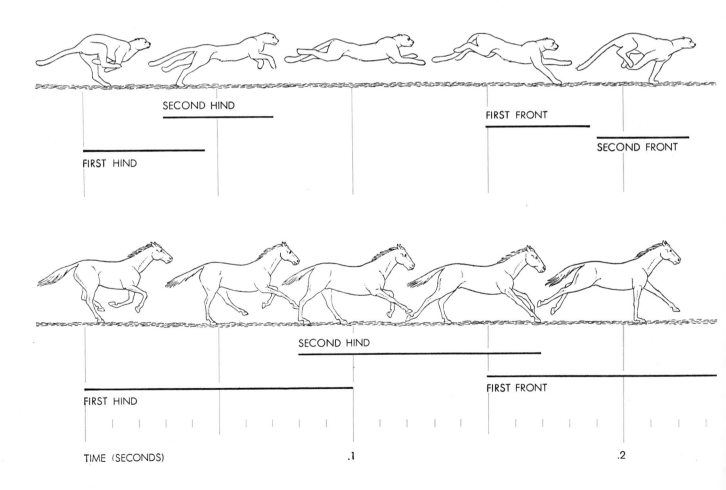

SECOND HIND

FIRST FRONT

FIRST HIND

SECOND FRONT

SECOND HIND

FIRST FRONT

FIRST HIND

TIME (SECONDS)   .1   .2

STRIDES OF THE CHEETAH AND THE HORSE in full gallop are contrasted in these illustrations. The sequence and duration of their footfalls, indicated by the horizontal lines under each animal, **relate to the time-scale at bottom, which is calibrated in 10ths of a second. The cheetah has two unsupported periods, which account for about half its stride; the horse has one unsupported period,**

ly increases stride length, but it also aggravates a problem. The body of the animal tends to rise when its shoulders and hips pass over its feet, and tends to fall when its feet extend to the front or rear. Carnivores offset this bobbing motion by flexing their ankles and wrists, thus shortening their legs. Ungulates do the same by sharply flexing the fetlock joint at the moment that the body passes over the vertical leg. The cheetah, a long-striding back-flexer, supplements its wrist-flexing by slipping its shoulder blade up its ribs about an inch, and thus achieves a smooth forward motion.

Since running is in actuality a series of jumps, the length of the jump must be reckoned as another important increment in the length of the stride. Hoofed runners have one major unsupported period, or jump, in each stride: when the legs are gathered beneath the body. The galloping carnivore has two major unsupported periods: when the back is flexed, and again when it is extended. In

the horse all of these anatomical and functional adaptations combine to produce a 23-foot stride. The cheetah, although smaller, has a stride of the same length.

Fast runners must take their long strides rapidly. The race horse completes about 2.5 strides per second and the cheetah at least 3.5. It is plain that the higher the rate of stride, the faster the runner must contract its muscles. One might infer that cursorial animals as a group would have evolved the ability to contract their muscles faster than other animals. Within limits that is true, but there is a general principle limiting the rate at which a muscle can contract. Assuming a constant load on the muscle fibers, the rate of contraction varies inversely with any of the muscle's linear dimensions; the larger muscle therefore contracts more slowly. That is why an animal with a larger body has a slower rate of stride and so loses the ad-

vantage of its longer length of stride.

The familiar mechanical principle of gear ratio underlies the fast runner's more effective use of its trim musculature. In the linkage of muscle and bone the gear ratio is equal to the distance between the pivot of the motion (the shoulder joint, for example) and the point at which the motion is applied (the foot) divided by the perpendicular distance between the pivot and the point at which the muscle is attached to the bone. Cursorial animals not only have longer legs; their actuating muscles are also attached to the bone closer to the pivot of motion. Their high-gear muscles, in other words, have short lever-arms, and this increases the gear ratio still further. In comparison, the anatomy of walking animals gives them considerably lower gear-ratios; digging and swimming animals have still lower gear ratios.

But while high gears enable an automobile to reach higher speed, they do

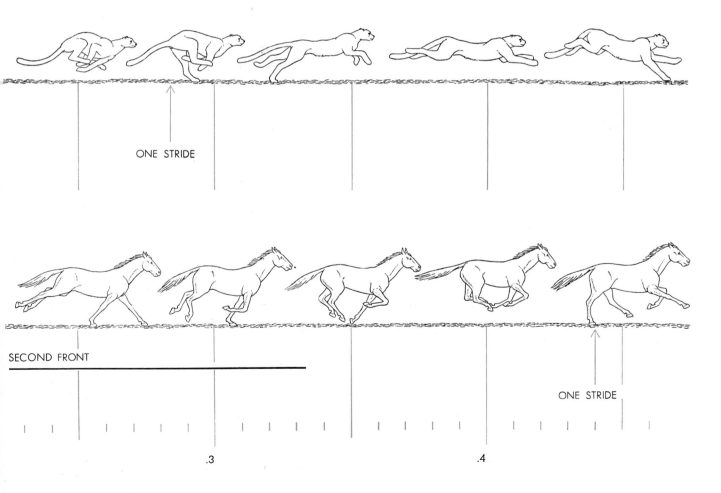

ONE STRIDE

SECOND FRONT

ONE STRIDE

.3   .4

which accounts for about a quarter of its stride. Although both the cheetah and the horse cover about 23 feet per stride, the cheetah attains speeds on the order of 70 miles per hour, to the horse's 43, because it takes about 3.5 strides to the horse's 2.5. The size of the horse has been reduced disproportionately in these drawings for the sake of uniformity in the stride-lines and time-scale.

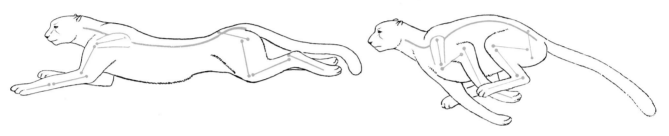

SWIVELING SHOULDER BLADES of the horse and the cheetah add several inches to their stride length. The faster cheetah gains a further advantage from the flexibility of the spine, which in addi- tion to adding the length of its extension to the animal's stride, adds the speed of its extension to the velocity of its travel. Horse's relatively longer leg partially compensates for its rigid spine.

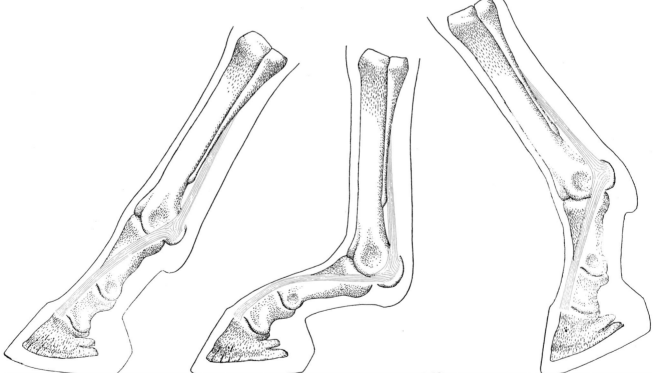

SPRINGING LIGAMENTS in the legs of horses, shown here, and other hoofed runners reduce the need for heavy muscles. Impact of the foot against the ground (*left*) bends the fetlock joint (*mid-* *dle*) and stretches an elastic ligament (*shown in color*) that snaps back when the foot leaves the ground (*right*). The springing action at once straightens the foot and gives the leg an upward impetus.

so at the expense of power. The cursorial animal pays a similar price, but the exchange is a good one for several reasons. Running animals do not need great power: air does not offer much resistance even when they are moving at top speed. Moreover, as the English investigators J. M. Smith and R. J. G. Savage have noted, the animal retains some relatively low-gear muscles. Probably the runner uses its low-gear muscles for slow motions, and then shifts to its high-gear muscles to increase speed.

Since the speed at which a muscle can contract is limited, the velocity of the action it controls must be correspondingly limited, even though the muscle speed is amplified by an optimum gear-ratio. A larger muscle, or additional muscles, applied to action around the same joint can produce increased power but not greater speed. Several men together can lift a greater weight than one can lift alone, but several equally skilled sprinters cannot run faster together than one of them alone. The speed of a leg can be increased, however, if different muscles simultaneously move different joints of the leg in the same direction. The total motion they produce, which is represented by the motion of the foot, will then be greater than the motion produced by any one muscle working alone. Just as the total speed of a man walking up an escalator is the sum of his own speed plus that of the escalator, so the independent velocities of each segment of the leg combine additively to produce a higher total velocity.

The trick is to move as many joints as possible in the same direction at the same time. The evolution of the cursorial body has produced just this effect. By abandoning the flat-footed plantigrade posture in favor of a digitigrade or unguligrade one, the cursorial leg acquired an extra limb-joint. In effect it gained still another through the altered functioning of the shoulder blade. The flexible back of the cursorial carnivore adds yet another motion to the compound motion of its legs; the back flexes in such a way that the chest and pelvis are always rotating in the direction of the swinging limbs.

The supple spine of the carnivore contributes to stride rate by speeding up the motion of its body as well as of its legs. The spine is flexed when the runner's first hind foot strikes the ground, and by the time its second hind foot leaves the ground the animal has extended its spine and thus lengthened its body. In the brief interval when its hind

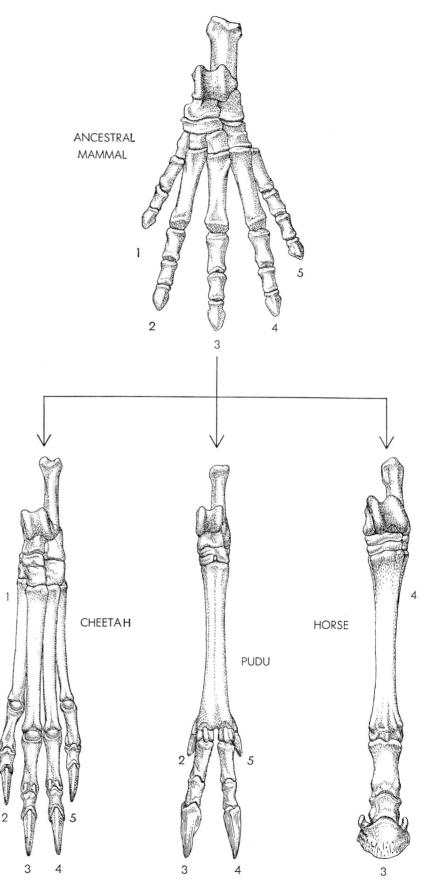

MODERN CURSORIAL FOOT EVOLVED from the broad, five-digited foot of an ancestral mammal (*top*). Lateral digits were lost and metatarsal bones, the longest in the foot, were further elongated. Resultant foot is lighter and longer. Pudu is a deer of the Andes.

feet are planted, the forequarters, riding on the extending spine, move farther and faster than the hindquarters. Similarly when the front feet are on the ground, the hindquarters move faster than the forequarters. So although the speed that the driving legs can impart to the forequarters or hindquarters is limited by their rate of oscillation, the body as a whole is able to exceed that limit. In a sense the animal moves faster than it runs. For the cheetah the advantage amounts to about two miles per hour—enough to add the margin of success in a close chase.

In addition to the obvious tasks of propelling the animal's body and supporting its weight, the locomotor muscles must raise the body to compensate for the falling that occurs during the unsupported phases of the stride. The load they must raise is proportional to the mass of the body, which is in turn proportional to the cube of any of its linear dimensions. A twofold increase in body length thus increases weight eightfold. The force that a muscle can exert, on the other hand, increases only as the square of its cross section. Thus against an eightfold increase of load, bigger muscles can bring only a fourfold increase of force. As body size increases, the capacity of the muscles to put the body in forward motion and to cause its legs to oscillate cannot quite keep up with the demands placed upon them. These factors in the nature of muscle explain why the largest animals can neither gallop nor jump, why small runners such as rabbits and foxes can travel as fast as race horses without having marked structural adaptations for speed and why the larger cursorial animals must be highly adapted in order to run at all.

If the bigger runners are to have endurance as well as speed, they must have not only those adaptations that increase the length and rate of their stride, but also adaptations that reduce the load on their locomotor structures and economize the effort of motion. In satisfying this requirement natural selection produced a number of large and fast runners that are able to travel for long distances at somewhat less than their maximum speed. In these animals the mass of the limbs is minimized. The muscles that in other animals draw the limbs toward or away from the midline of the body (the "hand-clapping" muscles in man) are smaller or adapted to moving the legs in the direction of travel, and the muscles that manipulate the digits or rotate the forearm have disappeared. The ulna in the forearm and the fibula in the shank —bones involved in these former motions—are reduced in size. The ulna is retained at the point where it completes the elbow joint, but elsewhere becomes a sliver fused to its neighbor; the fibula is sometimes represented only by a nubbin of bone at the ankle.

The shape of the cursorial limb embodies another load-reducing principle. Since the kinetic energy that must be alternately developed and overcome in oscillating the limb is equal to half the mass times the square of its velocity, the load on muscles causing such motions can be reduced not only by reducing the mass of the faster-moving parts of the limb but also by reducing the velocity of the more massive parts. Accordingly the fleshy parts of the limb are those close to the body, where they do not move so far, and hence not so fast, as the more distant segments. The lower segments, having lost the muscles and bones involved in rotation and in digit manipulation, are relatively light.

The rigor of design imposed by natural selection is especially evident in the feet of cursorial animals. The feet of other animals tend to be broad and pliable; the bones corresponding to those of the human palm and instep are rounded in cross section and well separated. In the foot of the cursorial carnivore, on the other hand, these bones are

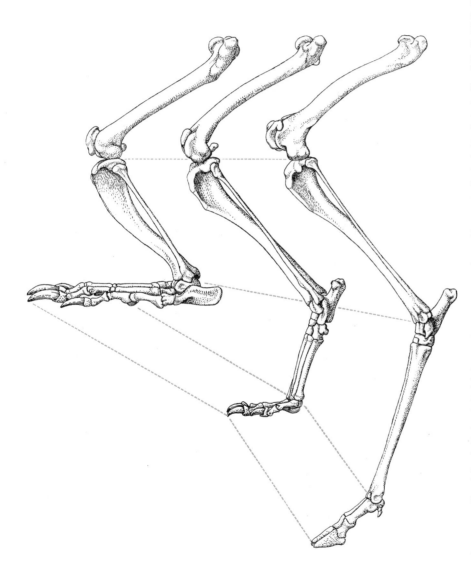

ADAPTATION OF THE LEG FOR SPEED is illustrated by the hind-leg bone of the slow badger (left), moderately fast dog (middle) and highly adapted deer (right). The lengthened metatarsus of the latter two has yielded a longer foot and an altered ankle posture that is better suited to running. The thigh bones of all three animals have been drawn to the same scale to show that the leg segments farthest from the body have elongated the most.

crowded into a compact unit, each bone having a somewhat square cross section. In the ungulates the ratio of strength to weight has been improved still further by reduction of the number of bones in the foot. The ungulates have tended to lose their lateral toes; sometimes the basal elements of the other toes are fused into a single bone. This process gave rise to the cannon bone: the shank of the hoofed mammals [*see illustration on page 35*]. In compensation for the bracing lost as the bones and muscles of their lower limbs were reduced or eliminated, these animals evolved joints that are modified to function as hinges and allow motion only in the line of travel.

The burden on the muscles of hoofed animals is relieved by an especially elegant mechanism built into the foot. When the hoof of the running animal strikes the ground, the impact bends the fetlock joint and stretches certain long ligaments called the suspensory or springing ligaments [*see bottom illustration on page 34*]. Because the ligaments are elastic, they snap back as the foot leaves the ground, thereby straightening the joint and giving the leg an upward push. Charles L. Camp of the University of California has found that these built-in pogo-sticks evolved from foot muscles at the time that the animals forsook river valleys for the open plains. The exchange was advantageous, for by means of this and the other adaptations, nature has reconciled the limitations of muscle mechanics with the exacting requirements of speed.

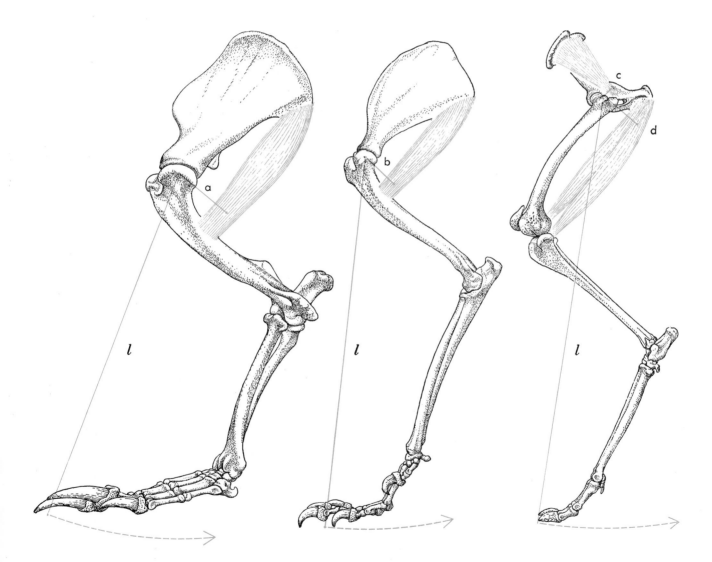

POWER AND SPEED are alternatively achieved in the badger (*left*) and the cheetah (*middle*) by placement of the teres major muscle. In the cheetah the small distance (b) between the muscle insertion and the joint it moves yields a higher rate of oscillation than in the badger, in which the distance (a) is greater. The higher oscillation rate, coupled with a longer leg (*l*), yields a faster stride. In the vicuña (*right*) the gluteus muscle (c) develops about five times the velocity but only a fifth the force of the larger semimembranosus muscle (d). The animal may use the latter to overcome inertia; the former, for high speed. Legs are not in same scale.

3

HOW SNAKES MOVE

CARL GANS
June 1970

*They have four modes of progression, termed lateral,
rectilinear, concertina and sidewinding. In lateral progression,
the commonest mode, the snake uses its loops to push not
downward but sideways*

"The way of a serpent upon a rock" is one of four phenomena described in the Book of Proverbs as being beyond understanding. A snake undulates in a smooth and asymmetrical fashion, so that the size and curvature of the loops formed by its body change as the animal glides along. Because the parts of the snake's body that exert motive forces are never as clearly defined as those in species with limbs it is not obvious to a casual observer how such movement propels the animal forward. It would seem that the snake's undulations would be just as likely to move it backward or leave it thrashing in one place. Such considerations make the movement of snakes a challenging subject for investigation.

Slithering and related forms of motion employed by snakes are the consequence of the snakes' evolution. Snakes presumably evolved from animals that lost their appendages when they took up a burrowing existence. Since they lacked limbs, the entire body had to become an organ of locomotion. This function favored elongation of the trunk and necessitated a repackaging of the viscera. Both of these characteristics were further developed by most snakes when they emerged to radiate into a broad spectrum of terrestrial, arboreal and even aquatic niches. As the trunk lengthened it remained flexible because the number of bony segments in the vertebral column increased. Thus snakes have between 100 and 400 trunk vertebrae, a number that contrasts strikingly with the 16 or so found in a typical mammal. Ophidian trunk vertebrae are also joined by at least one extra set of articulations, and each vertebra bears a pair of ribs that reach down to form a flexible yet strong support for the sides of the elongated visceral cavity.

The increase in the number of the seg-

ments gives the snake its flexibility; it also complicates the snake's control problems. Some 30 years ago the late Walter Mosauer of the University of California at Los Angeles showed that a snake's vertebrae are joined in such a way that one vertebra can bend up and down in relation to another through approximately 28 degrees and swing from side to side through about 50 degrees. Although the bulk of tissue probably reduces a living snake's flexibility, the number of degrees through which a snake can bend is still remarkable. Even twisting movements can be approximated by a combination of up-and-down and side-to-side bending, in spite of the fact that the joints between any pair of vertebrae do not allow one segment actually to twist in relation to the other. The posture assumed by a snake at any one moment is established by the muscles that together with the skeleton form the axis of the body. The most important of these muscles are arrayed in 12 or more bundles that connect one vertebra to another and to the ribs on each side of the trunk, and in two to four bands that attach the skin to each side of the snake's main mass.

How does a snake utilize all these elements in order to move? An exact description of slithering requires that the position of each element at any given instant be specified. It is easily seen that investigators who seek such a detailed description must solve a truly staggering number of simultaneous equations. The task is complicated by the fact that as the snake travels its parts move and accelerate at different rates. The solution gains several magnitudes of complexity when one wants to know which of the snake's thousands of muscles are actually exerting the forces that induce and maintain motion.

One method of achieving a simpler solution is to work inward from the outside. The snake is treated as a special kind of "black box" capable of exerting forces on areas within an envelope of space considerably greater than its own volume. One can then deal with the snake's external responses without, for the moment, concerning oneself with the way they are generated internally. One first asks where and how these external forces are exerted; then it can be asked what internal forces and mechanisms are responsible for the externally observed efforts. Ultimately this approach allows investigation of the function of the specific muscle groups, and of the motor units and their control.

Since the immediate area of contact between the snake and the environment is the skin it is important to consider the texture of the skin's surface. The sides and back of most snakes are covered with longitudinal rows of lozenge-shaped and more or less rigid scales. Such scales are generally mounted, so that their trailing edges are free, on a flexible and accordion-pleated intermediate layer of skin that is often lined by a layer of muscle. The skin of a snake has a remarkable capacity for extension, as is evidenced when the animal swallows prey more than twice the diameter of its body. The snake's bottom surface is covered by a series of scutes: wide scales whose free trailing edges overlap like plates of armor. Each scute tends to be associated with a single row of scales on the snake's sides and back, and to a single vertebra and a single set of muscles; the scutes thus reflect the snake's internal organization. The side and back scales often have sensory pits of unknown function and may be ridged; the scutes do not have pits and are normally smooth.

It is generally accepted that the mis-

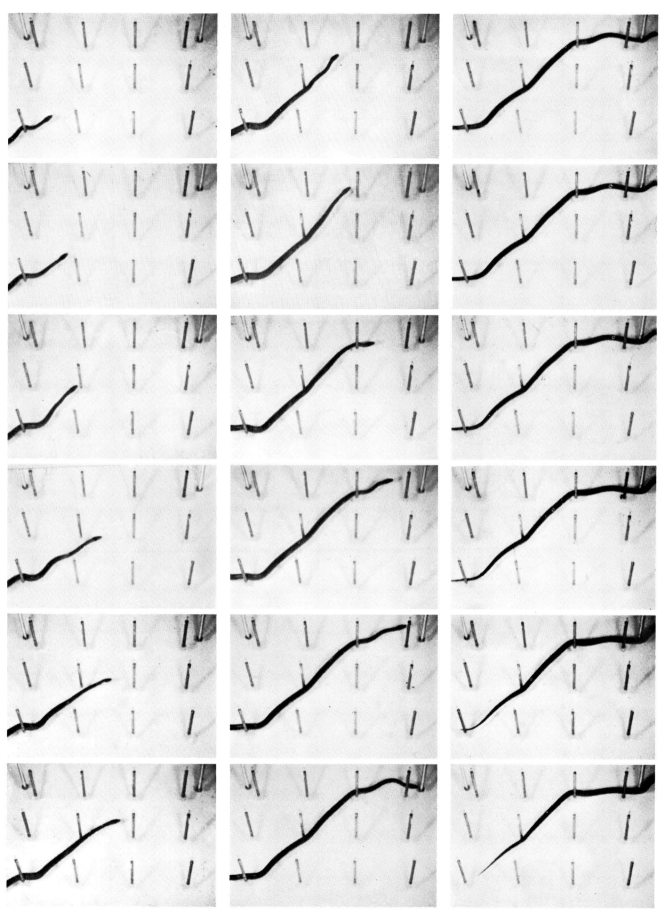

GARTER SNAKE slithering from left to right across an experimental surface demonstrates lateral undulation. These motion-picture frames show that the snake advances by forming curves with its body that it braces against the brass pegs projecting from surface.

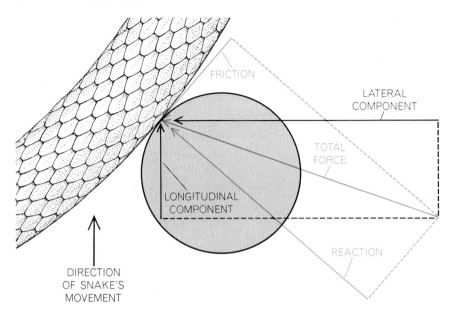

FORCES EXERTED by a snake's body against an object propel the animal. When the snake flexes its body, a reaction force is exerted that pushes the snake away from the peg, as indicated in diagram of forces (*color*). At the same time the snake slides past the peg because of pressure exerted elsewhere. This movement of its body against the peg produces a frictional force. Frictional force together with the reaction force equals the total force. This force consists of a propulsive longitudinal component and a lateral component.

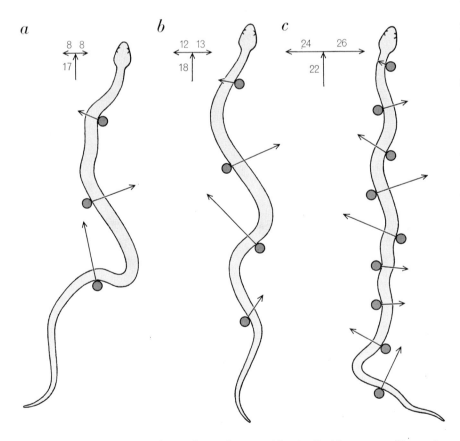

EFFICIENCY of lateral undulation, that is, the ratio of longitudinal forces propelling snake forward and lateral forces expended to the sides, declines as number of contact points increases. In *a* the snake propels itself forward by exerting force on two points at right and using one at left to balance the other two. Vector sum shows that forward-acting force is 17 grams and that 16 grams of force are expended to the sides. In *b* the snake uses four contact points and exerts 18 grams of force in a forward direction but expends about 25 grams of force to the sides. In *c* number of contact points has doubled and sum indicates that more force is expended to each side than is exerted in a forward direction.

cellaneous ways by which snakes propel themselves can be subdivided into four distinct patterns: lateral undulation, concertina progression, rectilinear locomotion and sidewinding. These categories are somewhat artificial. Most species of snakes can move in more than one of these patterns; indeed, a snake moves in more than one pattern at the same time. My colleagues and I at the State University of New York at Buffalo have investigated these patterns of movement. We are currently looking into their control mechanisms.

Lateral undulation is the kind of snake movement most frequently seen. It is the preferred movement not only of snakes but also of limbless lizards. Under certain conditions this kind of movement is even employed by limbed species, but their limbs are first folded against the body. The pattern can be shown to be the most primitive one in vertebrates. Lateral undulation involves the movement of the body along an S-shaped path. In an idealized situation the snake's track can be seen to be a single wavy line scarcely wider than the animal's body; each portion of the snake's trunk has traced this path. The snake begins to undulate laterally by bending the forward part of its body. This movement establishes a wavelike muscular contraction that travels down the snake's trunk.

How does this wave of muscular contraction cause the snake to move? Because human beings walk by exerting force directly on the surface under their feet, it is generally assumed that a snake too propels itself by somehow exerting a force directly on the ground. In actuality undulatory progression operates on a completely different principle. The wavelike contractions flowing down a snake's body do not propel it by acting on the ground under the snake's belly. Instead they cause the snake's body to exert force laterally on irregularities in the snake's path: small elevations and depressions, pebbles, tufts of grass and so on. The snake in effect pushes itself off from such points the way a man sitting in a chair with casters can push himself away from a desk. Since a snake can push only against a limited number of fixed objects, it sets its body in a corresponding pattern of loops, with the outside of each loop forming a contact point. The speed at which a loop progresses down the snake's body toward its tail equals the forward speed of the snake; the moving loops thus seem to be stationary with respect to the ground.

This analysis of lateral undulation has

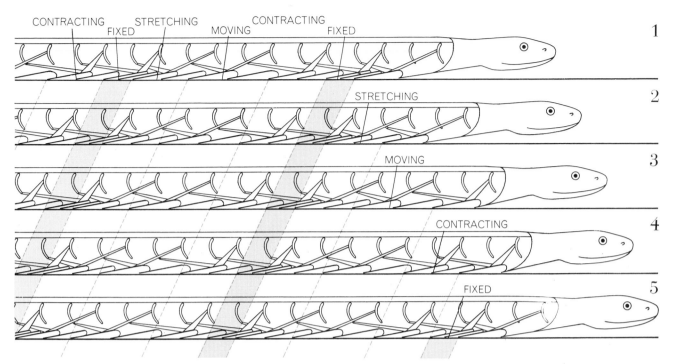

RECTILINEAR LOCOMOTION enables a snake to advance in a straight line as it stalks prey or crosses a flat surface. In *1* the snake pulls together two sets of scutes, its large abdominal scales, and fixes them against the ground. By pulling against these fixed zones that have established frictional contact with the ground the snake propels itself forward. For this task it tenses the bands of muscle running from the scutes back toward the ribs. In *2* the fixed zones have moved toward the snake's tail (*colored bars*). In order to move the fixed zones the advancing snake picks up the forward scutes in each zone, using muscles that run from scutes forward to ribs, and pulls them clear of the ground (*stretching areas*). Simultaneously it pulls additional scutes forward from behind the fixed zones (*contracting areas*). In *3* snake has begun to contract scutes near its head. In *4* these scutes are being gathered into a new fixed zone, while the hindmost fixed zone has disappeared. In *5* two zones are again fixed at more forward sites and cycle is complete.

been confirmed by Hans W. Lissmann and Sir James Gray in a series of experiments they conducted to characterize the forces exerted against the ground by certain snakes. Their snakes crawled across a smooth board over which hung a regular array of heavy, pendulum-like pegs. Any lateral force exerted against a peg would cause it to swing, indicating both the magnitude of the applied force and its direction. Calculations of these forces showed them to be consistent with the force and speed of a snake's movement. When the pegs were taken away, the snakes were totally unable to move by lateral undulation; they slipped about on the smooth board because they had no objects to exert lateral force against.

Although a snake propels itself by exerting force laterally, it does apply other forces to the ground. One of them is a vertical force induced by the weight of the animal. Another is the horizontal force due to friction. Only the frictional force is exerted by muscle activity; it is the result of the movement produced when a snake curves its trunk in order to apply propelling forces against surrounding objects.

What effect do these forces have on a snake's movement? Friction is produced by lateral undulation in two ways: by the underside of the body sliding along the ground and by the flanks of the body sliding along objects to the side. The magnitude of such sliding friction equals the force pressing the two surfaces together multiplied by a coefficient determined by their nature. In the first instance this amounts to the weight of the snake times the coefficient of friction between the snake's belly skin and the ground. In the second instance it is a function of the force exerted against the object to the side times the coefficient between the flank skin and the object.

In both instances the frictional forces act in opposition to the force inducing motion; thus in lateral undulation friction has to be overcome by muscular effort. This being so, there is a selective evolutionary pressure to reduce the frictional coefficient of the snake's skin, and various kinds of snake scales have been shown to have the low frictional coefficient of about .35. That friction is unnecessary to the locomotion of snakes can be demonstrated by placing a snake in a virtually frictionless experimental environment: a smooth surface lubricat-

ed with powder and studded with vertical pegs that are fixed in position but able to rotate. The snake will traverse such a surface even more rapidly than it can cross one studded with nonrotating pegs.

Continuous forward progression by a snake requires a minimum of three contact sites at all times. Moreover, the contact sites must have a particular spatial arrangement. The snake pushes against two of the sites to generate force and uses the third to balance the forces produced at the other two so that its body can move in a particular direction. If the snake is to move, the contact points need to be on both sides of its body and the forces on the contact points must be exerted at an acute angle to the rear. Such considerations suggest that as the snake moves from one contact point to another it shifts its position slightly, and this is confirmed in films we have made. The transition from one set of three contact points to the next must be made quickly; otherwise the snake must utilize four or more contact points at a time and will lose efficiency.

The minimum number of three contact points represents a compromise between two opposing factors. One factor is the snake's flexibility. Since a snake is notably flexible it must expend a certain amount of energy to make its body rigid in order to transmit force from one contact point to another. The amount of energy applied for this purpose can be reduced if the number of contact points is increased; less energy is needed to maintain rigidity over several short distances than is needed over a few long ones. Yet the greater the number of contact points, the shorter the amplitude of the snake's curves, the wider the angle of force application and the greater the waste of force in the snake's direction of motion. The snake balances the requirement for internal force transmission against the requirement for efficient force application by establishing a number of contact points that is consistent with both. The number of points is also affected by the "desired" speed, the rela-

tive elongation (diameter divided by length) of the body and the spacing of objects to the side.

Experiments in which snakes of varying sizes are chased through a fixed array of pegs suggest that the speed of a snake is not proportional to its size. In fact, the limitations on the speed of snakes remain obscure. There may well be an optimum peg-spacing for a particular size of snake, where optimum means the facilitation of a maximum velocity rather than a minimum energy expenditure per traverse. No one has yet done experiments with different peg-spacings to see if such spacings would actually affect the snakes' velocities. Nor have there been studies relating speed to work output or speed to the time or distance a snake is capable of traveling.

Such questions about peg-spacings and efficient movement lead one to ask

further questions about control mechanisms. As we have seen, the body of a snake can twist around its long axis to some extent. Thus a snake is capable of lifting a loop over one contact point and making use of another, an ability that enables the snake to select advantageous contact points in its environment. What environmental factors affect such loop placement? Experiments indicate that the key factor may well be the relative resistance of contact points. A snake moving by lateral undulation through grass or sand will increase the size of its loops in order to displace loose surface material until it can exert the maximum force possible under the circumstances. Yet how does the snake monitor information about such resistance, and what is the feedback sequence that shifts the information from one segment of the snake's body to the next as the snake

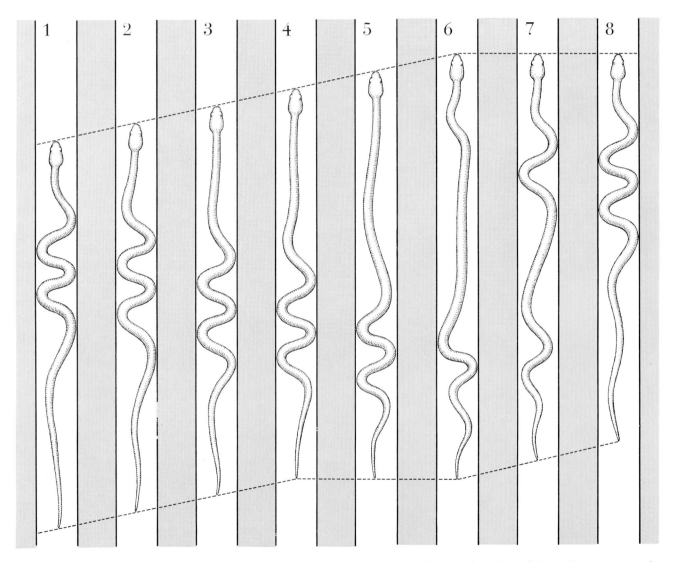

CONCERTINA MOVEMENT enables a snake to move in a narrow channel. In *1* the snake begins to move by bracing S-shaped loops against sides of channel. Snake then begins to extend its head and forebody (*2*) while pushing against passage walls until forward ex= tension is complete (*6*). In order to bring tail section up snake forms a new S-curve behind its head (*7*) and braces against the sides of channel. Snake can now bring tail section along by pulling it forward and forming more loops against the passage walls.

moves past a particular contact point?

The lateral undulatory progression of a snake has the inherent advantage that the capacity for establishing staggered, backward-traveling waves of muscular contraction is phylogenetically much older than snakes. Snakes have only had to develop a feedback sequence for the control of such waves. This kind of progression is also effective because it permits the snake's body to travel at a relatively constant speed with only limited changes in its velocity. Yet lateral undulation does not work under all circumstances. It is useless for traveling down a tunnel with parallel sides: no backward-directed forces can be exerted against the sides of the tunnel. It is also ineffective on a flat surface that lacks elevations and depressions, and on a rounded surface such as a branch or a fallen tree trunk. Indeed, some snakes find it difficult to use lateral undulation at all. This mode of locomotion is well suited to relatively slender snakes, but it is less effective for short, stout ones. When a snake species has the option of invading an environmental niche that requires a relatively stout body, there is immediate selective pressure for a substitute method of locomotion.

An alternative is rectilinear movement. This mode of locomotion differs from lateral undulation in two respects: it involves the application of force somewhat downward instead of laterally, and it is effective only if friction is established between the snake's skin and the ground. Rectilinear locomotion is made possible by the very loose skin found in many snakes (perhaps as a result of such a specialization as prey constriction). In order to move in this mode the snake fixes several series of scutes and starts to move the skin between them. For example, it pulls together a series of scutes near its head, fixes them against the ground and then moves the rest of its body with respect to this fixed zone and several similar zones behind it. As the body of the snake moves forward the skin is stretched, pulling the forwardmost scutes of each series out of contact with the ground, while additional scutes are continuously pulled up to the rear edge of the series. In this way a constant length of belly surface remains fixed to the ground. Normally a snake fixes two or three of these zones to the ground at once, and they can be seen to move continuously to the rear as the snake progresses [see illustration on page 41].

In rectilinear locomotion the two sides of the snake must move symmetrically rather than in alternation. X-ray motion pictures have confirmed an earlier observation that in this mode of movement the ribs and the vertebrae do not move with respect to one another. The scutes are shifted forward by slender muscles that stretch forward from the sides of each scute to points high on the side of each rib. A second set of muscles runs to the rear at a shallower angle to attach the skin to the bottom ends of the ribs. Contraction of the muscles in the first group pulls the skin forward and up, out of sliding contact with the ground; contraction of the stouter and mechanically better-placed muscles of the second group accelerates the snake's mass and maintains its constant forward movement.

The fact that rectilinear locomotion relies on friction places a certain limitation on this form of movement. The particular form of friction involved is called static friction. Static friction is defined as the force that must be applied parallel to the contact surface between two objects in order to start them sliding past each other. Then static friction becomes sliding friction. It follows that a snake can only exert a force parallel to the ground that is less than the force of static friction; otherwise the animal will slip. A snake in rectilinear movement must accelerate smoothly, because even a temporary imbalance may shift a group of scutes from static to sliding contact and make the effort ineffective.

Even those snakes that have achieved the necessary body form to move by rectilinear progression find this mode of locomotion limiting because of the need to keep some parts in static contact. Rectilinear progression enables the snake to cross flat surfaces and to advance in a straight line when it is stalking prey. Such progression, however, is slow because each scale must establish stationary contact with the ground before it can transmit horizontal forces. It seems that motor control is again of critical importance in keeping the contact zones from sliding.

Most snakes that do not have the muscle and bone structures necessary for rectilinear progression can still use static friction in locomotion by employing concertina progression. In order to move in this way the snake draws itself into an S-shaped curve similar to the posture assumed in lateral undulation, and sets the curved portion of its body in static contact with the ground. Motion begins when the head, the neck and the forward part of the body are extended by forces transmitted to the ground in the zone that remains in stationary contact. These forces produce movement by acting against the force of friction generated by the weight of the snake's body. This reserve of friction is called the static-friction reservoir. It should be noted that the area of contact between the snake's body and the ground does not affect the "capacity" of the reservoir. Friction is a function solely of the force pressing two surfaces together multiplied by a coefficient that represents

SCUTES on the belly of this snake, a Japanese species called the *awo daishu* (blue general), correspond to rows of back and side scales. This pattern is found in all species of snakes.

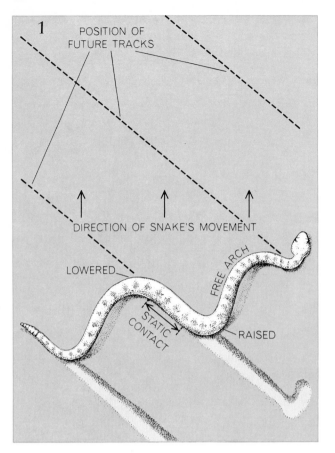

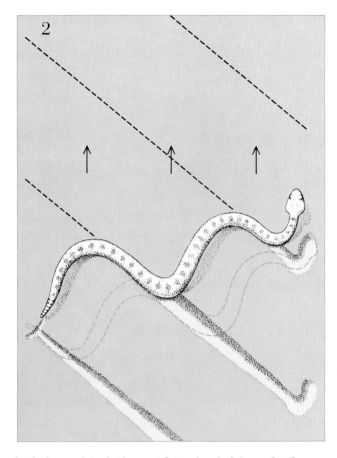

SIDEWINDING enables desert vipers to move rapidly. In *1* a snake with tail in first track (*lower left*) has lifted head and forebody from

hooked second track (*lower right*) and arched forward to begin a third track. In *2* head has risen from this new track at

their roughness. The force available for acceleration in the static-friction reservoir remains constant as long as the snake continues to support its weight on the stationary zone.

After the front end of such a snake has moved forward a short distance it stops. This establishes a new zone of stationary contact in which horizontal forces are exerted against the ground, and the rear end of the snake's body is pulled forward. The concertina progres-

sion of a snake is thus rather like the locomotion of an inchworm [*see illustration on page 42*].

As long as the moving portion of the snake's body can be lifted out of contact with the ground, the ratio of the moving portion to the stationary one depends only on the snake's stability. Yet if the moving parts slide forward or are dragged along they will induce sliding friction that must be overcome. Moreover, since each portion of the

snake's weight that induces sliding friction is unavailable for maintaining the static-friction reservoir, the snake will attempt to keep such areas to a minimum. It is therefore possible to predict the approximate fraction of a particular snake's body that can be kept in sliding motion over a particular kind of surface.

Snakes sometimes supplement weight-induced friction by muscular force, thereby enlarging the static-friction res-

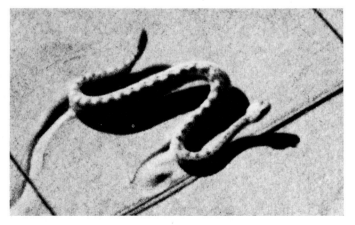

SIDEWINDER (*Crotalus cerastes*) undulates at an angle of about 60 degrees to its direction of travel. This angling helps the snake, a western

U.S. species, to avoid slipping. First the sidewinder sweeps out such a large area that it is likely to encounter objects to brace

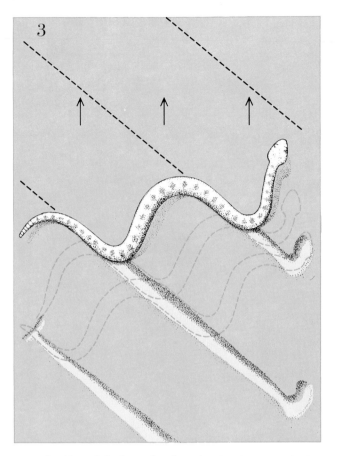

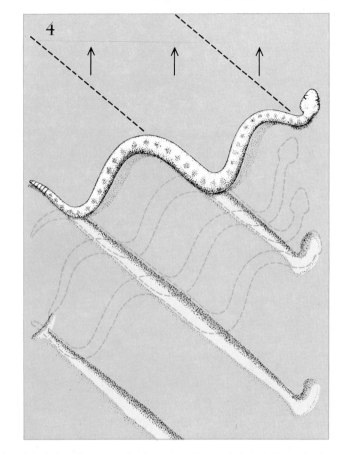

right. Meanwhile the snake's body has lifted clear of the first track and begun to move into the second one (*broken outline of snake's body indicates earlier position*). In 3 snake's body lies in both the second and the third track. In 4 a fourth track has been started.

ervoir. Such modified concertina progression is often used to traverse a straight-sided channel or to climb tree branches. In the first instance the snake widens the amplitude of its S-shaped loops, actively pushing them into contact with the channel's walls. In the second instance the snake forces its stationary surfaces into contact by constricting around a branch. Such enhancement of concertina progression probably arose among species that were already adapt-

ed for it through the accident of having an overdeveloped axial musculature for the constriction of prey.

This observation pays an unexpected literary dividend. In "The Adventure of the Speckled Band" Sherlock Holmes solves a murder mystery by showing that the victim has been killed by a Russell's viper that has climbed up a bell rope. What Holmes did not realize was that Russell's viper is not a constrictor. The snake is therefore incapable of concer-

tina movement and could not have climbed the rope. Either the snake reached its victim some other way or the case remains open.

Concertina progression is almost as common as lateral undulation, and one often sees a long snake combining the two kinds of movement. Various species of rat snake use this combination of movements to climb trees; they use the irregularities in the bark as channels. Many such snakes have a double keel on

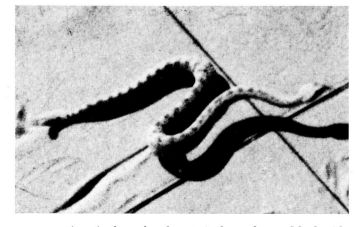

against. As the snake advances it also pushes sand back with its body until the sand piles up (*ridges in tracks*) and resists

further pressure. In this way the sidewinder is able to arrest the backward slippage of its loops and move its entire body forward.

SKELETON of *awo daishu* in this X-ray photograph has more than 300 vertebrae and 400 ribs. A large number of vertebrae enables a snake to bend and twist through many degrees.

can therefore only be sideways.

The series of actions producing these tracks can be visualized by assuming that the snake is lying so that its tail points about 60 degrees away from the direction in which the animal is going to move. The snake's head is raised and turned through an obtuse angle, so that it faces in the direction of travel. Only the bend of the snake's neck is in contact with the ground; this produces the hook at the rear end of each track.

As the snake starts to move forward it lifts its head and neck off the ground. In order to reach the next track the forward part of the body has to curve in a smooth loop. The extension of the snake's front end continues until approximately a quarter of the trunk is cantilevered out of contact with the ground. The head then arches downward as the cantilevered trunk remains off the ground; in making the first contact the neck bends at the next track of the sequence, so that another hook is produced. Successive sections of the body and then the tail follow along the new track, which parallels the preceding one. Considerably before the tail has been pulled into the second track the snake's front end starts into the third track. The body of a sidewinding snake thus lies on two or three separate tracks, with the body parts between tracks held off the ground.

Although the snake's body swings through loops that are reminiscent of lateral undulations, one can see that the force-transmission pattern is more like that of the concertina sequence. The force for the initial acceleration of the front part of the snake, and for maintaining the velocity of the moving parts, must be transmitted by static friction of the snake's belly. It is this pattern that gives sidewinding its peculiar advantages. When a nonsidewinding snake travels over loose sand, its lateral force shifts the sand and its track becomes a trough. Moreover, a snake that exerts a lateral force exceeding the resistance capacity of the force points tends to slip sideways instead of moving forward. Similarly, a snake in rectilinear locomotion will just dig in, slipping within its track, if the force exerted by the contact zones exceeds the limit of the static-friction reservoir. For the sidewinder such slippage proceeds sideways and causes the snake's body to dig in, piling up sand along its length and increasing the forces that produce forward movement. In other words, a slipping sidewinder quickly stabilizes itself by converting sliding friction into static friction.

H. Mendelssohn of the University of

their belly. The structure enables the snake to push laterally against irregularities in the surface it is climbing. Certain burrowing snakes increase their propulsive force by digging in their tail and then straightening their trunk.

Since concertina progression is effected by a bending of the body, it proceeds by asymmetrical contractions. Observation confirms that these intermittent contraction waves move along the body toward the tail. It thus appears that concertina progression is essentially a low-speed pattern of movement, because the animal must pause while it brings its rear end forward so that it can proceed again. Although concertina and rectilinear progression are suited to surfaces where lateral undulation would be ineffective, the price is lack of speed.

For pursuit—or escape—snakes need a mechanism that enables them to employ static friction without sacrificing speed. What is needed is a means of locomotion that keeps the zones that are in static contact from slipping, allows contraction

waves to pass continuously and regularly down the snake's body and prevents minor irregularities in the surface from affecting the pattern. Such a form of locomotion is sidewinding, which was first properly described by Walter Mosauer in 1932. A sidewinding snake achieves firm static contact by moving so that its body lies almost at right angles to the direction of its travel. In this orientation the snake is more likely to encounter rocks and other slight irregularities against which it can brace itself than it would if it were moving straight ahead, for the reason that it is sweeping out an area many times wider than the width of its loops.

The track of a sidewinding snake that has traversed a smooth flat surface appears as a series of straight parallel lines, each inclined some 60 degrees to the snake's direction of motion and each about as long as the snake itself. The rear end of each line is bent into a short, forward-pointing hook. The track of the scutes is well defined; slippage, if any,

Tel Aviv and I are currently analyzing the feedback mechanisms that control sidewinding. The energy required for sidewinding is apparently reflected in the height to which the traveling loops are cantilevered off the ground. The loop height is lowest for those surfaces that are smoothest and these therefore require the least amount of energy. Indeed, when the snakes are allowed to travel over a smooth, low-friction surface such as a polished terrazzo floor, they will slide rather than lift the moving parts of their body. If their tracks were visible, they would tend to be connected. When the snakes traverse a smooth but high-friction surface such as sandpaper, the tracks are quite separate. A rough surface such as a layer of crushed, sharp-edged aggregate causes a sidewinding snake to lift its loops quite high. Frame-by-frame analysis of films suggests that the loops continuously change their height.

Sidewinding has the further advantage that the contraction waves can start at the neck and pass down the snake's lateral musculature. Each motor sequence is thus continuous even though parts of the body stop and start. The snake is accordingly able to cross relatively flat areas, liberated from the need to search for lateral irregularities to propel itself.

Sidewinding seems to be a method of locomotion available to any kind of snake. Various conditions elicit it, and no special structural modification is required. Yet although many snakes sidewind under stress, only a few species do it effectively, and the beautiful control and minimal energy utilization described here were observed only in desert vipers. Such snakes seem to have found this method advantageous for traveling quickly over flat, sandy and rocky surfaces. Sidewinding would seem to be particularly useful when the crossing has to be made at midday, when speed and minimum contact may prevent overheating.

The various sidewinders differ in their track angles and in the number of tracks they occupy at one time. One of the most spectacular of the movement patterns that have been observed so far is seen in juvenile specimens of the Southwest African desert viper, a species that has evolved a special escape sequence when it is faced with heat stress. The muscle-contraction waves travel down the snake's body at very high speed, and some specimens jump completely off the ground from track to track [see illustration at right].

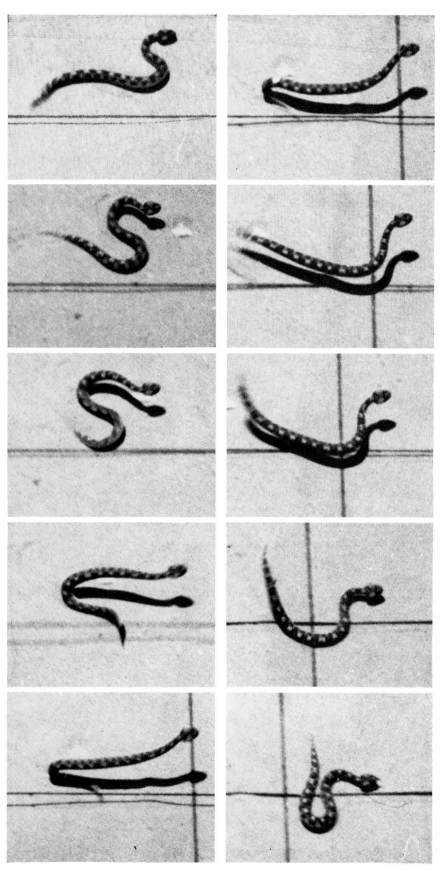

JUMPING SNAKE, the African desert viper *Bitis caudalis*, has developed this variation of the sidewinding pattern in order to escape intense heat. In the first of a series of frames from a motion-picture film this snake begins its leap by sending muscular contractions down its body toward its tail. In fourth frame down the impulses have lifted half of the snake from the ground. In the next two frames all of the snake except the tail is in the air. In the last frame the snake lands on a black line that is part of a measuring grid.

4

THE ANTIQUITY OF HUMAN WALKING

JOHN NAPIER

April 1967

*Man's unique striding gait may be the most significant
ability that sets him apart from his ancestors. A big-toe bone
found in Tanzania is evidence that this ability dates back
more than a million years*

Human walking is a unique activity during which the body, step by step, teeters on the edge of catastrophe. The fact that man has used this form of locomotion for more than a million years has only recently been demonstrated by fossil evidence. The antiquity of this human trait is particularly noteworthy because walking with a striding gait is probably the most significant of the many evolved capacities that separate men from more primitive hominids. The fossil evidence—the terminal bone of a right big toe discovered in 1961 in Olduvai Gorge in Tanzania—sets up a new signpost that not only clarifies the course of human evolution but also helps to guide those who speculate on the forces that converted predominantly quadrupedal animals into habitual bipeds.

Man's bipedal mode of walking seems potentially catastrophic because only the rhythmic forward movement of first one leg and then the other keeps him from falling flat on his face. Consider the sequence of events whenever a man sets out in pursuit of his center of gravity. A stride begins when the muscles of the calf relax and the walker's body sways forward (gravity supplying the energy needed to overcome the body's inertia). The sway places the center of body weight in front of the supporting pedestal normally formed by the two feet. As a result one or the other of the walker's legs must swing forward so that when his foot makes contact with the ground, the area of the supporting pedestal has been widened and the center of body weight once again rests safely within it. The pelvis plays an important role in this action: its degree of rotation determines the distance the swinging leg can move forward, and its muscles help to keep the body balanced while the leg is swinging.

At this point the "stance" leg—the leg still to the rear of the body's center of gravity—provides the propulsive force that drives the body forward. The walker applies this force by using muscular energy, pushing against the ground first with the ball of his foot and then with his big toe. The action constitutes the "push-off," which terminates the stance phase of the walking cycle. Once the stance foot leaves the ground, the walker's leg enters the starting, or "swing," phase of the cycle. As the leg swings forward it is able to clear the ground because it is bent at the hip, knee and ankle. This high-stepping action substantially reduces the leg's moment of inertia. Before making contact with the ground and ending the swing phase the leg straightens at the knee but remains bent at the ankle. As a result it is the

heel that strikes the ground first. The "heel strike" concludes the swing phase; as the body continues to move forward the leg once again enters the stance phase, during which the point of contact between foot and ground moves progressively nearer the toes. At the extreme end of the stance phase, as before, all the walker's propulsive thrust is delivered by the robust terminal bone of his big toe.

A complete walking cycle is considered to extend from the heel strike of one leg to the next heel strike of the same leg; it consists of the stance phase followed by the swing phase. The relative duration of the two phases depends on the cadence or speed of the walk. During normal walking the stance phase constitutes about 60 percent of the cycle and the swing phase 40 percent. Although

WALKING MAN, photographed by Eadweard Muybridge in 1884 during his studies of human and animal motion, exhibits the characteristic striding gait of the modern human.

the action of only one leg has been described in this account, the opposite leg obviously moves in a reciprocal fashion; when one leg is moving in the swing phase, the other leg is in its stance phase and keeps the body poised. Actually during normal walking the two phases overlap, so that both feet are on the ground at the same time for about 25 percent of the cycle. As walking speed increases, this period of double leg-support shortens.

Anyone who has watched other people walking and reflected a little on the process has noticed that the human stride demands both an up-and-down and a side-to-side displacement of the body. When two people walk side by side but out of step, the alternate bobbing of their heads makes it evident that the bodies undergo a vertical displacement with each stride. When two people walk in step but with opposite feet leading, they will sway first toward each other and then away in an equally graphic demonstration of the lateral displacement at each stride. When both displacements are plotted sequentially, a pair of low-amplitude sinusoidal curves appear, one in the vertical plane and the other in the horizontal [see illustrations on next page]. General observations of this kind were reduced to precise measurements during World War II when a group at the University of California at Berkeley led by H. D. Eberhart conducted a fundamental investigation of human walking in connection with requirements for the design of artificial legs. Eberhart and his colleagues found that a number of

functional determinants interacted to move the human body's center of gravity through space with a minimum expenditure of energy. In all they isolated six major elements related to hip, knee and foot movement that, working together, reduced both the amplitude of the two sine curves and the abruptness with which vertical and lateral changes in direction took place. If any one of these six elements was disturbed, an irregularity was injected into the normally smooth, undulating flow of walking, thereby producing a limp. What is more important, the irregularity brought about a measurable increase in the body's energy output during each step.

The Evidence of the Bones

What I have described in general and Eberhart's group studied in detail is the form of walking known as striding. It is characterized by the heel strike at the start of the stance phase and the push-off at its conclusion. Not all human walking is striding; when a man moves about slowly or walks on a slippery surface, he may take short steps in which both push-off and heel strike are absent. The foot is simply lifted from the ground at the end of the stance phase and set down flat at the end of the swing phase. The stride, however, is the essence of human bipedalism and the criterion by which the evolutionary status of a hominid walker must be judged. This being the case, it is illuminating to consider how the act of striding leaves its distinctive marks on the bones of the strider.

To take the pelvis first, there is a well-known clinical manifestation called Trendelenburg's sign that is regarded as evidence of hip disease in children. When a normal child stands on one leg, two muscles connecting that leg and the pelvis—the gluteus medius and the gluteus minimus—contract; this contraction, pulling on the pelvis, tilts it and holds it poised over the stance leg. When the hip is diseased, this mechanism fails to operate and the child shows a positive Trendelenburg's sign: the body tends to fall toward the unsupported side.

The same mechanism operates in walking, although not to the same degree. During the stance phase of the walking cycle, the same two gluteal muscles on the stance side brace the pelvis by cantilever action. Although actual tilting toward the stance side does not occur in normal walking, the action of the muscles in stabilizing the walker's hip is an essential component of the striding gait. Without this action the stride would become a slow, ungainly shuffle.

At the same time that the pelvis is stabilized in relation to the stance leg it also rotates to the unsupported side. This rotation, although small, has the effect of increasing the length of the stride. A familiar feature of the way women walk arises from this bit of anatomical mechanics. The difference in the proportions of the male and the female pelvis has the effect of slightly diminishing the range through which the female hip can move forward and back. Thus for a given length of stride women are obliged to rotate the pelvis through a greater

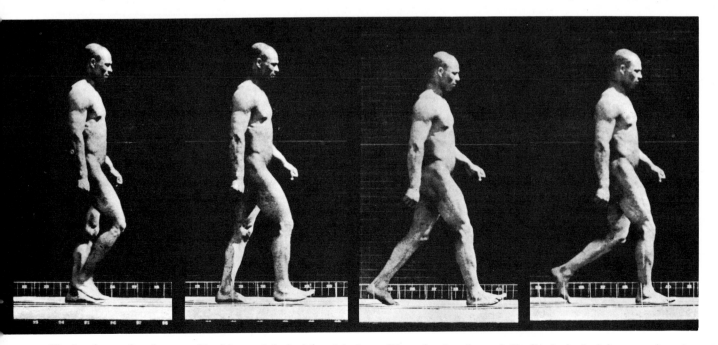

The free foot strikes the ground heel first and the body's weight is gradually transferred from heel to ball of foot as the opposite leg lifts and swings forward. Finally the heel of the stance foot rises and the leg's last contact with the ground is made with the big toe.

WALKING CYCLE extends from the heel strike of one leg to the next heel strike by the same leg. In the photograph, made by Gjon Mili in the course of a study aimed at improvement of artificial legs that he conducted for the U.S. Army, multiple exposures trace the progress of the right leg in the course of two strides. The ribbons of light allow analysis of the movement (*see illustration below*).

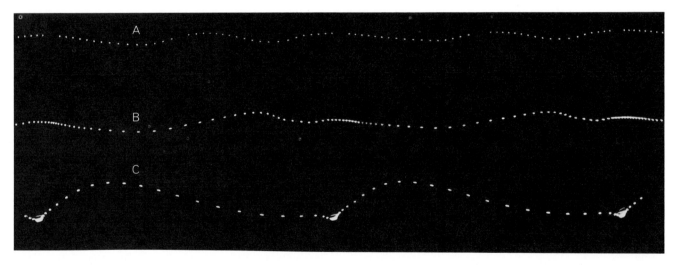

SINE CURVE described by the hip of a walking man was recorded on film by means of the experimental system illustrated above. An interrupter blade, passing in front of the camera lens at constant speed, broke the light from lamps attached to the walker into the three rows of dots. The speed of hip (*a*), knee (*b*) or ankle (*c*) during the stride is determined by measuring between the dots.

angle than men do. This secondary sexual characteristic has not lacked exploitation; at least in our culture female pelvic rotation has considerable erotogenic significance. What is more to the point in terms of human evolution is that both the rotation and the balancing of the pelvis leave unmistakable signs on the pelvic bone and on the femur: the leg bone that is joined to it. It is by a study of such signs that the walking capability of a fossil hominid can be judged.

Similar considerations apply to the foot. One way the role of the foot in walking can be studied is to record the vertical forces acting on each part of the foot while it is in contact with the ground during the stance phase of the walking cycle. Many devices have been built for this purpose; one of them is the plastic pedograph. When the subject walks across the surface of the pedograph, a motion-picture camera simultaneously records the exact position of the foot in profile and the pattern of pressures on the surface. Pedograph analyses show that the initial contact between the striding leg and the ground is the heel strike. Because the foot is normally turned out slightly at the end of the swing phase of the walking cycle, the outer side of the back of the heel takes the brunt of the initial contact [*see illustration on opposite page*]. The outer side of the foot continues to support most of the pressure of the stance until a point about three-fifths of the way along the sole is reached. The weight of the body is then transferred to the ball of the foot and then to the big toe. In the penultimate stage of push-off the brunt of the pressure is under the toes, particularly the big toe. Finally, at the end of the stance phase, only the big toe is involved; it progressively loses contact with the ground and the final push-off is applied through its broad terminal bone.

The use of pedographs and similar apparatus provides precise evidence about the function of the foot in walking, but every physician knows that much the same information is recorded on the soles of everyone's shoes. Assuming that the shoes fit, their pattern of wear is a true record of the individual's habitual gait. The wear pattern will reveal a limp that one man is trying to hide, or unmask one that another man is trying to feign, perhaps to provide evidence for an insurance claim. In any case, just as the form of the pelvis and the femur can disclose the presence or absence of a striding gait, so can the form of the foot bones, particularly the form and proportions of the big-toe bones.

The Origins of Primate Bipedalism

Almost all primates can stand on their hind limbs, and many occasionally walk in this way. But our primate relatives are all, in a manner of speaking, amateurs; only man has taken up the business of bipedalism intensively. This raises two major questions. First, how did the basic postural adaptations that permit walking—occasional or habitual—arise among the primates? Second, what advantages did habitual bipedalism bestow on early man?

With regard to the first question, I have been concerned for some time with the anatomical proportions of all primates, not only man and the apes but also the monkeys and lower primate forms. Such consideration makes it possible to place the primates in natural groups according to their mode of locomotion. Not long ago I suggested a new group, and it is the only one that will concern us here. The group comprises primates with very long hind limbs and very short forelimbs. At about the same time my colleague Alan C. Walker, now at Makerere University College in Uganda, had begun a special study of the locomotion of living and fossil lemurs. Lemurs are among the most primitive offshoots of the basic primate stock. Early in Walker's studies he was struck by the frequency with which a posture best described as "vertical clinging" appeared in the day-to-day behavior of living lemurs. All the animals whose propensity for vertical clinging had been observed by Walker showed the same proportions—that is, long hind limbs and short forelimbs—I had proposed as forming a distinct locomotor group.

When Walker and I compared notes, we decided to define a hitherto unrecognized locomotor category among the primates that we named "vertical clinging and leaping," a term that includes both the animal's typical resting posture and the essential leaping component

in its locomotion. Since proposing this category a most interesting and important extension of the hypothesis has become apparent to us. Some of the earliest primate fossils known, preserved in sediments laid down during Eocene times and therefore as much as 50 million years old, are represented not only by skulls and jaws but also by a few limb bones. In their proportions and details most of these limb bones show the same characteristics that are displayed by the living members of our vertical-clinging-and-leaping group today. Not long ago Elwyn L. Simons of Yale University presented a reconstruction of the lemur-

like North American Eocene primate *Smilodectes* walking along a tree branch in a quadrupedal position [see "The Early Relatives of Man," by Elwyn L. Simons; SCIENTIFIC AMERICAN Offprint 622]. Walker and I would prefer to see *Smilodectes* portrayed in the vertical clinging posture its anatomy unequivocally indicates. The fossil evidence, as far as it goes, suggests to us that vertical clinging and leaping was a major primate locomotor adaptation that took place some 50 million years ago. It may even have been the initial dynamic adaptation to tree life from which the subsequent locomotor patterns of all the living pri-

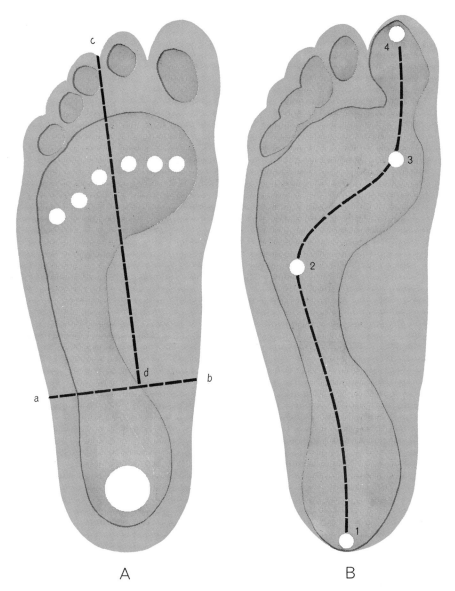

A   B

DISTRIBUTION OF WEIGHT in the human foot alters radically as action takes the place of rest. When motionless (A), the foot divides its static load (half of the body's total weight) between its heel and its ball along the axis *a–b*. The load on the ball of the foot is further divided equally on each side of the axis *c–d*. When striding (B), the load (all of the body's weight during part of each stride) is distributed dynamically from the first point of contact (1, *heel strike*) in a smooth flow via the first and fifth metatarsal bones (2, 3) that ends with a propulsive thrust (4, *push-off*) delivered by the terminal bone of the big toe.

mates, including man, have stemmed.

Walker and I are not alone in this view. In 1962 W. L. Straus, Jr., of Johns Hopkins University declared: "It can safely be assumed that primates early developed the mechanisms permitting maintenance of the trunk in the upright position.... Indeed, this tendency toward truncal erectness can be regarded as an essentially basic primate character." The central adaptations for erectness of the body, which have been retained in the majority of living primates, seem to have provided the necessary anatomical basis for the occasional bipedal behavior exhibited by today's monkeys and apes.

What we are concerned with here is the transition from a distant, hypothetical vertical-clinging ancestor to modern, bipedal man. The transition was almost

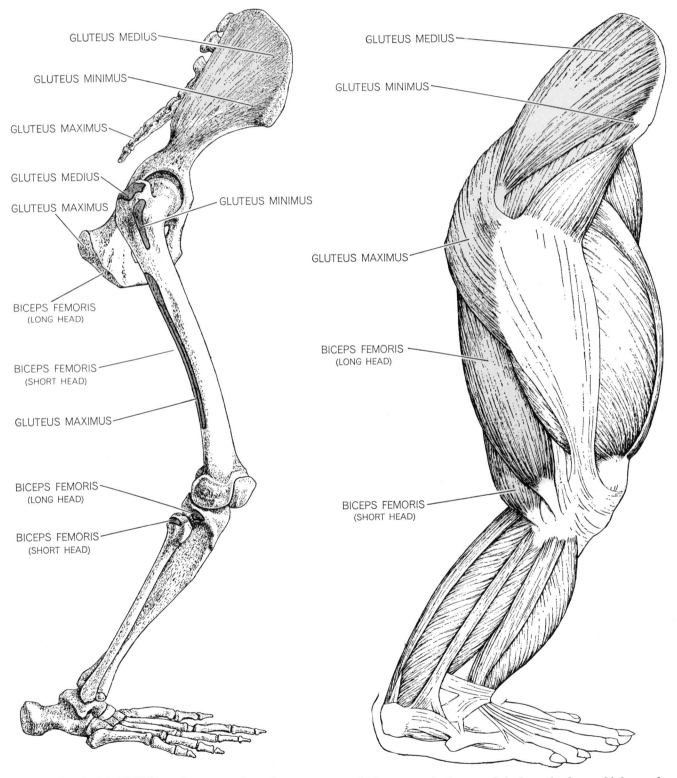

QUADRUPEDAL POSTURE needs two sets of muscles to act as the principal extensors of the hip. These are the gluteal group (the gluteus medius and minimus in particular), which connects the pelvis to the upper part of the femur, and the hamstring group. which connects the femur and the lower leg bones. Of these only the biceps femoris is shown in the gorilla musculature at right. The skeletal regions to which these muscles attach are shown in color at left. In most primates the gluteus maximus is quite small.

certainly marked by an intermediate quadrupedal stage. Possibly such Miocene fossil forms as *Proconsul,* a chimpanzee-like early primate from East Africa, represent such a stage. The structural adaptations necessary to convert a quadrupedal ape into a bipedal hom-inid are centered on the pelvis, the femur, the foot and the musculature associated with these bones. Among the nonhuman primates living today the pelvis and femur are adapted for four-footed walking; the functional relations between hipbones and thigh muscles are such that, when the animal attempts to assume a bipedal stance, the hip joint is subjected to a stress and the hip must be bent. To compensate for the resulting forward shift of the center of gravity, the knees must also be bent. In order to alter a bent-hip, bent-knee gait into

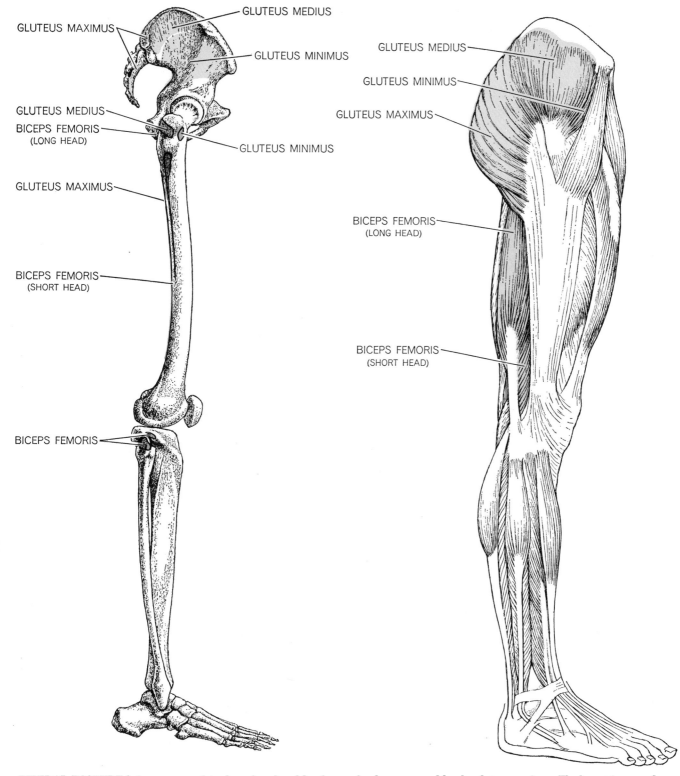

BIPEDAL POSTURE brings a reversal in the roles played by the same pelvic and femoral muscles. Gluteus medius and gluteus minimus have changed from extensors to abductors and the function of extending the trunk, required when a biped runs or climbs, has been assumed by the gluteus maximus. The hamstring muscles, in turn, now act mainly as stabilizers and extensors of the hip. At right are the muscles as they appear in man; the skeletal regions to which their upper and lower ends attach are shown in color at left.

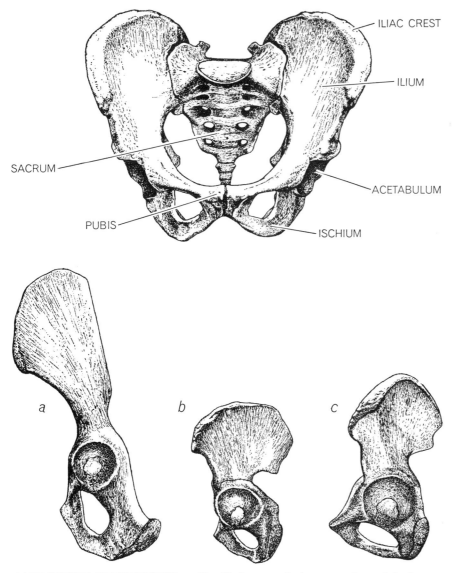

COMPONENTS OF THE PELVIS are identified at top; the bones are those of the human pelvis. Below, ilium and ischium of a gorilla (a), of *Australopithecus* (b) and of modern man (c) are seen from the side (the front is to the left in each instance). The ischium of *Australopithecus* is longer than man's; this almost certainly kept the early hominid from striding in the manner of *Homo sapiens*. Instead the gait was probably a kind of jog trot.

man's erect, striding walk, a number of anatomical changes must occur. These include an elongation of the hind limbs with respect to the forelimbs, a shortening and broadening of the pelvis, adjustments of the musculature of the hip (in order to stabilize the trunk during the act of walking upright), a straightening of both hip and knee and considerable reshaping of the foot.

Which of these changes can be considered to be primary and which secondary is still a matter that needs elucidation. Sherwood L. Washburn of the University of California at Berkeley has expressed the view that the change from four-footed to two-footed posture was initiated by a modification in the form and function of the gluteus maximus, a thigh muscle that is powerfully developed in man but weakly developed in monkeys and apes [see illustrations on preceding two pages]. In a quadrupedal primate the principal extensors of the trunk are the "hamstring" muscles and the two upper-leg muscles I have already mentioned: the gluteus medius and gluteus minimus. In man these two muscles bear a different relation to the pelvis, in terms of both position and function. In technical terms they have become abductor muscles of the trunk rather than extensor muscles of the leg. It is this that enables them to play a critical part in stabilizing the pelvis in the course of striding. In man the extensor function of these two gluteal muscles has been taken over by a third, the gluteus maximus. This muscle, insignificant in other primates, plays a sur-

prisingly unimportant role in man's ability to stand, or even to walk on a level surface. In standing, for example, the principal stabilizing and extending agents are the muscles of the hamstring group. In walking on the level the gluteus maximus is so little involved that even when it is paralyzed a man's stride is virtually unimpaired. The gluteus maximus comes into its own in man when power is needed to give the hip joint more play for such activities as running, walking up a steep slope or climbing stairs [see illustration on page 56]. Its chief function in these circumstances is to correct any tendency for the human trunk to jackknife on the legs.

Because the gluteus maximus has such a specialized role I believe, in contrast to Washburn's view, that it did not assume its present form until late in the evolution of the striding gait. Rather than being the initial adaptation, this muscle's enlargement and present function appear to me far more likely to have been one of the ultimate refinements of human walking. I am in agreement with Washburn, however, when he states that changes in the ilium, or upper pelvis, would have preceded changes in the ischium, or lower pelvis [see "Tools and Human Evolution," by Sherwood L. Washburn; SCIENTIFIC AMERICAN Offprint 601]. The primary adaptation would probably have involved a forward curvature of the vertebral column in the lumbar region. Accompanying this change would have been a broadening and a forward rotation of the iliac portions of the pelvis. Together these early adaptations provide the structural basis for improving the posture of the trunk.

Assuming that we have now given at least a tentative answer to the question of how man's bipedal posture evolved, there remains to be answered the question of why. What were the advantages of habitual bipedalism? Noting the comparative energy demands of various gaits, Washburn points out that human walking is primarily an adaptation for covering long distances economically. To go a long way with a minimum of effort is an asset to a hunter; it seems plausible that evolutionary selection for hunting behavior in man was responsible for the rapid development of striding anatomy. Gordon W. Hewes of the University of Colorado suggests a possible incentive that, acting as an agent of natural selection, could have prompted the quadrupedal ancestors of man to adopt a two-footed gait. In Hewes's view the principal advantage of bipedalism over quadrupedalism would be the free-

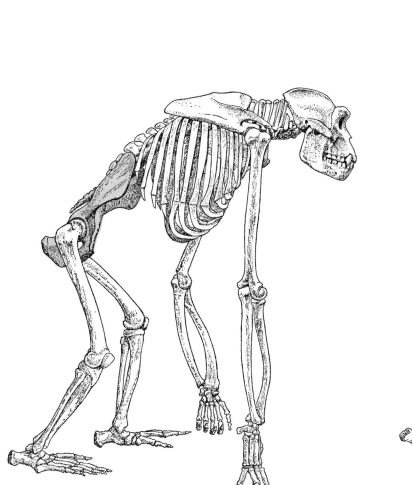

SHAPE AND ORIENTATION of the pelvis in the gorilla and in man reflect the postural differences between quadrupedal and bipedal locomotion. The ischium in the gorilla is long, the ilium ex- tends to the side and the whole pelvis is tilted toward the horizontal (*see illustration on opposite page*). In man the ischium is much shorter, the broad ilium extends forward and the pelvis is vertical.

TROPICAL FOREST   WOODLAND SAVANNA   OPEN GRASSLAND

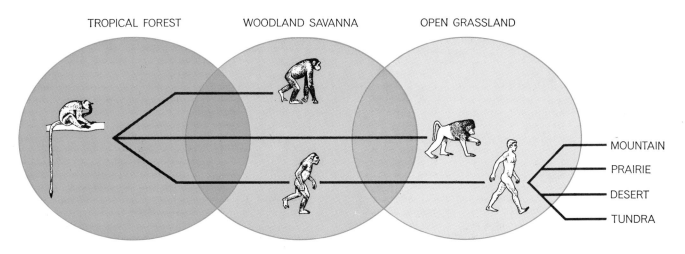

MOUNTAIN

PRAIRIE

DESERT

TUNDRA

ECOLOGICAL PATHWAY to man's eventual mastery of all environments begins (*left*) with a quadrupedal primate ancestor living in tropical forest more than 20 million years ago. During Miocene times mountain-building produced new environments. One, a transition zone between forest and grassland, has been exploited by three groups of primates. Some, for example the chimpanzees, have only recently entered this woodland savanna. Both the newly bipedal hominids and some ground-living quadrupedal monkeys, however, moved beyond the transition zone into open grassland. The quadrupeds, for example the baboons, remained there. On the other hand, the forces of natural selection in the new setting favored the bipedal hominid hunters' adaptation of the striding gait typical of man. Once this adaptation developed, man went on to conquer most of the earth's environments.

ing of the hands, so that food could be carried readily from one place to another for later consumption. To assess the significance of such factors as survival mechanisms it behooves us to review briefly the ecological situation in which our prehuman ancestors found themselves in Miocene times, between 15 and 25 million years ago.

The Miocene Environment

During the Miocene epoch the worldwide mountain-building activity of middle Tertiary times was in full swing. Many parts of the earth, including the region of East Africa where primates of the genus *Proconsul* were living, were being faulted and uplifted to form such mountain zones as the Alps, the Himalayas, the Andes and the Rockies. Massive faulting in Africa gave rise to one of the earth's major geological features: the Rift Valley, which extends 5,000 miles from Tanzania across East Africa to Israel and the Dead Sea. A string of lakes lies along the floor of the Rift Valley like giant stepping-stones. On their shores in Miocene times lived a fantastically rich fauna, inhabitants of the forest and of a new ecological niche—the grassy savanna.

These grasslands of the Miocene were the domain of new forms of vegetation that in many parts of the world had taken the place of rain forest, the dominant form of vegetation in the Eocene and the Oligocene. The savanna offered new evolutionary opportunities to a variety of mammals, including the expanding population of primates in the rapidly shrinking forest. A few primates—the ancestors of man and probably also the ancestors of the living baboons—evidently reacted to the challenge of the new environment.

The savanna, however, was no Eldorado. The problems facing the early hominids in the open grassland were immense. The forest foods to which they were accustomed were hard to come by; the danger of attack by predators was immeasurably increased. If, on top of everything else, the ancestral hominids of Miocene times were in the process of converting from quadrupedalism to bipedalism, it is difficult to conceive of any advantage in bipedalism that could have compensated for the added hazards of life in the open grassland. Consideration of the drawbacks of savanna living has led me to a conclusion contrary to the one generally accepted: I doubt that the advent of bipedalism took place in this environment. An environment neglected by scholars but one far better

suited for the origin of man is the woodland-savanna, which is neither high forest nor open grassland. Today this halfway-house niche is occupied by many primates, for example the vervet monkey and some chimpanzees. It has enough trees to provide forest foods and ready escape from predators. At the same time its open grassy spaces are arenas in which new locomotor adaptations can be practiced and new foods can be sampled. In short, the woodland-savanna provides an ideal nursery for evolving hominids, combining the challenge and incentive of the open grassland with much of the security of the forest. It was probably in this transitional environment that man's ancestors learned to walk on two legs. In all likelihood, however, they only learned to stride when they later moved into the open savanna.

Moving forward many millions of years from Miocene to Pleistocene times, we come to man's most immediate hominid precursor: *Australopithecus*. A large consortium of authorities agrees that the shape of the pelvis in *Australopithecus* fossils indicates that these hominids were habitually bipedal, although not to the degree of perfection exhibited by modern man. A few anatomists, fighting a rearguard action, contend that on the contrary the pelvis of *Australopithecus*

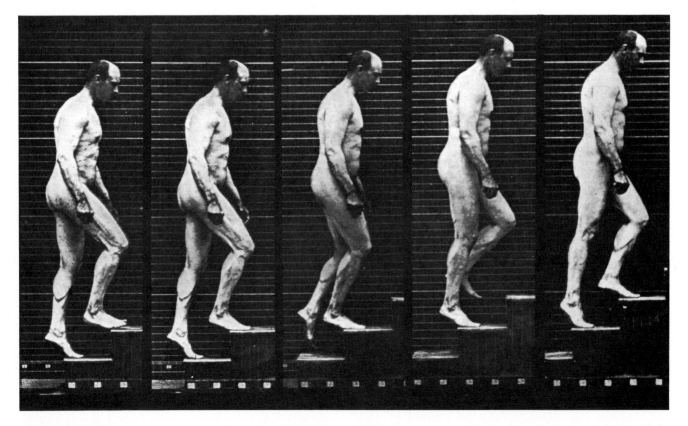

STAIR-CLIMBING, like running, is a movement that brings the human gluteus maximus into play. Acting as an extensor of the **trunk, the muscle counteracts any tendency for the body to jackknife over the legs. Photographs are from Muybridge's collection.**

shows that these hominids were predominantly quadrupedal. I belong to the first school but, as I have been at some pains to emphasize in the past, the kind of upright walking practiced by *Australopithecus* should not be equated with man's heel-and-toe, striding gait.

From Bipedalist to Strider

The stride, although it was not necessarily habitual among the earliest true men, is nevertheless the quintessence of the human locomotor achievement. Among other things, striding involves extension of the leg to a position behind the vertical axis of the spinal column. The degree of extension needed can only be achieved if the ischium of the pelvis is short. But the ischium of *Australopithecus* is long, almost as long as the ischium of an ape [*see illustration on page 54*]. Moreover, it has been shown that in man the gluteus medius and the gluteus minimus are prime movers in stabilizing the pelvis during each stride; in *Australopithecus* this stabilizing mechanism is imperfectly evolved. The combination of both deficiencies almost entirely precludes the possibility that these hominids possessed a striding gait. For *Australopithecus* walking was something of a jog trot. These hominids must have covered the ground with quick, rather short steps, with their knees and hips slightly bent; the prolonged stance phase of the fully human gait must surely have been absent.

Compared with man's stride, therefore, the gait of *Australopithecus* is physiologically inefficient. It calls for a disproportionately high output of energy; indeed, *Australopithecus* probably found long-distance bipedal travel impossible. A natural question arises in this connection. Could the greater energy requirement have led these early representatives of the human family to alter their diet in the direction of an increased reliance on high-energy foodstuffs, such as the flesh of other animals?

The pelvis of *Australopithecus* bears evidence that this hominid walker could scarcely have been a strider. Let us now turn to the foot of what many of us believe is a more advanced hominid. In 1960 L. S. B. Leakey and his wife Mary unearthed most of the bones of this foot in the lower strata at Olduvai Gorge known collectively as Bed I, which are about 1.75 million years old. The bones formed part of a fossil assemblage that has been designated by the Leakeys, by Philip Tobias of the University of the Witwatersrand and by me as possibly the earliest-known species of man: *Homo*

habilis. The foot was complete except for the back of the heel and the terminal bones of the toes; its surviving components were assembled and studied by me and Michael Day, one of my colleagues at the Unit of Primatology and Human Evolution of the Royal Free Hospital School of Medicine in London. On the basis of functional analysis the resem-

blance to the foot of modern man is close, although differing in a few minor particulars. Perhaps the most significant point of resemblance is that the stout basal bone of the big toe lies alongside the other toes [*see upper illustration on next page*]. This is an essentially human characteristic; in apes and monkeys the big toe is not exceptionally robust and

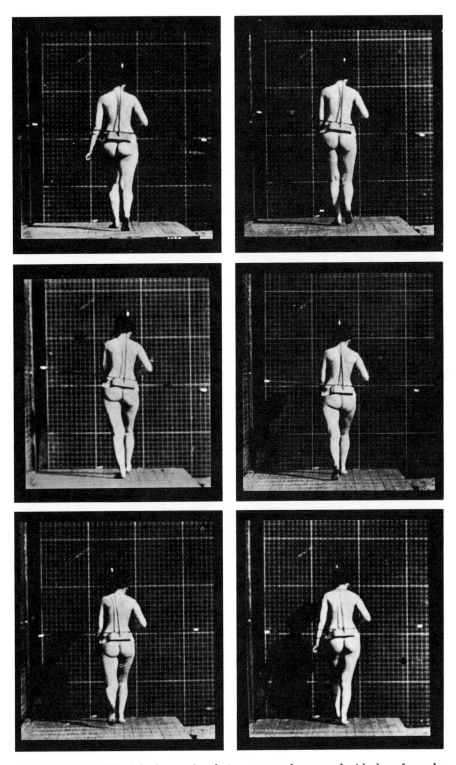

PELVIC ROTATION of the human female is exaggerated compared with that of a male taking a stride of equal length because the two sexes differ in pelvic anatomy. Muybridge noted the phenomenon, using a pole with whitened ends to record the pelvic oscillations.

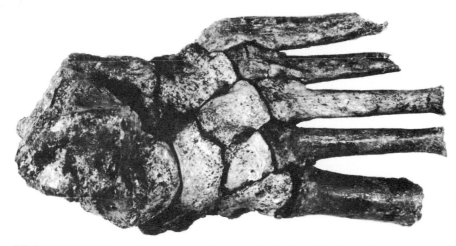

PRIMITIVE FOOT, complete except for the back of the heel and the tips of the toes, was unearthed from the lower level at Olduvai Gorge in Tanzania. Attributed to a very early hominid, Homo habilis, by its discoverer, L. S. B. Leakey, it is about 1.75 million years old. Its appearance suggests that the possessor was a habitual biped. Absence of the terminal bones of the toes, however, leaves open the question of whether the possessor walked with a stride.

BIG-TOE BONE, also discovered at Olduvai Gorge, is considerably younger than the foot bones in the top illustration but still probably more than a million years old. It is the toe's terminal bone (*bottom view at left, top view at right*) and bore the thrust of its possessor's push-off with each swing of the right leg. The tilting and twisting of the head of the bone in relation to the shaft is unequivocal evidence that its possessor walked with a modern stride.

diverges widely from the other toes. The foot bones, therefore, give evidence that this early hominid species was habitually bipedal. In the absence of the terminal bones of the toes, however, there was no certainty that *Homo habilis* walked with a striding gait.

Then in 1961, in a somewhat higher stratum at Olduvai Gorge (and thus in a slightly younger geological formation), a single bone came to light in an area otherwise barren of human bones. This fossil is the big-toe bone I mentioned at the beginning of this article [*see lower illustration at left*]. Its head is both tilted and twisted with respect to its shaft, characteristics that are found only in modern man and that can with assurance be correlated with a striding gait. Day has recently completed a dimensional analysis of the bone, using a multivariate statistical technique. He is able to show that the fossil is unquestionably human in form.

There is no evidence to link the big-toe bone specifically to either of the two recognized hominids whose fossil remains have been recovered from Bed I at Olduvai: *Homo habilis* and *Zinjanthropus boisei*. Thus the owner of the toe remains unknown, at least for the present. Nonetheless, one thing is made certain by the discovery. We now know that in East Africa more than a million years ago there existed a creature whose mode of locomotion was essentially human.

5

THE EVOLUTION OF THE HAND

JOHN NAPIER
December 1962

In 1960 tools were found together with the hand bones of a prehuman primate that lived more than a million years ago. This indicates that the hand of modern man has much later origins than had been thought

At Olduvai Gorge in Tanganyika two years ago L. S. B. Leakey and his wife Mary unearthed 15 bones from the hand of an early hominid. They found the bones on a well-defined living floor a few feet below the site at which in the summer of 1959 they had excavated the skull of a million-year-old man-ape to which they gave the name *Zinjanthropus*. The discovery of *Zinjanthropus* has necessitated a complete revision of previous views about the cultural and biological evolution of man. The skull was found in association with stone tools and waste flakes indicating that at this ancient horizon toolmakers were already in existence. The floor on which the hand bones were discovered has also yielded stone tools and a genuine bone "lissoir," or leather working tool. Hence this even older living site carries the origins of toolmaking still further back, both in time and evolution, and it is now possible for the first time to reconstruct the hand of the earliest toolmakers.

Research and speculation on the course of human evolution have hitherto paid scant attention to the part played by the hand. Only last year I wrote: "It is a matter of considerable surprise to many to learn that the human hand, which can achieve so much in the field of creative art, communicate such subtle shades of meaning, and upon which the pre-eminence of *Homo sapiens* in the world of animals so largely depends, should constitute, in a structural sense, one of the most primitive and generalized parts of the human body." The implication of this statement, which expresses an almost traditional view, is that the primate forebears of man were equipped with a hand of essentially human form long before the cerebral capacity necessary to exploit its potential had appeared. The corollary to this view is that the difference between the human hand and the monkey hand, as the late Frederic Wood Jones of the Royal College of Surgeons used to insist, is largely one of function rather than structure. Although broadly speaking it is true that the human hand has an extraordinarily generalized structure, the discovery of the Olduvai hand indicates that in a number of minor but nevertheless highly significant features the hand is more specialized than we had supposed.

Tool-using—in the sense of improvisa-

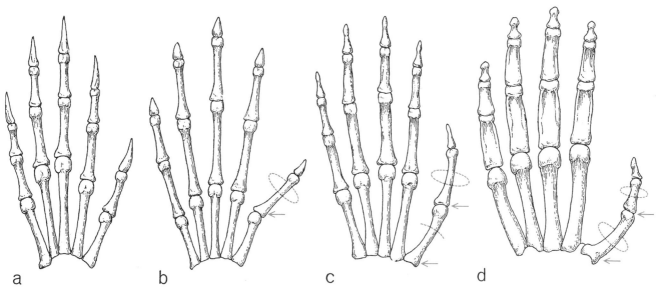

a b c d

HANDS OF LIVING PRIMATES, all drawn same size, show evolutionary changes in structure related to increasing manual dexterity. Tree shrew (*a*) shows beginnings of unique primate possession, specialized thumb (*digit at right*). In tarsier (*b*) thumb is distinct and can rotate around joint between digit and palm. In capuchin monkey (*c*), a typical New World species, angle between thumb and finger is wider and movement can be initiated at joint at base of palm. Gorilla (*d*), like other Old World species, has saddle joint at base of palm. This allows full rotation of thumb, which is set at a wide angle. Only palm and hand bones are shown here.

tion with naturally occurring objects such as sticks and stones—by the higher apes has often been observed both in the laboratory and in the wild and has even been reported in monkeys. The making of tools, on the other hand, has been regarded as the major breakthrough in human evolution, a sort of status symbol that could be employed to distinguish the genus *Homo* from the rest of the primates. Prior to the discovery of *Zinjanthropus,* the South African man-apes (Australopithecines) had been associated at least indirectly with fabri-

cated tools. Observers were reluctant to credit the man-apes with being tool-makers, however, on the ground that they lacked an adequate cranial capacity. Now that hands as well as skulls have been found at the same site with undoubted tools, one can begin to correlate the evolution of the hand with the stage of culture and the size of the brain. By the same token one must also consider whether the transition from tool-using to toolmaking and the subsequent improvement in toolmaking techniques can be explained purely in

terms of cerebral expansion and the refinement of peripheral neuromuscular mechanisms, or whether a peripheral factor—the changing form of the hand—has played an equally important part in the evolution of the human species. And to understand the significance of the specializations of the human hand, it must be compared in action—as well as in dissection—with the hands of lower primates.

In the hand at rest—with the fingers slightly curled, the thumb lying in the plane of the index finger, the poise of the

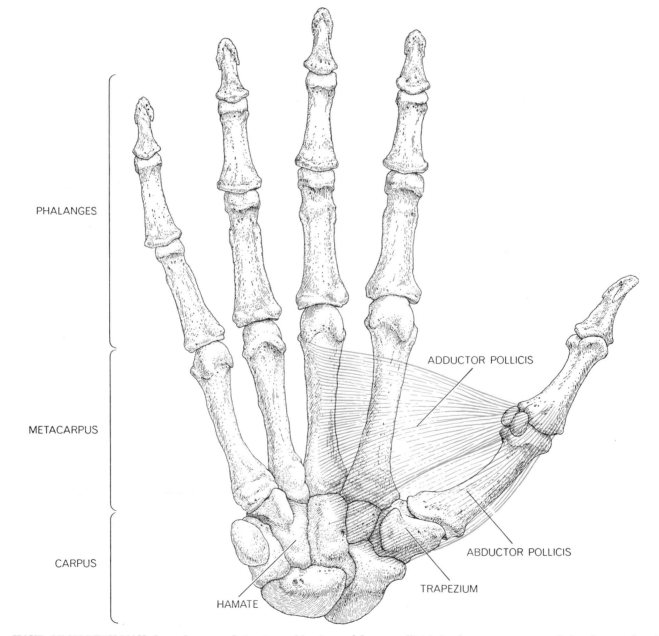

PHALANGES

METACARPUS

CARPUS

ADDUCTOR POLLICIS

ABDUCTOR POLLICIS

TRAPEZIUM

HAMATE

HAND OF MODERN MAN, drawn here actual size, is capable of precise movements available to no other species. Breadth of terminal phalanges (end bones of digits) guarantees secure thumb-to-finger grip. Thumb is long in proportion to index finger and is set at very wide angle. Strong muscles (*adductor pollicis* and *abductor pollicis*) implement movement of thumb toward and away from palm. Saddle joint at articulation of thumb metacarpal (a bone of the palm) and trapezium (a bone of the carpus, or wrist) enables thumb to rotate through 45 degrees around its own longitudinal axis and so be placed in opposition to all the other digits.

whole reflecting the balanced tension of opposing groups of muscles—one can see something of its potential capacity. From the position of rest, with a minimum of physical effort, the hand can assume either of its two prehensile working postures. The two postures are demonstrated in sequence by the employment of a screw driver to remove a screw solidly embedded in a block of wood [*see illustration below*]. The hand first grips the tool between the flexed fingers and the palm with the thumb reinforcing the pressure of the fingers; this is the "power grip." As the screw comes loose, the hand grasps the tool between one or more fingers and the thumb, with the pulps, or inner surfaces, of the finger and thumb tips fully opposed to one another; this is the "precision grip." Invariably it is the nature of the task to be performed, and not the shape of the tool or object grasped, that dictates which posture is employed. The power grip is the grip of choice when the full strength of the hand must be applied and the need for precision is subordinate; the precision grip comes into play when the need for power is secondary to the demand for fine control.

The significance of this analysis becomes apparent when the two activities are correlated with anatomical structure. The presence or absence of these structural features in the hands of a lower primate or early hominid can then be

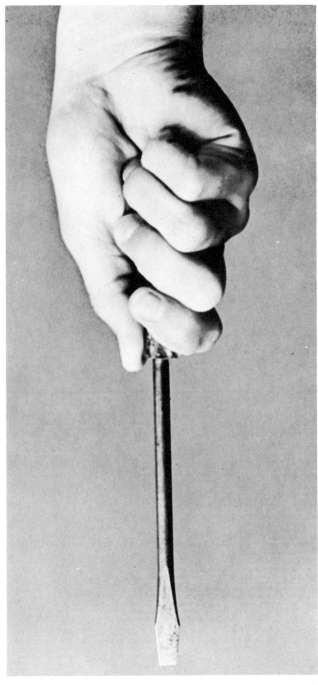

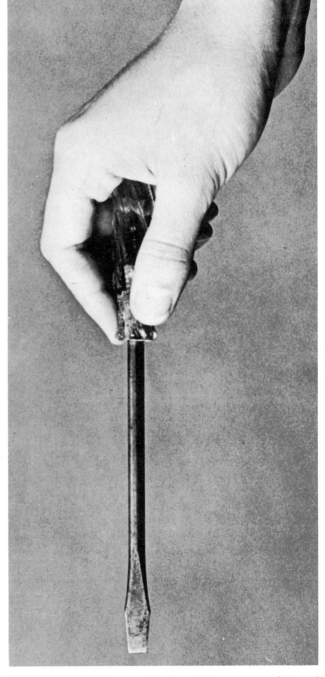

POWER GRIP is one of two basic working postures of human hand. Used when strength is needed, it involves holding object between flexed fingers and palm while the thumb applies counterpressure.

PRECISION GRIP is second basic working posture and is used when accuracy and delicacy of touch are required. Object is held between tips of one or more fingers and the fully opposed thumb.

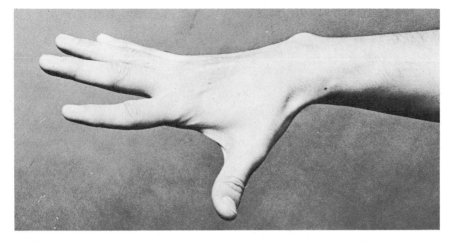

DIVERGENCE, generally associated with weight-bearing function of hand, is achieved by extension at the metacarpophalangeal joints. All mammalian paws are capable of this action.

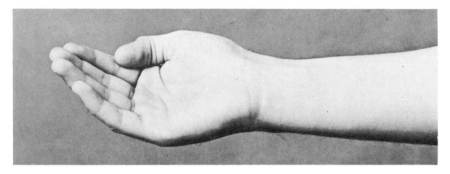

CONVERGENCE is achieved by flexion at metacarpophalangeal joints. Two convergent paws equal one prehensile hand; many mammals hold food in two convergent paws to eat.

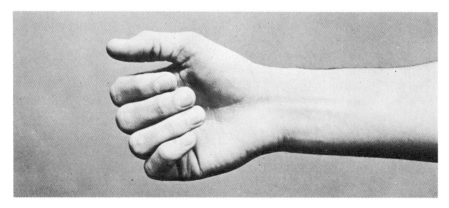

PREHENSILITY, the ability to wrap the fingers around an object, is a special primate characteristic, related to the emergence of the specialized thumb during evolutionary process.

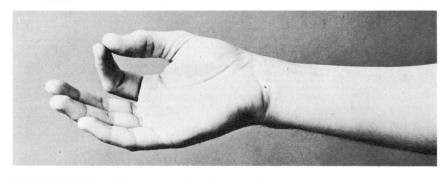

OPPOSABILITY is ability to sweep thumb across palm while rotating it around its longitudinal axis. Many primates can do this, but underlying structures are best developed in man.

taken to indicate, within limits, the capabilities of those hands in the cultural realm of tool-using and toolmaking. In the case of the hand, at least, evolution has been incremental. Although the precision grip represents the ultimate refinement in prehensility, this does not mean that more primitive capacities have been lost. The human hand remains capable of the postures and movements of the primate foot-hand and even of the paw of the fully quadrupedal mammal, and it retains many of the anatomical structures that go with them. From one stage in evolution to the next the later capability is added to the earlier.

The study of primate evolution is facilitated by the fact that the primates now living constitute a graded series representative of some of its principal chapters. It is possible, at least, to accept a study series composed of tree shrews, tarsiers, New World monkeys, Old World monkeys and man as conforming to the evolutionary sequence. In comparing the hands of these animals with one another and with man's, considerable care must be taken to recognize specializations of structure that do not form part of the sequence. Thus the extremely specialized form of the hand in the anthropoid apes can in no way be regarded as a stage in the sequence from tree shrew to man. The same objection does not apply, however, to certain fossil apes. The hand of the Miocene ancestral ape *Proconsul africanus* does not, for example, show the hand specializations of living apes and can legitimately be brought into the morphological sequence that branches off on the man-ape line toward man.

In the lowliest of the living primates —the tree shrew that inhabits the rain forests of the East Indies and the Malay Archipelago—the hand is little more than a paw. It exhibits in a primate sense only the most rudimentary manual capability. This is the movement of convergence that brings the tips of the digits together by a flexion of the paw at the metacarpophalangeal joints, which correspond in man to the knuckles at the juncture of the fingers and the rest of the hand. The opposite movement— divergence—fans the digits outward and is related to the pedal, or weight-bearing, function of the paw. With its paws thus limited the tree shrew is compelled to grasp objects, for example its insect prey, in two-handed fashion, two convergent paws being the functional equivalent of a prehensile hand. For purposes of locomotion in its arboreal

STONE TOOLS to left of center are similar to those found at Olduvai Gorge, Tanganyika, in conjunction with the hand bones of an early hominid. Such crude tools can be made by using the power grip, of which the Olduvai hand was capable. Finely flaked Old Stone Age tools at right can be made only by using the precision grip, which may not have been well developed in Olduvai hand.

habitat, this animal does not require prehensility because, like the squirrel, it is small, it has claws on the tips of its digits and is a tree runner rather than a climber. Even in the tree shrew, however, the specialized thumb of the primate family has begun to take form in the specialized anatomy of this digit and its musculature. Occasionally tree shrews have been observed feeding with one hand.

The hand of the tarsier, another denizen of the rain forests of the East Indies, exhibits a more advanced degree of prehensility in being able to grasp objects by bending the digits toward the palm. The thumb digit also exhibits a degree of opposability to the other digits. This is a pseudo opposability in that the movement is restricted entirely to the metacarpophalangeal joint and is therefore distinct from the true opposability of man's thumb. The movement is facilitated by the well-developed abductor and adductor muscles that persist in the hands of the higher primates. With this equipment the tarsier is able to support its body weight on vertical stems and to grasp small objects with one hand.

The tropical rain forests in which these animals live today are probably not very different from the closed-canopy forests of the Paleocene epoch of some 70 million years ago, during which the first primates appeared. In the wide variety of habitats that these forests provide, ecologists distinguish five major strata, superimposed like a block of apartments. From the top down these are the upper, middle and lower stories (the last being the main closed canopy), the shrub layer and the herb layer on the ground. To these can be added a sixth deck: the subterrain. In the emergence of prehensility in the primate line the three-dimensional arrangement of this system of habitats played a profound role. Prehensility is an adaptation to arboreal life and is related to climbing. In animals that are of small size with respect to the branches on which they live and travel, such as the tree shrew, mobility is not hampered by lack of prehensility. They can live at any level in the forest, from the forest floor to the tops of the tallest trees, their stability assured by the grip of sharp claws and the elaboration of visual and cerebellar mechanisms.

The tree-climbing as opposed to the tree-running phase of primate evolution may not have begun until the middle of the Eocene, perhaps 55 million years ago. What environmental pressure brought about this adaptation can only be guessed at. Thomas F. Barth of the University of Chicago has suggested that the advent of the widely successful order of rodents in the early Eocene may have led to the displacement of the primates from the shrub strata to the upper three strata of the forest canopy. In any case little is known about the form of the primates that made this transition.

In *Proconsul*, of the early to middle Miocene of 20 million years ago, the fossil record discloses a fully developed tree-climbing primate. His hand was clearly prehensile. His thumb, however, was imperfectly opposable. Functionally this hand is comparable to that of some of the living New World monkeys.

True opposability appears for the first time among the living primates in the Old World monkeys. In these animals the carpometacarpal joint shows a well-developed saddle configuration comparable to that in the corresponding joint of the human hand. This allows rotation of the thumb from its wrist articulation. Turning about its longitudinal axis through an angle of about 45 degrees, the thumb can be swept across the palm, and the pulp of the thumb can be directly opposed to the pulp surfaces of one of or all the other digits. This movement is not so expertly performed by the monkeys as by man. At the same time, again as in man, a fair range of movement is retained at the metacarpophalangeal joint, the site of pseudo opposability in the tarsier.

The hands of anthropoid apes display many of these anatomical structures but do not have the same degree of functional capability. This is because of certain specializations that arise from the fact that these apes swing from trees by their hands. Such specializations would seem to exclude the apes from the evolutionary sequence that leads to man. In comparing the hand of monkeys with the hand of man one must bear in mind an obvious fact that is all too often overlooked: monkeys are largely quadrupedal, whereas man is fully bipedal. Variations in the form of the hand from one species of monkey to the next are related to differences in their mode of locomotion. The typical monkey hand is rather long and narrow; the metacarpal, or "palm," bones are short compared with the digits (except in baboons); the terminal phalanges, or finger-tip bones, are slender and the tips of the fingers are consequently narrow from side to side. These are only the most obvious differences between the foot-hand of the Old World monkey and that of man. They serve nonetheless to show how too rigid an application of Frederic Wood Jones's criterion of morphological similarity can mislead one into assuming that the only important difference between the hands of men and monkeys lies in the elaboration of the central nervous system.

It seems likely that the terrestrial phase of human evolution followed on the heels of *Proconsul*. At that time, it is well known, the world's grasslands expanded enormously at the expense of the forests. By the end of the Miocene, 15 million years ago, most of the prototypes of the modern plains-living forms had appeared. During this period, apparently, the hominids also deserted their original forest habitats to take up life on the savanna, where the horizons were figuratively limitless. Bipedal locomotion, a process initiated by life in the trees and the ultimate mechanism for emancipation of the hands, rapidly followed the adoption of terrestrial life. The use of the hands for carrying infants, food and even weapons and tools could not have lagged far behind. As Sherwood L. Washburn of the University of California has suggested on the basis of observations of living higher primates,

tool-using must have appeared at an early stage in hominid evolution. It is a very short step from tool-using to tool-modifying, in the sense of stripping twigs and leaves from a branch in order to improve its effectiveness as a tool or weapon. It is an equally short further step to toolmaking, which at its most primitive is simply the application of the principle of modification to a stick, a stone or a bone. Animal bones are a convenient source of tools; Raymond A. Dart of the University of Witwatersrand in South Africa has advanced the hypothesis that such tools

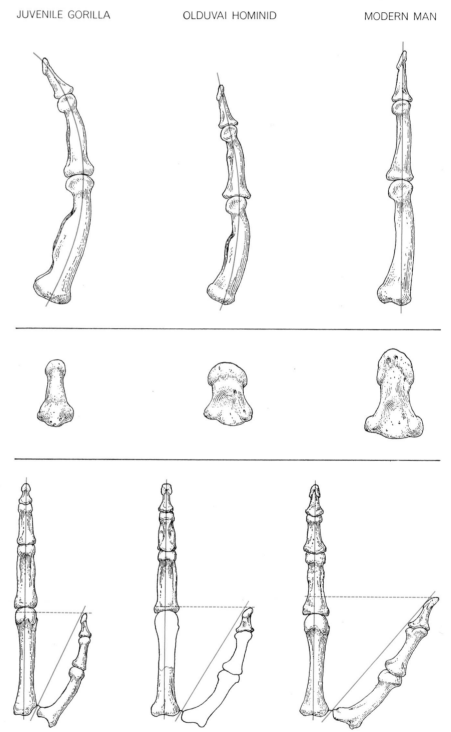

JUVENILE GORILLA   OLDUVAI HOMINID   MODERN MAN

HAND BONES of juvenile gorilla, Olduvai hominid and modern man are compared. Phalanges (*top row*) decrease in curvature from juvenile gorilla to modern man. Terminal thumb phalanx (*middle row*) increases in breadth and proportional length. Third row shows increase in length of thumb and angle between thumb and index finger. Olduvai bones in outline in third row are reconstructed from other evidence; they were not found.

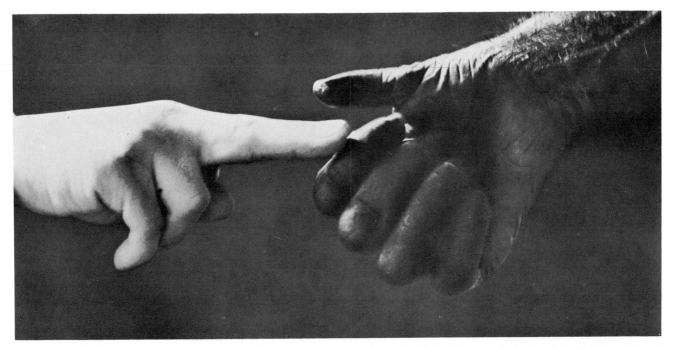

CHIMPANZEE, attempting to grasp experimenter's finger, uses an inefficient precision grip. Because animal's thumb is so short in proportion to the digits, it is compelled to bend the digits forward and grasp the object between the sides of index finger and thumb.

were used by early man-apes as part of an "osteodontokeratic" (bone-tooth-hair) culture.

The tools from the pre-*Zinjanthropus* stratum at Olduvai Gorge are little more than pebbles modified in the simplest way by striking off one or more flakes to produce a chopping edge. This technology could not have required either a particularly large brain or a hand of modern human proportions. The hand bones of the pre-*Zinjanthropus* individuals uncovered by the Leakeys in their more recent excavation of Olduvai Gorge are quite unlike those of modern *Homo sapiens*. But there seems to be no reason, on either geological or anthropological grounds, for doubting that the tools found with them are coeval. Modern man must recover from his surprise at the discovery that hands other than his own were capable of shaping tools.

At this point it may be useful to return to the analysis of the manual capability of modern man that distinguishes the power and the precision grip. When compared with the hand of modern man, the Olduvai hand appears to have been capable of a tremendously strong power grip. Although it was a smaller hand, the relative lengths of the metacarpals and phalanges indicate that the proportion of digits and palm was much the same as it is in man. In addition, the tips of the terminal bones of all the Olduvai fingers are quite wide and the finger tips themselves must therefore have been

broad—an essential feature of the human grip for both mechanical and neurological reasons. The curvature of the metacarpals and phalanges indicates that the fingers were somewhat curved throughout their length and were normally held in semiflexion. Unfortunately no hamate bone was found among the Olduvai remains. This wristbone, which articulates with the fifth metacarpal, meets at a saddle joint in modern man and lends great stability to his power grip.

It seems unlikely that the Olduvai hand was capable of the precision grip in its fullest expression. No thumb metacarpal was found in the Olduvai deposit; hence any inference as to the length of the thumb in relation to the other fingers must be derived from the evidence of the position of the wristbone with which the thumb articulates. This evidence suggests that the Olduvai thumb, like the thumb of the gorilla, was set at a narrower angle and was somewhat shorter than the thumb of modern man, reaching only a little beyond the metacarpophalangeal joint of the index finger. Thus, although the thumb was opposable, it can be deduced that the Olduvai hand could not perform actions as precise as those that can be undertaken by the hand of modern man.

Nonetheless, the Olduvai hand activated by a brain and a neuromuscular mechanism of commensurate development would have had little difficulty in making the tools that were found with it. I myself have made such pebble tools employing only the power grip to hold and strike two stones together.

The inception of toolmaking has hitherto been regarded as the milestone that marked the emergence of the genus *Homo*. It has been assumed that this development was a sudden event, happening as it were almost overnight, and that its appearance was coincidental with the structural evolution of a hominid of essentially modern human form and proportions. It is now becoming clear that this important cultural phase in evolution had its inception at a much earlier stage in the biological evolution of man, that it existed for a much longer period of time and that it was set in motion by a much less advanced hominid and a much less specialized hand than has previously been believed.

For full understanding of the subsequent improvement in toolmaking over the next few hundred thousand years of the Paleolithic, it is necessary to document the transformation of the hand as well as of the brain. Attention can now also be directed toward evidence of the functional capabilities of the hands of early man that is provided by the tools they made. These studies may help to account for the radical changes in technique and direction that characterize the evolution of stone implements during the middle and late Pleistocene epoch. The present evidence suggests that the stone implements of early man were as good (or as bad) as the hands that made them.

6

BIRDS AS FLYING MACHINES

CARL WELTY
March 1955

A sequel to the article on the aerodynamics of birds in the April, 1952, issue of Scientific American. *Among the remarkable adaptations birds have made to life in the air are high power and light weight*

The great struggle in most animals' lives is to avoid change. A chickadee clinging to a piece of suet on a bitter winter day is doing its unconscious best to maintain its internal status quo. Physiological constancy is the first biological commandment. An animal must eternally strive to keep itself warm, moist and supplied with oxygen, sugar, protein, salts, vitamins and the like, often within precise limits. Too great a change in its internal economy means death.

The spectacular flying performances of birds—spanning oceans, deserts and whole continents—tend to obscure the more important fact that the ability to fly confers on them a remarkably useful mechanism to preserve their internal stability, or homeostasis. Through flight birds can search out the external conditions and substances they need to keep their internal fires burning clean and steady. A bird's wide search for specific foods and habitats makes sense only when considered in the light of this persistent, urgent need for constancy.

The power of flight opens up to birds an enormous gaseous ocean, the atmosphere, and a means of quick, direct access to almost any spot on earth. They can eat in almost any "restaurant"; they have an almost infinite choice of sites to build their homes. As a result birds are, numerically at least, the most successful vertebrates on earth. They number roughly 25,000 species and subspecies, as compared with 15,000 mammals and 15,000 fishes.

At first glance birds appear to be quite variable. They differ considerably in size, body proportions, color, song and ability to fly. But a deeper look shows that they are far more uniform than, say, mammals. The largest living bird, a 125-pound ostrich, is about 20,000 times heavier than the smallest bird, a hummingbird weighing only one tenth of an ounce. However, the largest mammal, a 200,000-pound blue whale, weighs some 22 million times as much as the smallest mammal, the one-seventh-ounce masked shrew. Mammals, in other words, vary in mass more than a thousand times as much as birds. In body architecture, the comparative uniformity of birds is even more striking. Mammals may be as fat as a walrus or as slim as a weasel, furry as a musk ox or hairless as a desert rat,

long as a whale or short as a mole. They may be built to swim, crawl, burrow, run or climb. But the design of nearly all species of birds is tailored to and dictated by one pre-eminent activity—flying. Their structure, outside and inside, constitutes a solution to the problems imposed by flight. Their uniformity has been thrust on them by the drastic demands that determine the design of any flying machine. Birds simply dare not deviate widely from sound aerodynamic design. Nature liquidates deviationists much more consistently and drastically than does any totalitarian dictator.

Birds were able to become flying machines largely through the evolutionary gifts of feathers, wings, hollow bones, warm-bloodedness, a remarkable system of respiration, a strong, large heart and powerful breast muscles. These adaptations all boil down to the two prime requirements for any flying machine: high power and low weight. Birds have thrown all excess baggage overboard. To keep their weight low and feathers dry they forego the luxury of sweat glands. They have even reduced

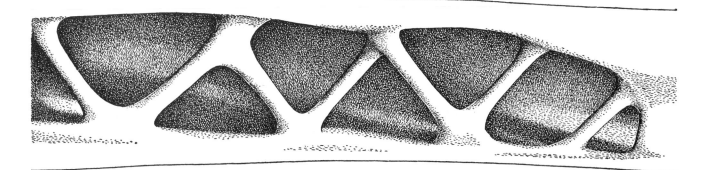

INTERNAL STRUCTURE of the metacarpal bone of a vulture's wing is shown in this drawing of a longitudinal section. The braces within the bone are almost identical in geometry with those of the Warren truss commonly used as a steel structural member.

their reproductive organs to a minimum. The female has only one ovary, and during the nonbreeding season the sex organs of both males and females atrophy. T. H. Bissonette, the well-known investigator of birds and photoperiodicity, found that in starlings the organs weigh 1,500 times as much during the breeding season as during the rest of the year.

As early as 1679 the Italian physicist Giovanni Borelli, in his *De motu animalium*, noted some of the weight-saving features of bird anatomy: ". . . the body of a Bird is disproportionately lighter than that of man or of any quadruped . . . since the bones of birds are porous, hollowed out to extreme thinness like the roots of the feathers, and the shoulder bones, ribs and wing bones are of little substance; the breast and abdomen contain large cavities filled with air; while the feathers and the down are of exceeding lightness."

The skeleton of a pigeon accounts for only 4.4 per cent of its total body weight, whereas in a comparable mammal such as a white rat it amounts to 5.6 per cent. This in spite of the fact that the bird must have larger and stronger breast bones for the muscles powering its wings and larger pelvic bones to support its locomotion on two legs. The ornithologist Robert Cushman Murphy has reported that the skeleton of a frigate bird with a seven-foot wingspread weighed only four ounces, which was less than the weight of its feathers!

Although a bird's skeleton is extremely light, it is also very strong and elastic—necessary characteristics in an air frame subjected to the great and sudden stresses of aerial acrobatics. This combination of lightness and strength depends mainly on the evolution of hollow, thin bones coupled with a considerable fusion of bones which ordinarily are separate in other vertebrates. The bones of a bird's sacrum and hip girdle, for example, are molded together into a thin, tube-like structure—strong but phenomenally light. Its hollow finger bones are fused together, and in large soaring birds some of these bones have internal trusslike supports, very like the struts inside airplane wings. Similar struts sometimes are seen in the hollow larger bones of the wings and legs.

To "trim ship" further, birds have evolved heads which are very light in proportion to the rest of the body. This has been accomplished through the simple device of eliminating teeth and the accompanying heavy jaws and jaw muscles. A pigeon's skull weighs about

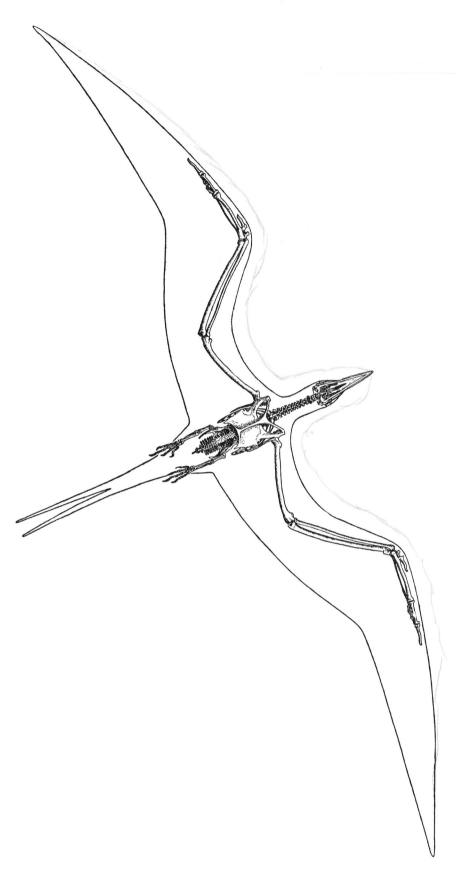

FRIGATE BIRD has a seven-foot wing span, but its skeleton weighs only four ounces. This is less than the weight of its feathers. The skeleton is shown against the outline of the bird.

one sixth as much, proportionately, as that of a rat; its skull represents only one fifth of 1 per cent of its total body weight. In birds the function of the teeth has been taken over largely by the gizzard, located near the bird's center of gravity. The thin, hollow bones of a bird's skull have a remarkably strong re-inforced construction [*see photograph on page 69*]. Elliott Coues, the 19th-century U. S. ornithologist, referred to the beautifully adapted avian skull as a "poem in bone."

The long, lizard-like tail that birds inherited from their reptilian ancestors has been reduced to a small plate of bone

at the end of the vertebrae. The ribs of a bird are elegantly long, flat, thin and jointed; they allow extensive movement for breathing and flying, yet are light and strong. Each rib overlaps its neighbor—an arrangement which gives the kind of resilient strength achieved by a woven splint basket.

Feathers, the bird's most distinctive and remarkable acquisition, are magnificently adapted for fanning the air, for insulation against the weather and for reduction of weight. It has been claimed that for their weight they are stronger than any wing structure devised by man. Their flexibility allows the broad trailing edge of each large wing-feather to bend upward with each downstroke of the wing. This produces the equivalent of pitch in a propeller blade, so that each wingbeat provides both lift and forward propulsion. When a bird is landing or taking off, its strong wingbeats separate the large primary wing feathers at their tips, thus forming wing-slots which help prevent stalling. It seems remarkable that man took so long to learn some of the fundamentals of airplane design which even the lowliest English sparrow demonstrates to perfection [see "Bird Aerodynamics," by John J. Storer; SCIENTIFIC AMERICAN Offprint 1115].

Besides all this, feathers cloak birds with an extraordinarily effective insulation—so effective that they can live in parts of the Antarctic too cold for any other animal.

The streamlining of birds of course is the envy of all aircraft designers. The bird's awkwardly angular body is trimmed with a set of large quill, or contour, feathers which shape it to the utmost in sleekness. A bird has no ear lobes sticking out of its head. It commonly retracts its "landing gear" (legs) while in flight. As a result birds are far and away the fastest creatures on our planet. The smoothly streamlined peregrine falcon is reputed to dive on its prey at speeds up to 180 miles per hour. (Some rapid fliers have baffles in their nostrils to protect their lungs and air sacs from excessive air pressures.) Even in the water, birds are among the swiftest of animals: Murphy once timed an Antarctic penguin swimming under water at an estimated speed of about 22 miles per hour.

A basic law of chemistry holds that the velocity of any chemical reaction roughly doubles with each rise of 10 degrees centigrade in temperature. In nature the race often goes to the metabolically swift. And birds have evolved the highest operating temperatures of all animals. Man, with his conservative 98.6

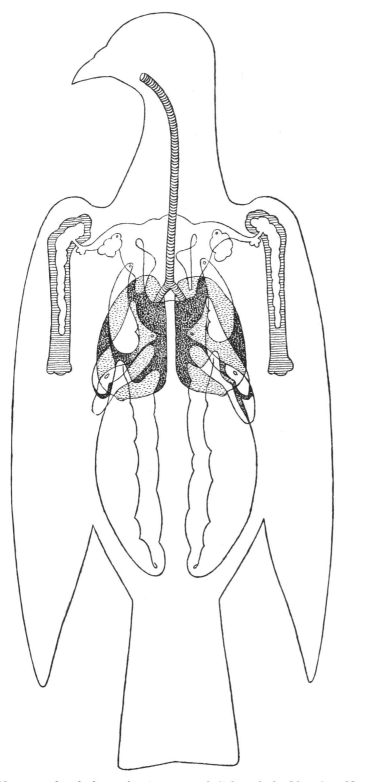

AIR SACS connected to the lungs of a pigeon not only lighten the bird but also add to the efficiency of its respiration and cooling. The lungs are indicated by the two dark areas in the center. Two of the air sacs are within the large bones of the bird's upper "arm."

degrees Fahrenheit, is a metabolic slow-poke compared with sparrows (107 degrees) or some thrushes (113 degrees). Birds burn their metabolic candles at both ends, and as a result live short but intense lives. The average wild songbird survives less than two years.

Behind this high temperature in birds lie some interesting circulatory and respiratory refinements. Birds, like mammals, have a four-chambered heart which allows a double circulation, that is, the blood makes a side trip through the lungs for purification before it is circulated through the body again. A bird's heart is large, powerful and rapid-beating [*see table of comparisons on page 70*]. In both mammals and birds the heart rate, and the size of the heart in proportion to the total body, increases as the animals get smaller. But the increases seem significantly greater in birds than in mammals. Any man with a weak heart knows that climbing stairs puts a heavy strain on his pumping system. Birds do a lot of "climbing," and their circulatory systems are built for it.

The blood of birds is not significantly richer in hemoglobin than that of mammals. The pigeon and the mallard have about 15 grams of hemoglobin per 100 cubic centimeters of blood—the same as man. However, the concentration of sugar in their blood averages about twice as high as in mammals. And their blood pressure, as one would expect, also is somewhat higher: in the pigeon it averages 145 millimeters of mercury; in the chicken, 180 millimeters; in the rat, 106 millimeters; in man, 120 millimeters.

In addition to conventional lungs, birds possess an accessory system of five or more pairs of air sacs, connected with the lungs, that ramify widely throughout the body. Branches of these sacs extend into the hollow bones, sometimes even into the small toe bones. The air-sac system not only contributes to the birds' lightness of weight but also supplements the lungs as a supercharger (adding to the efficiency of respiration) and serves as a cooling system for the birds' speedy, hot metabolism. It has been estimated that a flying pigeon uses one fourth of its air intake for breathing and three fourths for cooling.

The lungs of man constitute about 5 per cent of his body volume; the respiratory system of a duck, in contrast, makes up 20 per cent of the body volume (2 per cent lungs and 18 per cent air sacs). The anatomical connections of the lungs and air sacs in birds seem to provide a one-way traffic of air through most of the system, bringing in a constant stream of unmixed fresh air, whereas in the lungs

of mammals stale air is mixed inefficiently with the fresh. It seems odd that natural selection has never produced a stale air outlet for animals. The air sacs of birds apparently approach this ideal more closely than any other vertebrate adaptation.

Even in the foods they select to feed their engines birds conserve weight. They burn "high-octane gasoline." Their foods are rich in caloric energy—seeds, fruits, worms, insects, rodents, fish and so on. They eat no low-calorie foods such as leaves or grass; a wood-burning engine has no place in a flying machine. Furthermore, the food birds eat is burned quickly and efficiently. Fruit fed to a young cedar waxwing passes through its digestive tract in an average time of 27 minutes. A thrush that is fed blackberries will excrete the seeds 45

minutes later. Young bluejays take between 55 and 105 minutes to pass food through their bodies. Moreover, birds utilize a greater portion of the food they eat than do mammals. A three-weeks-old stork, eating a pound of food (fish, frogs and other animals), gains about a third of a pound in weight. This 33 per cent utilization of food compares roughly with an average figure of about 10 per cent in a growing mammal.

The breast muscles of a bird are the engine that drives its propellers or wings. In a strong flier, such as the pigeon, these muscles may account for as much as one half the total body weight. On the other hand, some species—*e.g.*, the albatross—fly largely on updrafts of air, as a glider does. In such birds the breast muscles are greatly re-

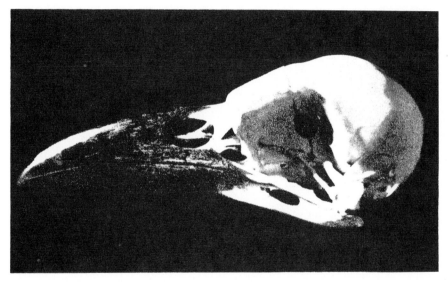

SKULL OF A CROW achieves the desirable aerodynamic result of making the bird light in the head. Heavy jaws are sacrificed. Their work is largely taken over by the gizzard.

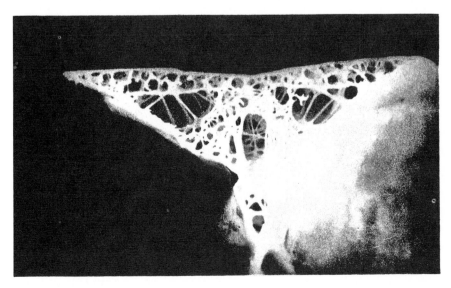

FRONTAL BONE in the skull of a crow is cut through to show its hollow and braced internal construction. The skull of the bird accounts for less than 1 per cent of its total weight.

HEART	PERCENT OF BODY WEIGHT	HEART BEATS PER MINUTE
FROG	.57	22
MAN	.42	72
PIGEON	1.71	135
CANARY	1.68	514
HUMMINGBIRD	2.37	615

HEART WEIGHT and pulse rate are compared for a number of animals. The hearts of birds are relatively large for body size.

duced, and there are well-developed wing tendons and ligaments which enable the bird to hold its wings in the soaring position with little or no effort.

A bird may have strong breast muscles and still be incapable of sustained flight because of an inadequate blood supply to these muscles. This condition is shown in the color of the muscles; that is the explanation of the "white meat" of the chicken and the turkey—their breast muscles have so few blood vessels that they cannot get far off the ground. The dark meat of their legs, on the other hand, indicates a good blood supply and an ability to run a considerable distance without tiring.

After a ruffed grouse has been flushed four times in rapid succession, its breast muscles become so fatigued that it can be picked up by hand. The blood supply is simply inadequate to bring in fuel and carry away waste products fast enough. Xenophon's *Anabasis* relates the capture of bustards in exactly this manner: "But as for the Bustards, anyone can catch them by starting them up quickly; for they fly only a short distance like the partridge and soon tire. And their flesh was very sweet."

In birds the active phase of the breathing cycle is not in inhaling but exhaling. Their wing strokes compress the rib case to expel the air. Thus instead of "running out of breath" birds "fly into breath."

Probably the fastest metabolizing vertebrate on earth is the tiny Allen's hummingbird [see "The Metabolism of Hummingbirds," by Oliver P. Pearson; SCIENTIFIC AMERICAN, January, 1953]. While hovering it consumes about 80 cubic centimeters of oxygen per gram of body weight per hour. Even at rest its metabolic rate is more than 50 times as fast as man's. Interestingly enough, the hovering hummingbird uses energy at about the same proportionate rate as a hovering helicopter. This does not mean that man has equalled nature in the efficiency of energy yield from fuel. To hover the hummingbird requires a great deal more energy, because of the aerodynamic inefficiency of its small wings and its very high loss of energy as dissipated heat. The tiny wings of a hummingbird impose on the bird an almost incredible expenditure of effort. Its breast muscles are estimated to be approximately four times as large, proportionately, as those of a pigeon. This great muscle burden is one price a hummingbird pays for being small.

A more obvious index of the efficiency of bird's fuel consumption is the high mileage of the golden plover. In the fall the plover fattens itself on bayberries in Labrador and then strikes off across the open ocean on a nonstop flight of 2,400 miles to South America. It arrives there weighing some two ounces less than it did on its departure. This is the equivalent of flying a 1,000-pound airplane 20 miles on a pint of gasoline rather than the usual gallon. Man still has far to go to approach such efficiency.

7

HORNS AND ANTLERS

WALTER MODELL
April 1969

They are commonly believed to be rather alike but in actuality they are quite different. Among other differences, the material of horns is related to skin and the material of antlers to bone

Some years ago a curator at the New York Zoological Society received a shipment of antelopes from South Africa to which he had been looking forward eagerly. It included rare specimens that were hardly known to Americans except in crossword puzzles: the hartebeest, the eland, the waterbuck, the impala and others. When the antelopes were unloaded from the ship, the curator was stricken with shock and horror. They had all been dehorned. In order to save space and prevent injury the shipper had polled the animals' horns with a cattle dehorner, assuming that they would regrow their "antlers" after settling down in their new home. Unfortunately the shipper had made a serious zoological error, failing to distinguish between horns and antlers. All the animals in the shipment had horns, which as every dairy farmer knows do not regenerate after polling. The curator rejected the mutilated animals as being unsuitable for exhibition in a zoo, and since no one else wanted them they were slaughtered and given to the large cats, which for the first time in a long while ate as they had been brought up to eat.

The hapless animal dealer who confused horns with antlers was not particularly ignorant but was a victim of a common misapprehension. Many people and even reference books are not entirely clear on the differences between horns and antlers. The differences are fundamental and complex, and they present interesting problems in biochemistry, physiology, animal behavior and evolution.

True horns, antlers and similar cephalic adornments are found today only in five families of ungulates (animals with hooves): (1) the Rhinocerotidae (rhinoceros), characterized by one or two permanent midline nasal horns; (2) the Bovidae (cattle, sheep, goats and antelopes), characterized by a pair of symmetrical permanent horns; (3) the Antilocapridae (pronghorn antelope), which annually renew their pair of symmetrical horns; (4) the Cervidae (moose, caribou, elk and deer), which annually renew a pair of antlers, and (5) the Giraffidae (the giraffe and the okapi), whose cephalic protuberances are permanent and paired but are neither horns nor antlers.

In functional terms horns make sense: they serve their possessors as effective weapons. The knobs on the giraffe's head are less understandable; borne some 18 feet above the ground, they are hardly in a position to attack anything except perhaps a low-flying airplane, and in any case they are short, blunt and cushioned with a tuft, so that they cannot inflict much damage. The function of antlers is even more mysterious; apart from giving the animal a noble appearance these headpieces have little utilitarian justification; indeed, they are an encumbrance. Antlers are too delicate to serve as weapons. When antlered animals really fight, they use their hooves and not their antlers. For several months of the year, between the annual shedding and the regrowth, the animal does without antlers and seems not to miss them at all.

The only observable function served by antlers is that during the mating season the males use them to tilt with other males in winning a harem. This unique application often ends unhappily; it is not uncommon for the two contestants to lock antlers (not horns!) so that both are immobilized and perish, and the herd loses the genes of what may well be its two best stags. As we shall see, the antler is a strange and uneconomic experiment of nature, extravagantly costly to its possessors in several ways, and it seems destined eventually to disappear.

In distinguishing antlers from horns we note to begin with that they are composed of entirely different materials. Horn consists mainly of keratin, the protein that is the principal constituent of hair, nails, hooves, scales, feathers, claws and other tough structures derived from epidermal tissue. Like hair and nails, horn is not a living, sensitive tissue: it has no nerves or blood supply and hence is insensible to pain and does not bleed when it is cut. Horns grow slowly and attain their definitive size and shape by extension from their source in an epidermal layer surrounding a bony core, the *os cornu* at the base of the frontal bone of the skull, as the animal grows to adulthood. If the horny material is cut off, it is not regenerated.

Antler, in contrast, is living tissue that resembles true bones of the body in physiology, chemical composition and cellular structure. During the antler's development it is covered with a hairy skin called velvet, which has a rich supply of blood vessels and nerves. While the antler is in velvet, it bleeds profusely when injured, and the skin is sensitive to touch and pain. At this stage the antlers are not only tender but also fragile. As the season progresses the antlers become ossified, the velvet is shed and the bare tines of bone are exposed. At the end of the mating season (usually in December in temperate regions of the Northern Hemisphere) the animal sheds its antlers, and four to five months later (in April or May) it begins to grow a new set. In short, the antler is a deciduous organ that is cast off and renewed annually like the leaves of a tree.

Horns and antlers differ significantly in their architecture. Antlers, at least those sought as trophies, tend to be large, complex and ornate. Antlers are paired,

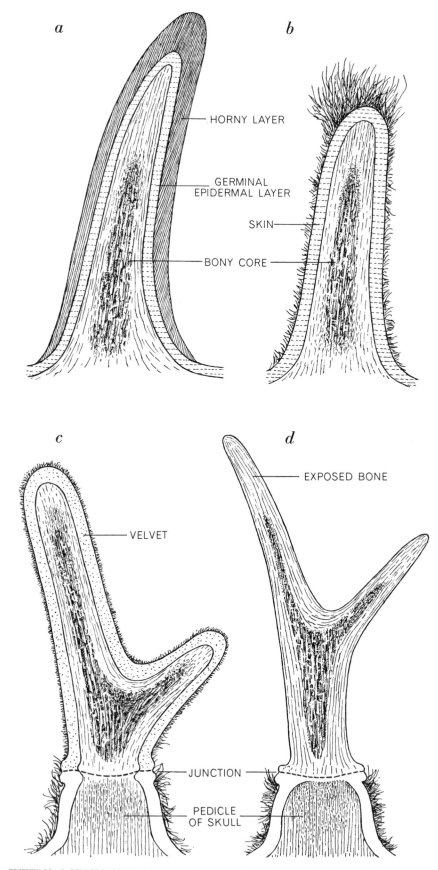

INTERNAL STRUCTURE of horns and antlers is shown for a bovine horn (*a*), the knob of a giraffe (*b*) and the antler of a deer, with its "velvet" covering (*c*) and with the bony structure exposed (*d*) after the velvet has been shed. The junction with the skull where growth begins in the spring and where the antler breaks off at the time of the annual shedding is indicated. Developing antlers have blood vessels and nerves, whereas horns entirely lack them.

although in some species they are not symmetrical. In certain animals, such as the reindeer and the moose, antlers have a winglike, palmate structure—like the palm of a hand with extended fingers. From year to year after a deer has reached maturity and until it reaches its prime each new crop of antlers becomes larger and more elaborate, adding branches and "points" that provide a measure of the animal's age and vigor. The number of points is also a measure of the magnificence of the deer; in Scotland, for example, a deer with 12 points is called a royal stag. In some species of Cervidae now extinct the antlers weighed more than the animal's entire internal skeleton. At the other extreme the tiny pudu of the Chilean Andes, a reddish deer only about a foot high, has simple pygmy antlers consisting of almost invisible spikes two or three inches long.

The horns of the Bovidae, although they too can be magnificent, are clearly structured for use rather than ornamentation. They exist as symmetrical pairs (an exception is the four-horned antelope with two symmetrical pairs) in a rich variety of forms: curved, twisting, coiled, helical and zigzag. They all end, however, in a single strong spear capable of impaling or tossing an adversary. (Testimony to the formidable power of horns as weapons has been erected outside the bullring in Madrid in the form of a statue of Sir Alexander Fleming, the discoverer of penicillin, whose wonder drug has reduced the death toll among gored matadors.) It also seems significant that in almost all species of Bovidae the female as well as the male possesses horns (although they are often smaller in the female), whereas among the Cervidae antlers are secondary sexual characteristics of the male (the only known exception being the reindeer and its identical American version, the caribou). This again suggests that the antler, unlike the horn, did not evolve primarily as a weapon of defense.

Let us examine the biological and evolutionary distinctions between horns and antlers in more detail. Horn is made up of filaments that closely resemble hair, and these filaments arise from papillae in the skin that are much like hair follicles. It is clear, however, that horn is not, as it has sometimes been said to be, simply a mass of agglutinized hair. On microscopic examination it can be seen to be made of distinct hollow filaments, whereas hair fibrils are solid. Furthermore, horns originated much earlier (probably at least 50 million years earlier) than hair, which apparently developed only after the arrival of mammals.

The simplest and most primitive horn of our day is the horn of the rhinoceros. It is made up of tubular, filamentous secretions from the skin that are cemented together to form a projection from the animal's nose. Having no bony core, the horn consists of solid keratin. As it grows it becomes cemented to the nasal bone, but if a dead rhinoceros's head is skinned, the horn often comes away with the skin, to which it is firmly attached.

In cattle and other animals of the bovid family the horns are hollow, as one can see in an antique powder horn. The horn is mounted like a shoe directly on the spikelike *os cornu*, which projects from the frontal bone of the skull. Part of the frontal sinus can often be found in the center of the *os cornu*. Horns arise from an inner epidermal layer (the *stratum germinativum*) immediately covering the *os cornu*. This layer lines the developing horn and produces its slow growth by continuing to secrete filaments. The reason a horned animal cannot regrow a polled horn is that the operation destroys the essential filament-secreting epidermal tissue.

The only horned animal that sheds its horns periodically is the pronghorn antelope (*Antilocapra americana*), a native of North America that was once numerous but was almost wiped out by hunters before conservationists took measures that have effectively preserved the species. Each year the pronghorn develops a set of true horns that consist of keratin and grow on an *os cornu*. The prong is an extra spike on the horn.

This animal's annual shedding process is not even remotely related to that of antler replacement. The new horn grows while the old one is still in place and pushes the old shoe off the *os cornu* as it achieves full development, so that the animal is never without a horn. Although such an experiment is not recorded, it is probable that, if the old horn were cropped at the base, a new one would not grow. The pronghorn's horn differs from the horns typical of the Bovidae in that it is often covered with a considerable growth of hair. Yet the pronghorn antelope is so like members of the family Bovidae in all its obvious physical features that I believe it belongs in that family.

In contrast to horn, antler is a unique anatomical object. It is far and away the fastest growing postnatal bone known. My interest in this unusual tissue goes back to my days as a second-

year medical student, when I began a study on seasonal changes in the elk at the laboratories of the New York Zoological Society. I showed one of my histologic sections of growing antler to my professor, James Ewing, who was then the world's outstanding student of malignant growths. On examining it under the microscope, he described the tissue, which was extraordinarily rich in mitotic figures and gave other signs of fierce growth, as a sample of malignant bone sarcoma. When I told Ewing that it was actually a slice of a normal growing antler, he urged me to pursue the study of the tissue because of its simulation of malignant growth.

Unfortunately neither I nor any other investigator since has been able to discover the nature of the mechanism that controls the exuberant growth of antler cells. Although under the microscope the

actively growing antler tissue cannot be distinguished from that of malignant neoplasm of bone, it is clear that its development is under rigid control. Instead of spreading out in all directions the cells produce a sharply defined structure growing away from the head. In the course of a few months the original cells grow into a large, bony tree of great complexity. Then, at the end of the mating season, special mechanisms of the body in effect cleanly amputate this structure and it is discarded. No surgeon has yet achieved such success in removing a bone sarcoma. It is noteworthy, however, that like a malignant neoplasm the antler recurs (every year until the stag becomes very old) because a few primordial cells are left behind.

Antlers have intrigued and mystified naturalists since ancient times. Until the beginning of the 19th century they were

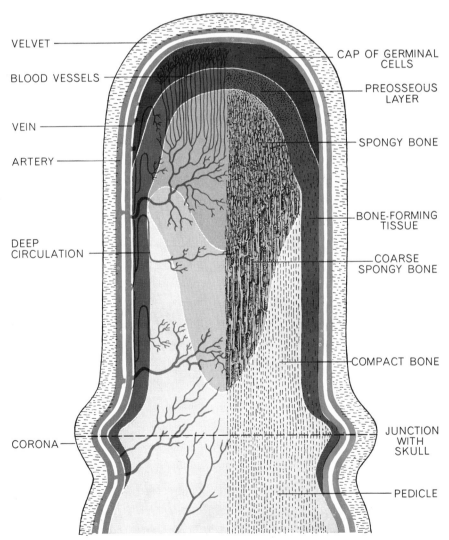

BLOOD SUPPLY of growing antlers is provided through the elaborate structure of arterioles and veins depicted at left. The major structural zones of the antler are shown at right.

a   *b*   *c*

ARCHITECTURE OF HEADPIECES is suggested by five horned animals and the giraffe. The horned animals are (*a*) the mouflon, (*b*) the springbok, (*c*) the wildebeest, (*d*) the rhinoceros, whose horn is made up of tubular filaments as other horns are but, unlike

generally thought to be composed of wood; indeed, French naturalists named them *bois,* and they attached the term *écorce,* meaning bark, to the velvet that came off when the antlers matured. The discovery that they were actually bone did not come until some 19th-century investigator boiled antlers and identified gelatin in the residue.

Not all naturalists took so naïve a view as the French. Since antlers are a male characteristic, some early investigators sought to relate them to male physiology. Aristotle correctly noted: "If stags are castrated before they are old enough to have antlers, these never appear, but if castrated after they have antlers, their size never varies nor are they subject to annual change." In recent years it has been shown that injections of the male hormone testosterone can cause a castrated male deer to develop antlers. The male hormone can also stimulate a spayed female deer to produce knobs, covered with skin resembling velvet, at the sites on the skull where antlers normally grow in the male. I have observed (as have others) elderly female deer with similar knobs after menopause.

How is the antler formed? Investiga-

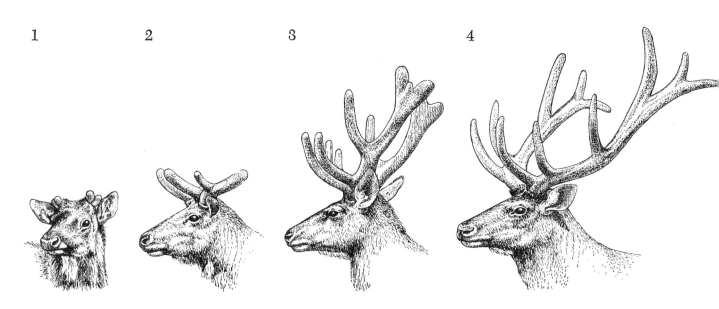

1   *2*   *3*   *4*

ANNUAL EVOLUTION of antlers, belonging in this case to the wapiti elk, begins (*1*) with the appearance of the velvet-covered buds of new antlers in April, about six weeks after the former antlers have been shed. Within two weeks (*2*) the characteristic branched pattern has appeared. By the end of May (*3*) the antlers are well developed and fully covered by velvet. During this

them, is solid rather than hollow and (*e*) the pronghorn antelope, which is the only horned animal with horns that are deciduous, or periodically shed. The giraffe's head (*f*) has growths that are not true horns but rather are protuberances covered with a hairy skin.

tors many years ago, using ordinary stains (hematoxylin and eosin) to study by microscope the intimate details of the growth of the tissue, concluded that antlers, like the long bones of the body, developed through an intermediate cartilage stage. I examined the microscopic details of antler development with special silver staining techniques and found that, on the contrary, the bone of antler,

like the bone of the skull, is formed by direct ossification of the framework of fibrous tissue that develops first. It becomes cancellous (spongy) bone with an internal cavity that is continuous with the cavity of the skull and shares its blood supply. Curiously, unlike other spongy bone in the body, such as the sternum and the pelvis, the bone of the antler does not manufacture blood, al-

though there is considerable blood (and a little fatty marrow) in it.

Under the influence of a hormonal rhythm the male deer begins to grow its antlers in the spring (late April or May in the U.S.). The developing antler bulges out of the velvet that covers the pedicle, a bony platform atop the frontal bone of the skull, soon after the old antler separates from it. From a few

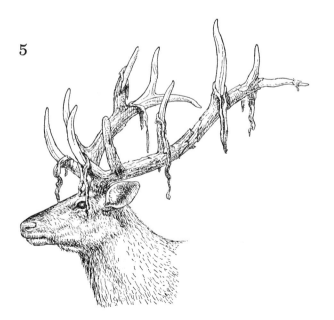

stage the animal is careful to avoid hard objects. By August (4) the antlers are mature. Growth has ceased and the velvet has begun to dry at the tips. When the bony material of the antler has become fully hardened, the velvet dies (5) and peels off in ragged shreds. After the antlers are mature and have lost their velvet (6) the wapiti, which is an American elk, becomes sexually aggressive.

ANCIENT HORNED ANIMAL was *Triceratops,* a reptile with three horns and also a horny shieldlike structure. Horns apparently originated with the early large reptiles.

ANCIENT ANTLERED ANIMAL, much more recent than *Triceratops* but also extinct, was the giant Irish elk. Its massive antlers weighed more than its entire internal skeleton.

fibroblasts (embryonic cells) left behind when the former antler was cast a mass of fibroblasts develops below the skin, and the velvet cover grows to conform with the developing antler. The fibroblasts rapidly form the armature of the antler, much as a branch and twigs grow out of a tree. Meanwhile osteoblasts (bone-forming cells) begin to stream into this framework and to lay down the bone-forming calcium. The resulting bone has a thinner cortex and consequently is not nearly as strong as the bone of the skull, but apart from the fact that it does not form blood it is indistinguishable from the cancellous bone of the skull and other spongy bones of the skeleton.

By September the antler has grown to full size and has firmly united itself to the pedicle. The bone at the base becomes progressively denser and eventually cuts off the flow of blood from the skull to the antler's interior. Some blood is still supplied, however, by arteries coursing through the velvet covering the antler. Soon, as a result of some mechanism that has not yet been satisfactorily explained, the velvet proceeds to degenerate. It dries up and is shed by the stag, coming off in strips when the animal rubs its antlers against trees or shrubs.

At this stage the antlers have no blood supply or nerves and are insensible to painful stimuli. The bare bone and sharp tines of the antler are exposed, and for a few weeks in the fall the many-spiked antler might be usable as a weapon. It is an awkward weapon at best, however, and its development is not particularly well timed, because by that season the fawns born of the stag's preceding matings are fleet enough to escape predators and no longer need the sire's protection. As I have noted, when stags tilt with each other to gain a harem for the new mating season, the antlers may be dangerous to both adversaries. Furthermore, how is one to explain the curious fact that sometimes the harem winners are stags (called hummels) that for some unknown reason have failed to develop antlers? Could they be better fighters because they lack antlers?

The antler's implausibility as a weapon suggests that it did not actually evolve for this function. One interesting current hypothesis is that antlers serve the deer and other Cervidae as a cooling device during the summer. The velvet covering the antler provides an admirable means of radiating body heat, because of its considerable surface area and its exten-

TYPICAL HORNS AND ANTLERS are shown in front and side views. At top are the horns of the kudu ram, a large antelope that is found in much of Africa. At bottom are the antlers of the caribou, which is the North American equivalent of the reindeer.

POSSIBLE EVOLUTIONARY TREND in antlered animals may be represented by the Chinese water deer (*left*) and the musk deer (*right*), which are the only animals of the family Cervidae that do not possess antlers. The two species, however, have evolved tusks.

sive apparatus of peripheral blood vessels and sebaceous glands. The fact that the velvet dies at the end of the summer and the antlers are shed not long afterward lends some support to this surmise about the antler's primary reason for being. It is difficult to imagine any other plausible reason why nature should have endowed animals with an elaborate superstructure that is used only during a few weeks of the year and is discarded annually after being produced at great metabolic expense. (To the question of why females lack these radiators the supporters of the hypothesis respond that females stay in the shade.)

In the late fall some of the cement holding the base of the antler to the pedicle is resorbed, the connection weakens and the antler is cast—not as an active process but by breaking off when it happens to strike something. The break occurs neatly at the junction with the skull where the antler started to grow in the spring; there is a little bleeding, but the blood promptly clots. The process is painless. A new growth of hairy velvet (rather than the hairless, nonsweating scar tissue that forms in the normal healing of wounds) quickly covers the wound on the exposed skull, and the organs and cells involved in the formation of antler go into a dormant period, lasting from late December until the fol-

lowing April or May. An injection of testosterone during this period, however, can trigger antler growth.

Notwithstanding the peculiarities and apparent frivolity of antlers, they cannot really be called a freak of nature. As a headpiece they are one version, albeit an exotic one, of a phenomenon that goes back to the early history of land animals. Horns are at least 100 million years old. In the Cretaceous period there were dinosaurs and crocodile-like reptiles with horns. Some of these growths beggar the imagination of science fictionists. *Triceratops* had three horns: a huge spike rising from the nose and one above each eye. *Styracosaurus* had an upright nasal spike nearly two feet long and a neck shield with six spikes thrust out from its edges. Skull protuberances have also been found among fossils of some early mammals, notably the elephantine ungulates. The horns were formidable weapons in some early mammals, such as *Arsinoitherium*, but in others, such as *Uintatherium* and *Brontotherium*, they were blunt extensions from the skull and were probably covered with skin. By the 'end of the Eocene epoch some 40 million years ago the rhinoceros, its naked horn as menacing a weapon as the horns of the ancient reptiles, had begun to appear.

The rhinoceros is unusual in several ways. It is the only odd-toed animal that has a horn. The Rhinocerotidae include the only nonmythical unicorn, but several surviving species of rhinoceroses have two horns, both positioned on the midline one behind the other. The rhinoceros's horn is also exceptional in that it is solid keratin, not a hollow shoe. In prehistoric times the animal must have made considerable use of its horn as a weapon, but today the rhinoceros's massive size and armor are sufficient protection to discourage attack by its less massive contemporary competitors.

Paradoxically, in our era the horn has been the rhinoceros's undoing. For more than 1,000 years man, now the animal's principal enemy, has been hunting down the rhinoceros for its horn. In China ground rhinoceros horn has long been prized for its supposed values as an aphrodisiac and a medicine for various ailments, and in the medical markets of China and Africa today the horn is said to be worth half its weight in gold. A thriving trade is also conducted in dried rhinoceros blood and in rhinoceros hide for use as a warrior's shield. A few rhinoceroses manage to survive in Asia by keeping out of sight, and in some parts of Africa the animal is protected from hunters by law.

During the Miocene epoch, beginning

some 25 million years ago, horns and antlers developed among many species of two important families of ungulates, the Bovidae and the Cervidae. These two families are so remarkably alike in many ways that they may well have had a common ancestry. The great differences between horn and antler, however, indicate that the two forms of head appendage had different origins. The Bovidae revived the keratinous horn growth that was already a 100-million-year-old carryover from the age of reptiles. The Cervidae introduced the antler as a basically new growth. According to the available fossil record, the first antlered cervid was *Dicrocerus*, an ungulate of the early Miocene that grew a very simple antler in the form of short spikes. Because antler is so skimpily constructed that it deteriorates about twice as fast as the skull bone under ordinary exposure, it may well be that its ancient history is not as well preserved in the paleontological record as that of skull and skeletal bone. The record does show, however, that in the Pleistocene epoch antlers became common among the Cervidae and some of them grew to monstrous size. Probably the most impressive antlers of all time were borne by the great stag (*Cervus megaceros*) of the Ice Age; its pair of antlers had a spread of three meters and weighed about 70 kilograms (154 pounds)! Among the living members of the family Cervidae today all but two species have antlers. The exceptions are the Chinese water deer and the musk deer [*see illustration on preceding page*].

There are reasons to believe the giraffe and the okapi are closely related to the Cervidae and evolved from the same group of ancestors. The permanent knobs on the head of the giraffe and the okapi are not made of keratin; hence they are not true horns. The giraffe's knobs are formed of a bony core extending straight up from the skull and are covered with a hairy skin that gives the knobs their tufted appearance. In the okapi a bit of bare bone is exposed at the tip. The knobs of these two strange beasts seem closer to antlers than to horns and perhaps are best compared to antlers permanently in velvet.

In a curious way horns and antlers are invariably associated with certain other apparently unrelated anatomical features. Headpieces of one kind or another (horns, antlers or knobs) are possessed by virtually all even-toed ungulates with a four-chambered (or true ruminant) stomach; this includes the Bovidae, the Cervidae, the pronghorn, the giraffe and the okapi. With the single exception of the rhinoceros no odd-toed animal (which includes the horse, the ass and the zebra) has a headpiece, and headpieces are also absent in all even-toed ungulates with false ruminant, or three-chambered, stomachs (camels, llamas and others), as well as in those with a single-chambered stomach (pigs, peccaries and hippopotamuses). What connection can there be between horns or antlers and a perfect ruminant stomach and even-toed hooves? By what odd quirk of evolution did these seemingly unrelated characteristics come to be associated with one another, if indeed the association is not mere coincidence?

Be that as it may, this enigma is much less intriguing than the mystery of why antlers evolved and why they have persisted so long. The evolutionary success of horns, which have proved their value over 100 million years of trials and are now firmly incorporated in a great number of species of ungulates, is quite understandable. The antler, on the other hand, is obviously an encumbrance with a very limited, if not entirely questionable, usefulness. Perhaps the main surprise is that it has lasted for upward of 25 million years and is still retained by almost the entire family Cervidae. There is evidence, however, both in the fossil record and in the hunting history of man, that antlers as well as the antlered animals are on the decline. Their often suicidal tilts with their antlers, by killing off some of the best and strongest stags, may have played a part in this decline. The preservation of the Cervidae is now aided by the U.S. program for the protection of wildlife, but this has led to a proliferation that has made them a true pest and necessitates large-scale seasonal hunting of the animals to spare them from starvation and our forests from ruination by debarking. Overprotection of the animals, with its consequent disturbance of ecology, may not be in our interest.

It may be that in the long run natural selection can save the Cervidae by eliminating those with elaborate antlers, so that the surviving members of the family come to be like the two nonantlered species, the Chinese water deer and the musk deer. In any case, it appears that antlers, if not the antlered animals themselves, are doomed by evolution, and that the eight arboreally ornamented reindeer of Santa Claus, like the handsome unicorn, will one day become strictly a legend.

II

VASCULAR SYSTEM BIOLOGY

The arteries have no sensation, for they even are without blood, nor do they all contain the breath of life; and when they are cut only the part of the body concerned is paralysed . . . the veins spread underneath the whole skin, finally ending in very thin threads, and they narrow down into such an extremely minute size that the blood cannot pass through them nor can anything else but the moisture passing out from the blood in innumerable small drops which is called sweat.

Pliny
NATURAL HISTORY, XI, lxxxix.

II

VASCULAR SYSTEM BIOLOGY

INTRODUCTION

Perhaps the most complex regulatory network in vertebrates involves the control of oxygen and carbon dioxide levels in the body. In homeotherms the temperature-regulating center and its connections impinge in many ways on the breathing and heart-rate systems. In "The Physiology of Exercise" Carleton Chapman and Jere Mitchell introduce the general subject of the vascular system with a general discussion of responses of higher vertebrates to exercise. They touch upon many topics that are treated in greater detail by subsequent articles in this section, and in the sections on respiratory adaptations and temperature.

We know a great deal about the evolution of the blood vascular system in vertebrates. Changes in the routes of flow, in the control machinery, and in the respiratory pigment complement each other throughout vertebrate history. Let us consider, first, some properties of the pigment, hemoglobin.

Each molecule of hemoglobin is a complex of the protein globin and four heme groups; each heme is a porphyrin molecule containing an iron atom. Oxygen binds reversibly to the iron in a manner that varies with the partial pressure of the gas in the neighborhood of the hemoglobin; thus, as partial pressure rises, binding rises, and vice versa. This relationship is described by a sigmoid curve (the oxygen dissociation curve, see page 83) that has a characteristic shape for each vertebrate species. The sigmoid shape is thought to refect the following relationship: it is somewhat difficult for the first heme group of a hemoglobin molecule to load the first molecule of oxygen; once the first oxygen is loaded, it is much easier for the second and third groups to load the second and third oxygen molecules, and very easy for the fourth group to load the fourth molecule. This loading pattern results from alterations in the three-dimensional shape of globin. When the first oxygen is bound, the globin changes shape, enabling the remaining hemes to interact with each other and to load more readily. The process is called heme-heme interaction.

The position of the dissociation curve at various partial pressures of oxygen defines the hemoglobin's affinity for oxygen. Curves for most fish bloods show complete loading at relatively low tensions. Such hemoglobin is said to have a "high" affinity for oxygen because it binds the oxygen avidly at low partial pressure. Terrestrial vertebrates (and, in particular, birds and mammals) have curves far to the right (see Figure 1) of those of fish; their hemoglobin has a "low" affinity for oxygen and loads only at the high partial pressure of oxygen in normal air. The low affinity of avian and mammalian hemoglobin for oxygen is of vital importance for high rates of metabolism and elevated body temperature. Low affinity in effect means that the hemoglobin *unloads* its oxygen in the tissues at relatively high partial pressures (pressures at which fish hemoglobin would remain completely loaded). This allows more oxygen to be present around and in cells, a necessity for the high rates of metabolism required for body temperatures of 37 to 40 degrees centigrade.

Dissociation curves are affected by carbon dioxide levels (the Bohr effect, see Chapman and Mitchell) and the protein packing within cells. An increase in carbon dioxide dissolved in blood lowers the pH (some other agents also produce hydrogen ions and lower the pH) and thus decreases hemoglobin's affinity for oxygen; consequently, dissociation occurs more readily in tissue spaces where carbon dioxide is produced by metabolic activity. But in lung alveoli, where carbon dioxide is lost to the air and pH rises, affinity is increased and oxygen is bound more readily to hemoglobin. The Bohr effect was formerly thought to be mediated by the number of reactive sulfhydryl groups (—SH) in the vicinity of the heme moieties on the globin molecule. This now appears to be an oversimplification at best, and it is safest to conclude for the moment that dissociable groups effected by hydrogen ion concentrations affect the heme-oxygen binding interaction in a way that is still undefined.

Oxygen dissociation curves are also effected by the presence of an organic molecule produced by erythrocytes themselves. This 2,3-diphosphoglycerate decreases the affinity of hemoglobin for oxygen. As one might predict, larger quantities of the substance are present in red cells as they pass through body tissues; smaller quantities are present as blood passes through the lungs. Thus, this intracellular substance acts like hydrogen ions and the Bohr effect. We will discuss adaptations to life at high altitude in the next section (see Hock, "The Physiology of High Altitude," in Part III). It is interesting that, even in the first day or two of acclimatization to high altitude, the overall quantities of 2,3-diphosphoglycerate rise in human blood, thereby shifting the dissociation curve to the right and allowing oxygen to be surrendered more easily in the tissues.

Alterations in hemoglobins of different vertebrates provide several interesting lessons about the mechanisms of evolution. In terrestrial mammals the degree of the Bohr effect is inversely proportional to the body weight: mouse hemoglobin responds dramatically to increased carbon dioxide by giving up its oxygen very easily, and, as a result, it is a more "efficient" respiratory pigment; the larger the mammal, the less the hemoglobin responds to increased partial pressures of carbon dioxide.

These differences are attributable to the problem of temperature control. The surface area of a mouse is large in relation to its mass; hence, the area from which heat can be lost is great in proportion to the quantity of tissue producing heat. Because the animal cannot remain active (and warm) if oxygen levels fall, the exaggerated Bohr effect works to insure that this emergency does not arise. The proportions are the opposite in an elephant, whose body mass is large in relation to surface area. As a result, there is less chance of precipitous heat loss in the event of subnormal oxygen supply to the tissues. Interestingly, a large aquatic mammal capable of prolonged diving has a much greater Bohr effect than would be predicted on the basis of mass. Humpback whale hemoglobin is the Bohr equivalent of a guinea pig's, for instance, and shows significant decrease in affinity for oxygen under conditions of increasing carbon dioxide. This is obviously a special adaptation that permits rapid oxygen loading while the animal is at the surface, complete unloading as carbon

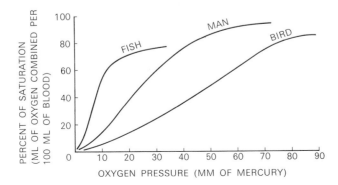

Figure 1. Oxygen dissociation curves for various vertebrates. Fish hemoglobin loads at low partial pressures of oxygen, whereas the pigment of the homeotherms requires much more oxygen in the environment before high percentages of saturation are reached.

dioxide accumulates during long dives, and tolerance of hydrogen ion production during such dives.

Quite a different and unexpected alteration has taken place in hemoglobins of extremely active pelagic fishes. So much muscular activity goes on in these creatures that large quantities of lactic acid continually tend to lower blood pH to the point where normal fish hemoglobin with its Bohr effect would be unable to load oxygen in the gills. As emphasized by Peter Hochachka and George Somera in their important book *Strategies of Biochemical Adaptation*, the ancestors of these fishes apparently evolved a hemoglobin lacking the Bohr effect, so that today's representatives can exploit a way of life, an ecological niche, not open to more lethargic relatives.

These and other physiological differences in the hemoglobins result from alterations in the amino acids of globin. Globin of mammals has a molecular weight of 64,500 and is composed of four subunits (two alpha units and two beta units). A heme is bound to each subunit. The hemoglobin of the most primitive living vertebrates, the cyclostomes, has a molecular weight of about 17,000, has one heme group, and shows a marked Bohr effect. Although the amino acid composition of cyclostome globin is very different from that of the globin in higher vertebrates, both types of molecules show the important property of undergoing conformational changes when oxygen binding occurs: thus heme-heme interactions take place. This conclusion is based upon the discovery of a *sigmoid* dissociation curve for lamprey hemoglobin. (Note that most texts quote the old finding that lamprey hemoglobin has a hyperbolic dissociation curve like that of the storage pigment myoglobin; this is incorrect and was probably the result of early experimentation with very dilute protein solutions.) At first glance, heme-heme interaction in lamprey hemoglobin seems paradoxical because only one heme is found on each molecule. The probable explanation is that hemoglobin aggregates as the protein concentration is raised, and that heme-heme interactions can occur in such aggregates, just as they do between the four subunits of higher vertebrate hemoglobin! It has been proposed that the single gene that produced a globin of 17,000 in cyclostomes, by processes of gene duplication and mutation, gave rise to the set of four genes that produces in higher vertebrates a four-unit globin with a molecular weight of almost 68,000.

In the course of vertebrate evolution, many amino acid substitutions have occurred, affecting the way that globin subunits bind to each other and to hemes, as well as altering dissociable groups near the hemes (the basis of the Bohr effect). For instance, the differences in Bohr effect shown by various hemoglobins may relate to the availability of histidine residues on the C-terminal end of the beta subunits and amino groups on the N-terminal end of the alpha subunits. Masking of the latter groups or alterations in the histidine can lead to alteration or disappearance of the Bohr effect. Such permutations of structure are the variations upon which natural selection has acted to yield hemoglobins adapted to different environments or ways of life.

The hemoglobin of vertebrates is, of course, normally contained within erythrocytes. This is not, as long assumed, because hemoglobin solutions are particularly viscous. Among the possible reasons for the retention of the respiratory pigment within red blood cells are: (1) the possibility that free hemoglobin would excessively increase the osmotic pressure of the blood, thereby necessitating a much higher blood pressure to counteract the tendency for fluid to enter the blood; (2) the fact that cells help to cause turbulence in flowing blood, thus increasing the efficiency of gas exchange; and (3) the desirability of keeping certain enzymes or small molecules (as 2,3-diphosphoglycerate) in close proximity to hemoglobin. Besides these features of red blood cells, a crucial property is their deformability. If

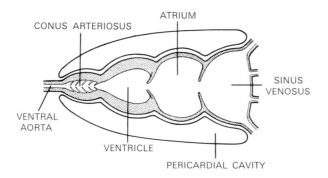

CONUS ARTERIOSUS   ATRIUM

SINUS VENOSUS

VENTRAL AORTA

VENTRICLE

PERICARDIAL CAVITY

Figure 2. The probable arrangement of the heart chambers in a primitive vertebrate. As blood flows from the right to the left through the heart, its pressure is raised by action of the increasingly thick muscular walls (stippled) of the chambers. Back flow is prevented by valves between the chambers.
[After Torrey, Theodore W., *Morphogenesis of the Vertebrates*, Second Edition, Wiley, 1967.]

erythrocytes are rigidified in shape with chemical fixatives, they cannot pass through the narrow capillary beds in some parts of the body.

As the blood pigment changed in the course of time, so did the vascular system and its controls. It seems likely that, in the earliest vertebrates, blood was originally propelled by rhythmic contractions along the main anterior ventral blood vessel. As these organisms increased in size, the pumping function apparently localized in a discrete region that was to become the heart. The walls of the blood vessel in that region acquired cardiac muscle cells and as a result thickened. Such a heart pump works most efficiently if its cavity is expanded somewhat beyond the resting volume before each contraction occurs. The four chambers of the primitive tubular vertebrate heart, such as that in some fish, appear to be stages for accomplishing this purpose by raising the blood pressure. When venous blood returns to the heart of a fish, it is under such low pressure that it cannot expand the thick-walled anterior pump, the ventricle. Therefore, the blood first enters a thin-walled chamber, the sinus venosus, where it is propelled forward into a somewhat thicker-walled space, the atrium, which finally raises blood pressure enough to expand the ventricle. This simplistic interpretation must be modified somewhat, because it does not take into account the elastic recoil of contracting cardiac muscle, a property that also tends to expand the chambers. The intricate workings of the modern heart are recounted in the article "The Heart" by Carl J. Wiggers. In an earlier *Scientific American* article (Offprint 1067, "The Heart's Pacemaker"), E. F. Adolph points out that the various chambers each beat at a characteristic rate during the embryonic development of a vertebrate; the anteriormost chamber, which develops first, beats at the slowest rate, the next most posterior chamber develops next and sets the pace with a higher beat, and lastly the most posterior chamber forms and sets the final fast pace. The derivative of this chamber—the sinoatrial node—is the pacemaker in higher vertebrate adult hearts.

If we look at primitive cyclostome adults, we find that the rate of heart beat is determined by the sinus venosus; the beat is wholly myogenic (intrinsic to the muscle cells) in origin and no nerves accelerate or slow the beat. Interestingly, the heart muscle does not respond to direct application of acetylcholine, the agent liberated by the vagus nerve of higher vertebrates to decrease cardiac output. In higher bony fish and sharks, vagal fibers do innervate the heart and their discharge of acetylcholine slows the beat rate and decreases the amplitude of contraction; consequently, less blood is pumped. Finally, in terrestrial vertebrates, the vagus functions as it does in fish, and, in addition, sympathetic nerve fibers discharge noradrenaline, which accelerates the beat and increases its amplitude. Thus, we see three stages of control: myogenic, myogenic plus inhibition, and myogenic plus inhibition and stimulation. Clearly, more precise regulation is possible with such increasingly complex control circuitry. In the embryos of today's birds and mammals, the basic myogenic pattern described above occurs before nerve fibers reach the heart and during that period the cardiac muscle tissue does not respond to application of acetylcholine or noradrenaline. In this way, the developmental sequence within a single embryo reflects the events that occurred during vertebrate evolution.

Although the control of heart beat is the core of vertebrate circulatory regulation, the actual cardiac output (volume of blood pumped) in a mammal is determined to a great degree by the volume of blood returning to the heart through the veins. This quantity is governed by the degree of vasoconstriction of peripheral blood vessels (arterioles, capillaries, veins; see in particular "The Microcirculation of Blood" by Benjamin Zweifach). Control of venous return is still not completely understood. For example, Edwin Wood ("The Venous System," *Scientific American*, January, 1968; Offprint 1093) has pointed out that when a man stands, return of blood upward through the veins of the leg is difficult because of gravity. Valves spaced along the veins prevent backflow and hold the blood in a series of stages. Nevertheless, a larger percent of total body blood accumulates in the legs. As this occurs, veins elsewhere in the body periphery contract and reduce their volume; this tends to maintain a constant volume of blood returning to the heart and so compensate for the temporary accumulation in the legs. In addition, the recent experiences of man in the prolonged, gravity-free conditions of space craft, where minimal use of leg musculature occurs, suggests the importance of actual compression of the veins by the muscles in aiding venous return and maintaining general health of the blood vessels. The mechanisms coordinating function in the venous system are not understood, although it seems likely that the sympathetic nervous network is involved. At the cell level, noradrenaline released by sympathetic nerves causes contraction of smooth muscle cells in the walls of veins, thus causing vasoconstriction (remember that the sympathetics also stimulate contraction of cardiac muscle). Other sympathetic fibers innervate smooth muscle cells in the capillary system, where they usually evoke vasoconstriction (but, for unknown reasons, some muscle cells respond differently and relax, yielding vasodilation, when noradrenaline is released). The vasomotor center of the medulla in the brain must be the ultimate coordinating switchboard for all of these processes.

Two specialized parts of the circulatory system are discussed in "The Lymphatic System" by H. S. Mayerson and in the article on the rete mirabile, " 'The Wonderful Net,' " by P. F. Scholander. Lymph vessels are found in all terrestrial vertebrates (and not just in mammals, as is implied by Mayerson). The higher blood pressures and rates of flow of the partially separate pulmonary circulation of amphibians and reptiles apparently produced leakage of plasma through the walls of blood capillaries. A collecting system evolved to drain the tissue spaces and to return the lost substances to the regular circulatory system. These lymphatic vessels are present in more complex form in birds and mammals.

Rete mirabile and countercurrent flow apparatus are used in many ways by vertebrates. Gases, heat, and some metabolites are commodities passed between the tiny vessels in which fluid flows in opposite directions. We can

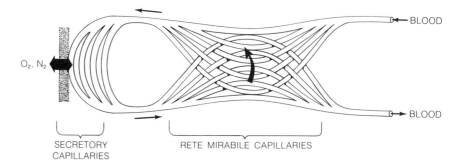

Figure 3. An idealized diagram demonstrating the physical separation of the secretory capillaries located near the epithelium of the gas gland in a swim bladder and the rete capillaries where countercurrent exchange takes place.

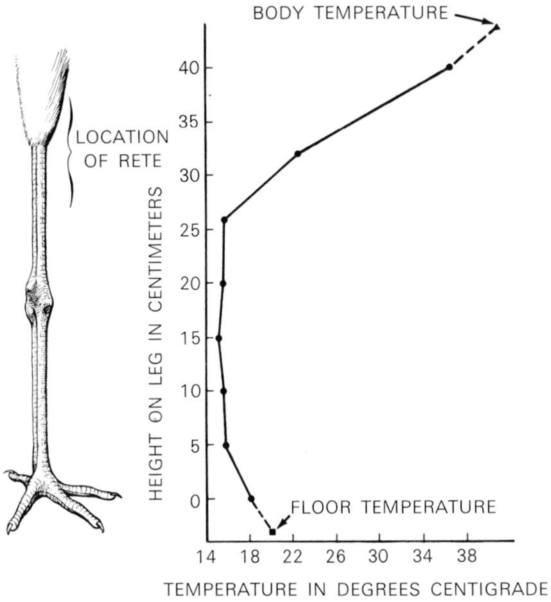

BODY TEMPERATURE

LOCATION OF RETE

FLOOR TEMPERATURE

HEIGHT ON LEG IN CENTIMETERS

TEMPERATURE IN DEGREES CENTIGRADE

Figure 4. The role of the rete mirabile in conserving body heat is demonstrated by these data on the leg of a stork. The bird was in an environment between 11.5° and 13.0° C. Without the rete bed, the leg would be a major source of heat loss because it lacks insulating feathers. [After Kahl, M. P., Jr., *Physiol. Zool.* 36, 1963. The University of Chicago Press.]

now add to Scholander's comments about the function of the rete mirabile in the fish swim bladder. The rete itself is responsible for the efficient exchange of gas (oxygen) from the capillary that carries blood away from the lumen of the bladder to the capillary that brings blood to the bladder. The idealized diagram of the rete in the article by Scholander (p. 125) may lead to some confusion in the mind of a reader about the anatomical and functional relationships involved. In fact, as seen in Figure 3, blood passes through the rete, into an artery, through the capillary bed beneath the gas-secreting epithelium, back into a vein, and once more through the rete capillaries. This diagram makes it clear that the business of the rete is *exchange*—exchange of commodities such as oxygen or organic molecules. The separate capillary bed, which is not a countercurrent exchanger, is the *source* of gas to be passed into the swim bladder.

How can oxygen be maintained or actually secreted into the swim bladder when a fish is at great depths and the bladder already contains nearly pure oxygen at hundreds of atmospheres pressure? One factor may be lactic acid, which is apparently produced by the secretory epithelial cells of the gas gland. This lactic acid enters the blood, is exchanged across the rete to inflowing blood, and therefore increases dramatically in concentration. Hydrogen ions apparently are produced and follow the same path. Both substances serve to lower the blood pH as they accumulate in the "closed" circuit. Because of the Bohr effect, hemoglobin in the blood gives up its oxygen more easily as this point is approached. The lactic acid is also thought to cause "salting out" of oxygen from the blood fluid. In other words, as the concentration of the acid rises, oxygen solubility decreases and transfer from blood fluid to swim bladder lumen is favored. In addition to these effects of lactic acid, the hemoglobin of many fishes

Table 1.	
DEPTH (in meters)	LENGTH RETE CAPILLARIES (in millimeters)
150–500	1–2.0
500–1000	2.5–6.0
750–1500	10–15

Table 1. Relationship between the depth at which given species of fish live in the ocean and the length of the capillaries in the rete mirabili of their swim bladders.
[After N. B. Marshall, *Explorations in the Life of Fishes*, Harvard University Press, 1971.]

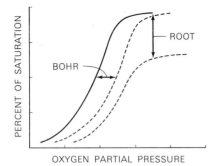

Figure 6. Oxygen dissociation curves showing the Bohr and Root effects. In the Bohr effect the affinity of hemoglobin for oxygen decreases (the curve is displaced to the right). In the Root effect, the maximal quantity of oxygen that can combine with the respiratory pigment and be transported is decreased.

with swim bladders shows a "Root effect," whereby the actual quantity of oxygen that can be bound at any one partial pressure of oxygen is decreased as pH falls (i.e., the oxygen dissociation curve does not move to the right as it does in the Bohr effect, but rather the upper segment of the curve is lowered, as shown in Figure 6). Thus, hemoglobin can unload oxygen, even though hundreds of atmospheres of oxygen may be present in the swim bladder.

Interestingly, the length of the capillary vessels in the rete mirabile is a direct function of the normal depth at which a fish lives. Obviously, such anatomical modifications must have evolved if the bony fish were to attain the ability to invade great depths, while maintaining buoyancy at the same time.

A survey of vertebrate vascular systems and hemoglobins provides good illustrations of the versatility of vertebrates in taking advantage of special features of the environment. For instance, Johan H. Ruud ("The Ice Fish," *Scientific American*; Offprint 1025) has found that the "ice" fishes produce no hemoglobin. Instead, they seem to have an extraordinarily large blood fluid volume that meets the demands of gas transport. Low rates of metabolism in very cold ocean waters demand far less oxygen and produce less carbon dioxide than would metabolism in warmer waters. Cold water also contains more dissolved oxygen than does warm water. This is probably important to ice fish who, because no hemoglobin passes through their gill capillaries, have no pigment to bind oxygen actively and to remove substantial quantities of it from the water. Instead, simple diffusion of oxygen into the blood fluid probably occurs. Consequently, the larger quantity of dissolved oxygen in sea water, the greater the amounts that can diffuse per unit of time into the flowing blood.

Another group of vertebrates has taken advantage of the high oxygen content of cold water in quite another way. In fast-running mountain streams, several types of lungless salamanders are found. These creatures have such extensive capillary nets in their skin that all gas exchange can occur there and no lungs are formed during embryonic development. A peripheral advantage of this modification is that the air-filled lungs are not present to act as buoyancy organs; therefore specific gravity of the body is greater and there is thus less chance of the animal being swept away in the rapidly flowing water currents. But unlike the ice fish, lungless salamanders do utilize hemoglobin and erythrocytes to transport oxygen.

8

THE PHYSIOLOGY OF EXERCISE

CARLETON B. CHAPMAN AND JERE H. MITCHELL
May 1965

*Muscular activity causes the body to mobilize an entire array
of adaptive mechanisms to meet rising respiratory and
circulatory demands. Even the prospect of exercise can initiate
the process*

When Peter Snell of New Zealand ran the mile in three minutes 54.4 seconds and Sixten Jernberg of Sweden covered 50 kilometers on skis in two hours 43 minutes 52.6 seconds, their record performances provided vivid testimony not only to their speed and endurance but also to the prodigious internal adjustments of which the human body is capable. An adult human being at rest gets along on about three-tenths of a liter of oxygen per minute; during maximal exertion the oxygen requirement increases to 10 or more times that amount. Since the body carries no appreciable reserve of oxygen (its entire store would be used up by the muscles in 20 seconds of strenuous exercise), this means that during exertion the body's machinery must step up enormously its intake and distribution of oxygen to the tissues, and it must be able to do so within seconds. On this basic fact hangs a tale that for nearly two centuries has provided physiologists with an absorbing realm of exploration.

The detailed study of muscular exercise began in the 18th century when Antoine Laurent Lavoisier and Pierre Simon de Laplace discovered that the process consumes oxygen and produces carbon dioxide. As investigation progressed it became clear that exercise involves not only the muscles but also many other tissues; that it depends, indeed, on an extraordinary coordination of the respiratory, circulatory and nervous systems, all working together under highly integrated controls. During the 19th century almost every physiologist of the first rank worked on muscular exercise at one time or another. During the 20th century three men—A. V. Hill of Britain, August Krogh of Denmark and Otto Meyerhof of Germany—

have received Nobel prizes largely for work relating to muscle or muscular exercise. Among the Americans who made notable contributions in the physiology of exercise were Edward Cathcart, Francis Benedict, Graham Lusk and D. Bruce Dill. In the years before World War II the Fatigue Laboratory at Harvard University became a world center for the experimental study of muscular exercise.

In recent years investigations of exercise have concentrated on processes in the cell and on physiological systems involved in control. This is not all there is to the study of exercise: in order to ensure an adequate supply of oxygen for the working muscle cell the body must coordinate the interaction of the lungs, the blood (specifically the oxygen-carrying pigment hemoglobin), the heart and circulatory system and finally the muscle cell itself. We shall discuss each component of the system in turn.

Oxygen enters the bloodstream—and carbon dioxide leaves it—by way of the alveoli: the tiny sacs that form the ultimate functional structure of the lungs. The total surface area of this multitude of sacs is normally so large that, as far as the body's needs go, the exchange of gases between the air on one side of the walls of the alveoli and the blood bathing the other side is virtually unlimited. A normal man at rest inhales between six and eight liters of air per minute, from which about .3 liter of oxygen is transferred in the alveoli to the blood. Simultaneously carbon dioxide is given off by the blood and exhaled. When the same man is engaged in maximal muscular activity, he may take in 100 liters of air per minute and extract five liters of oxygen.

The rate of oxygen intake provides an excellent measure of the physical work done. The term "maximal oxygen intake," introduced by Hill in 1924, characterizes the upper limit of performance of an individual in a remarkably predictable way and has proved to be an extremely useful physiological tool. In practice an experimental subject performs a series of work tests on a motor-driven treadmill or bicycle ergometer, each test being at a higher load than the one just preceding. The intake of oxygen per minute rises on a virtually linear curve with work load until maximal intake is attained. Thereafter most people can still increase the work load, but the oxygen intake can rise no further. The maximal oxygen intake in normal individuals, however, has to do not with the capacity of the lungs for ventilation or diffusion but with the maximal pumping capacity of the heart. Maximal oxygen intake is therefore a fair index to circulatory capacity but not to pulmonary capacity.

The outstanding question about the ventilation of the lungs is: How does the control system that so neatly balances oxygen supply against oxygen demand work? As long as 150 years ago it was suspected that mammals had a center for the control of breathing in the medulla oblongata; this center was believed to have the function of maintaining rhythmic respiratory movements regardless of the state of consciousness. Although the respiratory center has long been known to be influenced by higher centers in the brain, the identity of the chemical and hormonal factors that affect it has been elusive. The classical paper on the subject, published in 1905 by John Scott Haldane (father of the late J. B. S. Haldane) and John G. Priestley, held that the respiratory center is primarily subject to the influence of carbon dioxide in the arterial blood flowing into it. The amount of carbon dioxide in the arterial blood in turn is determined by the level of carbon dioxide in the alveoli of the lungs.

Haldane's experiments showed that ventilation can be doubled by increasing the amount of carbon dioxide in inspired air from the usual negligible amounts to about 3 percent. He also considered the possibility that lack of oxygen might stimulate respiration. Ultimately he concluded that the carbon dioxide effect is dominant except under circumstances of extreme lack of oxygen. As far as the Haldane school was concerned the primary function of the respiratory center is to govern ventilation in such a way that the level of carbon dioxide in the alveoli is held as constant as possible.

During physical exercise the production of carbon dioxide in the body rises very rapidly and carbon dioxide is brought to the alveoli in increasing quantities by the venous blood. It was Haldane's view that, as a result of increased amounts of carbon dioxide in the alveoli, the arterial blood becomes more acid. He thought that the acidity stimulates the respiratory center and that excess carbon dioxide is blown off in expired air.

More refined studies later showed that Haldane's chemical-control theory was a considerable oversimplification. It turned out that during exercise and recovery from exercise the rate of ventilation does not rise and fall in direct proportion to the acidity, the carbon dioxide pressure or the oxygen pressure of the arterial blood. Because of these findings various other hypotheses have been offered: that some secondary signal—chemical, hormonal or nervous— acts as the stimulus to the respiratory control center, or that small changes in the carbon dioxide and oxygen pressure can cause large changes in the breathing rate by a multiplying effect. The details of the system are still very much in doubt. Whatever its exact *modus operandi,* the respiratory control mechanism ordinarily prevents carbon dioxide from accumulating in any significant degree and virtually assures an adequate supply of oxygen over a range extending from rest to maximal exertion.

Some interesting possibilities were opened up when Hill studied the effect of breathing pure oxygen during exercise. There seems no reason to doubt that an immediate effect of switching during steady exercise from ordinary air to air enriched with oxygen is to lower considerably the rate of ventilation. The effect is cited to support the view that a change in arterial oxygen pressure has an influence on the regulation of breathing. Athletes who have breathed oxygen-enriched air during exercise have reported a pronounced relief of subjective distress and have evinced a decrease in ventilation. Moreover, the late C. G. Douglas of the University of Oxford and his colleagues showed that oxygen-breathing extended the work capacity of trained athletes. The topic is an important one, not only with respect to competitive athletics but also with respect to the deficient transport of oxygen in disease and at high altitudes. It is being actively investigated in many laboratories.

Another important mechanism employed by the body when it is under fairly severe stress from exercise is its ability to incur an oxygen debt. In theory the mechanism enables the exercising body to live temporarily beyond its capacity for transporting oxygen to active muscles and to compensate for doing so during rest after the exercise. The phenomenon is well documented by the fact that ventilation and oxygen intake do not immediately return to normal resting levels for some time after vigorous exercise stops.

The amount of oxygen debt that can be contracted is limited by the ability of the blood to absorb acid metabolic products without an unduly large change in its acidity. Some degree of oxygen debt is always built up during a short, hard burst of exercise such as a footrace over a distance of 100 yards or a quarter of a mile. Milers or marathon runners, particularly if they are in good physical condition, can cover their distances with a relatively small oxygen debt. The term "steady state" is used by exercise physiologists to define a rate of work that can be performed for considerable periods of time without oxygen debt. Actually most people— probably anyone other than an athlete in training—incur some oxygen debt during exercise at all levels but the very lowest. Waking activity, even if it is quite sedentary, usually involves a continuous process of acquiring and paying off oxygen debt at various rates. During continuous activity, such as walking steadily at three miles per hour for several hours, oxygen debt is built up rapidly at first and then levels off. Thereafter it may remain virtually un-

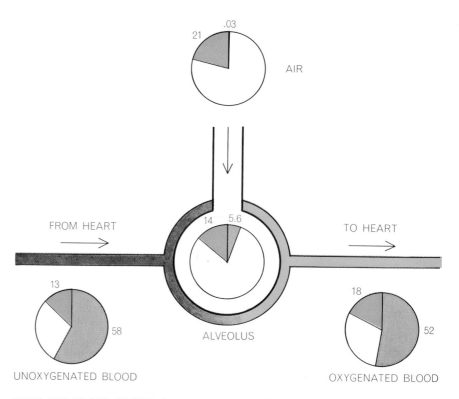

LUNG AND BLOOD GASES in human respiration under resting conditions are shown near and in one of the alveoli, the lung sacs where the blood acquires oxygen and yields carbon dioxide. Numbers give the percentage by volume of oxygen (*color*) and carbon dioxide (*gray*). The remaining percentages are accounted for by substances not involved in respiration.

changed for some hours. During violent activity it builds up rapidly and continuously until the activity ceases.

In any event, the contraction of an oxygen debt, like any other form of deficit financing, has its limits. Ideally the entire process of adaptation to physical stress, including the control of ventilation, is adjusted to the work load so elegantly that the oxygen debt incurred is kept at a minimum. The process of contracting a small oxygen debt is part of the normal mechanism of adaptation to exercise; a large oxygen debt is an emergency mechanism that is invoked only in case of dire need and in most sedentary individuals never comes into play at all.

After ventilation of the lungs the next major link in the system of adaptation to exercise is hemoglobin. Without this remarkable substance mammals would be completely incapable of muscular activity. The ability of hemoglobin to transport and deliver oxygen in the body resides in its iron atoms. Everyone knows that when metallic iron and oxygen combine, as they do in iron ore, it takes a blast furnace to separate them. In hemoglobin, however, iron combines with oxygen transiently and releases it readily to the tissue cells at body temperature.

As Christian Bohr of Denmark (father of the physicist Niels Bohr) first noted in 1904, the absorption and release of oxygen by the hemoglobin contained in the red blood cells are described by an S-shaped curve. Where the partial pressure of oxygen is that of ordinary air, as in the lungs, the hemoglobin of the blood becomes almost completely saturated with oxygen; when pure oxygen is breathed, the hemoglobin is 100 percent saturated. It gives up little of its oxygen as it flows through the arterial system, although the partial pressure of oxygen falls slightly there. When, however, the blood arrives at the capillaries and the surrounding tissues, where the oxygen pressure is substantially lower, the hemoglobin freely surrenders its oxygen and the curve of oxygen-binding bends sharply downward [see illustration on page 94].

Normally arterial blood contains about 18 percent of oxygen by volume, that is, about 18 milliliters of oxygen per 100 milliliters of blood. (Breathing pure oxygen raises this figure to 18.5 milliliters, and it also increases the amount of oxygen dissolved in the blood plasma from .2 milliliter to 1.9 milliliters.) When a person is at rest, the tissues absorb oxygen at a rate that reduces the 18 percent oxygen content of the arterial blood to about 12 percent in the venous blood. The drop is known as the "arteriovenous oxygen difference."

During exercise, on the other hand, the blood may have to give the tissues as much as 15 of its 18 percent of oxygen—about two and a half times more than it surrenders at rest. Several factors facilitate this increased surrender of oxygen by hemoglobin. One, of course, is the depletion of oxygen in the tissues by exercise; this results in a higher gradient between the oxygen pressure in the blood and that in the tissue cells. Another factor is the accumulation of carbon dioxide and acid metabolic products in the exercising muscle fibers, which acts to stimulate the release of oxygen by hemoglobin. Other factors, including the increased temperature of the active muscle cells, may also be involved.

The main point that emerges from this kind of investigation is that the biological properties of hemoglobin play a very important role in enabling the body to adapt to exercise. If the body had to depend solely on dissolving oxygen in a carrier fluid such as blood plasma, even at rest it would have to circulate nearly 300 pounds of fluid per minute—not far from twice the total weight of an average man! Thanks to hemoglobin, a flow of only five liters of blood per minute is enough to supply the body's oxygen needs at rest.

This brings us to the third link in the system of adjustment to exercise: the pumping and circulation of the blood. Comparatively recent developments make it possible to obtain accurate measurements of human cardiac output (the amount of blood pumped per minute) during exercise. It is now established that in a young man the heart can increase its output from about 5.5 liters of blood per minute at rest to nearly five times that figure during maximal exertion. At peak demand for oxygen the heart increases its output both by speeding up its rate of beating and by increasing the volume of blood pumped at each stroke. The pulse rate may double or triple; the stroke volume, which is about 60 to 80 milliliters at each beat when a person is standing at rest, may go as high as 120 milliliters.

The situation at levels of exercise less than maximal is somewhat uncertain. According to the available evidence, it seems likely that under moderate stress the heart may increase either its pulse rate or its stroke volume, depending on the individual's physical training and perhaps on other factors. There is good reason to believe that the heart of a trained athlete increases its stroke volume more readily than that of a sedentary person. Under the stress of emotions such as fear and anger the heart increases its output almost entirely by speeding up its rate of beating. If the nerves controlling the pulse rate are blocked experimentally by drugs, however, the heart will expand its stroke volume. The increase in cardiac output during emotional stress is considerably less than it is under physical stress; even during severe emotional crises it may not rise more than a third over the normal output at rest. The whole subject of the heart's output during stress still leaves much to be learned and is under wide investigation.

The facts we have reviewed so far make clear that the chief limitation on the body's capacity for physical exertion is the cardiac output: the ability of the heart to move blood to the tissues. The heart's capacity for increasing its output fivefold is not enough—even allowing for the ability of hemoglobin to release oxygen more freely—to account for the muscles' consumption of oxygen during violent exercise. An extra margin is provided, however, by a change in the pattern of blood flow. During heavy exercise most of the arterial blood is diverted to the active muscles, where the need for oxygen is most acute.

When the body is at rest, the muscles apparently account for no more than about 20 percent of its total oxygen consumption. Substantial amounts of oxygen go to the brain, the heart, the skin, the kidneys and other organs [see illustration on the next page]. If the muscles' share of the oxygen is about 20 percent, they use at rest only 60 to 70 milliliters of oxygen per minute (the body's total consumption being 300 milliliters per minute). Working at full tilt, as in running or swimming, however, the active muscles need about 3,000 milliliters per minute, or 50 times their resting requirement. The other organs do not require anything like this increase of oxygen; in fact, most of them actually use substantially less oxygen during maximal exertion. (The flow to the skin follows a unique pattern. In light or fairly strenuous exercise there is a substantial increase of blood flow through the skin to dissipate heat generated by

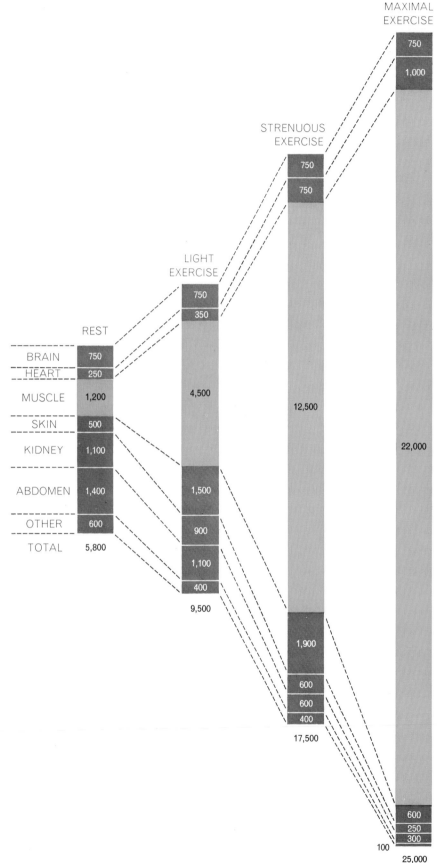

the exercise, but when the muscles' need for oxygen becomes acute in maximal exertion, the skin, like the other organs, surrenders most of its claim on the supply of blood.)

The result of this cooperative adaptation by the other organs is that all but a relatively small portion of the cardiac output can go to the active muscles. The absolute amount of oxygen diverted to the muscles by this means is not great: it is no more than about 400 to 500 milliliters per minute. This amount, however, apparently represents a critical margin for the muscles. Experimental drug treatments that block the preferential distribution of blood to the muscles bring about a considerable impairment of the subject's capacity for exertion.

What are the mechanisms that adapt the output of the heart to the muscles' needs during exercise? On this question only fragmentary information is available; obviously it is a difficult question to investigate experimentally. One well-established finding is that the heart muscle itself contains a mechanism for the control of its activity. An experimental preparation of an animal's heart, isolated from any nervous or hormonal influence, still responds to an increase in its rate of filling with blood by increasing the amount it ejects at each beat. This property is known as the Frank-Starling law of the heart, after Otto Frank of Germany and Ernest Henry Starling of Britain, the physiologists who discovered it. Starling called it the "central fortress" of the system whereby the heart adapts itself to varying demands.

Concerning the nervous and hormonal controls of cardiac output certain general facts are known. The heart is powerfully influenced by sympathetic nerves, which stimulate its activity; by parasympathetic nerves (principally the vagus nerves), which depress its activity, and by hormones that reach the heart muscle by way of the bloodstream. These influences, both nervous and hormonal, are in turn triggered by various receptors elsewhere in the circulatory system or by complex factors involving the central nervous system. Typically a receptor at a strategic point in the arterial system (for example in the arch of the aorta) responds to increased pressure by transmitting impulses to integrating centers in the central nervous system. These increase the number of depressor impulses to the heart via the vagus nerves and decrease the

DIVERSION OF BLOOD from certain organs to the muscles is one of the body's major mechanisms for adapting to exercise. Numbers show blood flow in milliliters per minute. The brain is the only organ to which the flow remains almost constant under all conditions.

number of augmentor impulses via the sympathetic nerves. The net effect is to reduce the rate and force of the heartbeat, thus slowing the flow of blood in the arteries. Conversely, a fall of pressure in the aortic arch gives rise to a signal that stops the sending of impulses to the central nervous system, and so the depression of the heart's activity is diminished.

The eyes and ears may operate in the same way as a blood-pressure receptor. They too, in response to perceptions of the external world, send information in the form of impulses that control the heart's output. The impulses go to the cortex of the brain, which responds by sending appropriate directives to lower centers, whence augmentor or depressor messages are forwarded to the heart. In this way the heart is directed to gear its activity to anticipated needs of the body or to emotional stress.

The ultimate element of the system of adaptation to exercise, and the beneficiary of the marvelously adaptable machinery for supplying oxygen, is of course the muscle cell itself. Compared with many other cell types, the muscle cell is unique in one respect: its tolerance for oxygen deprivation. Certain other types of cell are highly sensitive to even a temporary shortage of oxygen. The cells of the brain, for instance, quickly suffer irreversible damage if the partial pressure of oxygen in the venous blood draining from them falls below 20 millimeters of mercury. Muscle cells, on the other hand, are known to survive even when the venous blood draining from their vicinity shows a partial pressure of oxygen at or near zero!

The extraordinary tolerance of muscle cells for a temporary shortage in the oxygen supply may be accounted for by the presence in these cells of myoglobin. This substance, like hemoglobin, carries oxygen, but it differs from hemoglobin in that it surrenders the oxygen much less readily. The S-shaped curve picturing hemoglobin's uptake and release of oxygen signifies that this behavior varies with the partial pressure of oxygen in hemoglobin's surroundings. As the oxygen pressure rises, the hemoglobin molecules absorb more oxygen; as the pressure falls, hemoglobin gives up its oxygen, and it does so with increasing rapidity as the oxygen pressure in the surroundings drops below 60 millimeters of mercury. (Normally, even during violent exercise, the

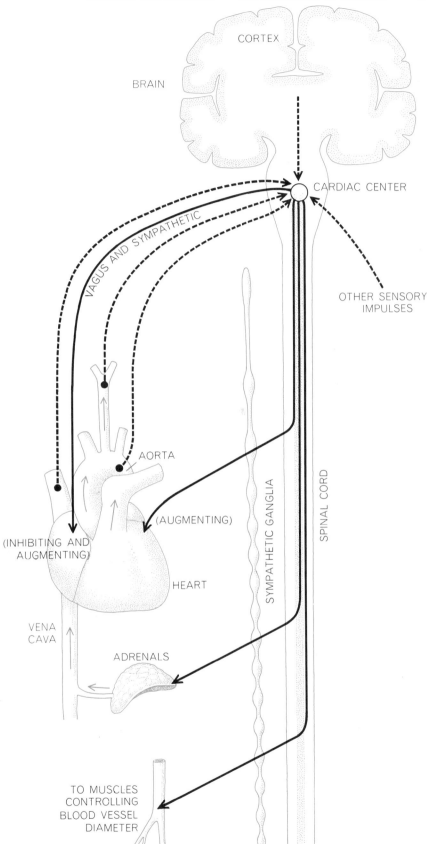

NERVE CONTROLS affecting the action of the heart are depicted schematically. Broken lines represent afferent, or input, nerves; solid lines, efferent, or output, nerves. Some efferent nerves have a sympathetic, or stimulating, effect, in which the sympathetic ganglia participate. Others, such as the vagus, have a parasympathetic, or slowing, effect. Stimuli from cortex, resulting from visual or auditory signals or just from thought, can cause circulatory and muscular systems to begin preparing for exercise before it starts.

oxygen pressure in the venous blood coming from active muscles does not fall much below 20 millimeters, at which pressure hemoglobin is about 30 percent saturated with oxygen.) Myoglobin's behavior contrasts sharply with this. Its curve of uptake and retention of oxygen is not S-shaped but hyperbolic, which is to say that after becoming saturated with oxygen the myoglobin molecule does not give up its oxygen until the pressure has fallen to a low level—15 millimeters or less.

Hemoglobin seems designed, therefore, to perform its vital function in surroundings where partial pressures of oxygen range from about 100 down to 20 millimeters (at which level hemoglobin is about 30 percent saturated). Myoglobin, on the other hand, would be of little physiological use (and would relinquish little or no oxygen) if it were confined to the range of partial pressures suitable for hemoglobin. Since myoglobin is found only inside muscle cells, and since there is fair evidence that partial pressures of oxygen in such cells may approach zero during very heavy activity, myoglobin may conceivably serve as a special oxygen store for the cell. According to this view, an intermittently contracting muscle cell might recharge its myoglobin during the resting phase and then, during a succeeding contraction, call on myoglobin oxygen stores when intercellular oxygen pressures fall below 10 millimeters of mercury.

This phenomenon may well account for the fact that mammalian muscles can work more smoothly and effectively when they intermittently contract and relax than when they are kept contracted for an extended period. Actually continuous contraction is probably rare except in laboratory experiments; even in an activity such as weight lifting, which might be thought to keep the muscles under unremitting tension, some of the individual muscle fibrils probably have a chance to relax with each slight shift in grasp and body position.

Myoglobin may also have something to do with the fact that short periods of work alternated with short periods of rest seem to be more efficient—metabolically as well as in work accomplished —than long work periods followed by long rest periods. Another interesting point inviting investigation is the indication that the amount of myoglobin in muscle tissue can be increased by physical training. The physiological significance and functions of myoglobin are a

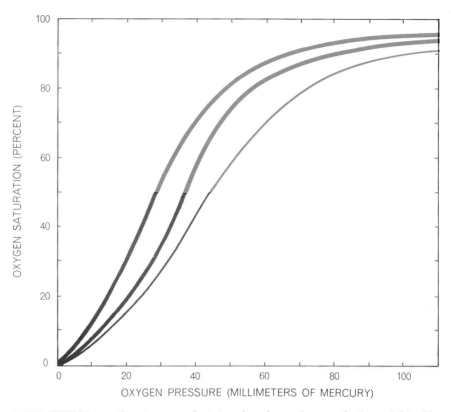

BOHR EFFECT is evident in curves depicting the release of oxygen by hemoglobin. The curve at left represents a condition of rest; the middle curve, exercise. Arterial blood is indicated in color; venous, in gray. Hemoglobin's release of oxygen increases with declining oxygen pressure in the tissues, a condition that is intensified as exercising muscles use oxygen. Such muscles also produce acid wastes, and acidity entering the blood helps to stimulate the release of oxygen by hemoglobin. This is the Bohr effect, reflected in the shift of the curves to the right. A pH reading of 7.4, slightly alkaline, is normal for a man at rest; a slightly acid reading of 6.8 (*light curve*) can occur with heavy exercise.

largely unexplored subject and represent one of the most promising frontiers of research on muscular exercise.

Having looked separately at the main components of the system that effects adaptation to physical stress, let us now put them together and see how the system works as a whole. Nature demands that for survival an organism must be able in an emergency to rise to maximal activity on an instant's notice. Usually, however, the anticipatory warning allows enough time for efficient mobilization of the body's machinery. In man and other mammals a great deal of physiological preparation begins as soon as the individual senses that vigorous physical activity is imminent. The autonomic nervous system becomes active and the adrenal glands release the hormones epinephrine and norepinephrine into the bloodstream. The pulse rate quickens, the heart muscle contracts more forcibly and the output of the heart probably begins to increase slightly. The respiration deepens and may speed up. The voluntary muscles

become tense. The switching of blood flow from other areas to the muscles begins.

When exercise itself starts, there is an immediate jump in the rate of ventilation of the lungs. The cardiac output, pulse rate and stroke volume rise to their maximal level, probably within no more than a minute or two. Simultaneously the hemoglobin starts to surrender more of its oxygen to the muscle tissues, widening the difference between the oxygen content of arterial and venous blood. By the end of the first minute most of the known adaptive mechanisms are working nearly to capacity. Thereafter, with continuation of the exercise, the muscle cells begin to run up an oxygen debt.

If the exercise is pushed to the point of exhaustion, various mechanisms begin to fail. The rhythm of the heart becomes less regular; cardiac output declines; the oxygen content of the arterial blood may fall, and marked pallor —sometimes even blueness—of the lips and fingernails develops. Finally an inordinate rise in the pulse rate and

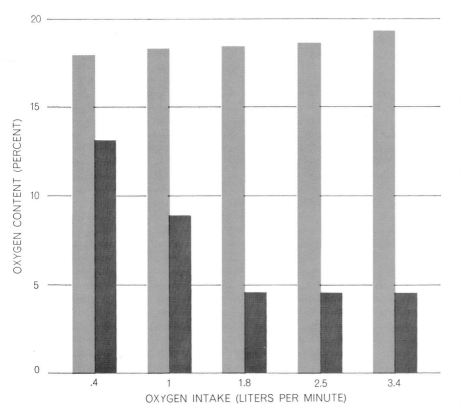

ARTERIOVENOUS OXYGEN DIFFERENCE is shown as oxygen intake rises with increasing exercise. Arterial blood is represented in color; venous, in gray. The oxygen content of arterial blood remains almost constant. That of venous blood drops rapidly at first, then levels off as the intake of oxygen and the output of the heart become maximal.

the heart beat faster, by attempting to increase the ventilation of the lungs and by surrendering more oxygen from each liter of blood pumped by the heart to the tissues. By acquiring a better understanding of the mechanisms involved it may be possible to improve the body's adaptation to its needs under a wide variety of conditions.

Could special training techniques and other devices, including drugs, raise man's physical performances to greater heights? The training methods of athletes, from the time of the Greeks to the present, have been largely empirical and rule of thumb. With methods based on fuller physiological knowledge it may well be possible to achieve performances now considered superhuman. One must pay due regard, however, to the limits of the human organism. Once the organism has reached its maximal oxygen intake and cardiac output it cannot go further without drawing on its vital reserves, and this must have its limits of safety.

According to Herodotus, Pheidippides, "a trained runner," covered the 150 miles from Athens to Sparta in two days and two nights. Legend (but not Herodotus) says that a few days later he ran the 22 miles from Marathon to Athens to report the Greek victory over the Persians and fell dead at the end of his run. It may become possible to achieve by artificial means the kind of performance that extreme motivation is said to have made possible for Pheidippides, but only at the same cost. In all likelihood, however, there is room for improvement of human physical achievement, well within the bounds of safety, by rationally based methods.

clumsiness of the muscular movements give warning that collapse is imminent.

Research on the physiology of exercise of course has important practical objectives, particularly with respect to what can be accomplished by physical training. The problem is not simply a matter of enabling astronauts to cope

with a landing on the moon. Probably a more useful goal, from the standpoint of the human population at large, is to find training regimes and other measures that will help the body to cope with diseases of the lungs and heart. In many ways the symptoms of chronic heart failure are like those accompanying heavy exercise. The body responds by making

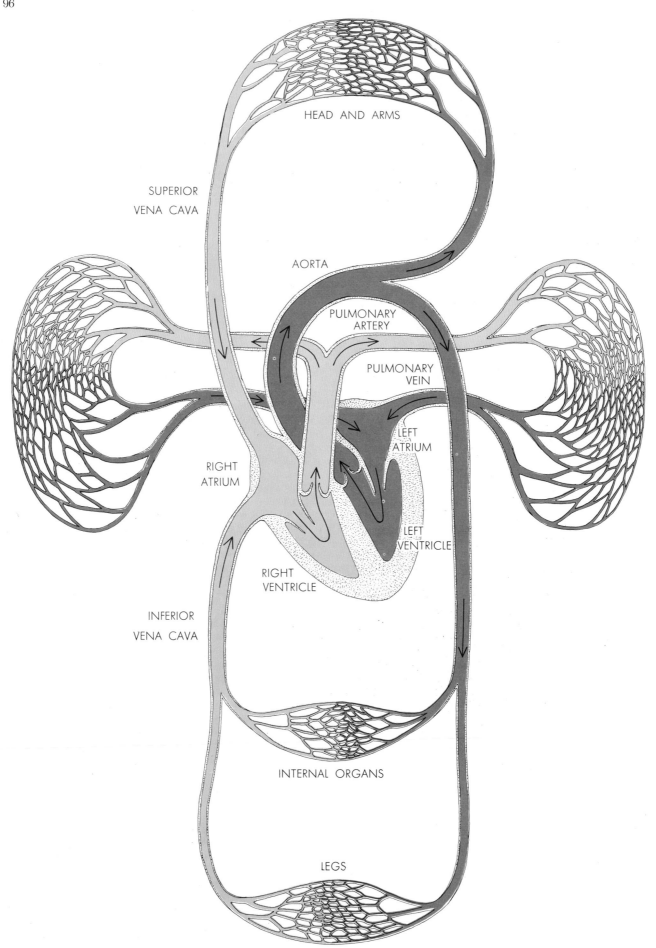

HEAD AND ARMS

SUPERIOR
VENA CAVA

AORTA

PULMONARY
ARTERY

PULMONARY
VEIN

LEFT
ATRIUM

RIGHT
ATRIUM

LEFT
VENTRICLE

RIGHT
VENTRICLE

INFERIOR
VENA CAVA

INTERNAL ORGANS

LEGS

9

THE HEART

CARL J. WIGGERS
May 1967

It pumps five quarts of blood in a minute, 75 gallons in an hour, 70 barrels in a day and 18 million barrels in 70 years. It does this by means of the most intricately woven muscle in the body

The blood bathes the tissues with fluid and preserves their slight alkalinity; it supplies them with food and oxygen; it conveys the building stones for their growth and repair; it distributes heat generated by the cells and equalizes body temperature; it carries hormones that stimulate and coordinate the activities of the various organs; it conveys antibodies and cells that fight infections—and of course it carries drugs administered for therapeutic purposes. No wonder that William Harvey, the discoverer of the circulation, ardently defended the ancient belief that the blood is the seat of the soul.

The blood cannot support life unless it is kept circulating. If the blood flow to the brain is cut off, within three to five seconds the individual loses consciousness; after 15 to 20 seconds the body begins to twitch convulsively; and if the interruption of the circulation lasts more than nine minutes, the mental powers of the brain are irrevocably destroyed. Similarly the muscles of the heart cannot survive total deprivation of blood flow for longer than 30 minutes. These facts emphasize the vital importance of the heart as a pump.

The work done by this pump is out of all proportion to its size. Let us look at some figures. Even while we are asleep the heart pumps about two ounces

ANATOMY OF THE HEART and its relationship to the circulatory system is schematically depicted on the opposite page. The arterial blood is represented in a bright red; venous blood, in a somewhat paler red. The capillaries of the lungs are represented at left and right; the capillaries of the rest of the body, at top and bottom. The term atrium is now used in preference to auricle.

of blood with each beat, a teacupful with every three beats, nearly five quarts per minute, 75 gallons per hour. In other words, it pumps enough blood to fill an average gasoline tank almost four times every hour just to keep the machinery of the body idling. When the body is moderately active, the heart doubles this output. During strenuous muscular efforts, such as running to catch a train or playing a game of tennis, the cardiac output may go up to 14 barrels per hour. Over the 24 hours of an average day, involving not too vigorous work, it amounts to some 70 barrels, and in a lifetime of 70 years the heart pumps nearly 18 million barrels!

The Design

Let us look at the design of this remarkable organ. The heart is a double pump, composed of two halves. Each side consists of an antechamber, formerly called the auricle but now more commonly called the atrium, and a ventricle. The capacities of these chambers vary considerably during life. In the human heart the average volume of each ventricle is about four ounces, and of each atrium about five ounces. The used blood that has circulated through the body—low in oxygen, high in carbon dioxide, and dark red (not blue) in color—first enters the right half of the heart, principally by two large veins (the superior and inferior venae cavae). The right heart pumps it via the pulmonary artery to the lungs, where the blood discharges some of its carbon dioxide and takes up oxygen. It then travels through the pulmonary veins to the left heart, which pumps the refreshed blood out through the aorta and to all regions of the body [*see diagram on opposite page*].

The thick muscular walls of the ventricles are mainly responsible for the pumping action. The wall of the left ventricle is much thicker than that of the right. The two pumps are welded together by an even thicker dividing wall (the septum). Around the right and left ventricles is a common envelope consisting of several layers of spiral and circular muscle [*see diagrams on page 99*]. This arrangement has a number of mechanical virtues. The blood is not merely pushed out of the ventricles but is virtually wrung out of them by the squeeze of the spiral muscle bands. Moreover, it is pumped from both ventricles almost simultaneously, which insures the ejection of equal volumes by the two chambers—a necessity if one or the other side of the heart is not to become congested or depleted. The effectiveness of the pumping action is further enhanced by the fact that the septum between the ventricles becomes rigid just before contraction of the muscle bands, so that it serves as a fixed fulcrum at their ends.

The ventricles fill up with blood from the antechambers (atria). Until the beginning of the present century it was thought that this was accomplished primarily by the contractions of the atria, *i.e.*, that the atria also functioned as pumps. This idea was based partly on inferences from anatomical studies and partly on observation of the exposed hearts of frogs. But it is now known that in mammals the atrium serves mainly as a reservoir. The ventricles fill fairly completely by their elastic recoil from contraction before the atria contract. The contraction of the latter merely completes the transfer of the small amount of blood they have left. Indeed, it has been found that the filling of the ventricles is not significantly impaired when

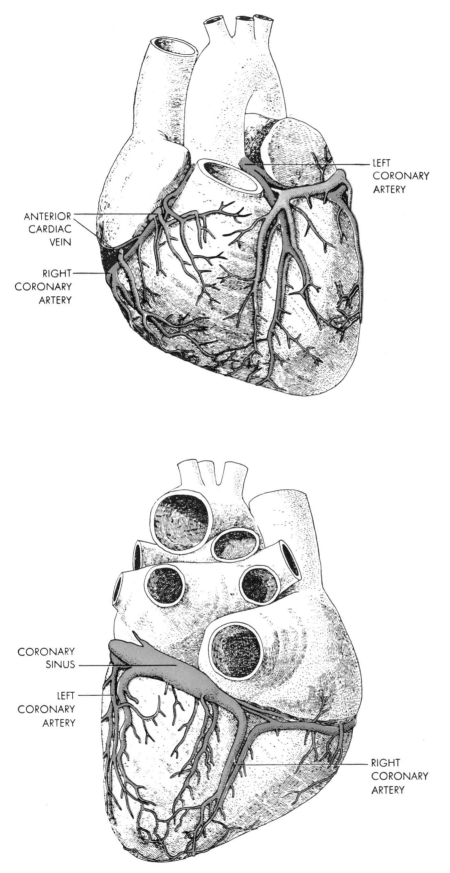

ANTERIOR
CARDIAC
VEIN

RIGHT
CORONARY
ARTERY

LEFT
CORONARY
ARTERY

CORONARY
SINUS

LEFT
CORONARY
ARTERY

RIGHT
CORONARY
ARTERY

ARTERIES AND VEINS which carry blood to and from the muscles of the heart are shown
from the front (*top*) and back (*bottom*). The arteries are bright red; the veins, pale red.

disease destroys the ability of the atria
to contract.

Factors of Safety

Since most of us believe that every
biological mechanism must have some
purpose, the question arises: Why com-
plicate the cardiac pump with contractions of the atria if the ventricles alone
suffice? The answer is that they provide
what engineers call a "factor of safety."
While the atrial contractions make only
a minor contribution to filling the ventricles under normal circumstances, they
assume an important role when disease
narrows the valve openings between the
atria and the ventricles. Their pumping
action then is needed to drive blood
through the narrowed orifices.

The ventricular pumps also have their
factors of safety. The left ventricle can
continue to function as an efficient pump
even when more than half of its muscle
mass is dead. Recently the astounding
discovery was made that the right ventricle can be dispensed with altogether
and blood will still flow through the
lungs to the left heart! An efficient circulation can be maintained when the
walls of the right ventricle are nearly
completely destroyed or when blood is
made to by-pass the right heart. Obviously the heart is equipped with large
factors of safety to meet the strains of
everyday life.

This applies also to the heart valves.
Like any efficient pump, the ventricle
is furnished with inlet and outlet valves;
it opens the inlet and closes the outlet
while it is filling, and closes the inlet
when it is ready to discharge. The pressure produced by contraction of the
heart muscles mechanically closes the
inlet valve between the atrium and ventricle: shortly afterward the outlet valve
opens to let the ventricle discharge its
blood—into the pulmonary artery in the
case of the right ventricle and into the
aorta in the case of the left. Then as
the muscles relax and pressure in the
chambers falls, the outlet valves close
and shortly thereafter the inlet valves
open. The relaxation that allows the
ventricles to fill is called diastole; the
contraction that expels the blood is
called systole.

While it might seem that competent
valves are indispensable for the forward
movement of blood, they are in fact not
absolutely necessary. The laws of hy-
draulics play some peculiar tricks. As
every farmhand knew in the days of
hand well-pumps, if the valves of the
pump were worn and leaky, one could

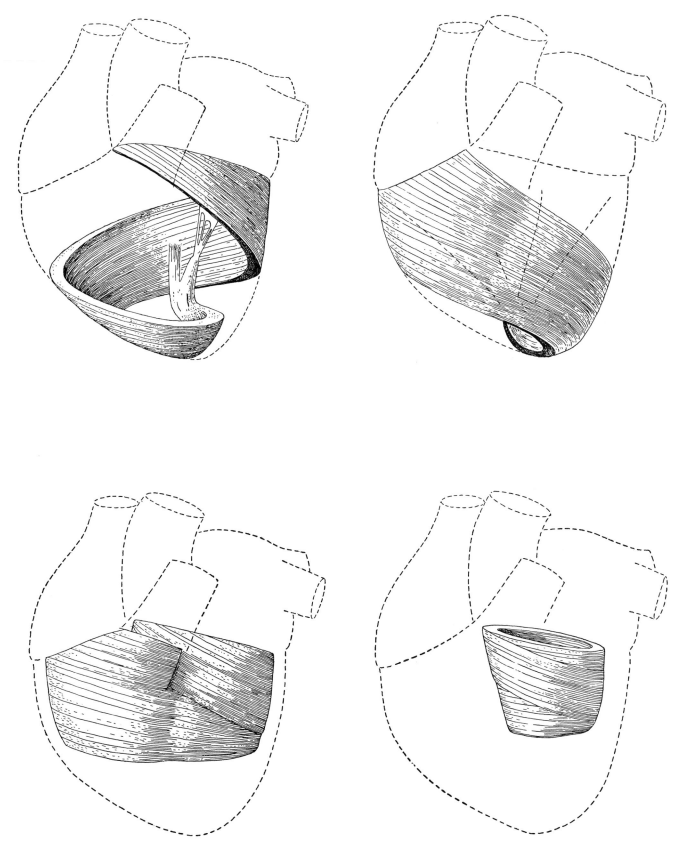

MUSCLE FIBERS of the ventricles are divided into four groups, one of which is shown in each of these four drawings. Two groups of fibers (*two drawings at top*) wind around the outside of both ventricles. Beneath these fibers a third group (*drawing at lower left*) also winds around both ventricles. Beneath these fibers, in turn, a fourth group (*drawing at lower right*) winds only around the left ventricle. The contraction of all these spiral fibers virtually wrings, rather than presses, blood out of the ventricles.

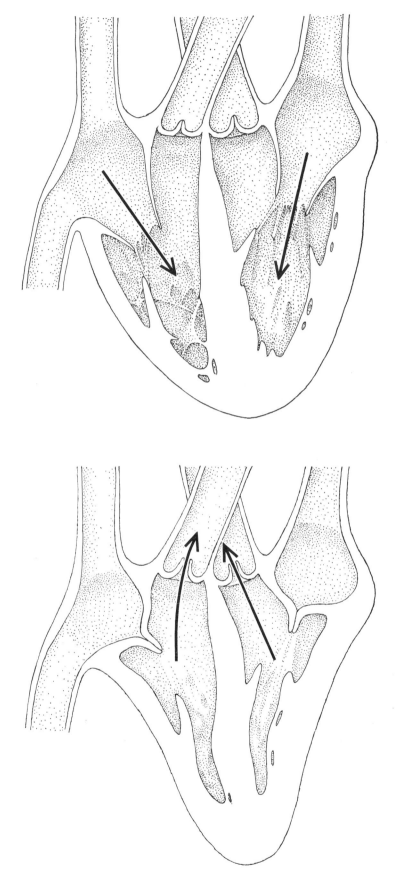

SYSTOLE AND DIASTOLE is the pumping rhythm of the heart. At the top is diastole, in which the ventricles relax and blood flows into them from the atria. The inlet valves of the ventricles are open; the outlet valves are closed. At the bottom is systole, in which the ventricles contract, closing the inlet valves and forcing blood through the outlet valves.

still draw water from the well by pumping harder. Similarly doctors have long been aware that patients can maintain a good circulation despite serious leaks in the heart valves. The factors of safety concerned are partly physical and partly physiological. The physical factor is more vigorous contraction of the heart muscles, aided by a structural arrangement of the deep muscle bands which tends to direct the blood flow forward rather than backward through the leaky valve. The physiological factor of safety is the mechanism known as "Starling's Law." In brief the rule is that, the more a cardiac muscle is stretched, the more vigorously it responds, of course within limits. The result is that the more blood the ventricles contain at the end of diastole, the more they expel. Of course they will fill with an excess of blood when either the inlet or the outlet valves leak. By pumping an extra volume of blood with each beat, the ventricles compensate for the backward loss through the atrial valves. In addition, the sympathetic nerves or hormones carried in the blood may spur the contractile power of the muscles. Under certain circumstances unfavorable influences come into play that depress the contractile force. Fortunately drugs such as digitalis can heighten the contractile force and thus again restore the balance of the circulation even though the valves leak.

Like any sharp closing of a door, the abrupt closings of the heart valves produce sounds, which can be heard at the chest wall. And just as we can gauge the vigor with which a door is slammed by the loudness of the sound, so a physician can assess the forces concerned in the closing of the individual heart valves. When a valve leaks, he hears not only the bang of the valves but also a "murmur" like the sigh of a gust of wind leaking through a broken window pane. The quality and timing of the murmur and its spread over the surface of the chest offer a trained ear considerable additional information. Sometimes a murmur means that the inlet and outlet orifices of the ventricles have been narrowed by calcification of the valves. In that case there is a characteristic sound, just as the water issuing from a hose nozzle makes a hissing sound when the nozzle is closed down.

Blood Supply

In one outstanding respect the heart has no great margin of safety: namely, its oxygen supply. In contrast to many other tissues of the body, which use as little as one fourth of the oxygen brought to them by the blood, the heart uses 80 per cent. The amount of the blood supply is therefore all-important to the heart, particularly when activity raises its demand for oxygen.

Blood is piped to the heart muscles via two large coronary arteries which curl around the surface of the heart and send branchings to the individual muscle fibers [*see diagrams on page 98*]. The left coronary artery is extraordinarily short. It divides almost immediately into two branches. A large circumflex branch runs to the left in a groove between the left atrium and ventricle and continues as a large vessel which descends on the rear surface of the left ventricle. It supplies the left atrium, the upper front and whole rear portion of the left ventricle. The other branch circles to the left of the pulmonary artery and then runs downward in a furrow to the apex. It supplies the front wall of the left ventricle and a small part of the rear right ventricle. Close to its origin the left coronary artery gives off several twigs which carry blood to the septum. The right coronary artery, embedded in fat, runs to the right in a groove between the right atrium and ventricle. It carries blood to both of them.

From the surface branches vessels run into the walls of the heart, dividing repeatedly until they form very fine capillary networks around the muscle elements. Eventually three systems of veins return the blood to the right heart to be pumped back to the lungs.

In the normal human heart there is little overlap by the three main arteries. If one of them is suddenly blocked, the area of the heart that it serves cannot obtain a blood supply by any substitute route. The muscles deprived of arterial blood soon cease contracting, die and become replaced by scar tissue. Now while this is the ordinary course of events, particularly in young persons, the amazing discovery was made some 20 years ago that the blocking of a main coronary artery does not always result in death of the muscles it serves. It has since been proved that new blood vessels grow in, from other arteries, if a main branch is progressively narrowed by atherosclerosis over a period of months or years. In other words, if the closing of a coronary artery proceeds slowly, a collateral circulation may develop. This biological process constitutes another factor of safety. Recent experiments on dogs indeed indicate that exercise will accelerate the development of collaterals when a major coronary artery is constricted. If this indication is con-

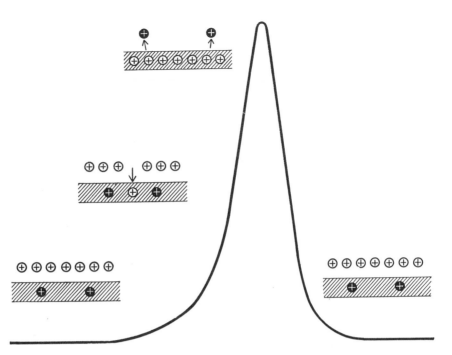

ELECTRICAL IMPULSE IS GENERATED by a cell in the "pacemaker" of the heart, a system of specialized muscle tissue (*see diagram on next page*). At lower left is a cross section of the cell; positively charged potassium ions (*black*) are inside it and a larger number of positively charged sodium ions (*white*) are outside. Because there are more positive charges outside the cell than inside, the inside of the cell is negatively charged with respect to the outside. When the cell is stimulated (*second cross section*), a sodium ion leaks across the membrane. Then many sodium ions rush across the membrane and potassium ions rush out (*third cross section*); this reverses the polarity of the cell and gives rise to an action potential (*peak in curve*). The original situation is then restored (*fourth cross section*)

firmed, it may well be that patients with atherosclerosis will be encouraged to exercise, rather than to adopt a sedentary life.

We have seen that there are many structural and functional factors of safety which enable the heart to respond, not only to the stress and strain of everyday life, but also to unfavorable effects of disease. Their existence has long been recognized; modern research has now thrown some light on the fundamental physiological and chemical processes involved.

The heart's transformation of chemical energy into the mechanical energy of contraction has certain similarities to

the conversion of energy in an automobile engine; but there are also essential differences. In both cases a fuel is suddenly exploded by an electric spark. In both the fuel is complex, and the explosion involves a series of chemical reactions. In each case some of the energy is lost as unusable heat. In each the explosions occur in cylinders, but in the heart these cylinders (the heart muscle cells) not only contain the fuel but are able to replenish it themselves from products supplied by the blood. The mechanical efficiency of these cells, *i.e.*, the fraction of total energy that can be converted to mechanical energy, has not been equaled by any man-made

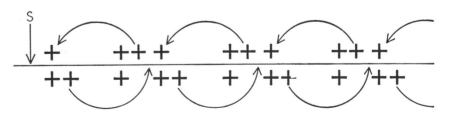

ELECTRICAL IMPULSE IS TRANSMITTED through the pacemaker system not as an electric current but as an electrical chain reaction. When a pacemaker cell is stimulated (S), it discharges and generates a local current which causes the depolarization and discharge of adjacent cells. In effect a wave of positive charge passes through the system (*curved arrows*).

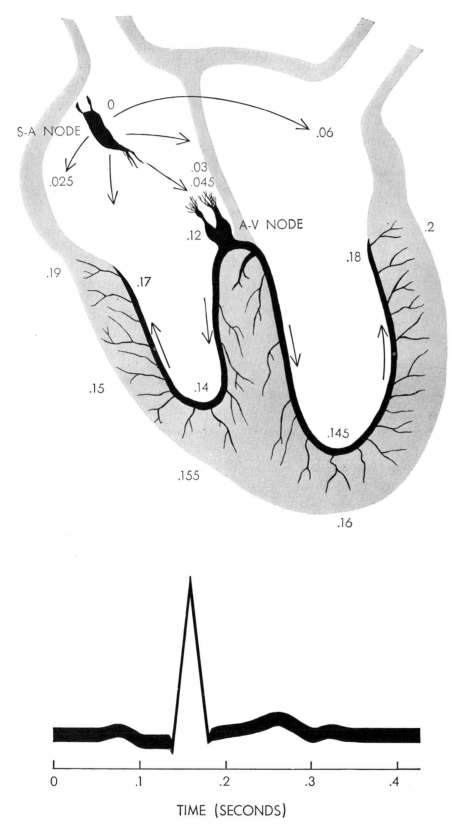

S-A NODE

0

.025

.06

.03
.045

.12   A-V NODE

.18   .2

.19   .17

.15   .14

.145

.155

.16

TIME (SECONDS)

0   .1   .2   .3   .4

PACEMAKER SYSTEM generates and transmits the impulses which cause the contraction of the heart muscles. The impulse is generated in the sino-atrial (S-A) node and spreads across the atria, causing their contraction and stimulating the atrio ventricular (A-V) node. This in turn stimulates the rest of the system, causing the rest of the heart to contract. The numbers indicate the time (in fractions of a second) it takes an impulse to travel from the S-A node to that point. The electrocardiogram curve at bottom indicates the change in electrical potential that occurs during the spread of one impulse through the system.

machine designed in the pre-atomic age. The mechanism responsible for this efficiency is unique and very complex.

Under the microscope we can see that cardiac muscle consists of long, narrow networks of fibers, with connective tissue and tiny blood vessels filling the spaces between. Each muscle fiber is made up of innumerable fibrils embedded in a matrix. It has been demonstrated that these fibrils are responsible for the contraction of the muscle as a whole. By special and clever techniques the fibrils can be washed free of the matrix, and it has been shown that when brought into contact with the energy-rich substance ATP, the fibrils shorten.

Examinations with the polarizing and electron microscopes and with X-rays have produced a fairly good picture of the ultimate design of these microscopic fibers. Each fibril is composed of many smaller filaments, or "protofibrils," just distinguishable under the highest microscopic magnification. The fibrils of a single muscle fiber may contain a total of some 10 million such filaments. The filaments are the smallest units known to stiffen and shorten. It has been possible to extract the actomyosin of which they are composed and to reconstitute filaments by squirting the extracted protein into a salt solution. These synthetic filaments can be made to contract.

The filaments themselves remain straight during contraction; therefore the kinking or coiling necessary for their contraction must take place at a still lower level—the level of molecules. Here the picture is clouded. We know from X-ray diffraction analyses that the molecules composing myosin filaments are arranged as miniature stretched spiral springs or stretched rubber bands. But just how they effect the filaments' contraction can only be guessed. Regardless of the mechanism, there is no doubt that the stiffness and shortening which are the features of contraction are mediated by changes in the molecular arrangement. This rearrangement requires energy. The consensus is that a tiny electric spark delivered to each individual cell causes the explosion of ATP. Not all the energy released is used for shortening of the actomyosin filaments. Some of it is converted to heat and some is used to initiate a series of complex chemical reactions which replenishes the fuel by reconstituting ATP. The explosion of ATP differs from that of gasoline in that no oxygen is required. But oxygen is indispensable for the rebuilding of ATP.

The millions of cardiac cylinders, as

MUSCLE FIBRILS of the heart, which make up its muscle fibers, are revealed in this electron micrograph made by Bruno Kisch of the American College of Cardiology in New York. The fibrils, which are from the heart of a guinea pig, are the long bands running diagonally across the micrograph. The round bodies between the fibrils are sarcosomes, which are found in large numbers in heart muscle and which appear to supply the enzymes that make possible its tireless contractions. The fibrils themselves are made up of protofibrils, which may barely be seen in the micrograph. The micrograph enlarges these structures some 50,000 diameters.

in an automobile engine, must fire in proper sequence to contract the muscle effectively. When they fire haphazardly, there is a great liberation of energy but no coordinated action. This chaotic condition is called fibrillation—*i.e.*, independent and uncoordinated activity of the individual fibrils.

The Beat

What causes the heart to maintain its rhythmic beat? The ancients, performing sacrificial rites, must have noticed that the heart of an animal continues to beat for some time after it has been removed from the body. That the beat must originate in the heart itself was apparently clear to the Alexandrian anatomist Erasistratus in the third century B.C. But anatomists ignored this evidence for the next 20 centuries because they were convinced that the nerves to the heart must generate the heartbeat. In 1890, however, Henry Newell Martin at the Johns Hopkins University demonstrated that the heart of a mammal could be kept beating though it was completely separated from the nerves, provided it was supplied with blood. And many years before that Ernst Heinrich Weber of Germany had made the eventful discovery that stimu-

lation of the vagus nerve to the heart does not excite it but on the contrary stops the heart. In short, it was established that the beat is indeed generated within the heart, and that the nerves have only a regulating influence.

The nature and location of the heart's "pacemaker" remained enigmatic until comparatively recent times. Within the span of my own memory there was considerable evidence for the view that the pacemaking impulses were generated by nerve cells in the right atrium and transmitted by nerve fibers to the heart-muscle cells. At present the evidence is overwhelming that the impulses are actually generated and distributed by a system of specialized muscle tissue consisting of cells placed end to end. Seventy-two times per minute—more or less—a brief electric spark of low intensity is liberated from a barely visible knot of tissue in the rear wall of the right atrium, called the sino-atrial or S-A node. The electric impulse spreads over the sheet of tissue comprising the two atria and, in so doing, excites a succession of muscle fibers which together produce the contraction of the atria. The impulse also reaches another small knot of specialized muscle known as the atrioventricular or A-V node, situated between the atria and ventricles. Here

the impulse is delayed for about seven hundredths of a second, apparently to allow the atria to complete their contractions; then from the A-V node the impulse travels rapidly throughout the ventricles by way of a branching transmission system, reaching every muscle fiber of the two ventricles within six hundredths of a second. Thus the tiny spark produces fairly simultaneous explosions in all the cells, and the two ventricles contract in a concerted manner.

If the heart originates its own impulses, of what use are the two sets of nerves that anatomists have traced to the heart? A brief answer would be that they act like spurs and reins on a horse which has an intrinsic tendency to set its own pace. The vagus nerves continually check the innate tempo of the S-A node; the sympathetic nerves accelerate it during excitement and exercise.

Normally, as I have said, the S-A node generates the spark, but here, too, nature has provided a factor of safety. When the S-A node is depressed or destroyed by disease, the A-V node becomes the generator of impulses. It is not as effective a generator (its maximum rate is only 40 or 50 impulses per minute, and its output excites the atria and ventricles simultaneously), but it suffices to keep the heart going. Patients

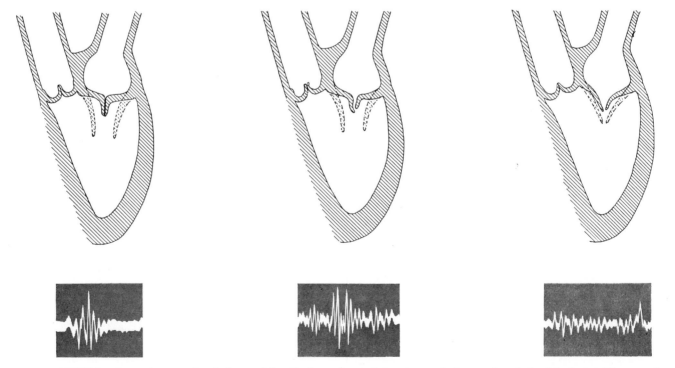

HEART SOUNDS indicate the normal and abnormal functioning of the heart valves. The three drawings at the top show the left side of the heart in cross section. The aortic, or outlet, valve is at upper left in each drawing; the mitral, or inlet, valve is at upper right. The first drawing shows the normal closing of the mitral valve; the trace below it records the sound made by this closing. The second drawing shows the partial closing of a leaky mitral valve; the trace below it records the murmur of blood continuing to flow through the valve. The third drawing shows the partial closing of a mitral valve with stiff leaves; the trace below it records a fainter murmur.

have survived up to 20 years with the A-V pacemaker substituting for a damaged S-A node.

There are still lower pacemakers which can maintain a slow heartbeat when the higher ones fail. When all the pacemakers are so weakened that, like an old battery, they are barely able to emit impulses, an anesthetic administered during an operation may stop the heart. In that case the beat can often be restored by rhythmic electric shocks —a system now incorporated in an apparatus for revival of the heart.

The Spark

What sort of mechanism exists in nodal tissues that is able to emit electric sparks with clockwork regularity 104,-000 times a day? We must look first to the blood. The fact that an excised heart does not long continue to beat unless supplied artificially with blood shows that the blood must supply something essential for preservation of its beat. In the 19th century physiologists began to experiment on isolated hearts, first of frogs and turtles and later of rabbits and cats, to determine what constituents of blood could be spared without halting the rhythmic heart contractions. They found that the serum (the blood fluid without cells) could maintain the beat of a mammalian heart, provided the serum was charged with oxygen under pressure. What constituents of the serum, besides the oxygen, were necessary? Attention first focused on the proteins, on the theory that the beating heart required them for nourishment. The heart was, in fact, found to be capable of maintaining its beat fairly well on a "diet" of blood proteins or even egg white or oxygenated milk whey. But the nourishment idea received a blow when it was discovered that the heartbeat could be maintained on a solution of gum arabic! It was then suggested that serum proteins act by virtue of their viscosity. This is an example of how experimenters are sometimes led astray in trying to uncover nature's secrets.

An eventful discovery in 1882 by the English physiologist Sydney Ringer changed the direction of thinking. He showed that a solution containing salts of sodium, potassium and calcium and a little alkali, in the concentrations found in the blood, would sustain the beat of a frog's heart. It was but a step to show that Ringer's solution, when oxygenated, also keeps the mammalian heart beating for a short time. Later it was found that the addition of a biological fuel—glucose

or, better yet, lactic acid—would extend the heart's performance.

Summing up the evidence, it was known at the beginning of the present century that the beat of the mammalian heart, and obviously also the generation of the spark, depends primarily on a balanced proportion of sodium, potassium and calcium plus a supply of oxygen and an energy-yielding substance such as glucose.

During the present century the scientific minds have sought to learn how these inorganic elements are involved in the initiation and spread of impulses. In order to understand the intricate mechanisms we must recall what most of us learned in high school: viz., the theory that, when a salt is dissolved in water, the elements are dissociated and become ions charged with positive and negative electricity.

The delicate enclosing membrane of all cells is differentially permeable: that is, ordinarily (at rest) it allows potassium ions to enter the cell but excludes sodium ions. We may say that the potassium ions have admission tickets, while the sodium ions do not. Since sodium ions predominate in the body fluids, the positively charged potassium ions within a cell are greatly outnumbered by positively charged sodium ions around the outside of the cell; the net result is that the outside is more positive than the inside, and the interior can therefore be regarded as negative with respect to the exterior. The potential difference is about one tenth of a volt. Each cell thus becomes a small charged battery. Now in the case of cells of the S-A node, the membrane leaks slightly, allowing some sodium ions to sneak in. This slowly but steadily reduces the potential difference between the inside and the outside of the membrane. When the difference has diminished by a critical amount (usually about six hundredths of a volt), the tiny pores of the membrane abruptly open. A crowd of sodium ions then rushes in, while some of the imprisoned potassium ions escape to the exterior. As a result the relative charges on the two sides of the membrane are momentarily reversed, the inside being positive with respect to the outside. The action potential thus created is the release of the electric spark.

As soon as activity is over, the membrane repolarizes, i.e., reconstitutes a charged battery. How this is accomplished is little understood, beyond the fact that oxidation of glucose or its equivalent is required. The mechanism

is pictured as a kind of metabolic pump which ejects the sodium ions that have gained illegal admittance, allows potassium ions to re-enter and closes the pores again. Then the cells are ready to be discharged again.

A little reflection should make it evident that the frequency with which such cells discharge depends on at least two things: (1) the rate at which sodium ions leak into the cell, and (2) the degree to which the potential across the membrane must be reduced in order to discharge it completely. The rate of sodium entry is known to be increased by warming and decreased by cooling, which accounts in part for the more rapid firing of the pacemakers in a patient who has a fever. The magnitude of the potential difference required to discharge the cell depends on the characteristics of the membrane. In this the concentration of calcium ions plays a basic role. Calcium favors stability of the membrane: if its concentration falls below a certain critical value, the rate of discharge by the cells increases; if calcium ions are too abundant, the rate is slowed. The rate of discharge is also affected by other factors. The vagus nerves tend to reduce it, the sympathetic nerves to increase it. The blood's content of oxygen and carbon dioxide, its degree of alkalinity, hormones and drugs—these and other influences can change the stability of the membranes and thus alter the rate of cell discharge.

Transmission of Impulses

The spark from a pacemaker is transmitted to the myriads of cylinders constituting the ventricular pumps by way of the special conducting system. When we say that the impulse travels over this system, this does not mean that electricity flows, as over wires to automobile cylinders. The electric impulse spreads by a kind of chain reaction involving the successive firing of the special transmitting cells. When pacemaker cells discharge, they generate a highly localized current which in turn causes the depolarization and discharge of an adjacent group of cells, and thus the impulse is relayed to the muscle cells concerned with contraction. An advantage of this mechanism is that the strength of the very minute current reaching the contracting fibers is not reduced.

Such a mode of transmission is not unknown in the inanimate world. There is a classic experiment in chemistry which illustrates an analogous process. An iron wire is coated with a microscopic film of iron oxide and suspended in a cylinder of strong nitric acid. Protected by this coating, the iron does not dissolve. But if the coat is breached (by a scratch or by an electric current) at a spot at one end of the immersed wire, a brown bubble immediately forms at this spot and a succession of brown bubbles then traverses the whole length of the wire. An electrical recorder connected to a number of points along the wire shows that a succession of local electric currents is generated down the wire as it bubbles. At each spot the current breaks the iron oxide film and the ensuing chemical reaction generates a new action potential. The contact between bare iron and nitric acid is only momentary, because the break in the film is quickly repaired.

Summarizing, the passage of electric impulses over the conduction system of the heart represents a series of local bio-electric currents, relaying the impulses step by step over special tissue to the contracting cells. On arrival at these cells the electric charges trigger the breakdown of ATP and so release the chemical energy needed for contraction.

Diagnosis

Considering the complexity of the cardiac machinery, it is remarkable that the heartbeat does not go wrong more often. Like a repairman for an automobile or a television set, a physician sometimes has to make an extensive hunt for the source of the disorder. It seems appropriate to close this article with a list of the points at which the machinery is apt to break down.

1. The main (S-A) pacemaker, or in rare instances all the pacemakers, may fail.

2. There may be too many pacemakers. The secondary pacemakers occasionally spring into action and work at cross purposes with the normal one, producing too rapid, too slow, or ill-timed beats.

3. The system for conduction of the pacemaker impulses may break down, leading to "heart block."

4. The heart muscle cylinders may respond with little power because of poor fuel, lack of enough oxygen for building fuel, fatigue or lack of adequate vitamins, hormones or other substances in the blood.

5. Some of the cylinders may be put out of commission by blockage of a coronary artery.

6. The heart valves may leak, and in its gallant effort to compensate, the heart may be overworked to failure.

10

THE MICROCIRCULATION OF THE BLOOD

BENJAMIN W. ZWEIFACH

January 1959

The primary purpose of the circulatory system is served by the microscopic vessels in which the blood flows from the arteries to the veins and thereby nourishes all the tissues of the body

When we think of the circulatory system, the words that first occur to us are heart, artery and vein. We tend to forget the microscopic vessels in which the blood flows from the arteries to the veins. Yet it is the microcirculation which serves the primary purpose of the circulatory system: to convey, to the cells of the body the substances needed for their metabolism and regulation, to carry away their products—in short, to maintain the environment in which the cells can exist and perform their interrelated tasks. From this point of view the heart and the larger blood vessels are merely secondary plumbing to convey blood to the microcirculation.

To be sure, the entire circulatory system is centered on the heart. The two chambers of the right side of the heart pump blood to the lungs, where it is oxygenated and returned to the chambers of the left side of the heart. Thence the blood is pumped into the aorta, which branches like a tree into smaller and smaller arteries. The smallest twigs of the arterial system are the arterioles, which are too small to be seen with the unaided eye. It is here that the micro- circulation begins. The arterioles in turn branch into the capillaries, which are still smaller. From the capillaries the blood flows into the microscopic tributaries of the venous system: the venules. Then it departs from the microcirculation and is returned by the tree of the venous system to the chambers in the right side of the heart.

The vessels of the microcirculation permeate every tissue of the body; they are never more than .005 inch from any cell. The capillaries themselves are about .0007 inch in diameter. To give the reader an idea of what this dimension means, it would take one cubic centimeter of blood (about 14 drops) from five to seven hours to pass through a capillary. Yet so large is the number of capillaries in the human body that the heart can pump all the blood in the body (about 5,000 cubic centimeters in an adult) through them in a few minutes. The total length of the capillaries in the body is almost 60,000 miles. Taken together, the capillaries comprise the body's largest organ; their total bulk is more than twice that of the liver.

If all the capillaries were open at one time, they would contain all of the blood in the body. Obviously this does not happen under normal circumstances, whereby hangs the principal theme of this article. How is it that the flow of blood through the capillaries can be regulated so as to meet the varying needs of all the tissues, and yet not interfere with the efficiency of the circulatory system as a whole?

It was William Harvey, physician to Charles I of England, who first demonstrated that the blood flows continuously from the arterial system to the venous. In 1661, 33 years after Harvey had published his famous work *De Motu Cordis* (*Concerning the Motion of the Heart*), the Italian anatomist Marcello Malpighi

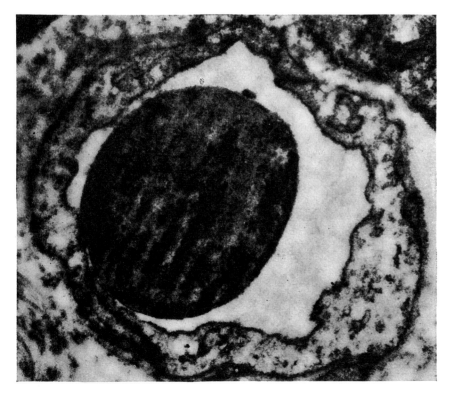

CAPILLARY from a cat's leg muscle is shown in cross section by this electron micrograph, which enlarges the structure some 20,000 diameters. The band running around the picture is the wall of the capillary. The large, dark object in the center is a single red blood cell. The micrograph was made by George D. Pappas and M. H. Ross of Columbia University.

PULMONARY CAPILLARIES

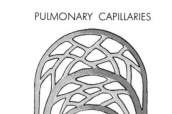

PULMONARY ARTERY

PULMONARY VEIN

RIGHT HEART

LEFT HEART

VEIN

ARTERY

CAPILLARIES
OF BODY AND ORGANS

CIRCULATORY SYSTEM is schematically outlined. The blood is pumped by the right heart through the pulmonary artery into the capillaries of the lungs. It returns from the lungs through the pulmonary vein to the left heart, which pumps it through the arteries to the capillaries of the internal organs and of the rest of the body. It finally returns to the right heart through the veins.

first observed through his crude microscope the fine conduits which link the two systems. These vessels were named capillaries after the Latin word *capillus*, meaning hair. Since Malpighi's time the capillaries have been intensively examined by a host of microscopists. Their work has established that not all the vessels in the network lying between the arterioles and the venules are the same. Indeed, we must regard the network as a system of interrelated parts. Hence it is preferable to think not of capillaries, but of a functional unit called the capillary bed.

The capillary bed, unlike muscle or liver or kidney, cannot be removed from an experimental animal and studied as an intact unit outside the body of the animal. By their very nature the capillaries are interwoven with other tissues. It is possible, however, to examine the capillary bed in a living animal. For example, one can open the abdomen of an anesthetized rat and carefully expose a thin sheet of mesentery: the tissue that attaches the intestine to the wall of the abdominal cavity. In this transparent sheet the capillary bed is displayed in almost diagrammatic form.

The tube of a capillary is made of a single layer of flat cells resembling irregular stones fitted together in a smooth pavement. The wall of the tube is so thin that even when it is viewed edge-on at a magnification of 1,000 diameters it is visible only as a line. When the wall is magnified in the electron microscope, it may be seen that the wall is less than .0001 inch thick. This so-called endothelium not only forms the walls of the capillaries but also lines the larger blood vessels and the heart, so that all the blood in the body is contained in a single envelope.

In a large blood vessel the tube of endothelium is sheathed in fibrous tissue interwoven with muscle. The fibrous tissue imparts to the vessel a certain amount of elasticity. The muscle is of the "smooth" type, characterized by its ability to contract slowly and sustain its contraction. The muscle cells are long and tapered at both ends; they coil around the vessel. In the tiny arterioles, in fact, a single muscle cell may wrap around the vessel two or three times. When the muscle contracts, the bore of the vessel narrows; when the muscle relaxes, the bore widens.

The muscular sheath of the larger blood vessels does not continue into the capillary bed. Yet as early as the latter part of the 19th-century experimental

physiologists reported that the smallest blood vessels could change their diameter. Moreover, when the flow of blood through the capillary bed of a living animal is observed under the microscope, the pattern of flow constantly changes. At one moment blood flows through one part of the network; a few minutes later that part is shut off and blood flows through another part. In some capillaries the flow even reverses. Throughout this ebb and flow, however, blood passes steadily through certain thoroughfares of the capillary bed.

If the capillaries have no muscles, how is the flow controlled? Some investigators suggested that although the endothelium of the capillaries was not true muscle, it could nonetheless contract. Indeed, it was demonstrated that in many lower animals blood vessels consisting only of endothelium contract and relax in a regular rhythm. However, contractile movements of this kind have not been observed in mammals.

Another explanation was advanced by Charles Rouget, a French histologist. He had discovered peculiar star-shaped cells, each of which was wrapped around a capillary, and he assumed that they were primitive muscle cells which opened and closed the capillaries. Many investigators agreed with him, among them the Danish physiologist August Krogh, who in 1920 won a Nobel prize for his work on the capillary system. It was not possible, however, to prove or disprove the contractile function of the Rouget cells by simple observation.

There the matter rested until methods were developed for performing microsurgical operations on single cells [see "Microsurgery," by M. J. Kopac; SCIENTIFIC AMERICAN, October, 1950]. Now it was possible to probe the cell with extremely fine needles, pipettes and electrodes. Microsurgery established that in mammals neither the capillary endothelium nor the Rouget cells could control the circulation by contraction. The endothelium did not contract when it was stimulated by a microneedle, or by the application with a micropipette of substances that cause larger blood vessels to contract. When one of the star-shaped Rouget cells was stimulated, it became thicker but did not occlude the capillary. When the same stimulus was applied to the recognizable muscle cell of an arteriole, on the other hand, the cell contracted and the arteriole was narrowed.

The microsurgical experiments established an even more significant fact: not

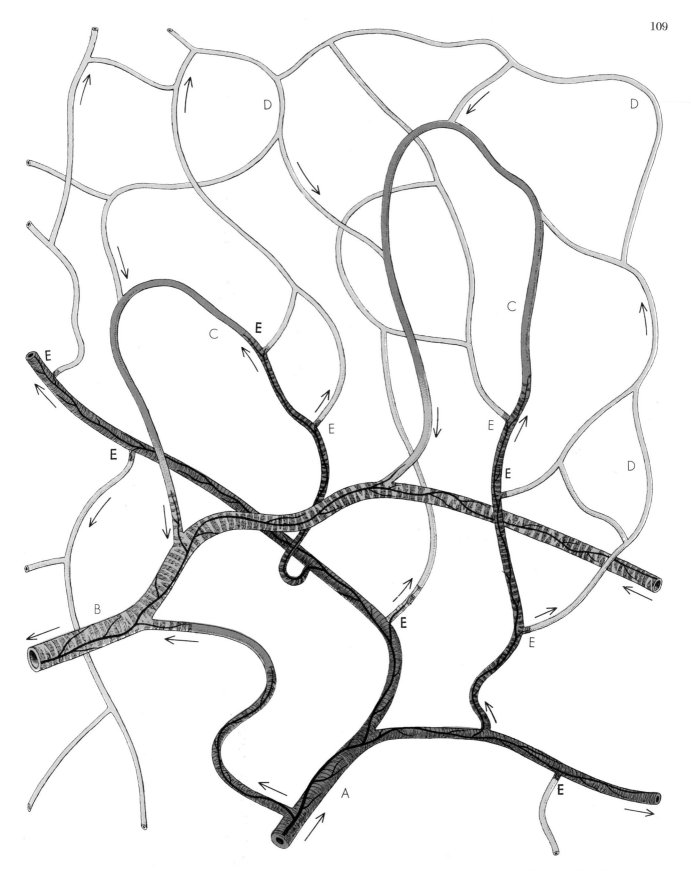

TYPICAL CAPILLARY BED is depicted in this drawing. The blood flows into the bed through an arteriole (A) and out of it through a venule (B). Between the arteriole and the venule the blood passes through thoroughfare channels (C). From these channels it passes into the capillaries proper (D), which then return it to the channels. The arteriole and venule are wrapped with muscle cells; in the thoroughfare channels the muscle cells thin out. The capillaries proper have no muscle cells at all. The flow of blood from a thoroughfare channel into a capillary is regulated by a ring of muscle called a precapillary sphincter (E). The black lines on the surface of the arteriole, venule and thoroughfare channels are nerve fibers leading to muscle cells. At lower left, between the arteriole and venule, is a channel which in many tissues shunts blood directly from the arterial system to the venous when necessary.

all the vessels in the capillary bed entirely lack muscle. For example, if epinephrine, which causes larger blood vessels to contract, is injected into the capillary bed with a micropipette, some of the vessels in the bed become narrower. Even when no stimulating substances are added, the same vessels open and close with the ebb and flow of blood in the capillary bed. It is these vessels, moreover, through which the blood flows steadily from the arterial to the venous system.

So the arterial system, with its muscular vessels, does not end at the capillary bed. The blood is continuously under muscular control as it flows into the venous system. To be sure, the muscle cells along the thoroughfare are sparsely distributed. As the arterial tree branches into the tissues the muscular sheath of the endothelium becomes thinner and thinner until in the smallest arterioles it is only one cell thick. In the thoroughfare channel of the capillary bed the muscle cells are spaced so far apart that the channel is almost indistinguishable from the true capillaries. The major portion of the capillary network arises as abrupt side branches of the thoroughfare channels, and at the point where each of the branches leaves a thoroughfare channel there is a prominent muscle structure: the muscle cells form a ring around the entrance to the capillary. It is this ring, or precapillary sphincter, which acts as a floodgate to control the flow of blood into the capillary network

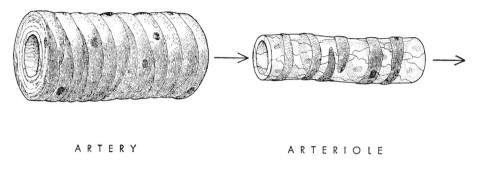

ARTERY   ARTERIOLE

WALLS OF BLOOD VESSELS of various kinds reflect their various functions. The wall of an artery consists of a single layer of endothelial cells sheathed in several layers of muscle cells interwoven with fibrous tissue. The wall of an arteriole consists of a single layer of

from the thoroughfare channel.

The muscular specialization of the circulatory system is illuminated by its embryonic development. In the early embryo the circulatory system is a network of endothelial tubes through which the primitive blood cells flow in an erratic fashion. The tubes are at first just large enough to pass the blood cells in single file. Attached to the outer wall of the tubes are numerous star-shaped cells which have wandered in from the surrounding tissue.

As the development of the embryo proceeds, those tubes through which the blood flows most rapidly are transformed into heavy-walled arteries and veins. In the process the star-shaped cells evolve through several stages into typical muscle cells. The outer reaches of the adult circulatory system possess a graded series of muscle-cell types, which are a direct representation of this developmental process. Thus the star-shaped cells of the capillary bed—the Rouget cells—are primitive muscle elements which have no contractile function.

From this point of view the capillary bed can be considered the immature part of the circulatory system. Like embryonic tissue, it has the capacity for growth, which it exhibits in response to injury. It also ages to some extent, and ultimately becomes less capable of dealing with the diversified demands of the tissue cells.

When we put these various facts together, we see the capillary bed not as a simple web of vessels between the arterial and venous systems, but as a

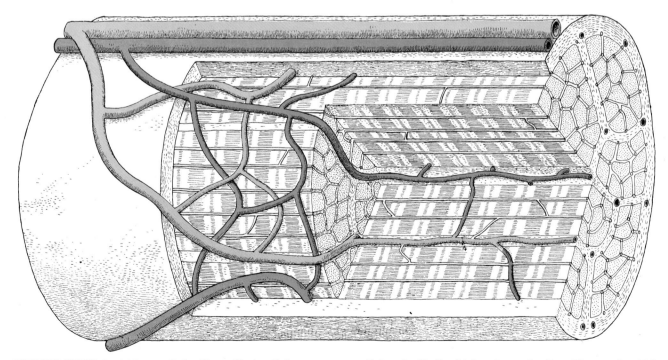

MUSCLE FIBER is richly supplied with capillaries. Lying atop this dissected muscle fiber are two blood vessels, the smaller of which is an artery and the larger a vein. Most of the capillaries run parallel to the fibrils which make up the fiber. The vessels which cut across two or more capillaries are thoroughfare channels. The system is shown in cross section at the right end of the drawing.

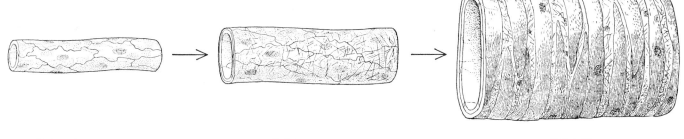

CAPILLARY   VENULE   VEIN

endothelial cells sheathed in a single layer of muscle cells. The wall of a capillary consists only of a single layer of endothelial cells. The wall of a venule consists of endothelial cells sheathed in fibrous tissue. The wall of a vein consists of endothelial cells sheathed in fibrous tissue and a thin layer of muscle cells. Thus a layer of endothelial cells lines the entire circulatory system.

physiological unit with two specialized components. One component is the thoroughfare channel, into which blood flows from the arteriole. The other is the true capillaries, which form a secondary network connected to the thoroughfare channel. The precapillary sphincters along the channel open and close periodically, irrigating first one part of the capillary network, then another part. When the sphincters are closed, the blood is restricted to the thoroughfare channel in its movement toward the venous system.

The structure of the physiological unit varies from one tissue to another in accordance with the characteristic needs of the tissues. For example, striated muscle, which unlike the smooth muscle of the blood vessels and other organs contracts rapidly and is under voluntary control, requires over 10 times more blood when it is active than when it is at rest. To meet this wide range of needs each thoroughfare channel in striated muscle gives rise to as many as 20 or 30 true capillaries. Glandular tissues, on the other hand, require only a steady trickle of blood, and each of their thoroughfare channels may give rise to as few as one or two capillaries. In the skin, which shields the body from its outer environment, there are special shunts through which blood can pass directly from the arteries to the veins with minimum loss of heat. Still other tissues require specialized capillary beds. The capillary beds of all the tissues, however, have the same basic feature: a central channel whose muscle cells control the flow of blood into the true capillaries.

But what controls the muscle cells? To answer this question we must draw a distinction between the control of the larger blood vessels and the control of the microcirculation. The muscle cells of the arteries and veins are made to contract and relax by two agencies: (1) the nervous system and (2) chemical "messengers" in the blood. These influences not only cause the vessels to constrict and dilate but also keep the muscle cells in a state of partial contraction. This muscle "tone" maintains the elasticity of the vessels, which assists the heart in maintaining the blood pressure. The operation of the system as a whole is supervised by special regulatory centers in the brain, working in collaboration with sensory monitoring stations strategically located in important vessels.

In the capillary bed, on the other hand, the role of the nervous system is much less significant. Most of the muscle cells in the capillary bed have no direct nerve connections at all. A further circumstance sets the response of the microscopic vessels apart from that of the larger vessels. Whereas the muscle cells of the large vessels are isolated from the surrounding tissues in the thick walls of the vessels, the muscle cells of the arterioles and the thoroughfare channels are immersed in the environment of the very tissues which they supply with blood. This feature introduces another chemical regulatory mechanism: the continuous presence of substances liberated locally by the tissue cells. As a consequence the contraction and relaxation of muscle cells in the microcirculation are under the joint control of messenger substances in the blood and specific chemical products of tissue metabolism.

The chemical substances that influence the function of the blood-vessel muscle cells comprise a subtly orchestrated system which is still imperfectly understood. Among the more important messengers are those released into the bloodstream by the cortex of the adrenal gland. These corticosteroids are essential to all cells in the body, notably maintaining the cells' internal balance of water and salts. (They have also been used with spectacular results in the treatment of degenerative diseases such as arthritis.) When the corticosteroids are deficient or absent, the muscles of the blood vessels lose their tone and the circulation collapses.

Another substance of profound importance to the circulatory system is epinephrine, which is secreted by the core of the adrenal gland (as distinct from its cortex). Epinephrine is one of two principal members of a family of substances called amines; the other principal member is norepinephrine, which is released both by the adrenal gland and by the endings of nerves in the muscles. All the amines cause the contraction of the muscle cells of the blood vessels, with the exception of certain vessels such as the coronary arteries of the heart. Also liberated at the nerve endings is acetylcholine, the effect of which is directly opposite that of the amines: it causes muscle cells to relax.

Many workers have suggested that it is norepinephrine and acetylcholine which control the flow of blood through the small vessels. Our own work at the New York University–Bellevue Medical Center leads us to conclude that such an explanation is too simple. The mechanism could not by itself account for the behavior of the small vessels.

In our view the function of the muscle cells of the small vessels is regulated not only by substances that directly cause them to contract and relax, but also by other substances that simply modify the capacity of the cells to react to stimuli and do work. It is known that a wide variety of substances extracted from tissues cause the small vessels to dilate. We postulate that when the metabolism of tissue cells is accelerated, the cells produce substances of this sort. When such substances accumulate in the vicinity of a precapillary sphincter, they depress the capacity of its muscle cells to respond to stimuli. As a result the sphincter relaxes, and blood flows from the thoroughfare channel into the capillary which nourishes the tissue.

The reaction limits itself, because the blood flow increases to the point where

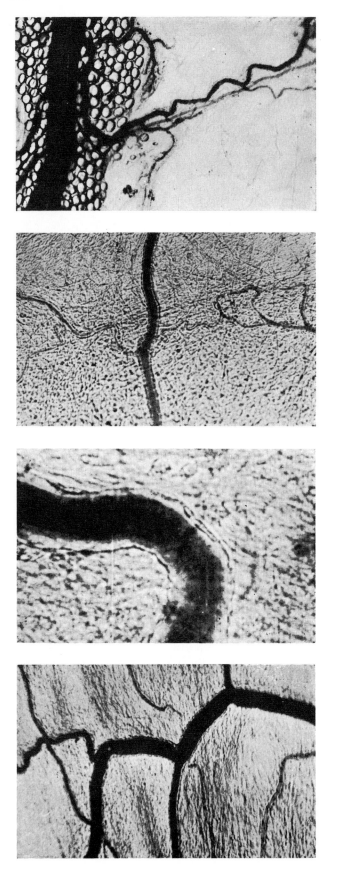

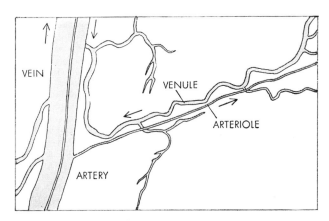

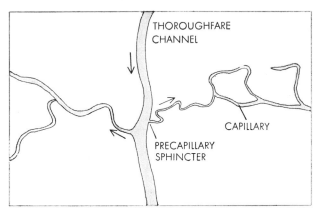

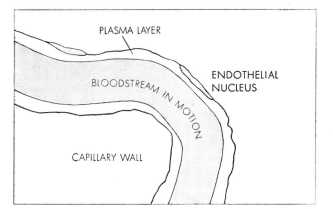

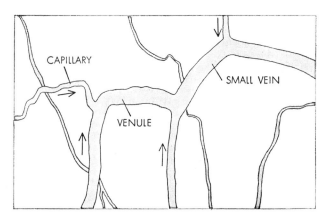

MESENTERY of a rat is photographed at various magnifications to show the characteristic structures of the microcirculation. The drawings at right label the structures. The magnification of the photomicrograph at top is 100 diameters; of the photomicrograph second from top, 200 diameters; of the third photomicrograph, 1,000 diameters; of the photomicrograph at bottom, 200 diameters.

it is sufficient to meet the nutritional requirements of the tissue cells. This leads to a gradual disappearance of the substances liberated by accelerated metabolism, and to a gradual lessening of the inhibition of the precapillary sphincter. As the muscle cells regain their tone, the sphincter shuts off the capillary.

The muscle cells of the arterioles and the capillary bed are extraordinarily sensitive to chemical stimuli, so sensitive that they respond to as little as a hundredth of the amount of substance required to constrict or dilate a large blood vessel. This sensitivity is dramatically demonstrated by microsurgical experiments on the capillary bed of a living rat. As little as .000000001 gram (.001

microgram) of epinephrine, injected into the capillary bed by means of a micropipette, is sufficient to close its capillary sphincters completely. Such substances reduce the flow of blood through the capillary bed by an orderly sequence of events: first the precapillary sphincters are narrowed, then the thoroughfare channels, then the arterioles, and finally the venules. Substances that cause the blood vessels to dilate, such as acetylcholine, set in motion a similar sequence: first the precapillary sphincters are opened, then the thoroughfare channels, and so on. The sensitivity of the arterioles and the capillary bed to such stimuli contributes to their independent behavior. An amount of substance sufficient to cause dramatic changes in the micro-

circulation simply has no effect on the larger vessels.

The tone of the muscle cells of the microcirculation may well be maintained by norepinephrine continuously discharged from the nerve endings, and by the level of epinephrine circulating in the blood. Our work indicates that the tone is also influenced by the local release of sulfhydryl compounds, which are key substances in the regulation of the oxidations conducted by cells. Now epinephrine and norepinephrine lose their activity when they are oxidized. Thus the actual level of these substances in the vicinity of muscle cells is not only dependent on their formation but also on their removal or destruction. Sulfhydryl compounds have been found to reduce the rate at which epinephrine and norepinephrine are oxidized. In this way the local release of such compounds could regulate the tone of smooth muscle.

Recently it has been suggested that a role in the local control of the microcirculation is played by the so-called mast cells, large numbers of which adjoin the small blood vessels. Various investigators have shown that the mast cells release at least three substances that strongly affect blood vessels: histamine, serotonin and heparin. It has been proposed that these substances, working alone or in certain combinations, are local regulatory factors.

It must be borne in mind that, even though the control of the microcirculation is largely independent of the rest of the circulatory system, the small blood vessels depend upon the nervous controls of the larger blood vessels for the shifting of blood from one organ to another as it is needed. Obviously the nervous controls of the larger vessels and the chemical controls of the microcirculation must be linked in some fashion. Under normal circumstances tissues that are inactive are perfused with a minimal amount of blood to allow the flow to be diverted to the tissues that need it most. During shock and acute infections, on the other hand, the demands of the tissues may be so great that the circulatory system cannot meet them, and the circulation collapses. In such conditions the effect of substances released locally to relax the muscle cells of the capillary bed has superseded the efforts of the nervous system to restrict the blood flow by the release of substances such as norepinephrine. It is ironic that this primitive response, in striving to insure the survival of individual cells, frequently overtaxes the circulation and brings about the death of the organism.

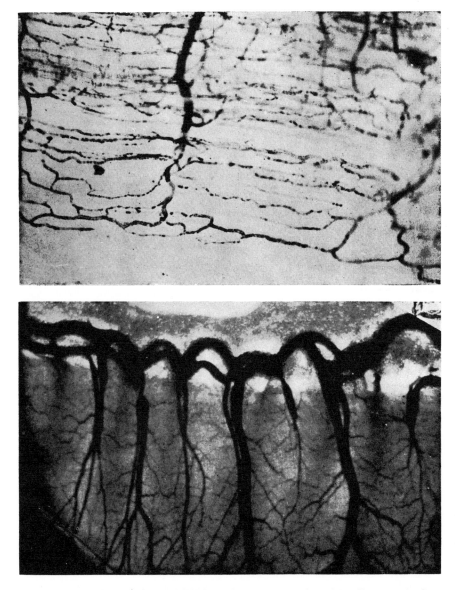

CAPILLARY BEDS OF TWO TISSUES in a living rat are enlarged 200 diameters in these photomicrographs. At top is the capillary bed of a striated muscle; the capillaries run parallel to underlying muscle fibers. At bottom are capillaries in the surface of intestine.

11

THE LYMPHATIC SYSTEM

H. S. MAYERSON
June 1963

This second circulation plays an essential role in maintaining the body's steady state, draining from the spaces between cells fluid, protein and other substances that leak out of the blood

Living tissue is for the most part a collection of cells bathed in a fluid medium. This interstitial fluid constitutes what the French physiologist Claude Bernard named the *milieu intérieur:* the internal environment of the organism that is the true environment of its cells. The interstitial fluid brings nutrients to the cells and carries away waste products; its composition varies in space and time under the control of the co-ordinated physiological processes that maintain homeostasis, the remarkably steady state that characterizes the internal environment of a healthy organism. In the maintenance of the homeostasis of the interstitial fluid the circulation of the blood is obviously of fundamental importance. In the higher vertebrates there is a second circulation that is equally essential: the lymphatic system. Its primary function is to recirculate the interstitial fluid to the bloodstream, thereby helping to create a proper cellular environment and to maintain the constancy of the blood itself. It also serves as a transport system, conducting specialized substances from the cells that make them into the bloodstream. In recent years physiologists, biochemists, physicians and surgeons have been studying the lymphatic system intensively, in health and in disease. Their investigations are providing much new information on how the body functions, explaining some heretofore poorly understood clinical observations and even suggesting new forms of treatment.

The fact that the lymphatic system is an evolutionary newcomer encountered only in the higher vertebrates is significant. In lower animals there is no separation between the internal and external environments; all the cells of a jellyfish, for example, are bathed in sea water. With progression up the evolutionary scale the cells become separated from the external environment, "inside" is no longer identical with "outside" and rudimentary blood circulatory systems make their appearance to conduct the exchange of nutrients and waste products. As the organism becomes more complex the blood system becomes more specialized. The system develops increasing hydrostatic pressure until, in mammals, there is a closed, high-pressure system with conduits of diminishing thickness carrying blood to an extensive, branching bed of tiny capillaries.

At this point in evolution a snag was encountered: the high pressures made the capillaries leaky, with the result that fluid and other substances seeped out of the bloodstream. A drainage system was required and lymphatic vessels evolved (from the veins, judging by embryological evidence) to meet this need.

In man the lymphatic system is an extensive network of distensible vessels resembling the veins. It arises from a fine mesh of small, thin-walled lymph capillaries that branch through most of the soft tissue of the body. Through the walls of these blind-end capillaries the interstitial fluid diffuses to become lymph, a colorless or pale yellow liquid very similar in composition to the interstitial fluid and to plasma, the liquid component of the blood. The lymphatic capillaries converge to form larger vessels that receive tributaries along their length and join to become terminal ducts emptying into large veins in the lower part of the neck. The largest of these great lymphatics, the thoracic duct, drains the lower extremities and all the organs except the heart, the lungs and the upper part of the diaphragm; these are drained by the right lymphatic duct. Smaller cervical ducts collect fluid from each side of the head and neck. All but the largest lymph vessels are fragile and difficult to trace, following different courses in different individuals and even, over a period of time, in the same individual. The larger lymphatics, like large veins, are equipped with valves to prevent backflow.

Along the larger lymphatics are numerous lymph nodes, which are of fundamental importance in protecting the body against disease and the invasion of foreign matter. The lymph nodes serve, first of all, as filtering beds that remove particulate matter from the lymph before it enters the bloodstream; they contain white cells that can ingest and destroy foreign particles, bacteria and dead tissue cells. The nodes are, moreover, centers for the proliferation and storage of lymphocytes and other antibody-manufacturing cells produced in the thymus gland; when bacteria, viruses or antigenic molecules arrive at a lymph node, they stimulate such cells to make antibodies [see "The Thymus Gland," by Sir Macfarlane Burnet; SCIENTIFIC AMERICAN Offprint 138].

Starling's Hypothesis

The present view of the lymphatic circulation as a partner of the blood system in maintaining the fluid dynamics of the body stems from the investigations early in this century by the British physiologist Ernest H. Starling. "Starling's hypothesis" stated that the exchange of fluid between the capillaries and the interstitial space is governed by the relation between hydrostatic pressure and osmotic pressure. Blood at the arterial end of a capillary is still under a driving pressure equivalent to some 40 millimeters of mercury; this constitutes a "filtration pressure" that tends to make plasma seep out of the capillary. Starling

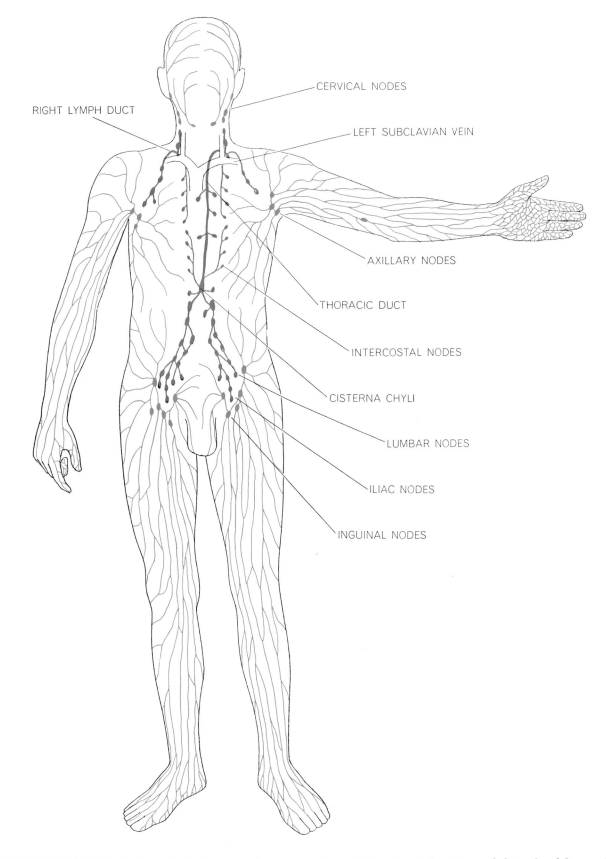

CERVICAL NODES

RIGHT LYMPH DUCT

LEFT SUBCLAVIAN VEIN

AXILLARY NODES

THORACIC DUCT

INTERCOSTAL NODES

CISTERNA CHYLI

LUMBAR NODES

ILIAC NODES

INGUINAL NODES

LYMPHATIC VESSELS drain the entire body, penetrating most of the tissues and carrying back to the bloodstream excess fluid from the intercellular spaces. This diagram shows only some of the larger superficial vessels (*light color*), which run near the surface of the body, and deep vessels (*dark color*), which drain the interior of the body and collect from the superficial vessels. The thoracic duct, which arises at the cisterna chyli in the abdomen, drains most of the body and empties into the left subclavian vein. The right lymph duct drains the heart, lungs, part of the diaphragm, the right upper part of the body and the right side of the head and neck, emptying into the right subclavian vein. Lymph nodes interspersed along the vessels trap foreign matter, including bacteria.

visualized the wall of the capillary as being freely permeable to plasma and all its constituents except the plasma proteins albumin, globulin and fibrinogen, which could leak through only in very small amounts. The proteins remaining in the capillary exert an osmotic pressure that tends to keep fluid in the capillary, countering the filtration pressure. Similar forces are operative in the tissue spaces outside the capillary. At the arterial end of the capillary the resultant of all these forces is ordinarily a positive filtration pressure: water and salts leave the capillary. At the venous end, however, the blood pressure is decreased, energy having been dissipated in pushing the blood through the capillary. Now the osmotic force exerted by the proteins is dominant. The pressure gradient is reversed: fluid, salts and the waste products of cell metabolism flow into the bloodstream [see top illustration on page 118].

It follows, Starling observed, that if the concentration of plasma proteins is decreased (as it would be in starvation), the return of fluid to the bloodstream will be diminished and edema, an excessive accumulation of fluid in the tissue spaces, will result. Similarly, if the capillaries become too permeable to protein, the osmotic pressure of the plasma decreases and that of the tissue fluid increases, again causing edema. Capillary poisons such as snake venoms have this effect. Abnormally high venous pressures also promote edema, by making it difficult for fluid to return to the capillaries; this is often one of the factors operating in congestive heart disease.

A fundamental tenet of Starling's hypothesis was that not much protein leaves the blood capillary. In the 1930's the late Cecil K. Drinker of the Harvard Medical School challenged this idea. Numerous experiments led him to conclude "that the capillaries practically universally leak protein; that this protein does not re-enter the blood vessels unless delivered by the lymphatic system; that the filtrate from the blood capillaries to the tissue spaces contains water, salts and sugars in concentrations found in blood, together with serum globulin, serum albumin and fibrinogen in low concentrations, lower probably than that of tissue fluid or lymph; that water and salts are reabsorbed by blood vessels and protein enters the lymphatics together with water and salts in the concentrations existing in the tissue fluid at the moment of lymphatic entrance." In other words, Drinker believed that protein is continuously filtering out of the blood; the plasma-protein level is maintained only because the lymphatic system picks up protein and returns it to the bloodstream.

Unfortunately Drinker had no definitive method by which to prove that the protein in lymph had leaked out of the blood and was not somehow originating in the cells. Perhaps for this reason his conclusions were not generally accepted. Teachers and the writers of textbooks continued to maintain that "healthy" blood capillaries did not leak protein. It was in an effort to clarify this point that I undertook an investigation of lymph

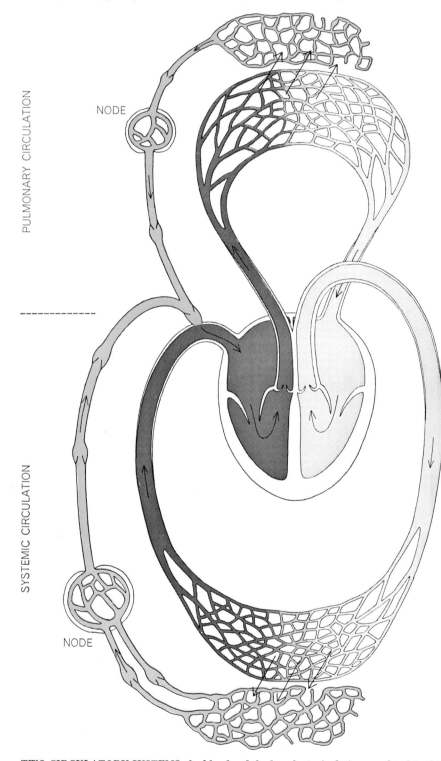

PULMONARY CIRCULATION

NODE

SYSTEMIC CIRCULATION

NODE

TWO CIRCULATORY SYSTEMS, the blood and the lymphatic (color), are related in this schematic diagram. Oxygenated blood (light gray) is pumped by the heart through a network of capillaries, bringing oxygen and nutrients to the tissue cells. Venous blood (dark gray) returns to the heart and is oxygenated in the course of the pulmonary (lung) circulation. Fluid and other substances seep out of the blood capillaries into the tissue spaces and are returned to the bloodstream by the lymph capillaries and larger lymphatic vessels.

and the lymphatics some 15 years ago. At that time I was working with a clinical group measuring the retention of blood by patients given large infusions. We saw that the patients were retaining the cellular components of the blood quite well but were "losing" the plasma. The loss was clearly into the tissue spaces, not by way of excretion from the kidneys.

If blood capillaries did indeed leak plasma, together with its proteins and other large molecules, then Drinker was correct. If the proteins entered the interstitial fluid, they would stay there, since Starling's measurements and Drinker's findings made it clear that large molecules could not get back into the blood capillaries—unless they were picked up by the lymphatic system. If they leaked from the blood vessels and were in fact returned by the lymphatic vessels, the evolutionary reason for the development of the lymphatic system would be established beyond question. I decided to return to my laboratory at the Tulane University School of Medicine and investigate the problem.

Over the years I have had the enthusiastic assistance of several colleagues—notably Karlman Wasserman, now at the Stanford University Medical School, and Stephen J. LeBrie—and of many students. Time had provided us with two tools not available to Drinker. One was flexible plastic tubing of small diameter, which we could insert into lymphatic vessels much more effectively than had been possible with the glass tubing available earlier. And we now had radioactive isotopes with which to label proteins and follow their course.

Experiments with Proteins

We injected the blood proteins albumin and globulin, to which we had coupled radioactive iodine atoms, into the femoral veins of anesthetized dogs. The proteins immediately began to leave the bloodstream. By calculating the slope of the disappearance curve in each experiment we could arrive at a number expressing the rate of disappearance [see top illustration on page 119]. The average rate of disappearance of albumin, for example, turned out to be about .001; in other words, a thousandth of the total amount of labeled albumin present at any given time was leaking out of the capillaries each minute. If we infused large amounts of salt solution, plasma or whole blood into our dogs, the disappearance rate increased significantly. The same thing happened in animals subjected to severe hemorrhage. In some experiments we simultaneously collected and analyzed lymph from the thoracic duct [see bottom illustration on page 119]. As before, labeled protein left the blood; within a few minutes after injection it appeared in the lymph, at first in small quantities and then at a faster rate. It leveled off, in equilibrium with the blood's protein, seven to 13 hours after injection.

We were able to calculate from our data that in dogs the thoracic duct alone

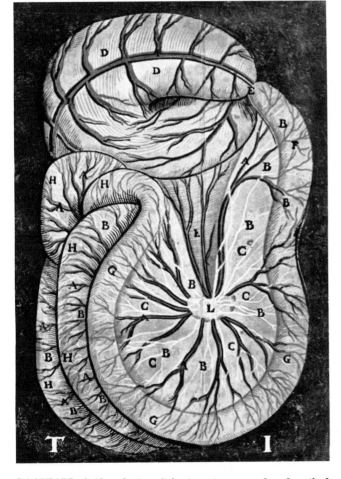

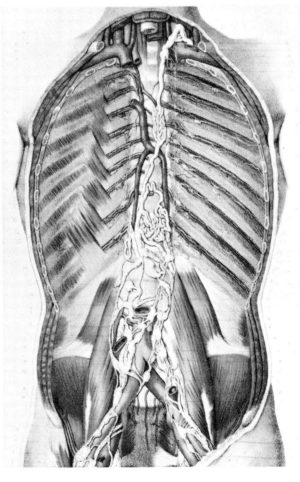

LACTEALS, the lymphatics of the intestine, were first described by the Italian anatomist Gasparo Aselli in 1622. They were pictured in his *De Lactibus*, the first anatomical work with color plates. This plate shows veins (*A*), *lacteals* (*B*), *mesentery* (*C*), stomach (*D*), small intestine (*F, G, H*) and a lymph node (*L*).

THORACIC DUCT and the major lymph vessels of the lower extremities and trunk that contribute to it are seen in this plate from a French book of 1847, *Atlas d'Anatomie Descriptive du Corps Humain*. The duct arises from a plexus of abdominal vessels and arches up into the lower neck before entering the subclavian vein.

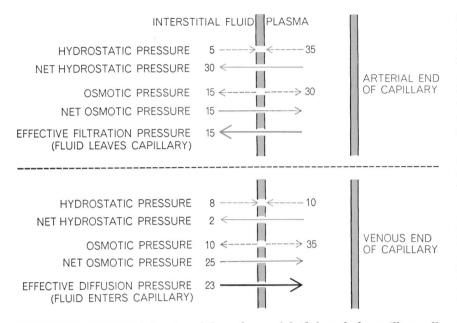

INTERSTITIAL FLUID | PLASMA

HYDROSTATIC PRESSURE	5	→ ←	35
NET HYDROSTATIC PRESSURE	30	←	
OSMOTIC PRESSURE	15	← →	30
NET OSMOTIC PRESSURE	15	→	
EFFECTIVE FILTRATION PRESSURE (FLUID LEAVES CAPILLARY)	15	←	

ARTERIAL END OF CAPILLARY

HYDROSTATIC PRESSURE	8	→ ←	10
NET HYDROSTATIC PRESSURE	2	←	
OSMOTIC PRESSURE	10	← →	35
NET OSMOTIC PRESSURE	25	→	
EFFECTIVE DIFFUSION PRESSURE (FLUID ENTERS CAPILLARY)	23	→	

VENOUS END OF CAPILLARY

STARLING'S HYPOTHESIS explained the exchange of fluid through the capillary wall. At the arterial end of the capillary the hydrostatic pressure (*given here in centimeters of water*) delivered by the heart is dominant, and fluid leaves the capillary. At the venous end the osmotic pressure of the proteins in the plasma dominates; fluid enters the capillary.

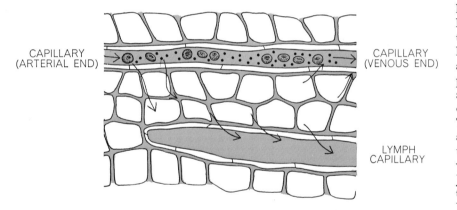

CAPILLARY (ARTERIAL END)

CAPILLARY (VENOUS END)

LYMPH CAPILLARY

FLUID EXCHANGE is diagramed as postulated by Starling. He believed that fluid and salts (*arrows*) left the blood capillaries, mixed with the interstitial fluid and for the most part were reabsorbed by the capillaries. Excess fluid was drained by the lymph vessels. He took it for granted that most of the protein in the blood stayed inside the blood capillaries.

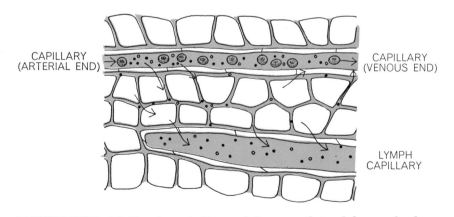

CAPILLARY (ARTERIAL END)

CAPILLARY (VENOUS END)

LYMPH CAPILLARY

PRESENT VIEW of fluid exchange is diagramed. It appears that such large molecules as proteins (*black dots*) and lipids (*open circles*) leave the blood capillaries along with the fluid and salts. Some of the fluid and salts are reabsorbed; the excess, along with large molecules that cannot re-enter the blood capillaries, is returned via the lymphatic system.

returned about 65 per cent of the protein that leaked out of the capillaries. Extension of this kind of experiment to man showed similar rates of leakage. In the course of a day 50 per cent or more of the total amount of protein circulating in the blood is lost from the capillaries and is returned to the bloodstream by the lymphatic system.

The importance of lymphatic drainage of protein becomes clear if one considers its role in lung function, which was elucidated by Drinker. The pulmonary circulation, in contrast to the general circulation, is a low-pressure system. The pulmonary capillary pressure is about a quarter as high as the systemic capillary pressure and the filtration pressure in the pulmonary capillaries is therefore considerably below the osmotic pressure of the blood proteins. As a result fluid is retained in the bloodstream and the lung tissue remains properly "dry."

When pulmonary capillary pressure rises significantly, there is increased fluid and protein leakage and therefore increased lymph flow. For a time the lymph drainage is adequate and the lungs remain relatively dry. But when the leakage exceeds the capacity of the lymphatics to drain away excess fluid and protein, the insidious condition pulmonary edema develops. The excessive accumulation of fluid makes it more difficult for the blood to take up oxygen. The lack of oxygen increases the permeability of the pulmonary capillaries, and this leads to greater loss of protein in a vicious circle. Some recent findings by John J. Sampson and his colleagues at the San Francisco Medical Center of the University of California support this concept. They found that a gradual increase in lymphatic drainage occurs in dogs in which high pulmonary blood pressure is produced and maintained experimentally. This suggests that the lymphatic system attempts to cope with the abnormal situation by proliferating, much as blood capillaries do when coronary circulation is impaired.

Leakage from blood capillaries and recirculation by the lymphatic system is, as I indicated earlier, not limited to protein. Any large molecule can leak out of the capillaries, and it cannot get back to the bloodstream except via the lymphatics. All the plasma lipids, or fatty substances, have been identified in thoracic-duct lymph. Even chylomicrons, particles of emulsified fat as large as a micron (a thousandth of a millimeter) in diameter that are found in blood during the digestion of fat, leak out of the bloodstream and are picked up and recirculated to the vascular system by the lymphatics. As a matter of fact, there is evidence that they may leak out even faster than proteins. The significance of these findings remains to be explained. Aaron Kellner of the Cornell University Medical College has suggested that atherosclerosis, a form of hardening of the arteries in which there is infiltration of the walls of the arteries by lipids, may have its origin in the fact that under normal conditions there is a constant flow of fluid containing lipids and proteins across the blood-vessel lining into the vessel wall. Ordinarily this fluid is removed by the small blood vessels of the wall itself and by the lymphatics. It is conceivable that something may interfere with the removal of lipids and cause them to accumulate in the blood-vessel wall. It is even conceivable that the high capillary filtration that accompanies hypertension may increase the leakage of lipids from capillaries to a level exceeding their rate of removal from the interstitial fluid, which would then bathe even the outer surfaces of the arteries in lipids.

In addition to demonstrating that the lymph returns large molecules from the tissue spaces to the bloodstream, recent investigation has confirmed the importance of lymphatic drainage of excess fluid filtered out of the capillaries but not reabsorbed. Experiments with heavy water show that blood is unquestionably the chief source of the water of lymph. In dogs the amount of lymph returned to the bloodstream via the thoracic duct alone in 24 hours is roughly equivalent to the volume of the blood plasma. Most of this fluid apparently comes from the blood. In some of our experiments we drained the thoracic-duct lymph outside the dog's body and found that the plasma volume dropped about 20 per cent in eight hours and the plasma-protein level some 16 per cent. Translated to a 24-hour basis, the loss would be equivalent to about 60 per cent of the plasma volume and almost half of the total plasma proteins circulating in the blood. Thus the return of lymph plays an essential role in maintaining the blood volume.

One situation in which this function can be observed is the "lymphagogue" effect: the tendency of large infusions into the vascular system to increase the flow of lymph. As we increased the size of infusions in dogs, lymph flow increased proportionately; with large infusions (2,000 milliliters, about the normal blood volume of a large dog) the thoracic-duct lymph flow reached a peak value about 14 times greater than that

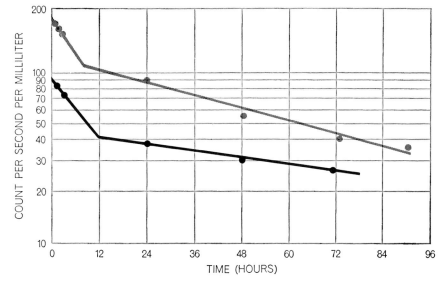

LABELED PROTEINS were injected into dogs' veins. Measuring the radioactivity per milliliter of blood withdrawn from an artery showed how quickly the globulin (*gray*) and albumin (*black*) disappeared. The steep slopes represent disappearance, the shallow slopes subsequent metabolism of the proteins. The radioactivity is plotted on a logarithmic scale.

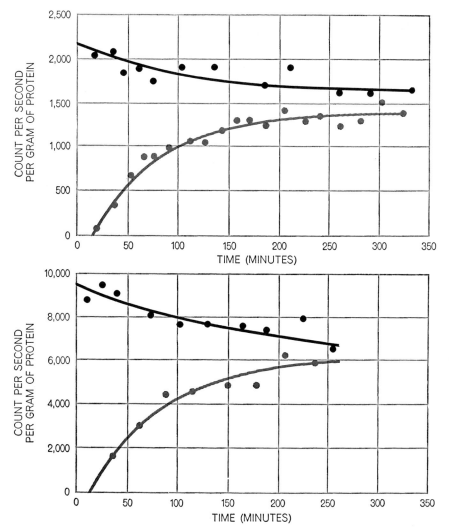

PROTEIN appeared in thoracic-duct lymph (*gray curves*) and increased in the lymph as it disappeared from the blood (*black curves*). The upper graph is for albumin, the lower one for globulin. In time the labeled protein in blood and lymph reached equilibrium.

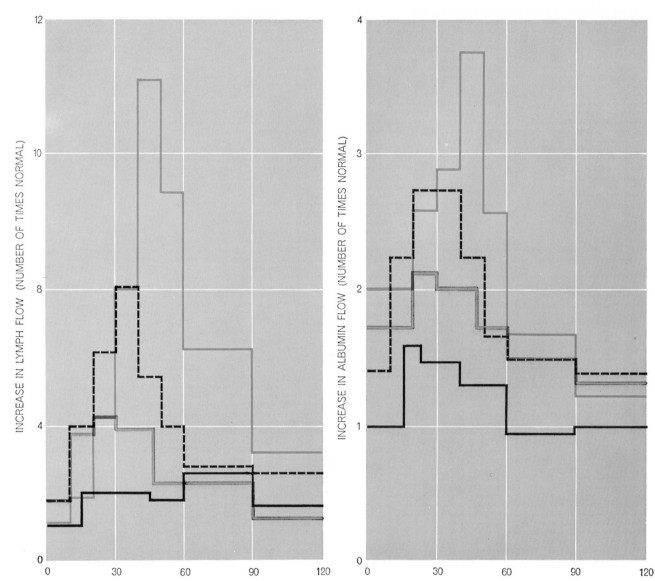

EFFECT OF INFUSIONS on lymph flow (*graph at left*) and on albumin flow in lymph (*graph at right*) in dogs is illustrated. Each curve shows the ratio of the flow after an infusion of a given size to the flow before the infusion. The curves are for infusions of 250 (*solid black*), 500 (*gray*), 1,000 (*broken black*) and 2,000 (*colored*) milliliters. The larger the infusion, the more leakage.

of the preinfusion level [*see illustration above*]. Most of the excess fluid is excreted by the kidneys in increased urine flow. But the displacement of fluid from the blood circulation into the lymph "saves" some of the fluid. In other words, it can be considered as being a fine adjustment of the blood volume so that not all the fluid is irrevocably lost from the body. Large infusions also increase protein leakage, but again the fact that the protein goes to the lymph and slowly returns to the bloodstream minimizes changes in total circulating protein and the loss of its osmotic effect.

The Mechanism of Filtration

The exact processes or sites of the filtration of large molecules through the capillary wall, and through cell membranes in general, are still unclear. As Arthur K. Solomon has pointed out in these pages [see "Pores in the Cell Membrane," by Arthur K. Solomon; SCIENTIFIC AMERICAN Offprint 76], "some materials pass directly through the fabric of the membrane, either by dissolving in the membrane or by interacting chemically with its substance. But it seems equally certain that a large part of the traffic travels via holes in the wall. These are not necessarily fixed canals; as the living membrane responds to changing conditions inside or outside the cell, some pores may open and others may seal up."

This last point appears to explain our results with infusions in dogs. The massive infusions overfill the closed blood system, raise filtration pressure in the capillaries and result in increased leakage through the capillary walls; lymph flow is copious and the lymph contains more large molecules. Small infusions do not do this. The reason, then, that patients did not do as well as expected after receiving large infusions or transfusions was that the plasma and proteins leaked out through stretched capillary pores. A similar effect accounted for the case of animals subjected to severe hemorrhage: there was not enough blood to oxygenate the capillary walls adequately, the walls became more permeable and the protein molecules passed through.

We found that the rate of leakage for any molecule depends on its size. Globulin, which has a molecular weight of

250,000, leaked more slowly than albumin, which has a weight of 70,000. The third plasma protein, fibrinogen, has a molecular weight of about 450,000 and leaves the blood still more slowly. By introducing into the blood various carbohydrate molecules, which unlike protein molecules do not carry a charge, we were able to demonstrate that it is size and not electrical charge that determines a molecule's rate of movement through the wall.

When, after infusing the carbohydrates, we collected lymph from different parts of the body, we found larger molecules in intestinal lymph than in leg lymph, and still larger molecules in lymph from the liver. This indicated that capillaries in the liver have substantially larger openings than leg capillaries do, and that the vessels of the intestine probably have both large and small openings. There are other indications that liver capillaries are the most permeable of all; for example, apparently both red cells and lymphocytes pass between the blood and the lymph in the liver. Recent studies with the electron microscope confirm the indirect evidence for variations in the size of capillary openings; the structure of the capillaries seems to vary with the organ, and these differences may be related to differences in function.

Transport by the Lymph

When Gasparo Aselli first described lymphatic vessels in 1622, the ones he noted were the lacteals: small vessels that drain the intestinal wall. The dog Aselli was dissecting had eaten recently; the lacteals had absorbed fat from the intestine, which gave them a milky-white appearance and made them far more visible than other lymphatics. The transport of certain fats from the intestine to the bloodstream by way of the thoracic duct is one of the lymphatic system's major functions. Studies in which fatty acids have been labeled with radioactive carbon show that the blood capillaries of the intestine absorb short-chain fatty acid molecules directly, together with most other digested substances, and pass them on to the liver for metabolism. But the lacteal vessels absorb the long-chain fats, such as stearic and palmitic acids, and carry them to the bloodstream via the thoracic duct. The lymphatic system is also the main route by which cholesterol, the principal steroid found in tissues, makes its way into the blood.

Since the lymphatics are interposed between tissue cells and the blood sys-

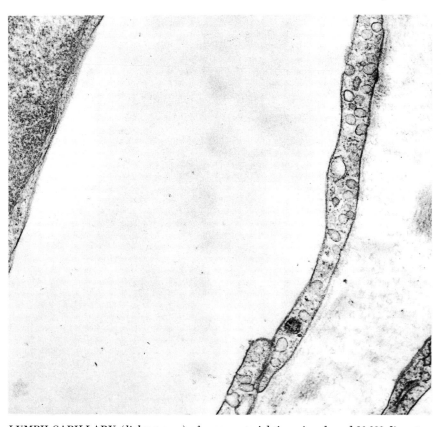

LYMPH CAPILLARY (*lightest area*) of mouse arterial tissue is enlarged 30,000 diameters in this electron micrograph made by Johannes A. G. Rhodin of the New York University School of Medicine. At upper left is part of the nucleus of an endothelial cell of the vessel wall. The thin ends of two other cells overlap near the bottom. The faint shadow in the connective tissue outside the wall is a slight indication of a basement membrane.

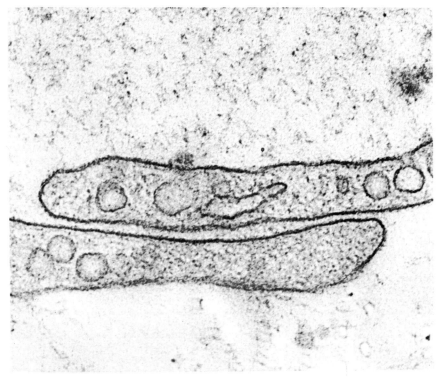

"LOOSE JUNCTION" between the overlapping ends of two endothelial cells of a vessel wall is seen in this electron micrograph, also made by Rhodin, of a lymphatic in mouse intestinal tissue. The lymph vessel is at the top, connective tissue at the bottom. In this case there is no sign of a basement membrane outside the wall. The enlargement is 90,000 diameters.

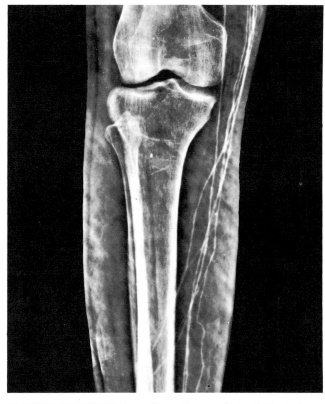 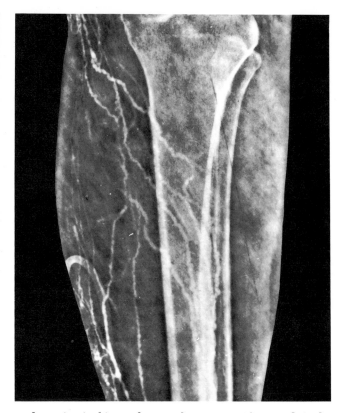

LYMPHANGIOGRAM is an X-ray photograph in which the lymphatic vessels are made visible by injecting into them a radiopaque dye. In the normal leg (*left*) the vessels are straight and well defined. In lymphedema (*right*) there is insufficient drainage of fluid and proteins, in this case because there were too few vessels in the thigh. The extra pressure on the lower-leg vessels increased their number and made them tortuous. These pictures were made by Carl A. Smith of the New York University School of Medicine.

tem it is not surprising to find that they serve as the channel for transport to the bloodstream of substances that originate in tissue cells. The lymph is probably the route by which at least some hormones, many of which are very large molecules, are carried to the blood from the endocrine glands where they are synthesized. Some enzymes, found in lymph in small concentrations, may merely have leaked out of the capillaries. But others are apparently picked up from their cells of origin and carried to the blood by the lymph. Certain enzymes, including histaminase and renin, are present in greater concentrations in lymph than in the blood. The finding on renin, reported by A. F. Lever and W. S. Peart of St. Mary's Hospital Medical School in London, is of particular interest to investigators working on the problem of hypertension. One concept ascribes high blood pressure in some individuals to the production of renin by a kidney suffering from inadequate blood circulation; the renin is thought to combine with a globulin in the plasma to form hypertensin, an enzyme that narrows the arterioles and results in high blood pressure. It has been difficult to establish this concept because no one has been able consistent-

ly to demonstrate the presence of renin in the blood of hypertensive patients. Now that it has been discovered in lymph coming from the kidney it is clear that renin is indeed being formed in these patients, although in amounts so small that it usually escapes detection after being diluted in the blood.

Recently Samuel N. Kolmen of the University of Texas Medical Branch in Galveston has provided what may be a confirmation of the renin concept. He produced hypertension in dogs by removing one kidney and partially constricting the artery supplying the other one. When he shunted the thoracic-duct flow into the dog's gullet or allowed it to escape, the hypertension diminished. The implication is that renin was being kept out of the bloodstream. When Kolmen stopped lymph flow in the shunt, presumably inhibiting the diversion of the renin, the hypertension returned.

The Lymphatic Circulation

To the investigators of the 19th century the lymphatic system was "open-mouthed": its capillaries were assumed to be open to the tissue spaces. More recent evidence has shown that the

lymphatics form a closed system, that fluid enters not through the open ends of vessels but through their walls. The walls of the terminal lymph capillaries, like those of blood capillaries, consist of a single layer of platelike endothelial cells. This layer continues into the larger vessels as a lining but acquires outer layers of connective tissue, elastic fibers and muscle. Although the capillaries of the blood and lymphatic systems are structurally very similar, recent electron micrographs show differences in detail that may help to explain the ease with which the lymph vessels take up large molecules. Sir Howard Florey of the University of Oxford and J. R. Casley-Smith of the University of Adelaide in Australia believe that the most important difference is the poor development or absence in lymph capillaries of "adhesion plates," structures that hold together the endothelial cells of the blood capillaries. They suggest that as a result there are open junctions between adjacent cells in lymph capillaries that allow large molecules to pass through the walls. Johannes A. G. Rhodin of the New York University School of Medicine puts more emphasis on the apparent absence or poor development in lymph capillaries of the "basement membrane" that surrounds blood capillaries.

Certainly it is clear that very large particles do enter the lymphatic vessels: proteins, chylomicrons, lymphocytes and red cells—the last of which can be as much as nine microns in diameter. Bacteria, plastic spheres, graphite particles and other objects have been shown to penetrate the lymphatics with no apparent difficulty. Yet we have found that when we introduce substances directly into the lymphatic system, anything with a molecular weight greater than 2,000 is retained almost completely within the lymphatics, reaching the blood only by way of the thoracic duct. If large particles can get into the lymphatic vessels, why do substances with a molecular weight of 2,000 not get out by the same channels?

I have spent many hours trying to formulate an answer to this question without arriving at a sophisticated concept, and have had to be content with a simple explanation that is at least consistent with the current evidence. Assume that the smallest terminal lymphatics are freely permeable to small and large molecules and particles moving in either direction through intercellular gaps. Compression of these vessels in any way would tend to force their contents in all directions. At least some of the contents would be forced along into the larger lymph vessels, where the presence of valves would prevent backflow. And once the lymph reaches a larger vessel it can no longer lose its large particles through the thick and relatively impermeable wall of the lymphatic.

One can argue that this seems to be a rather inefficient and even casual way of getting the job done. Indeed it is, and this physiological casualness is a characteristic of the lymphatic system as a whole. There is no heart to push the lymph, and although lymphatic vessels do contract and dilate like veins and arteries this activity does not seem to be an important factor in lymph movement. The flow of lymph depends almost entirely on forces external to the system: rhythmic contraction of the intestines, changes in pressure in the chest in the course of breathing and particularly the mechanical squeezing of the lymphatics by contraction of the muscles through which they course.

Lymphatic Malfunction

In spite of the casualness of the lymphatic system, its development, as I have tried to show, was an absolute necessity for highly organized animals. Its importance is most visibly demonstrated in various forms of lymphedema, a swelling of one or more of the extremities due to the lack of lymphatic vessels or to their malfunction. In some individuals the lymphatic system fails to develop normally at birth, causing gradual swelling of the affected part. Lymphangiographic studies, in which the vessels are injected with radiopaque dyes to make them visible in X-ray photographs, show that the lymphatics are scarce, malformed or dilated. Insufficient drainage causes water and protein to accumulate in the tissues and accounts for the severe and often disabling edema. There is evidence that genetic factors may play a role in this condition. Surgical procedures that destroy lymph vessels may have a similar effect in a local area. Elephantiasis is a specific form of lymphedema resulting from the obstruction of the vessels. It can be caused by infection of the lymphatics or by infestation with a parasitic worm that invades and blocks the vessels.

The lymphedemas have been recognized as such for many years. Recently the view that the lymphatic system is essential to homeostasis—which is to say "good health"—has led to a number of investigations of its role in conditions in which no lymphatic involvement was previously suspected. Our group at Tulane has found that lymphatic drainage of the kidneys is essential in order to maintain the precise osmotic relations

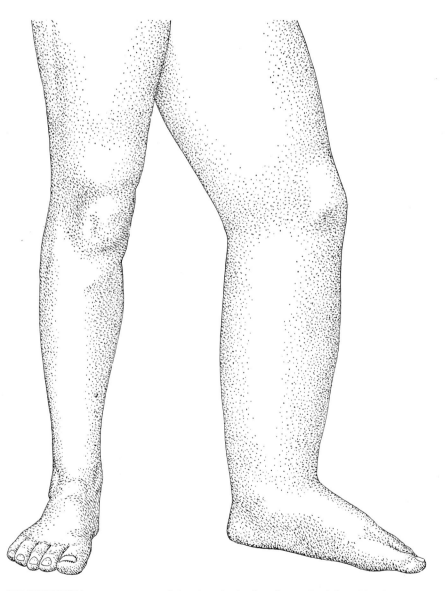

LYMPHEDEMA can cause gross deformity of a limb and even disability. The drawing is based on a photograph of an 11-year-old girl whose leg began to swell at the age of seven, probably because of an insufficiency of lymphatic vessels. The patient's condition was greatly improved by an operation in which the tissue between skin and muscles was removed.

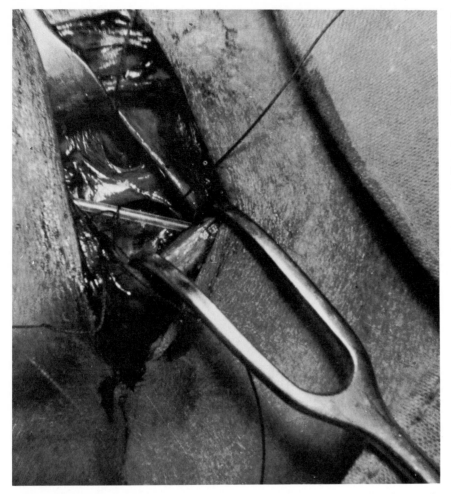

DILATED THORACIC DUCT of a patient with cirrhosis is seen in this photograph made by Allan E. Dumont and John H. Mulholland of the New York University School of Medicine and reprinted from *Annals of Surgery*. The plastic tube just below the duct is a tenth of an inch in diameter. A normal duct would be smaller than the tube.

on which proper kidney function depends. This may explain the dilution of the urine observed in some patients after kidney operations: the lymph vessels may have been damaged, decreasing their capacity for draining proteins and interfering with the reabsorption of water by the kidney tubules.

At the New York University School of Medicine, John H. Mulholland and Allan E. Dumont have been investigating the relation between thoracic-duct flow and cirrhosis of the liver. Their results suggest that the cirrhosis may be associated with increased lymph flow in the liver and that the inability of the thoracic duct to handle the flow may bring on the accumulation of fluid, local high blood pressure and venous bleeding that are frequently seen in cirrhosis patients; drainage of the duct outside the body temporarily relieves the symptoms.

These and other clinical observations are consistent with my feeling that the lymphatic system does a capable job when all is going well but that its capacity for dealing with disturbances is limited. As a phylogenetic late-comer it may simply not have evolved to the point of being able to cope with abnormal stresses and strains. The role of the second circulation in disease states is currently under intensive investigation. As more and more is learned about its primary functions and its reactions to stress, the new knowledge should be helpful in diagnosis and perhaps eventually in the treatment of patients.

12

"THE WONDERFUL NET"

P. F. SCHOLANDER

April 1957

*These words are a translation of rete mirabile, an
arrangement of blood vessels in which animals can conserve
heat and oxygen pressure by applying the principle of
counter-current exchange*

A man standing barefoot in a tub of ice water would not survive very long. But a wading bird may stand about in cold water all day, and the whale and the seal swim in the arctic with naked fins and flippers continually bathed in freezing water. These are warm-blooded animals, like man, and have to maintain a steady body temperature. How do they avoid losing their body heat through their thinly insulated extremities? The question brings to light a truly remarkable piece of biological engineering. It seems that such animals block the loss of heat by means of an elementary physical mechanism, familiar enough to engineers, which nature puts to use in a most effective way. In fact, the same principle is employed for several very different purposes by many members of the animal kingdom from fishes to man.

The principle is known as counter-current exchange. Consider two pipes lying side by side so that heat is easily transmitted from one to the other. Suppose that fluids at different temperatures start flowing in opposite directions in the two pipes: that is, a cold stream flowing counter to a warm [*see diagram on next page*]. The warmer stream will lose heat to the colder one, and if the transfer is efficient and the pipes long enough, the warm stream will have passed most of its heat to the counter current by the time it leaves the system. In other words, the counter current acts as a barrier to escape of heat in the direction of the warm current's flow. This method of heat exchange is, of course, a common practice in industry: the counter-flow system is used, for example, to tap the heat of exhaust gases from a furnace for preheating the air flowing into the furnace. And the same method apparently serves to conserve body heat for whales, seals, cranes, herons and other animals with chilly extremities.

Claude Bernard, the great 19th-century physiologist, suggested many years ago that veins lying next to arteries in the limbs must take up heat from the arteries, thus intercepting some of the body heat before it reaches the extremities. Recent measurements have proved that there is in fact some artery-to-vein transfer of heat in the human body. But this heat exchange in man is minor compared to that in animals adapted to severe exposure of the extremities. In those animals we find special networks of blood vessels which act as heat traps. This type of network, called *rete mirabile* (wonderful net), is a bundle of small arteries and veins, all mixed together, with the counter-flowing arteries and veins lying next to each other. The retes are generally situated at the places where the trunk of the animal deploys into extremities—limbs, fins, tail and so on. There the retes trap most of the blood heat and return it to the trunk. The blood circulating through the extremities is therefore considerably cooler than in the trunk, but the limbs can function perfectly well at the lower temperature. It has been found that many arctic animals have a leg temperature as low as 50 degrees Fahrenheit or even less.

Anatomists have confirmed that the whale, the seal and the long-legged wading birds possess such retes. However, these networks have also been discovered in the extremities of many tropical animals. It is not surprising to find heat-trapping retes in a water-dwelling animal such as the Florida manatee, for even tropical waters are chilling to a constantly immersed body, but why should the retes appear in tropical land animals like the sloth, the anteater and

the armadillo? The answer may be that these animals are hypersensitive to cool air. The sloth, for instance, begins to shiver when the air temperature drops below 80 degrees F. It has to adjust to this situation almost every night, and the retes may well be the means by which it makes the adjustment: that is, the sloth may let its long arms and legs cool to the temperature of the night air, as a reptile does, to preserve its body heat. Recent measurements have shown that it takes a sloth two hours to rewarm a chilled arm from 59 to 77 degrees, whereas an animal without retes, such as a monkey, accomplishes this in 10 minutes.

There is another finding, however, which at first sight is more puzzling. Many animals that spend a great deal of time in cold water or live in the arctic snow seem to lack retes to sidetrack body heat from their poorly insulated legs or feet: Among them are ducks, geese, sea gulls, the fox and the husky (the Eskimo dog). The absence of retes in these animals is not difficult to explain, however, when we consider that all of them are heavily insulated over most of their bodies. Their principal problem lies in getting rid of body heat rather than in conserving it. Consider, for instance, the situation of a husky. It is so well insulated that it can sleep on the snow at 40 degrees below zero without raising its normal rate of metabolism at rest. When this animal gets up after a cold night and begins to run in the warm sun, increasing its metabolic rate 10- or 20-fold because of the exercise, it is immediately faced with the problem of dissipating a good deal of excess heat. Because of its thick fur covering, it can lose heat only through exposed surfaces such as its tongue, face and legs. An arteriovenous network impeding the transport of heat

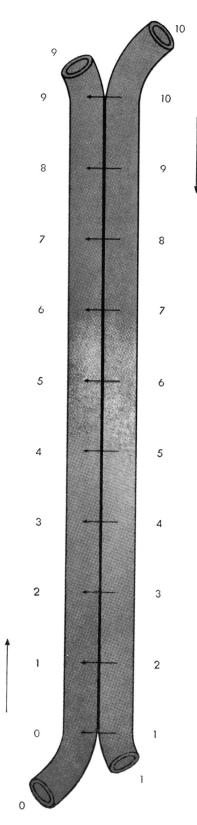

THE PRINCIPLE of counter-current exchange is demonstrated in two pipes lying side by side. Hot water enters one pipe (*top right*); cold water enters the other (*bottom left*). The fluids flow in opposite directions. Under ideal conditions heat will diffuse almost completely from one to the other.

to its legs would be a severe handicap. The same is true of the duck and other extremely well-insulated birds, which probably depend upon their webbed feet for heat dissipation.

As I mentioned at the beginning, heat conservation is only one of the functions performed by counter-current networks such as I have described. Indeed, there are more dramatic manifestations of this sort of system in the animal world. Nowhere in nature is counter-current exchange more strikingly developed nor more clearly illustrated than in the swim-bladder wall of deep-sea fishes. Here the function of the "wonderful network" is to prevent the loss of oxygen from the fish's air bladder.

A deep-sea fish keeps its swim bladder filled with gas which is more than 90 per cent oxygen. At depths of 9,000 feet or so it must maintain an oxygen pressure amounting to 200 to 300 atmospheres—nearly double the pressure in a fully charged steel oxygen cylinder. On the other hand, the oxygen pressure in the bladder's surroundings—in the fish's bloodstream and in the sea water outside—is no more than a fifth of an atmosphere. So the oxygen pressure difference across the thin swim-bladder wall is some 200 atmospheres. What is more, blood is constantly streaming along this wall through myriads of blood vessels embedded in it. Oxygen from the bladder, under the enormous pressure of 200 atmospheres, must diffuse into these blood vessels. How is it, then, that the streaming blood does not quickly drain the oxygen from the bladder? The answer, of course, is a counter-current exchange system. Very little oxygen escapes from the swim-bladder wall to the rest of the fish's body, because the outgoing veins, highly charged with oxygen, give it up to adjacent incoming arteries. There is a network of thousands of looping capillaries, so closely intermingled that diffusion of oxygen from veins to arteries goes on at a high rate.

What would be the most efficient arrangement of veins and arteries to give the maximum surface for transfer from one type of vessel to the other? We can treat this as a problem in topology and ask: How can we arrange black and white polygons (representing the cross sections of the blood vessels) so that black always borders white? If we allow only four polygons to meet at each corner, there are two different possible solutions: a checkerboard of squares or a pattern of hexagons with triangles filling the open corners. Under the microscope we observe that evolution has

produced precisely these two patterns in the swim-bladder retes of deep-sea fishes [*see photographs and diagrams on page 128*].

From the number and dimensions of the capillaries, the speed of the blood flow and other information we can calculate the amount of the oxygen-pressure drop across the rete, or, in other words, how effectively the rete traps oxygen. The calculation indicates that across a rete only one centimeter long, the oxygen pressure is reduced by a factor of more than 3,000. That is to say, the leak of oxygen through the rete is insignificant: translating the situation into terms of heat, if boiling water were to enter such a rete from one end and ice water from the other, the exchange of heat would be all but complete—to within one 10,000th of a degree!

To put it another way, the counter-current exchange in the swim-bladder rete is so efficient that in a single pass a rete one centimeter long is capable of raising a given concentration 3,000-fold, which leaves industrial engineering far behind. In speaking of concentration we include several different kinds: heat, gas pressure, the concentration of a solution and so on. A counter-current exchange system can establish a steep gradient in any of these quantities.

It has recently been suggested that counter-current exchange may be involved in the process whereby the kidneys filter the blood and produce urine. During the process of conversion of blood fluid to urine, the concentration of salts and urea in the fluid may be increased three- or four-fold. Just how is this concentration carried out?

The machines that perform the transformation are the units called nephrons, of which a kidney contains several millions. A nephron consists of a glomerulus capsule (a small ball of capillaries), a long, twisting tubule and a collecting duct [*see diagram on page 131*]. From the blood in the glomerulus capillaries a filtered fluid passes through the capsule wall into the tubule. The fluid travels along the tortuous course of the tubule,

RETE MIRABILE in the wall of the swim bladder of the deep-sea eel is enlarged 100 times in this photomicrograph. The rete is a bundle of small blood vessels; each of the small light areas in the photomicrograph is a blood vessel seen in cross section. The veins and arteries in the bundle are arranged in such a way that the blood in one vessel flows in a direction opposite that of the blood in an adjacent vessel.

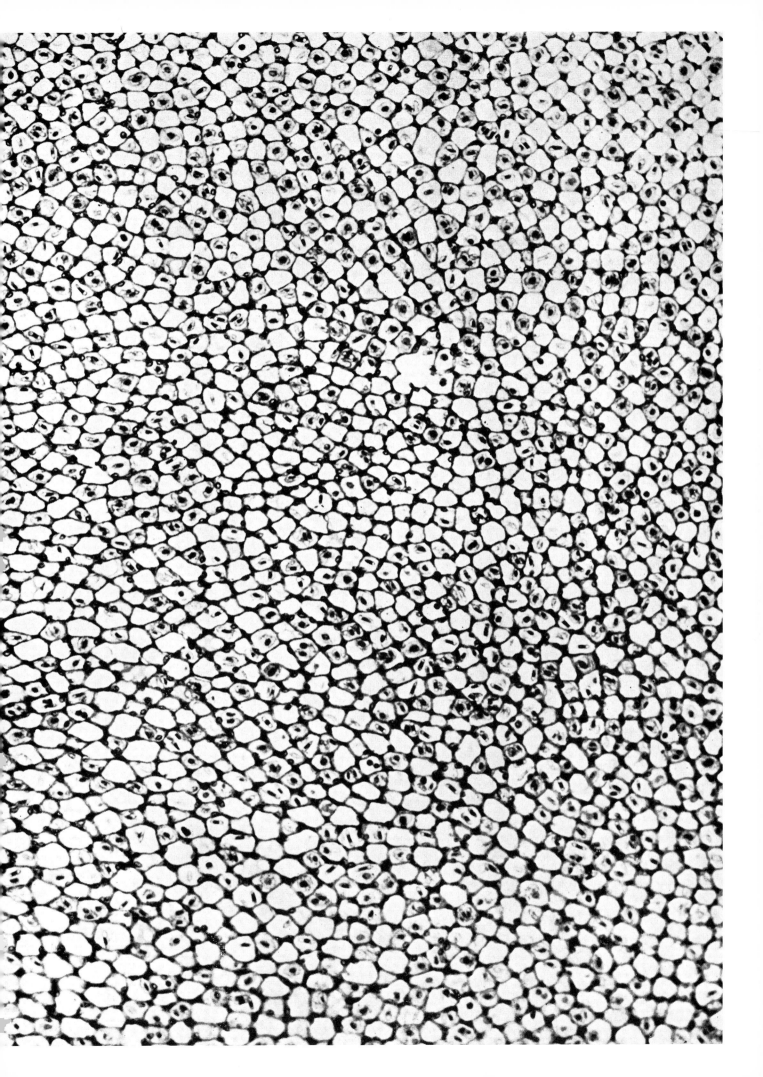

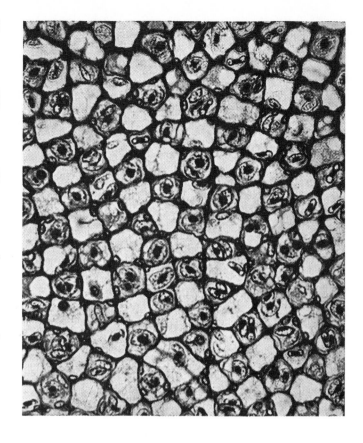

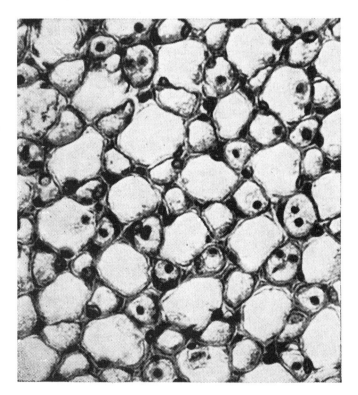

THE APPLICATION of counter-current exchange reaches the ultimate in the swim-bladder retes of deep-sea fishes. Drawings at left show two ways in which veins (*white*) and arteries (*black*) might be arranged to gain the greatest possible area of exchange. One pattern is non-staggered; it gives rise to a checkerboard (*top left*). The other is staggered; it gives rise to stars (*bottom left*). Photomicrographs show the checkerboard in the deep-sea eel (*top right*) and the star pattern in the rosefish (*bottom right*).

doubles back at the loop of Henle [*see diagram*], and by the time it leaves the collecting duct it has become concentrated urine. The conversion does not take place in the glomerulus capsule; it has been established that the fluid emerging from the capsule has essentially the same salt concentration as the blood. The problem is to determine exactly where in the system the change in concentration is produced.

B. Hargitay and Werner Kuhn of the University of Basel recently proposed that it takes place primarily in the loop of Henle. They pointed out that only animals with a Henle loop in the kidney nephron—namely, mammals and birds—can produce a concentrated urine.

The loop of Henle, with its two arms running parallel and fairly close together, is a structure which reminds us of the arteriovenous capillary loops that make up the rete of a fish [*see diagram on page 130*]. Hargitay and Kuhn reasoned that salts or water might migrate from one arm of Henle's loop to the other, and that as a result salts might be concentrated in the bend of the loop. This part of the loop is situated in an internal structure of the kidney called the papilla, which also contains the collecting ducts and adjacent blood capillaries. The investigators assumed that the fluid concentrated in the loop would be transmitted to the ducts and capillaries and become concentrated urine. To test their idea, they first froze rat kidneys and examined small sections of the tissue under a microscope. The sections of tissue in the papilla, around the loop of Henle, proved to have the same melting point as the rats' urine, while the tissue in the cortex (outer part) of the kidney had the same melting point as frozen blood. Since the melting point depends on the salt concentration this finding tended to confirm the idea that the primary site for the concentration of urine is located in the bend of Henle's loop. But it was possible that the brutal freezing process had damaged the tissues so that urine escaped from the collecting ducts and diffused to the area around the loop. To make a clearer test, H. Wirz, a colleague of Hargitay and Kuhn, developed a technique for drawing samples of blood from the capillaries around Henle's loop. This blood proved to have the same melting point as the urine, *i.e.*, its salt concentration was as much as three times higher than that of blood in other parts of the animal's body.

Thus Hargitay and Kuhn seem to have strong support for their thesis that a counter-current exchange in Henle's

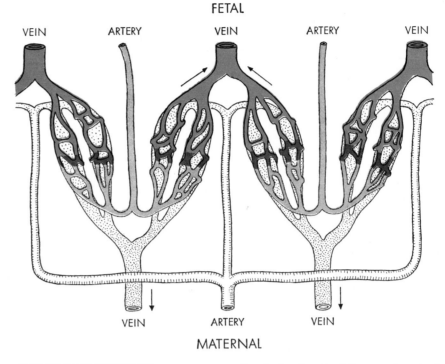

FETAL

VEIN   ARTERY   VEIN   ARTERY   VEIN

VEIN   ARTERY   VEIN

MATERNAL

OXYGEN IS EXCHANGED by counter-current flow in the placenta of the ground squirrel. Oxygen-rich maternal blood (*white*) enters the arteries (*bottom*) and flows counter to the oxygen-poor fetal blood (*gray*). The fetal blood picks up oxygen (*red*) and leaves through the fetal vein (*top*). Oxygen-poor maternal blood (*stippled*) returns to the maternal heart.

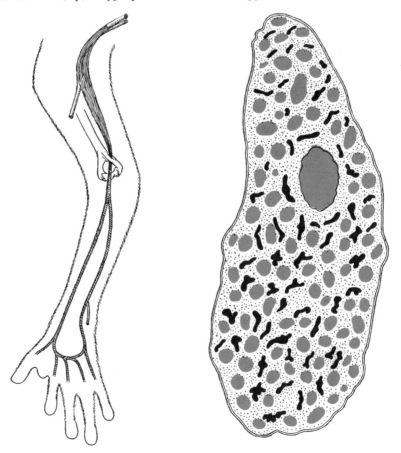

HEAT IS CONSERVED by rete bundles in the loris. They occur where the limbs join the body (*upper left*). Arterial blood enters the upper end of the rete and flows counter to the venous blood. Heat diffuses from the arteries to the veins and is returned to the body. A cross section of the rete (*right*) shows the arrangement of arteries (*red*) and veins (*black*).

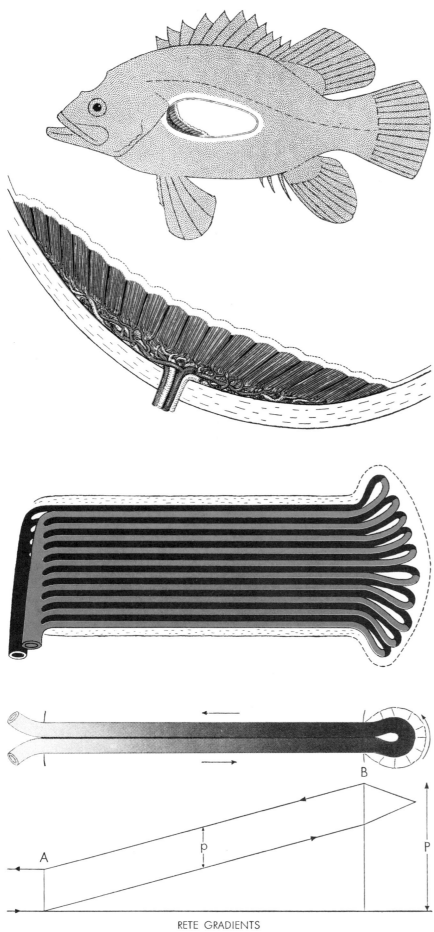

RETE GRADIENTS

loop plays a part in the formation of urine by the kidney. But the question still needs further research.

Various other examples of counter-current exchange in animals have been discovered. One of them concerns the breathing of fishes. A fish requires a far more efficient and resourceful breathing apparatus than an animal that lives in the air. Each quart of air contains some 200 cubic centimeters of oxygen, but a quart of sea water has only about five cubic centimeters, and oxygen diffuses through water slowly. The fish therefore has to be remarkably efficient in extracting oxygen from the water that flows over its gills. It can, in fact, take up as much as 80 per cent of the oxygen in the water. Anatomical studies and experiments have proved that fishes employ a counter-current system in this process. The blood in the capillaries of the fish's gill plates flows in the direction counter to the flow of water over its gills. When experimenters reversed the direction of the water current over the gills, fishes extracted only one fifth as much oxygen as they did normally!

In many species of animals the mother and the fetus she is carrying share their blood substances by means of a counter-current exchange system. This apparently is not true of the human animal, for the fetus's capillaries are bathed directly by the mother's blood. But in the rabbit, the sheep, the squirrel, the cow, the cat, the dog and other animals, the mother's blood vessels are intermingled with the fetus's in the placenta, and by counter-current flow they exchange oxygen, nutrients, heat and wastes.

In sum, the principle of counter-current exchange is employed in many and various ways in the world of living

THE WRECKFISH swims at great depths and must therefore keep its swim bladder filled with oxygen at tremendous pressures. It does this by means of counter-current bundles (*solid red*) in the swim-bladder wall (*first and second drawings*). One of these bundles (*third drawing*) and a single counter-current capillary (*fourth drawing*) are schematically depicted. Very little oxygen (*red shading*) escapes beyond the swim-bladder wall because it diffuses (*small arrows*) from the outgoing vein into the adjacent incoming artery. The diagram (*bottom*) represents build-up of pressure (P) by means of a pressure difference (p) between the counter-current artery and vein.

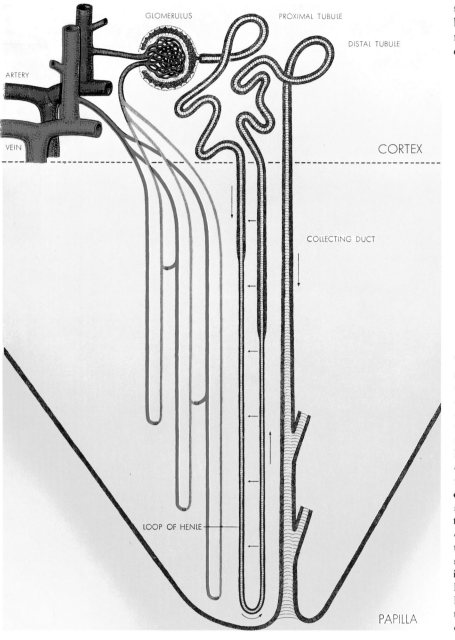

ARTERY

VEIN

GLOMERULUS

PROXIMAL TUBULE

DISTAL TUBULE

CORTEX

COLLECTING DUCT

LOOP OF HENLE

PAPILLA

things. We cannot fail to be impressed by the marvels of bio-engineering that nature has achieved in its development of "the wonderful net."

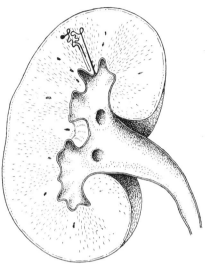

THE NEPHRON filters urea and other dissolved substances out of the blood and concentrates them in the urine. It may accomplish this with counter-current exchange. The drawing above is a cross section of the human kidney. At its upper left is a schematic representation of one of the kidney's several million nephrons. In the drawing at left the nephron is enlarged. Arterial blood (*red*) is depicted entering the capillaries of the glomerulus. The dissolved substances filter out of the glomerulus into the proximal tubule. The solution then travels down the tubule and doubles back at the loop of Henle. If the dissolved substances diffused from the ascending tubule into the descending tubule, the urine would be concentrated (*brown shading*) near the bend of the loop. The concentration might then be transmitted to the portion of the collecting ducts located in the papilla.

III

GAS EXCHANGE AND THE LUNGS: ADAPTATIONS FOR DIVING AND HIGH ALTITUDE

. . . It is universally admitted that a very few other creatures in the sea (other than whales) also breathe, those whose internal organs include a lung, since it is thought that no animal is able to breathe without one. Those who hold this opinion believe that the fishes possessing gills do not alternatively expire and inspire air. . . . Nor do I pretend that I do not myself immediately accept this view of theirs, since it is possible that animals may also possess other respiratory organs in place of lungs, if nature so wills, just as also many possess another fluid instead of blood. . . . Undoubtedly to my mind there are additional facts that make me believe that in fact all creatures in the water breath . . . in the first place a sort of panting that has often been noticed in fishes during summer heat, and another form of gasping, so to speak, in calm weather . . .

Pliny
NATURAL HISTORY, IX, vi.

III

GAS EXCHANGE AND THE LUNGS: ADAPTATIONS FOR DIVING AND HIGH ALTITUDE

INTRODUCTION

Almost all vertebrates use gills or lungs as respiratory organs. Except for rare cases (see Johansen's "Air Breathing Fishes"), individual gill filaments of fishes cannot function adequately in air. The reason is simply that adjacent filaments tend to adhere to each other, making it impossible for air to flow between the sheets of tissue and near the blood capillaries where gas exchange occurs. Consequently, all terrestrial vertebrates and many fish that obtain oxygen from the air possess lungs or lung-like sacs in which oxygen and carbon dioxide may be exchanged.

It seems likely that the simple lungs of modern lungfish are good facsimiles of early vertebrate lungs that appeared long ago in the history of fishes. The circumstance most conducive to the appearance of lungs was probably a fresh-water habitat that was subject to heating and drying; under such conditions, the amount of dissolved oxygen would decrease, creating an absolute necessity for ability to breathe air. Johansen presents data pertinent to these considerations and also discusses specializations other than lungs that permit various fishes to breathe air.

The steps by which lungs originally developed remain mysterious. All complex evolutionary changes occur in the course of many generations, with those organisms that happen to possess certain mutations being differentially successful at surviving to reproduce; this is natural selection, and the offspring of such animals are more likely to survive to pass on the mutated genes, assuming that the selective forces remain constant. This reasoning suggests many stages in the appearance of any structure as large and complex as a lung. What came first? A respiratory surface on the gut wall? A diverticulum that secondarily became vascularized? We cannot say. Furthermore, development of the lungs in modern animals tells us little. For many years it was thought that the developmental sequence observed in an animal alive today was a reasonable portrait of the evolutionary history of the species. For instance, in the introduction to Part II we mentioned the similarity between development of the heart and the stages of heart structure and physiology seen in various vertebrates. But in lungs, such a relationship is not so evident. A hollow endodermal diverticulum from the gut forms and grows into special condensed lung mesoderm. The initial single diverticulum branches at its tip; then each branch bifurcates again, and the process continues until the thousands of alveoli (in mammals) or parabronchi (in birds) are formed. Only the terminal product, with thin alveolar walls or minute air capillaries, is capable of normal function and gas exchange. Can this sort of process reflect the evolutionary history of lungs? It is difficult to imagine a function for any organs that would correspond to the early developmental stages of modern lungs. Hence we cannot guess why such early "nonfunctional" stages would be preserved by selection. Nevertheless, we must be cautious, for the developmental sequence may be misleading. The early stages we see today might correspond to former adult stages, if blood vessel formation was precocious in those early vertebrates. If this were so, a single diverticulum might have functioned as a primitive lung. Then in the course of subsequent evolution, a *delay* in vascularization would have provided time for development of a more complex surface area and for the construction of the complex mammalian or avian lung.

Once vertebrates began living on land, lungs began to increase in surface area. With the appearance of the dry skin in reptiles, all respiratory exchange was restricted to the lungs, so that the necessity for more surface area was greater. The two groups of temperature-regulating animals have radically different gas-exchange structures: mammals have large numbers of minute blind sacs, alveoli; birds, tiny tubes (air capillaries with a diameter of only 10 microns), which have the same function as the alveoli.

Birds also fill their lungs in a different manner than mammals. The expandable rib cage, first seen in reptiles, is still present in birds, but additional filling and emptying is caused by movement of the abdominal muscles. Mammals have a diaphragm—a muscular sheet of tissue that closes off the lung cavity from the abdominal cavity—as well as the expanding rib cage system. Three articles in this section, "The Lung" by Julius Comroe, "How Birds Breathe" by Knut Schmidt-Nielsen, and "Surface Tension in the Lungs" by John Clements, explain in detail the function of the avian and mammalian lung systems.

Recent work on the lipoprotein "surfactant" responsible for controlling the surface tension of mammalian lungs indicates the following: The substance is apparently stored in alveolar cells in special inclusion bodies (which are not mitochondria as implied by early studies and reported by Clements). This conclusion is based on the following kind of correlation: If a guinea pig is exposed to an atmosphere of 15 percent carbon dioxide, the inclusion bodies in the alveolar cells disappear, and concurrently the minimal surface tension of extracts taken from alveolar spaces rises. Then, as recovery from the carbon dioxide treatment occurs, the bodies increase in number and surface tension falls again. Thus, the presence of the surfactant is correlated with the presence of its apparent storage sites in the cells. Other recent work shows that the lipid portion of the lipoprotein surfactant is dipalmityl lecithin. The evolutionary origin of this compound identified in mammalian lungs is still not clear. Surfactant has been found in bird lungs—a surprising finding since birds have tiny tubes that are always open, unlike the mammalian alveolar cavities, which tend to collapse as air is exhaled from the lungs and the diameter of the cavities diminishes. At present it is thought that the surfactant of birds tends to reduce fluid transport into the capillary lumen, and so contributes to keeping the little tubes open for passage of air. Recent experiments have also demonstrated that extracts from some amphibian and reptilian lungs contain fatty substances that lower the surface tension of fluid. How these agents are related to avian and mammalian surfactants is not yet known.

The crucial role of the lung surfactant is nowhere better illustrated than in cases of hyaline membrane disease, a potentially lethal condition found at birth in some human infants. Because of insufficient surfactant, the alveoli in the lungs of these newborn infants collapse each time air is expelled from the body. For the baby, each new breath is an exhausting fight to re-expand the alveoli. In the absence of intensive care for at least five or six days, such infants usually die. It now seems likely that the disease results from the fact that the alveolar cells which make surfactant (so-called Type II cells) fail to differentiate during the latter part of gestation. In exciting experiments reported by Avery, Wang, and Taeusch (*Scientific American*, April 1973), it is shown that early injection of adrenal steroids such as cortisone into other mammals (rabbits and sheep) causes the precocious maturation of Type II cells; the result is production of surfactant. These observations have led to the first attempts in pregnant human females to accelerate fetal lung development and thus prevent hyaline membrane disease from appearing at birth.

The interrelated control of breathing and heart rate is treated superbly by the various articles in this section. But, because of some very recent results in studies of respiratory control, modification of the material on that subject is necessary here: the respiratory chemosensitive cells of the brain are not located in the inspiratory and expiratory centers themselves. Instead, they appear to be in the ventrolateral walls of the medulla of the brain. In these walls, the cells are sensitive to the level of hydrogen ions (H^+) in the cerebrospinal fluid. (The acid level is, of course, related to the partial pressure of carbon dioxide (CO_2) in the fluid by the reactions, $CO_2 + H_2O \rightleftarrows H_2CO_3 \rightleftarrows H^+ + HCO_3^-$. Thus, elevated carbon dioxide produces hydrogen ions.) In fact, the pH of the cerebrospinal fluid is normally kept very constant by active transport regulation. However, the pH apparently does change in localized regions near the receptor cells to modify their activity. As a result the main control of breathing can occur. Complex exchange of Cl^- ions for the HCO_3^- that is a byproduct of H^+ production is an additional ingredient in the process.

The relative roles of the CO_2 sensors of the medulla and O_2 sensors of the great vessels is demonstrated by the following exercise: a student breathes into a paper bag. When the CO_2 level rises to 3.5 percent and the O_2 level falls to 17 percent, breathing increases in speed. However, if a container of hydroxide is placed in the bottom of the bag so that CO_2 is absorbed, then the O_2 level must fall to 14 percent before breathing accelerates. From these and other types of observations, it is concluded that the "fine" control of respiratory rate is carried out in the medulla.

Reptiles, birds, and mammals have all given rise to lines of organisms that reentered the water to make it their prime habitat. All of these organisms retained their lungs, failed to develop new gill systems, and continued to breathe air. It seems safe to presume that all such animals possessed the sorts of adaptations outlined by P. F. Scholander in "The Master Switch of Life" and by Suk Ki Hong and Hermann Rahn in "The Diving Women of Korea and Japan." Although the heart beat has been slowed, and the blood flow has been restricted to the brain and heart, there is simply not enough oxygen dissolved in sea water to allow a whale or a penguin to change from lung to gill respiration. And, as was pointed out in the Introduction to Part II, the hemoglobin of a mammal or a bird could not load and serve as a respiratory carrier under such conditions, because the partial pressure of oxygen in water is too low. Quite simply, the high metabolic rate of homeotherms can be supported only by the respiratory efficiency of lungs and air breathing. In spite of the handicap of having to come to the surface periodically, several mammals and a few birds have come to compete successfully with the fishes.

Organisms that have ventured into the sea have given rise to some species that live at great depths. The pressures encountered hundreds or thousands of meters deep do unexpected things to proteins, cell membranes and organelles, and other animal components. The turnover number of some enzymes increases at very high pressures, whereas that of other enzymes decreases. Some molecules, such as lactate dehydrogenase, tend to aggregate as pressure rises. Others, such as actin and myosin, tend to remain in the monomer form. In fact, electron micrographs of muscle from bathypelagic fishes reveal that actin and myosin filaments are unusually short. There is still too little information available on pressure effects in pressure-adapted organisms to allow us to understand the ways that vertebrates have adapted to the dense, cold world deep in the sea.

An interesting example of parallel evolution is seen in the prehistoric reptilian ichthyosaurs and in the mammalian dolphins. Of the many reptiles that reentered the water, the ichthyosaurs were the ones that changed the most from the terrestrial body plan. They resembled a dolphin in shape: their hind limbs were reduced to small paddles, and their pelvic girdle no longer articulated with the vertebral column; the forelimbs were

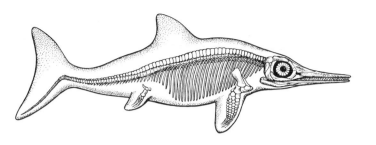

Figure 1. The skeleton of an ichthyosaur reptile. Note the fishlike tail in which the vertebral column turns ventrally. Fishes that gave rise to the first land vertebrates probably had tails in which the column turned upward. The condition seen here reflects the secondary return of this line of reptiles to the water, and an apparent different solution to the problem of tail support. Note that the small hind fin and its supporting pelvic girdle bones no longer articulate with the vertebral column, as they do in land vertebrates. [After Young, J. Z., *The Life of Vertebrates*, Oxford, 1962.]

flat-steering fins containing a large number of "finger" bones that extended outward into the fin blades; the vertebral column consisted of discs similar to those of fish and unlike those of land vertebrates. Ichthyosaurs swam by sweeping their tails back and forth through the water as fishes do. Finally, as might be expected, a shelled egg was not passed from the body, but rather the young were gestated within the female reproductive tract.

Dolphins are streamlined in a similar manner. External protrusions—hair, ear pinnae, penis, scrotum, nipples—are either absent or sunken beneath the body surface. Blubber, besides being the prime insulator, is a contouring material that is important for laminar flow and fast swimming (as discussed in Part I). Unlike ichthyosaurs, dolphins do not swim by moving the tail back and forth; rather, they sweep it up and down with large muscles that are inserted on spines on the tail vertebrae. The centers of the vertebrae are enlarged discs like those of ichthyosaurs, so that the column acts as a compression strut (like that of fish and unlike the "suspension girder" backbone of land mammals). New structures, the tail flukes, serve as propulsive blades that amplify the effects of tail movements. The hind limb is not visible externally, and only remnants of the pelvic girdle remain embedded in the lateral body musculature. As in ichthyosaurs, many finger bones extend outward in the large pectoral fins for support. Both groups of animals have a shortened neck region and nostrils that are located well back from the snout and rather dorsal on the head. Interestingly, the basic sensory systems are different: ichthyosaurs had large eyes and presumably used them as their main sense organ; dolphins and other cetaceans have less prominent eyes and a small optic lobe of the brain, and they use, instead, pulses of high-frequency sound for navigation and communication (discussed in detail in the Introduction to Part VI). In summary, the overwhelming similarities of these groups, which reentered the water millions of years apart, attest to the stringent demands imposed by the aqueous environment as well as to the remarkable plasticity of the vertebrate organization in adapting to a new environment.

The last article in this section, "The Physiology of High Altitude" by Raymond Hock, provides an interesting contrast to the study of pearl diving women and also reemphasizes much of what has been discussed in Parts II and III of this book. The cardiovascular, pulmonary, and heat-generating systems all must be integrated and modified in special ways if mammals or other vertebrates are to survive near the boundaries of an acceptable environment for life.

13

AIR-BREATHING FISHES

KJELL JOHANSEN
October 1968

Lack of oxygen in water forced certain Devonian fishes to develop air-breathing organs. Some left the water and colonized the land; descendants of others are today's remarkable air-breathing species

The transition from breathing water to breathing air was perhaps the most significant single event in the evolution of vertebrate life. How did it come about? The change in respiration is commonly identified with the emergence of animals from the water to dry land. Actually the fossil and living evidence shows that air-breathing by vertebrates began long before that development. Well before the evolution of amphibians some fishes had begun to use air instead of water for respiration. The crucial steps toward the development of lungs came about not as an adaptation to living out of the water but in response to changes in the aquatic environment itself.

During the Devonian period of some 350 million years ago a respiration crisis developed for much of the vertebrate life inhabiting the freshwater basins of the earth. The oxygen content of the waters gradually declined, as a result of high temperatures and the oxygen-consuming decay of dead organic material in shallow lakes, rivers and swamps. Throughout a vast portion of the aquatic environment the oxygen supply dropped to a marginal level. For the fishes, equipped only with gills for obtaining oxygen from water, life became quite precarious. Yet they were mere inches away from the limitless reservoir of oxygen in the atmosphere above the surface of the water. This saving circumstance led to the evolution of a variety of organs that enabled fishes to obtain oxygen from the air.

We can see the adjustments that were made to the oxygen problem in many fishes today, among both modern and archaic species. Certain species living in stagnant tropical pools, for instance, spend their entire existence close to the surface, within ready reach of the thin top layer of water that contains more oxygen than the waters below. Some of these species come to the surface layer to replenish their oxygen at frequent intervals; others visit the layer less frequently but stay in it longer each time. During a recent expedition to the Amazon River area I witnessed a dramatic demonstration of the latter performance. We had placed a freshwater stingray in a tank containing only a small amount of water. We noted that as the oxygen content of the water declined the fish pumped water over its gills more and more rapidly. Suddenly, when the oxygen pressure dropped below 20 millimeters of mercury, the fish swam straight to the surface, began to draw the surface film of water into the intake openings on its upper side and continued this performance at the surface for more than 15 minutes. The whole act was beautiful, and we were excited to discover this illuminating behavior—which the Amazon stingray probably has been performing routinely for millions of years. All the close relatives of this stingray are marine or brackish-water fishes with the flattened shape of bottom-dwellers and with water-intake openings on their upper surface, where the openings cannot be clogged by debris from the bottom. In the amazing freshwater stingray the

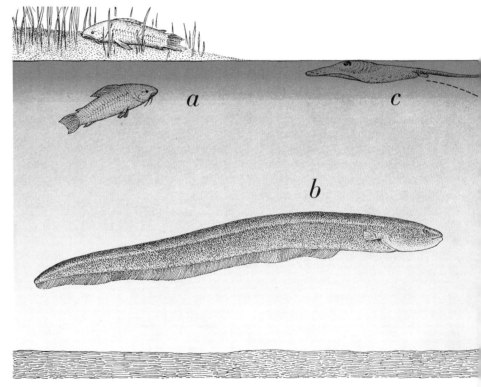

OXYGEN-DEFICIENT TROPICAL SWAMP is the habitat of air-breathing fishes and some that breathe near the surface where the oxygen supply is best. The relative oxygenation is shown by the color intensity. *Hoplosternum littorale* (*a*) is an intestinal breather that gets

same morphology exploits the opposite side of the hydrosphere: the interface with the air! It is also significant that the gills of the stingray (and of other fishes with similar habits) are large and well developed, maximizing gas exchange so that the fish can make efficient use of the limited supply of oxygen in the surface water.

The step from tapping surface water for oxygen to obtaining oxygen from the air above the surface was of course a drastic one for fishes, requiring radical modification of their respiratory structures. The typical gills of a fish, although well suited for gas exchange in the water, are totally unsuited to performance of that function in the air. The thin lamellae of the gills, with their fine blood vessels, collapse under gravity in the air. In order to breathe air fishes had to develop air-holding chambers of one kind or another. Some species became air-breathers by adapting the stomach or a segment of the intestine to this purpose; they swallowed air and expelled it again at the mouth or the cloaca. Others developed elaborate air-holding structures in the mouth or throat. The most common adaptation for breathing air was the development of the special organ now recognized in many fishes as the swim bladder. This chamber serves other purposes in water-breathing fishes: it

maintains the fish's buoyancy by adjusting to the hydrostatic pressure, and it may also act as an organ for detecting and producing sounds. The fossil evidence indicates, however, that the swim bladder of fishes originally came into being as an organ for respiration, and in some species it eventually evolved into a true lung.

A surprisingly large number of modern species of teleosts (bony fishes) are capable of breathing air. Evidently conditions like those that existed in Devonian times—swampy, oxygen-deficient waters—have given rise again and again to air-breathing fishes. Apart from these latter-day examples, we have considerable direct information on the respiratory revolution that led to the liberation of vertebrates from the aquatic environment during the Devonian. The information is provided not only by fossils from that time but also by living representatives of several archaic groups of air-breathing fishes that still exist today. These include two African genera in the primitive order Chondrostei, two American genera in the order Holostei and three genera of lungfishes (Dipnoi) in Africa, Australia and South America. These descendants from Devonian times inhabit oxygen-deficient waters and are so remarkably little changed from their ancestors that we can consider them representative of the early vertebrate forms

that made the first crucial step toward air-breathing. The amphibians are believed to have arisen from the archaic ancestors of the lungfishes.

The lungfishes have an air bladder that developed originally as a diverticulum, or pouch, from the gut. The acquisition of this incipient lung was of course only a first step toward effective air-breathing. It had to be accompanied by changes in the fishes' system of blood circulation that would provide efficient transportation to the body tissues of blood oxygenated in the lung. In the lungfishes we see the circulation transformed in that way. In short, the lungfishes are truly lung breathers, with a respiratory and circulatory system that forecasts the system later developed by the higher vertebrates—the birds and the mammals. Fortunately the three living genera of lungfishes depict three stages in the transition from water-breathing to air-breathing, so that we are able to trace the anatomical changes that effected this transition.

Curiously, although the lungfishes offer an inviting opportunity to investigate the evolution of air-breathing by experimental studies, it is only within the past decade that such studies have been undertaken. With the collaboration of Claude Lenfant and other colleagues I have been pursuing this inquiry with lungfishes and certain other

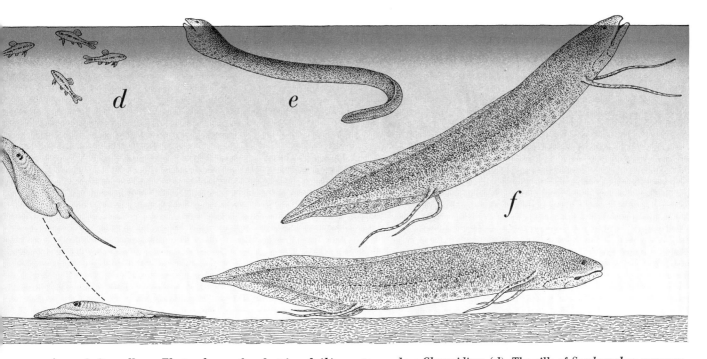

oxygen from air it swallows. *Electrophorus*, the electric eel (*b*), gulps air and obtains oxygen through the walls of the mouth. The freshwater stingray (*c*) visits the surface to breathe oxygenated wa-

ter, as does *Characidium* (*d*). The gills of *Symbranchus marmoratus* (*e*) function in air as well as water. In lungfishes such as *Protopterus* (*f*) the swim bladder has become a well-developed lung.

air-breathing fishes in Africa, Australia, South America, Norway and in my laboratory at the University of Washington. The following is an account of our findings.

The lungfishes have long been known to man; they have been an important source of food for primitive peoples, in part because many of the species go into estivation (a dormant state) during dry seasons and therefore are easy to catch. Some of the lungfishes grow to considerable size; one African specimen displayed in a Nairobi museum measures seven feet in length. It was only about 130 years ago that lungfishes began to attract the attention of scientists. The discovery at that time of their remarkable ability to breathe air evoked considerable excitement and confusion in zoological societies. Reluctant to accept an air-breather as a fish, the zoologists at first classified the lungfishes as amphibians, and the surprise occasioned by the discovery is still preserved in the species name of a South American lungfish: *paradoxa*.

In the three lungfish genera we can observe a sequence of development from very little dependence on air-breathing to a stage where air-breathing became the principal means of respiration. The

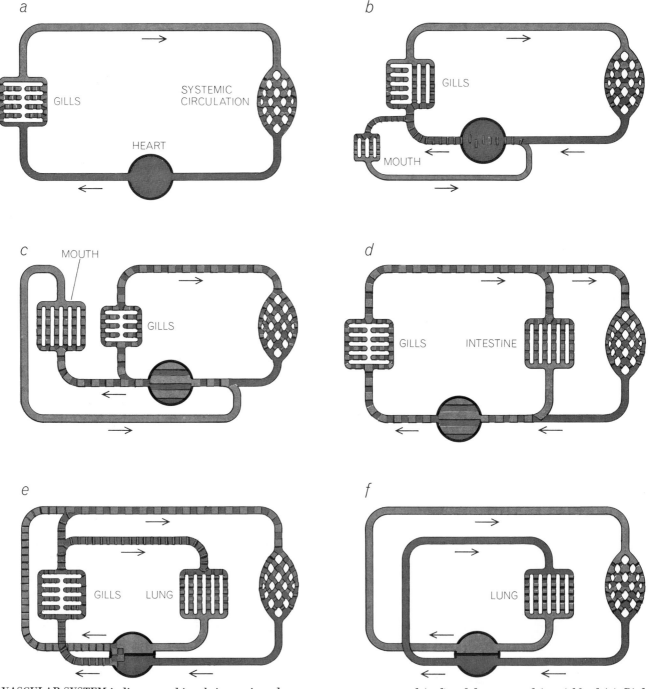

VASCULAR SYSTEM is diagrammed in relation to air- and water-breathing organs in typical fishes (*a*), birds and mammals (*f*) and various "in-between" air-breathing fishes: *Symbranchus* (*b*), *Electrophorus* (*c*), *Hoplosternum* (*d*) and the lungfishes (*e*). Water-breathing fishes have a single circulation with no mixing between oxygenated (*red*) and deoxygenated (*gray*) blood (*a*). Birds and mammals have a double circulation with no mixing (*f*). In the others there is more or less mixing of blood. The lungfish arrangement (*e*) was conducive to development toward the vertebrate condition: two circuits are arranged in parallel, with variable mixing.

Australian lungfishes (genus *Neoceratodus*), living today in rivers that are never severely deficient in oxygen, are primarily water-breathers, with well-developed and functioning gills. They do, however, possess lungs and can raise their heads out of the water to breathe air when necessary. The South American lungfishes (*Lepidosiren*) and the African (*Protopterus*) live in swamps whose waters are severely deoxygenated and periodically dried up by droughts. In dry periods these fishes entomb themselves in burrows or cocoons in the mud and go into a profound state of suspended animation; one might say their metabolic furnace is turned down all the way to the pilot light. Within minutes or hours after the swamp is flooded again by rainfall they revert to active life. Although water is their natural habitat, they depend mainly on their lungs and air-breathing for the absorption of oxygen. In the adult South American and African lungfishes large portions of the gills have degenerated to the point where they are little more than vestiges.

For examination of the functioning of the lungfishes' respiratory apparatus we have used catheters and transducers that enable us to determine blood flow and blood pressure in the vessels supplying the respiratory organs. We implant the instrumentation in the important blood vessels and in the lung itself, and after the fishes' recovery from the surgery we are able to obtain a continuous record of blood flow and pressure as the animals swim about in an aquarium, and to measure the oxygen tension in the blood and lung by analyzing samples from those organs. With this technique we have studied the lungfishes' respiratory behavior under normal conditions and under experimental variations of their environment.

The observations that are thus made possible show clearly the extent of the difference between the Australian lungfish and, say, the African lungfish with regard to breathing behavior [*see illustration on this page*]. With the fish swimming in well-oxygenated water, the blood coming from the gills of the Australian lungfish is almost fully saturated with oxygen. The lung then has little or no importance; there is no need to add to the oxygen supply by means of air in the lung. Furthermore, the carbon dioxide in the blood is kept at a very low level, because carbon dioxide has a relatively high solubility in water and the gills are remarkably efficient in exchanging this gas. In the African lungfish, on the other hand, the situation is some-

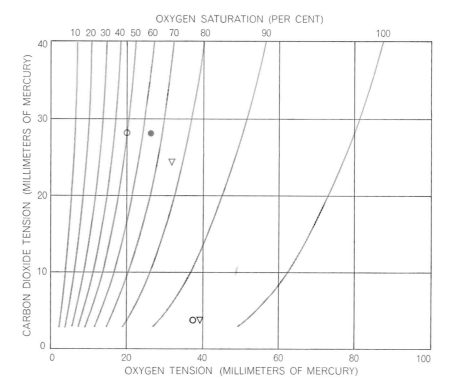

MEASUREMENTS of the oxygen and carbon dioxide content of the blood in the Australian lungfish *Neoceratodus* (*black*) and the African lungfish *Protopterus* (*color*) show that the latter depends more on its lung even in the water. In *Neoceratodus*, blood going to the lung in the pulmonary artery (*circle*) is about as well saturated with oxygen (*gray curves*) as blood coming from the lung in the pulmonary vein (*triangle*). In *Protopterus*, however, oxygen saturation is higher and carbon dioxide tension lower in blood coming from the lung (*triangle*) than in blood going to it (*open circle*); the oxygenated blood is channeled preferentially to the systemic arteries (*dot*) rather than to the lung. The lower carbon dioxide tensions in *Neoceratodus* reflect the dominance of water-breathing with efficient gills.

what reversed. The blood delivered to the lung from this fish's vestigial gills is not nearly as rich in oxygen, and the lung makes a large additional contribution to the oxygen tension, raising the oxygen saturation of the blood to about 80 percent. The lung also serves to eliminate some carbon dioxide. It is less efficient in this function than gills are, however, and the carbon dioxide tension in the blood is therefore considerably higher in the African lungfish than it is in the Australian lungfish.

Now let us drain the water from the tank and leave both types of fish exposed to air. The Australian lungfish frantically searches for a return to water and at the same time begins to gulp air into its lung at a rapid rate. This effort succeeds in saturating the blood flowing through the lung with oxygen, but the rate of blood flow is not sufficient to maintain the oxygen level in the arterial system. The oxygen tension in the arteries of the fish's body rapidly declines. Moreover, the lung does not eliminate carbon dioxide at the requisite rate, and in less than 30 minutes the carbon dioxide tension in

this fish's arterial blood rises more than fivefold. In contrast, the African lungfish fares much better when it is left completely exposed to air. The intensification of this fish's air-breathing succeeds in keeping the oxygen tension in the arteries at a high level. The carbon dioxide concentration in the blood does rise for a time, but the elimination of that gas through the lung is such that the carbon dioxide tension soon levels off [*see top illustration on page 143*]. Thus the African lungfish clearly shows that its air-breathing system is sufficiently developed to enable it to survive out of water, as it must do during periods of drought, whereas the Australian lungfish normally never leaves the water.

Obviously the African lungfish's superior performance in air-breathing must depend on a circulatory advantage that enables it to make more effective use of the oxygen taken up through the lung. The possession of a lung, even one well supplied with blood vessels, cannot efficiently provide oxygen to an animal for its metabolic needs unless it is accom-

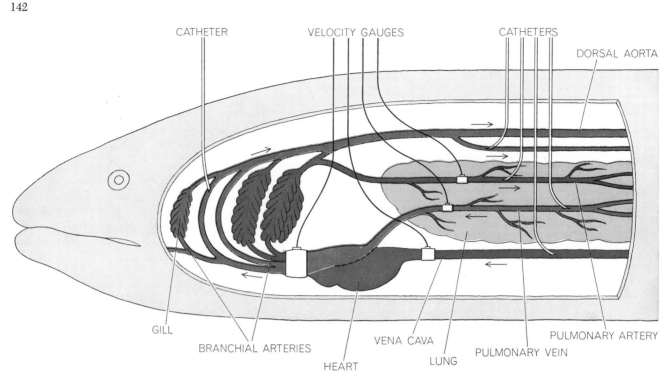

CATHETER   VELOCITY GAUGES   CATHETERS

DORSAL AORTA

GILL

BRANCHIAL ARTERIES

HEART

VENA CAVA

LUNG

PULMONARY VEIN

PULMONARY ARTERY

LUNGFISH was the major subject of the author's investigations. The schematic diagram shows how blood-velocity gauges and catheters for the measurement of blood pressure and blood-gas tensions were attached; animals swam freely after the instruments were implanted. In the lungfish blood is shunted either to the gills and lung for gas exchange or to the dorsal aorta for systemic circulation.

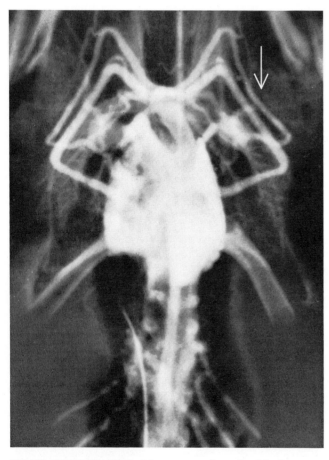

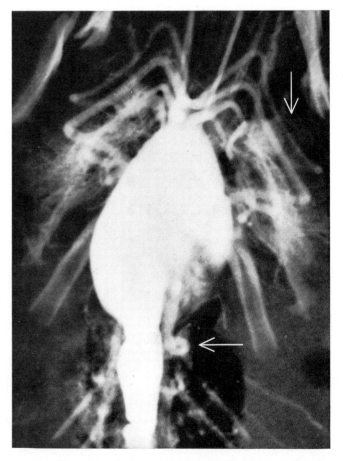

X-RAY ANGIOGRAPHY of *Protopterus* shows that blood returning to the heart from the lung is preferentially sent to the systemic circulation, with slight recirculation to the lung. When contrast medium is injected in the pulmonary vein (*left*), it appears mainly in two gill-less branchial arteries (*arrow*) and does not enter the pulmonary artery. Injected in the vena cava, however (*right*), it appears in all the branchial arteries and fine gill capillaries (*vertical arrow*) and also enters the pulmonary artery (*horizontal arrow*).

panied by an appropriate circulation for effective direct delivery of the oxygen to the metabolizing tissues. The circulatory requirement for an air-breather is fundamentally different from that for a water-breather. Hence a principal key to the transition from water-breathing to air-breathing lies in the evolution of the cardiovascular apparatus. The circulatory systems of the air-breathing fishes, differing one from another, show the steps in this evolution.

In ordinary water-breathing fishes the blood issuing from the pumping heart is not oxygenated. It flows to the gills, picks up oxygen there, emerges into a dorsal aorta and then flows through arteries to the various parts of the body, returning to the heart by way of the venous system. In short, the water-breathing fish has a single circulation, in contrast to the double circulation of a mammal, which first sends the venous blood from the right ventricle of the heart to the lungs and then, on return of the oxygenated blood to the left ventricle, pumps it out into the systemic circulation. For the fish the single circulation is perfectly adequate; the gills are highly efficient gas exchangers, and the blood flowing freely through them needs no extra input of energy in order to travel on through the arterial system.

When a fish acquires a lung and breathes air, complications begin to arise. The fish tends to retain its gills, at least at first, as an escape hatch for carbon dioxide, which is eliminated much more readily through gills than through a lung. Now, with gas exchange taking place both in the lung and in the gills, and with new avenues of blood circulation developing, the oxygenated blood from the lung becomes mixed with deoxygenated blood from the veins [*see illustration on page 140*]. Many of the air-breathing fishes still retain a single-circulation system in spite of this disadvantage. In the lungfishes, however, we see the beginnings of development of a double circulation. The two circulations are parallel, one passing through the lung, the other through the gills. They are not distinctly separated, and consequently there is some mixing of blood between them.

The Australian lungfish, as we have noted, is principally a water-breather and still has well-developed gills. It resorts to air-breathing only when the oxygen content of the water falls below normal. This fish could not survive long in severely deoxygenated water because the oxygen it gained by breathing air would be lost to the water as its oxy-

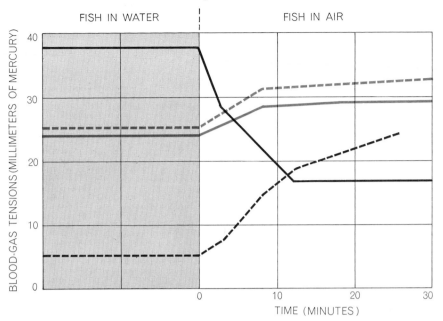

ADAPTABILITY TO AIR exposure varies in *Protopterus* (*color*) and *Neoceratodus* (*black*). When the tank is drained, both fish intensify air-breathing. *Protopterus* increases arterial oxygen tension (*solid curve*) and controls carbon dioxide (*broken curve*). *Neoceratodus* cannot get enough oxygen into its circulation or eliminate carbon dioxide efficiently.

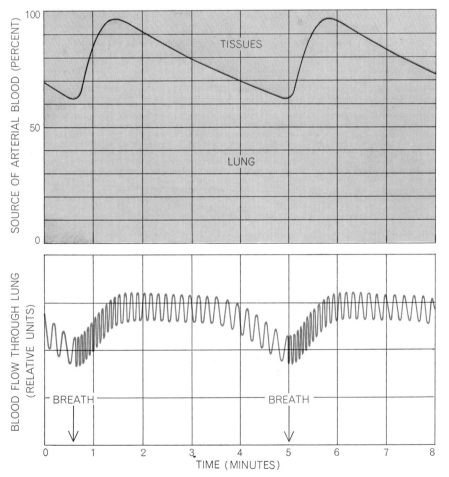

SHUNTING of oxygenated blood changes with the breathing cycle in *Protopterus*. Right after each breath almost all blood entering the systemic arteries is oxygenated blood from the lung; later in the cycle the proportion is lower. Blood flow through the lung also tends to be highest just after a breath and then to diminish until the next breath is taken.

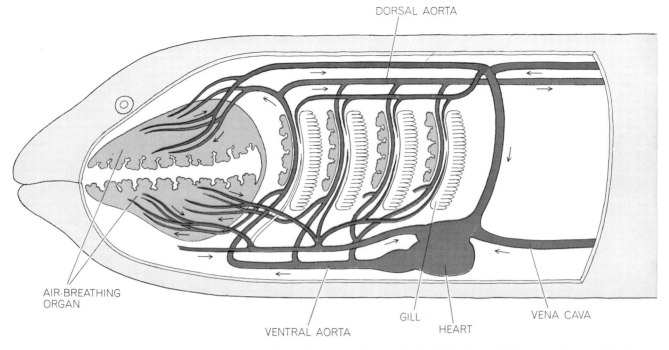

DORSAL AORTA

AIR-BREATHING
ORGAN

VENTRAL AORTA

GILL   HEART

VENA CAVA

ELECTRIC EEL is a mouth breather. The mucous membrane lining its mouth is creased and wrinkled, providing a large surface richly supplied with blood vessels for gas exchange. Blood moves past the vestigial gills, which are unimportant in gas exchange.

genated blood circulated through the well-developed gills. The African lungfish, however, gets along without stress in oxygen-depleted waters partly because its gills are degenerated and even bypassed by the two largest arteries from the heart, which go directly to the dorsal aorta, and partly because its lung circulation is more fully developed.

By means of X-ray angiography, using special film changers capable of exposing several frames per second, we have examined the pattern of circulation in the African lungfish in considerable detail. We find that the channeling of venous and arterial blood is highly selective in this fish; the recirculation, or mixing, of blood is very slight in the pulmonary circuit and only moderate in the systemic circuit. The X-ray studies of blood flow and oxygen analysis of the blood also show that the blood flow through the lung is greatest immediately after the intake of an air breath (when the lung gas is richest in oxygen) and that the blood in the arteries issuing from the heart to the tissues reflects a similar cycle: it is richest in oxygen right after a breath and gives way to deoxygenated blood from the tissues in the interval before the next breath.

Evidently the African lungfish has a control system that coordinates the respiratory and circulatory mechanisms to achieve a maximal yield in gas exchange and gas transport. Interestingly enough, an interaction of the two mechanisms such as we observe in the African lungfish can also be seen in man. When part of the human lung is blocked from contact with air or is poorly ventilated for

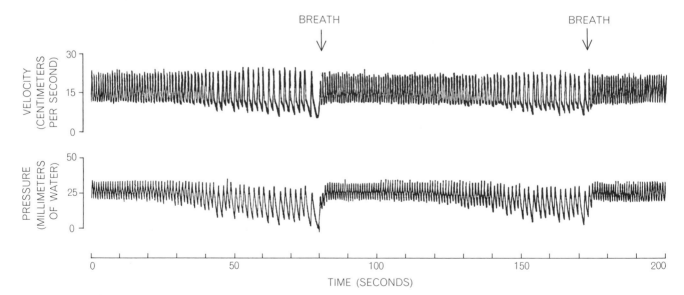

BREATH

BREATH

BLOOD FLOW in the electric eel is coordinated with breathing. Right after each breath (*arrows*) the cardiac output is high; the heart rate (pulse) and the average blood velocity and pressure then decline until the next breath, when they return to the high level.

some other reason, the vascular system reduces the flow of blood to that part. That is to say, the blood flow is adjusted to the availability of oxygen at the lung interface, as it is in the lungfish.

An even more basic parallel between the lungfish and human respiratory systems can be noted. Human respiration is regulated by an elaborate process of negative feedback in which the respiratory activities are controlled by changes in the oxygen and carbon dioxide tensions and the acidity of the blood. A rise in the blood's carbon dioxide tension, for instance, stimulates more rapid breathing to reduce it to the normal level. In air the African lungfish displays a similar sensitivity to specific internal cues. If it is exposed to air deficient in oxygen, the lungfish responds by breathing more rapidly than it would on taking normal atmospheric air into its lungs. Evidently its internal receptors sense hypoxia (oxygen deficiency) within the blood, and this acts as negative feedback to stimulate increased respiratory activity. In water, however, the response is different. If an adult African lungfish is placed in even severely deoxygenated water, the rate of its air-breathing, the gas tensions in its arterial blood and the normally slow rate of water-pumping across its gills all remain unchanged. The fish has been relieved of the threat of internal deoxygenation from oxygen-poor water; the gills no longer offer effective communication between the blood and the external environment and the respiratory control system monitors predominantly the activity of what is now the principal organ for oxygen extraction: the lung.

(It is easy to see that a feedback response prompting increased water-breathing in severely oxygen-depleted water would be harmful rather than advantageous to a fish that depends primarily on water-breathing. For an inhabitant of a tropical swamp it would lead only to an energy-consuming effort to obtain oxygen that is simply not available. The few truly water-breathing fishes living in such waters depend on a high tolerance for lack of oxygen in their blood and an ability to stay away from water masses most deficient in oxygen.)

What causes a lungfish to intensify its breathing efforts when it is totally exposed to air? This response is so prompt that it seems unlikely it is initiated by the operation of some internal chemical mechanism. My own surmise is that the response is probably triggered by the sudden translation of the fish from its weightless condition in water to the ex-

posure to net gravitational force in the air. This represents a massive physical stimulus that may well jolt the animal into a burst of rapid breathing. We may have here a phylogenetic parallel to mammalian ontogeny. The newborn baby's first breath after it emerges from the amniotic pool of its mother's womb seems to be triggered by the massive impingement of physical stimuli to which it is suddenly exposed in the air.

We come now to certain fishes in which the mouth serves as a lung. One example is the famous climbing perch of India (which, contrary to its mythology, does not actually climb trees). Probably the most remarkable of the mouth breathers is the electric eel (*Electrophorus electricus*) that lives in shallow, muddy pools along creeks and rivers in tropical South America. This fish is an obligate air-breather; it could not survive if it did not come up frequently for air. At regular intervals—every minute or two—the fish rises to the surface, gulps air into its mouth and then sinks back to the bottom. It expels the air through flap-covered opercular openings.

The electric eel has markedly degenerate gills that play no significant part in its respiration. The uptake of oxygen occurs in the mouth, which is almost entirely lined with papillae (protuberances) that provide a large total surface area for gas exchange. Comparatively little

carbon dioxide is released through the mouth; most of it is discharged into the water by way of the skin and the vestigial gills.

A rich network of blood vessels is embedded in the lining of the mouth for the absorption of oxygen. For the carnivorous electric eel the presence of these fragile vessels would be a serious drawback if it had to kill and chew its prey with its mouth, but it avoids that problem by stunning its victim with a powerful electric charge (up to 500 volts) and swallowing the food whole. The electric eel's respiration is handicapped, however, by a relatively inefficient circulatory system. The blood, after its oxygenation in the mouth, is not dispatched directly to the body tissues by way of the arteries but goes into veins that carry it to the heart [*see illustration on page 140*]. Consequently the oxygenated blood is mixed with the deoxygenated blood. As a result of the circulatory pattern, in which the blood is shunted through the respiratory organ between the arterial side and the venous side of the circulation, the arteries receive only part of the cardiac output, and the blood delivered to the tissues is a mixture of oxygenated and deoxygenated blood.

Our studies of blood flow in the electric eel showed that it pulses in a cyclic pattern like the pattern in the lungfish. Immediately after the eel has taken a breath the heart steps up its output; the

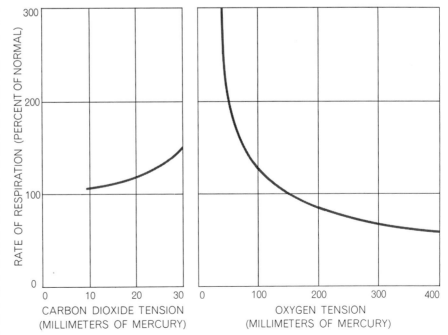

RESPONSE of electric eel to changes in the ambient atmosphere is clear and prompt. If it surfaces in an atmosphere low in oxygen (or high in carbon dioxide), its breathing rate is accelerated. If there is more oxygen present than in normal air, the breathing rate declines

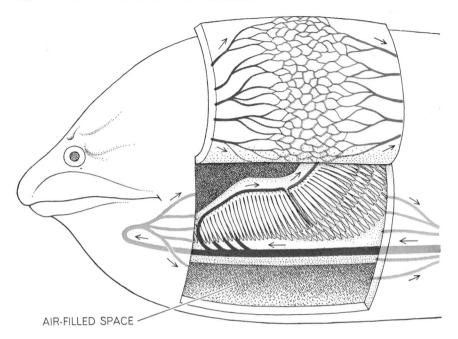

AIR-FILLED SPACE

SYMBRANCHUS **is unusual in that it can use the same organ, its gills, for air- and water-breathing, and can do both quite successfully. In addition to gills modified for air-breathing, it has a richly vascularized mouth lining (*flap*) that participates in gas exchange.**

heart rate, output and blood pressure then decline until the next air breath. During the accelerated phase of the cardiac output a high proportion of the blood flow goes to the mouth, where it can be oxygenated; then, as the oxygen in the mouth is used up, the proportion of the cardiac output delivered there declines. Thus the operation of the circulatory system partly compensates for the inefficiency of its structure by delivering more blood to the respiratory organ when the availability of oxygen in the organ rises to a peak. The increase in cardiac output evidently stems from stimulation of mechanical receptors in the mouth by the act of inhalation; we have found that it can be elicited by in-

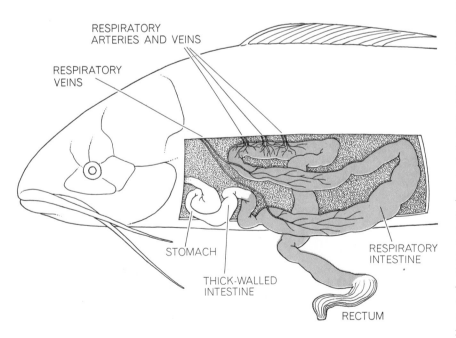

RESPIRATORY
ARTERIES AND VEINS

RESPIRATORY
VEINS

STOMACH

THICK-WALLED
INTESTINE

RESPIRATORY
INTESTINE

RECTUM

HOPLOSTERNUM **is one of several kinds of intestinal-tract breathers. It swallows air, which fills a long, coiled portion of the intestine that is thin-walled and richly supplied with blood vessels. Only a short segment of the intestine remains to function in digestion.**

flating the eel's mouth with pure oxygen or with nitrogen. Our experiments have also shown that the electric eel, like the African lungfish, is responsive to the oxygen content of the air it breathes. When it is subjected to air that is low in oxygen, the eel speeds up its breathing efforts; when the air is abnormally high in oxygen, the fish's breathing slows down.

A few remarkable fishes are able to use their gills for breathing air. An outstanding example is an eellike freshwater fish of South America named *Symbranchus marmoratus*. The gills of this fish are so organized that their fine blood vessels do not collapse under the force of gravity in the air; hence the blood flow through the gills remains normal. In *Symbranchus* the gills serve as an effective breathing organ both in the water and in the air. In air they are assisted by the lining of the mouth cavity, which also is well vascularized and provides gas exchange between the air and the blood. As long as the water contains an adequate oxygen content the fish spends all its time underwater. When the oxygen content drops, however, as commonly happens in tropical waters during the heat of the day, the fish rises to the surface, inflates its mouth and throat with air, closes its mouth and uses the trapped air for respiration. The inflation of the head makes the fish buoyant, and it may float on the surface for a considerable time, sometimes for hours, opening its mouth at intervals to expel the spent air and take a fresh breath. *Symbranchus'* air-breathing ability is so well developed that during a dry season it can survive for several months out of the water in the dormant state.

According to our measurements of the gases in this fish's blood, the oxygen tension of the arterial blood rises to higher levels when the fish is breathing air than when it breathes water. The carbon dioxide tension of the blood, however, also rises during air-breathing, both because the gills are less efficient in eliminating carbon dioxide in air than in water and because, as the fish rebreathes the air trapped in its closed mouth, waste carbon dioxide accumulates there. We found that when we removed *Symbranchus* from the water, it changed its breathing behavior so that the mouth was ventilated more or less continually, which allowed more carbon dioxide to escape. The fish kept its mouth open and made intermittent inhaling movements with its lower jaw. This change in breathing behavior on dry land may also be related to the fact that the fish no longer needs to maintain positive buoyancy.

Several species of tropical freshwater fishes use rhythmically ventilated air pockets in the gastrointestinal tract for air respiration. Air is swallowed and oxygen is delivered to the blood at the specialized, thin-walled and richly vascular portions of the gut. The circulatory system in these cases is generally inefficient for delivery of the oxygen to the body tissues because the arrangement is such that the oxygenated and deoxygenated blood is completely mixed and the arteries carry this mixture to the tissues [*see illustration on page 140*].

For the gastrointestinal-breathing species air-breathing by way of the gut is accessory to water-breathing with the gills. Nonetheless, the intestinal breathers are vitally dependent on air; most of them cannot survive if they are denied occasional access to the air, and for some species air-breathing has become the dominant method of respiration. A case in point is *Hoplosternum littorale*, a common fish of tropical South America that cannot live by water-breathing alone even if the water is saturated with oxygen, but that on the other hand is able to survive indefinitely in waters very severely deficient in oxygen by breathing air. Incidentally, this fish, like most other intestinal breathers, presents a puzzling question. Most of its intestine, a coiled tube that occupies nearly the entire body cavity, is modified for air-breathing. How can it carry out the usual intestinal functions, or, for that matter, how can it convey foodstuffs without disturbing the gas-exchange process of respiration? In our examinations we have always found the intestine full of gas and empty of solid food. It appears that the fish must feed only at long intervals and perform the digestive and absorptive functions with the short sections of the intestine that have not been modified for breathing.

The air-breathing fishes give a striking exhibition of the combination of exploratory and conservative forces that shaped the course of animal evolution. Changes in the oxygen tension of their aquatic environment forced the fishes to develop an apparatus for breathing air, yet they clung to the advantages that living primarily in water had to offer. At the same time, the crucial acquisition of the ability to breathe air opened a new world that inexorably led some fishes to develop the structural modifications that enabled vertebrates to step out of the water and walk onto dry land.

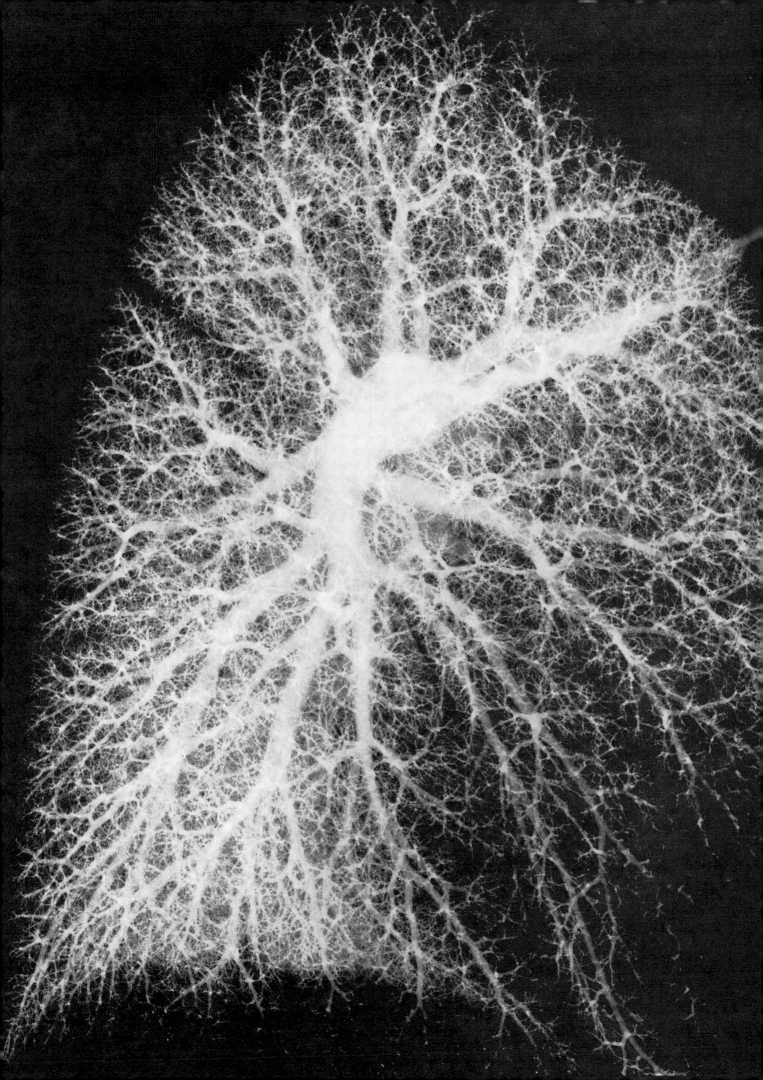

14

THE LUNG

JULIUS H. COMROE, JR.
February 1966

This elaborately involuted tissue of air sacs and blood vessels serves to exchange gases between the air and the blood. In man the total area of the membrane between the two systems is 70 square meters

Each year an adult human inhales and exhales between two million and five million liters of air. Each breath consists of about half a liter of air, 20 percent of which is molecular oxygen (O_2); the air swirls briefly through a maze of branching ducts leading to tiny sacs that comprises a gas-exchange apparatus in which some of the oxygen is added to the blood. The apparatus I am describing is of course the lung: the central organ in the system the larger land animals have evolved as a means of supplying each of their cells with oxygen.

Oxygen is essential to most forms of life. A one-celled organism floating in water requires no complex apparatus to extract oxygen from its surroundings; its needs are satisfied by the process of diffusion—the random movement of molecules that results in a net flow of oxygen from regions of abundance to regions of scarcity. Over very short distances, such as the radius of a cell, the diffusion of oxygen is rapid. Over larger distances diffusion is a much slower process; it cannot meet the needs of any many-celled organism in which the distance between the source of oxygen and the most remote cell is greater than half a millimeter.

The evolution of large animals has therefore required the development of various special systems that deliver oxygen from the surrounding medium to each of the animal's cells. In higher vertebrates such as man this system consists of a gas pump (the thorax) and two fluid pumps (the right and left ventricles of the heart). The fluid pumps are linked to networks of capillaries, through the walls of which the actual exchange of gas molecules takes place. The capillary network that receives its blood from the left ventricle distributes oxygen throughout the body. The network that receives its blood from the right ventricle serves a different purpose. In it the actions of the fluid pump and the gas pump are combined to obtain the oxygen needed by the body from the surrounding atmosphere.

This integrated system works in the following manner. The expansion of the thorax allows air to flow into ducts that divide into finer tubes that terminate in the sacs called alveoli. The rhythmic contraction of the right ventricle of the heart drives blood from the veins through a series of vessels that spread through the lung and branch into capillaries [*see illustration on opposite page*]. These blood-filled pulmonary capillaries surround the gas-filled alveoli; in most places the membrane separating the gas from the blood is only a thousandth of a millimeter thick. Over such a short distance the molecules of oxygen, which are more abundant in the inhaled air than in the venous blood, diffuse readily into the blood. Conversely the molecules of the body's waste product carbon dioxide, which is more abundant in the venous blood than in the inhaled air, diffuse in the opposite direction. In the human lung the surface area available for this gas exchange is huge: 70 square meters—some 40 times the surface area of the entire body. Accordingly gas can be transferred to and from the blood quickly and in large amounts.

The gas exchange that occurs in the lungs is only the first in a series of events that meet the oxygen needs of the human body's billions of cells. The second of the two liquid pumps—the left ventricle of the heart—now distributes the oxygenated blood throughout the body by means of arteries and arterioles that lead to a second gas-exchange system. This second system is in reality composed of billions of individual gas-exchangers, because each capillary is a gas-exchanger for the cells it supplies. As the arterial blood flowing through the capillary gives up its oxygen molecules to the adjacent cells and absorbs the cells' waste products, it becomes venous blood and passes into the collecting system that brings it back to the right ventricle of the heart [*see illustration on next page*].

The customary use of the term "pulmonary circulation" for that part of the circulatory system which involves the right ventricle and the lungs, and of the term "systemic circulation" for the left ventricle and the balance of the circulatory system, seems to imply that the body has two distinct blood circuits. In actuality there is only one circuit; the systemic apparatus is one arc of it and the pulmonary apparatus is the other. The systemic part of the circuit supplies arterial blood, rich in oxygen and poor in carbon dioxide, to all the capillary gas-exchangers in the body; it also collects the venous blood, poor in oxygen and rich in carbon dioxide, from these exchangers and returns it to the right ventricle. The pulmonary part of the circuit delivers the venous blood to the pulmonary gas-exchanger and then sends it on to the systemic part of the circuit.

This combination of two functions not only provides an adequate exchange of oxygen and carbon dioxide but also

ARTERIAL SYSTEM of the human lung (*opposite page*) is revealed by X-ray photography following the injection of a radio-opaque fluid. The finest visible branches actually subdivide further into capillaries from five to 10 microns in diameter that surround the lung's 300 million air sacs.

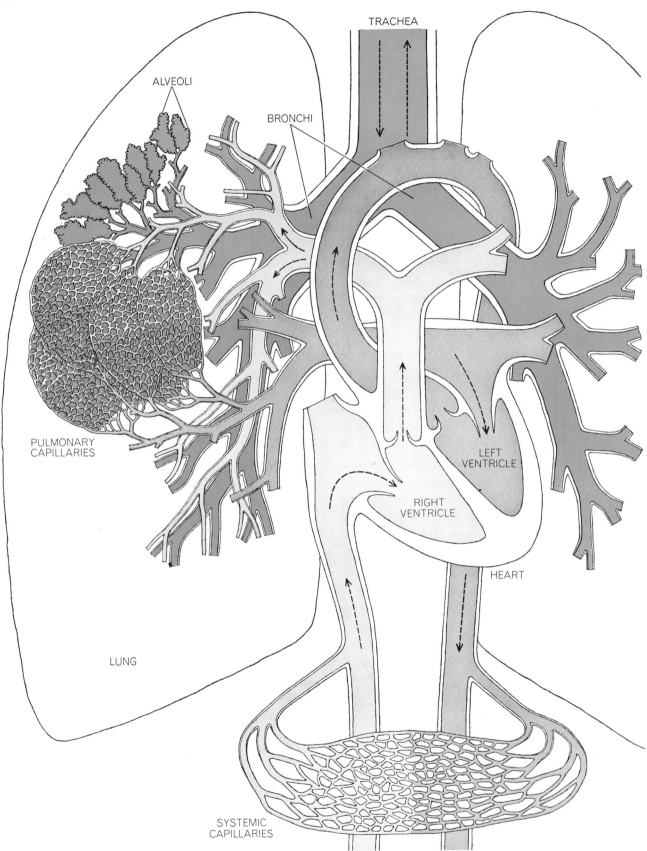

TRACHEA

ALVEOLI

BRONCHI

PULMONARY
CAPILLARIES

LEFT
VENTRICLE

RIGHT
VENTRICLE

HEART

LUNG

SYSTEMIC
CAPILLARIES

THREE PUMPS operate the lungs' gas-exchange system. One gas pump, the thorax (*see illustration on opposite page*), moves from five to seven liters of air (*color*) in and out of the lungs' air sacs every minute. At top left these alveoli are shown without their covering of pulmonary capillaries, which are shown at center left. One fluid pump, the heart's right ventricle, forces from 70 to 100 cubic centimeters of blood into the pulmonary capillaries at each contraction. This blood (*light gray*) is low in oxygen and high in carbon dioxide; oxygen, abundant in the air, is diffused into the blood while carbon dioxide is diffused from the blood into the air. Oxygenated and low in carbon dioxide, the blood (*dark gray*) then reaches the third pump, the left ventricle, which sends it on to the systemic capillaries (*example at bottom of illustration*) that deliver oxygen to and collect carbon dioxide from all the body's cells.

supplies a variety of nutrients to the tissues and removes the products of tissue metabolism, including heat. Complex regulatory mechanisms ensure enough air flow and blood flow to meet both the body's overall needs and the special needs of any particular part of the body according to its activity.

The Respiratory System

To the physiologist respiration usually means the movement of the thorax and the flow of air in and out of the lungs; to the biochemist respiration is the process within the cells that utilizes oxygen and produces carbon dioxide. Some call the first process external respiration and the second internal respiration or tissue respiration. Here I shall mainly discuss those processes that occur in the lung and that involve exchanges either between the outside air and the gas in the alveoli or between the alveolar gas and the blood in the pulmonary capillaries.

The structure of the respiratory system is sometimes shown in an oversimplified way that emphasizes only the conducting air path and the alveoli. The system is far more complex. It originates with the two tubes of the nose (the mouth can be regarded as a third tube), which join to become one: the trachea. The trachea then subdivides into two main branches, the right bronchus and the left. Each of the bronchi divides into two, each of them into two more and so on; there are from 20 to 22 bronchial subdivisions. These subdivisions give rise to more than a million tubes that end in numerous alveoli, where the gas exchange occurs. There are some 300 million alveoli in a pair of human lungs; they vary in diameter from 75 to 300 microns (thousandths of a millimeter). Before birth they are filled with fluid but thereafter the alveoli of normal lungs always contain gas. Even at the end of a complete exhalation the lungs of a healthy adult contain somewhat more than a liter of gas; this quantity is known as the residual volume. At the end of a normal exhalation the lungs contain more than two liters; this is called the functional residual capacity. When the lungs are expanded to the maximum, a state that is termed the total capacity, they contain from six to seven liters.

More important than total capacity, functional residual capacity or residual volume is the amount of air that reaches the alveoli. An adult human at rest inhales and exhales about half a liter of gas with each breath. Ideally each

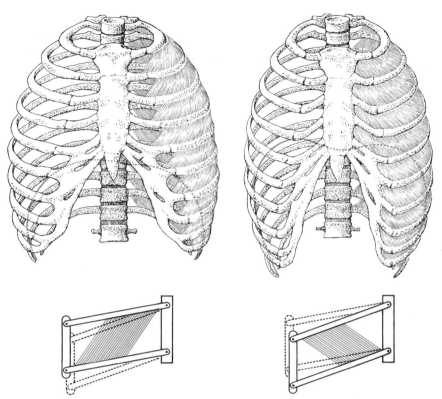

INHALATION takes place when the thorax expands, thus forcing the lungs to enlarge and bringing the pressure of the gas within them below atmospheric pressure. The expansion (*top left*) is principally the work of the diaphragm but also involves the muscles of the rib cage (*detail at bottom left*). Exhalation occurs when passive relaxation reduces the size of the thorax (*top right*), thereby raising the gas in the lungs to greater than atmospheric pressure. Only under conditions of stress do muscles (*detail at bottom right*) aid exhalation.

breath supplies his two-liter reservoir of gas—the functional residual capacity—with a volume of oxygen equal to the volume absorbed from the reservoir by the blood flowing through the pulmonary capillaries. At the same time each breath removes from the reservoir a volume of carbon dioxide equal to the volume produced by the body's cells and yielded to the lungs by the venous blood. The resting adult breathes from 10 to 14 times a minute, with the result that his ventilation—the volume of air entering and leaving the lungs—is from five to seven liters per minute. The maximum human capacity for breathing is about 30 times the resting ventilation rate, a flow of from 150 to 200 liters per minute. Even during the most intense muscular exercise, however, ventilation averages only between 80 and 120 liters per minute; man obviously has a great reserve of ventilation.

The volume of ventilation that is important in terms of gas exchange is less than the total amount of fresh air that enters the nose and mouth. Only the fresh air that reaches the alveoli and mixes with the gas already there is useful. Unlike the blood in the circulatory system, the air does not travel only in

one direction. There are no valves in the bronchi or their subdivisions; incoming and outgoing air moves back and forth through the same system of tubes. Little or no gas is exchanged in these tubes. They represent dead space, and the air that is in them at the end of an inhalation is wasted because it is washed out again during an exhalation. Thus the useful ventilation obtained from any one breath consists of the total inhalation—half a liter, or 500 cubic centimeters—minus that part of the inhalation which is wasted in the dead space. In an adult male the wasted volume is about 150 cubic centimeters.

From an engineering standpoint this dead space in the air pump represents a disadvantage: more ventilation, which necessitates more work by the pump, is required to overcome the 30 percent inefficiency. The dead space may nonetheless represent a net advantage. The use of a single system of ducts for incoming and outgoing air eliminates the need for a separate set of ducts to carry each flow. Such a dual system would certainly encroach on the lung area available for gas exchange and might well result in more than a 30 percent inefficiency.

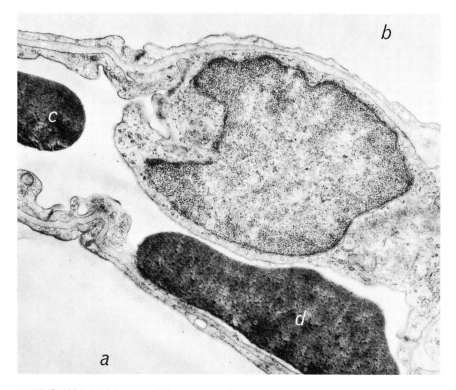

DIFFUSION PATH from gas-filled air sac to blood-filled capillary can be extremely short. In this electron micrograph of mouse lung tissue, *a* and *b* are air spaces; between these lies a capillary in which there are two red blood cells, *c* and *d*. The large light gray mass is the nucleus of an endothelial cell. The distances from air space *a* to red cell *d* and from air space *b* to red cell *c* are less than half a micron; the diffusion of gases across these gaps is swift.

Ventilation of the alveoli is not enough to ensure an adequate supply of oxygen. The incoming air must also be distributed uniformly so that each alveolus receives its share. Some alveoli are very close to the main branchings of the bronchi; others at the top and bottom of the lung are 20 or 30 centimeters away from such a branch. It is a remarkable feat of engineering to distribute the proper amount of fresh air almost simultaneously to 300 million alveoli of varying sizes through a network of a million tubes of varying lengths and diameters.

By the same token, ideal gas exchange requires that the blood be evenly distributed to all the pulmonary capillaries—that none of it should escape oxygenation by flowing through shunts in the capillary bed. Under abnormal circumstances, however, some alveoli are poorly ventilated, and venous blood flowing through the adjacent capillaries is not properly oxygenated. Conversely, there are situations when the flow of blood through certain capillaries is inadequate or nonexistent; in these cases good ventilation of the neighboring alveoli is wasted.

In spite of such deficiencies, the gas exchange can remain effective if the defects are matched (if the same re-gions that have poor ventilation, for example, also have poor blood flow), so that regions with increased ventilation also have increased blood flow. Two mechanisms help to achieve this kind of matching. First, a decrease in the ventilation of a group of alveoli results in a constriction of the blood vessels and a decrease in the blood flow of the affected region. This decrease is not the result of a nerve stimulus or a reflex but is a local mechanism, probably initiated by oxygen deficiency. Second, a local deficiency in blood flow results in a constriction of the pathways that conduct air to the affected alveoli. The resulting increase in airway resistance serves to direct more of the air to alveoli with a normal, or better than normal, blood supply.

The Pulmonary Circulation

The system that distributes blood to the lungs is just as remarkable as the system that distributes air. As I have noted, the right ventricle of the heart receives all the venous blood from every part of the body; the contraction of the heart propels the blood into one large tube, the pulmonary trunk. Like the trachea, this tube divides and subdivides, ultimately forming hundreds of millions of short, narrow, thin-walled capillaries. Each capillary has a diameter of from five to 10 microns, which is just wide enough to enable red blood cells to pass one at a time. The wall of the capillary is less than .1 micron thick; the capillary's length ranges from .1 to .5 millimeter.

If the pulmonary capillaries were laid end to end, their total length would be hundreds of miles, but the overall capillary network offers surprisingly little resistance to the flow of blood. In order to pump from five to 10 liters of blood per minute through the pulmonary system the right ventricle needs to provide a driving pressure of less than 10 millimeters of mercury—a tenth of the pressure required of the left ventricle for systemic circulation. With only a small increase in driving pressure the pulmonary blood flow can be increased to 30 liters per minute.

Although the total surface area provided by the pulmonary capillaries is enormous, at any instant the capillary vessels contain only from 70 to 100 cubic centimeters of blood. This volume is almost identical with the volume of blood the right ventricle ejects with each contraction. Thus with each heartbeat the reoxygenated blood in the pulmonary capillaries is pushed on toward the left ventricle and venous blood refills the capillaries. In the human body at rest each red blood cell remains in a pulmonary capillary for about three-quarters of a second; during vigorous exercise the length of its stay is reduced to about a third of a second. Even this brief interval is sufficient, under normal conditions, for gas exchange.

Electron microscopy reveals that the membrane in the lung that separates the gas-filled alveolus from the blood-filled capillary consists of three distinct layers. One is the alveolar epithelium, the second is the basement membrane and the third is the capillary endothelium. The cellular structure of these layers renders between a quarter and a third of the membrane's 70 square meters of surface too thick to be ideal for rapid gas exchange; over the rest of the area, however, the barrier through which the gas molecules must diffuse is very thin—as thin as .2 micron. Comparison of the gas-transfer rate in the body at rest with the rate during vigorous exercise provides a measure of the gas-exchange system's capacity. In the body at rest only 200 to 250 cubic centimeters of oxygen diffuse per minute; in the exercising body the system can deliver as much as 5,500 cubic centimeters per minute.

The combination of a large diffusion area and a short diffusion path is responsible for much of the lung's efficiency in gas exchange, but an even more critical factor is the remarkable ability of the red blood pigment hemoglobin to combine with oxygen. If plasma that contained no red cells or hemoglobin were substituted for normal blood, the adult human heart would have to pump 83 liters per minute through the pulmonary capillaries to meet the oxygen needs of a man at rest. (Even this assumes that 100 percent of the oxygen in the plasma is delivered to the tissues, which is never the case.) In contrast, blood with a normal amount of hemoglobin in the red cells picks up 65 times more oxygen than plasma alone does; the heart of a man at rest need pump only about five and a half liters of blood per minute, even though the tissues normally extract only from 20 to 25 percent of the oxygen carried to the cells by the red blood corpuscles.

The Mechanics of Breathing

Just as water inevitably runs downhill, so gases flow from regions of higher pressure to those of lower pressure. In the case of the lung, when the total gas pressure in the alveoli is equal to the pressure of the surrounding atmosphere, no movement of gas is possible. For inhalation to occur, the alveolar gas pressure must be less than the atmospheric pressure; for exhalation, the opposite must be the case. There are two ways in which the pressure difference required for the movement of air into the lungs can be created: either the pressure in the alveoli can be lowered below atmospheric pressure or the pressure at the nose and mouth can be raised above atmospheric pressure. In normal breathing man follows the former course; enlarging the thorax (and thus enlarging the lungs as well) enables the alveolar gas to expand until its pressure drops below that of the surrounding atmosphere. Inhalation follows automatically.

The principal muscle for enlarging the thorax is the diaphragm, the large dome-shaped sheet of tissue that is anchored around the circumference of the lower thorax and separates the thoracic cavity from the abdominal cavity. When the muscle of the diaphragm contracts, the mobile central portion of the sheet moves downward, much as a piston moves in a cylinder. In addition, skeletal muscles enlarge the bony thoracic cage by increasing its circumference [see *illustration on page 151*]. The lungs, of course, lie entirely within the thorax. They have no skeletal muscles and cannot increase their volume by their own efforts, but their covering (the visceral pleura) is closely linked to the entire inner lining of the thorax (the parietal pleura). Only a thin layer of fluid separates the two pleural surfaces; when the thorax expands, the lungs must follow suit. As the pressure of the alveolar gas drops below that of the atmosphere, the outside air flows in through the nose, mouth and tracheobronchial air paths until the pressure is equalized.

This kind of pulmonary ventilation requires work; the active contraction of the thoracic muscles provides the force necessary to overcome a series of opposing loads. These loads include the recoil of the elastic tissues of the thorax, the recoil of the elastic tissue of the lungs, the frictional resistance to the flow of air through the hundreds of thousands of ducts of the tracheobronchial tree, and the surface forces created at the fluid-gas interfaces in the alveoli [see "Surface Tension in the Lungs," by John A. Clements, beginning on page 169].

In contrast to inhalation, exhalation is usually a passive process. During the active contraction of muscles that causes the enlargement of the thorax, the tissues of thorax and lungs are stretched and potential energy is stored in them. The recoil of the stretched tissue and the release of the stored energy produce the exhalation. Only at very high rates of ventilation or when there is an obstruction of the tracheobronchial tree is there active contraction of muscles to assist exhalation.

Artificial ventilation can be produced either by raising external gas pressure or by lowering internal pressure. Body respirators of the "iron lung" type lower the pressure of the air surrounding the thorax in part of their cycle of operation. As a consequence the volume of the thorax increases, the alveolar pressure falls and, since the patient's nose and mouth are outside the apparatus, air at atmospheric pressure flows into his lungs. Later in the cycle the pressure within the respirator rises; the volume of the thorax decreases and the patient exhales.

Other types of artificial ventilation depend on raising the external pressure at the nose and mouth above the atmospheric level. Some mechanical respirators operate by supplying high-pressure

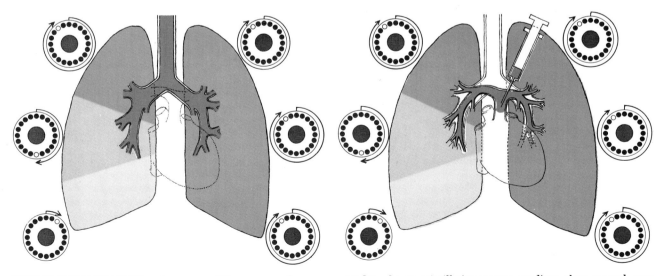

RADIOACTIVE TRACERS such as xenon 133 can assess the performance of the two components in the lungs' gas-exchange system. Inhaled as a gas (*left*), the xenon will be unevenly distributed if the air ducts are blocked; zones that are poorly ventilated will produce lower scintillation-counter readings than normal ones will. Injected in a solution into the bloodstream (*right*), the xenon will diffuse unevenly into the air sacs if the blood vessels are blocked. Such faulty blood circulation also causes low readings.

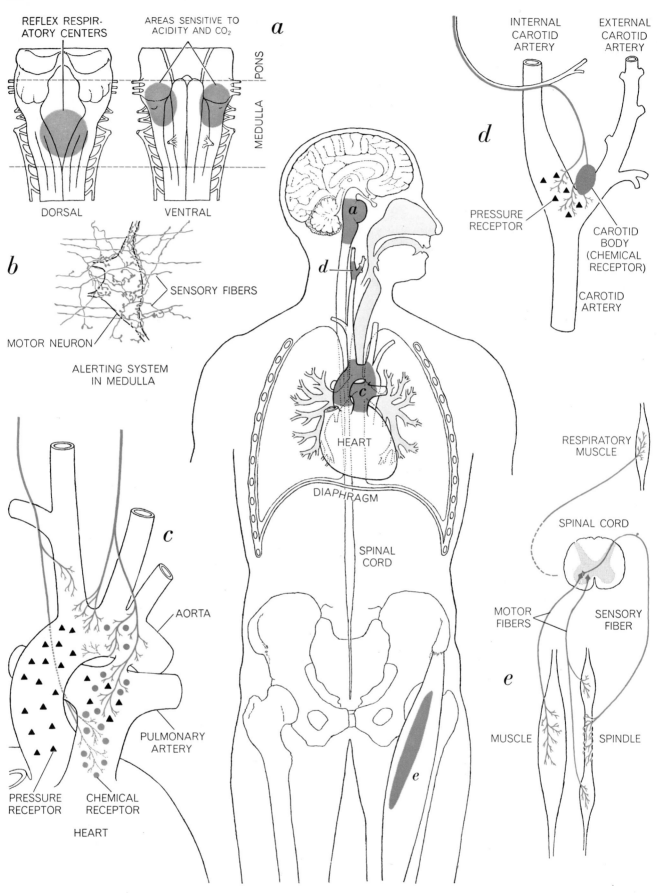

REFLEX RESPIR-
ATORY CENTERS

AREAS SENSITIVE TO
ACIDITY AND CO_2

a

DORSAL

VENTRAL

PONS

MEDULLA

INTERNAL
CAROTID
ARTERY

EXTERNAL
CAROTID
ARTERY

d

PRESSURE
RECEPTOR

CAROTID
BODY
(CHEMICAL
RECEPTOR)

CAROTID
ARTERY

b

SENSORY FIBERS

MOTOR NEURON

ALERTING SYSTEM
IN MEDULLA

d

HEART

c

DIAPHRAGM

SPINAL
CORD

c

AORTA

PULMONARY
ARTERY

PRESSURE
RECEPTOR

CHEMICAL
RECEPTOR

HEART

RESPIRATORY
MUSCLE

SPINAL CORD

MOTOR
FIBERS

SENSORY
FIBER

e

MUSCLE

SPINDLE

REGULATION OF BREATHING is controlled not only by the lower brain but also by a variety of other receptor and reflex centers. Portions of the pons and medulla of the lower brain (*a*) react to increases in acidity and carbon dioxide pressure. Other sensitive cells are attached to the aorta and pulmonary artery (*c*) and the carotid artery (*d*). The spindle receptors of skeletal muscles (*e*) also affect breathing, as do such external factors as temperature and vibration and such psychological ones as anxiety and wakefulness.

air at rhythmic intervals to the nose or mouth or more directly to the trachea. In mouth-to-mouth resuscitation the person who administers the air conveys it to the person who receives it by contracting his thorax and pushing air into the other person's lungs. Frogs can push air into their lungs by first filling their cheeks and then contracting their cheek muscles; some people whose respiratory muscles are paralyzed but whose head and neck muscles are not can learn to do this "frog breathing."

Still another way to increase the amount of oxygen available to human tissues is to place a hospital patient in an oxygen tent, in which the concentration of oxygen is increased, or to supply oxygen to him in a hyperbaric chamber, in which the total gas pressure is increased to two or three atmospheres. The extra oxygen that is taken up by the blood under these circumstances may be of help if the patient's clinical problem requires for its correction only more oxygen in the blood and tissues; such is the case, for example, when a patient has an infection caused by anaerobic bacteria. But supplying oxygen in greater than normal amounts does not increase ventilation and therefore cannot eliminate more carbon dioxide, nor does it increase the amount of blood being circulated. In the first instance, help in eliminating carbon dioxide is of prime importance in cases of pulmonary or respiratory disease; in the second, the tissues of patients with circulatory problems need not only more oxygen but also the added glucose, amino acids, lipids, white blood cells, blood platelets, proteins and hormones that can only be obtained from adequate blood flow.

The Measurement of Lung Function

Before the 1950's only a few tests of specific lung function had been devised. These included measurements of the amount of air in the lungs at the end of a full inhalation, the amount in the lungs at the end of a full exhalation, and the maximum amount exhaled after a full inhalation; these three volumes were known respectively as the total lung capacity, the residual volume and the vital capacity. In addition, some measurements had been made of the way in which gas was distributed throughout the tracheobronchial tree during inhalation by means of observing the dilution of a tracer gas such as helium or the exhalation of nitrogen following the inhalation of pure oxygen. Since the 1950's, however, pulmonary physiologists have developed a number of new instruments and test techniques with which to measure objectively, rapidly and accurately not only lung volume and the distribution and diffusion of gases but also the pulmonary circulation and the physical properties of the lung and its connecting air paths.

One such new instrument is the plethysmograph (also known as the body box), which can measure the volume of gas in the lungs and thorax, the resistance to air flow in the bronchial tree and even the blood flow in the pulmonary capillaries. For the first measurement the subject sits in the box (which is airtight and about the size of a telephone booth) and breathes the supply of air in the box through a mouthpiece fitted with a shutter and a pressure gauge. To measure the volume of gas in the subject's lungs at any moment an observer outside the box triggers a circuit that closes the shutter in the mouthpiece and then records the air pressure both in the subject's lungs and in the body box as the subject attempts to inhale; Boyle's law yields a precise measurement of the volume of gas in the subject's lungs.

In order to measure the blood flow in the pulmonary capillaries the subject in the body box is provided with a bag that contains a mixture of 80 percent nitrous oxide and 20 percent oxygen. At a signal the subject inhales a single breath of this mixture and holds the breath for a few seconds. Nitrous oxide dissolves readily in the blood; as its molecules diffuse from the alveoli to enter the blood flowing through the pulmonary capillaries, the total number of gas molecules in the alveoli obviously decreases. But the nitrous oxide that dissolves in the blood does not increase the volume of the blood; therefore the total gas pressure must decrease as the nitrous oxide molecules are subtracted. Knowing the total volume of gas in the lungs, the volume of nitrous oxide and the solubility of nitrous oxide in the blood, one can calculate the flow of blood through the pulmonary capillaries instant by instant. These calculations can be used both to measure the amount of blood pumped by the heart—and thus to arrive at an index of cardiac performance—and to detect unusual resistance to blood flow through the pulmonary capillaries.

To measure resistance to air flow in the bronchial tree the subject in the body box is instructed to breathe the air about him without any interruption while the observer continuously records changes in pressure in the box. One would expect that in a closed system the mere movement of 500 cubic centimeters of gas from the supply in the box into the lungs of a subject also in the box would not bring about any overall change in pressure. In actuality the pressure in the box increases. The reason is that gas can flow only from a region of higher pressure to a region of lower pressure. The subject cannot inhale unless the pressure of the gas in his lungs is lower than the pressure of the gas in the box; therefore the molecules of gas in his lungs during inhalation must occupy a greater volume than they did in the box before inhalation. The expansion of the subject's thorax, in turn, compresses—and thus is the equivalent of adding to—the rest of the gas in the body box; the effect is reversed as the thorax contracts during exhalation. An appropriately calibrated record of these pressure changes makes it possible for the pressure of the gas in the subject's alveoli to be calculated at any moment in the respiratory cycle. This test can be used to detect the increased resistance to air flow that arises in patients with bronchial asthma, for example, and to evaluate the effectiveness of antiasthmatic drugs.

Another useful instrument for the study of lung function is the nitrogen meter, developed during World War II by John C. Lilly at the University of Pennsylvania in order to detect leaks in or around aviators' oxygen masks. The instrument operates as an emission spectroscope, continuously sampling, analyzing and recording the concentration of nitrogen in a mixture of gases; its lag is less than a tenth of a second. Pulmonary physiologists use the nitrogen meter to detect uneven distribution of air within the lung. Assume that when one breathes ordinary air, the lungs contain 2,000 cubic centimeters of gas, 80 percent of which is nitrogen. If one next inhales 2,000 cubic centimeters of pure oxygen, and if this oxygen is distributed uniformly to the millions of alveoli, each alveolus should now contain a gas that is only 40 percent nitrogen instead of the former 80 percent. If, however, the 2,000 cubic centimeters of oxygen are not evenly distributed, some alveoli will receive less than their share, others will receive more, and the composition of the alveolar gas at the end of the oxygen inhalation will be decidedly nonuniform. In the alveoli that receive the most oxygen the proportion of nitrogen may be reduced to 30 percent; in those that receive little oxygen the proportion of nitrogen may remain as high as 75 percent.

It is impossible to put sampling

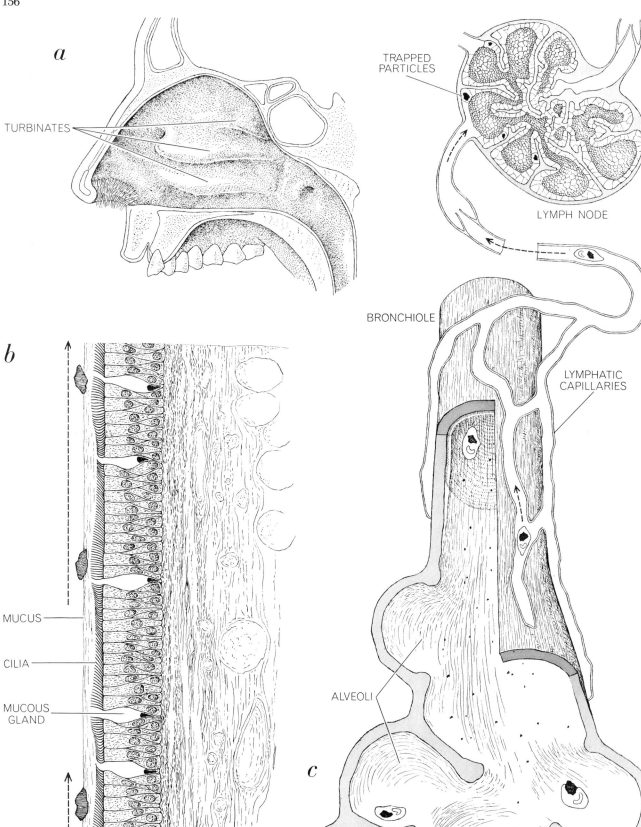

a

TURBINATES

TRAPPED PARTICLES

LYMPH NODE

b

BRONCHIOLE

LYMPHATIC CAPILLARIES

MUCUS

CILIA

MUCOUS GLAND

ALVEOLI

c

BAFFLE SYSTEM protects the lung's interior from intrusion by particles of foreign matter. Hairs in the nose and the convolutions of the turbinate bones (*a*) entrap most particles larger than 10 microns in diameter. Particles of from two to 10 microns in diameter usually settle on the walls of the trachea, the bronchi or the bronchioles, where the escalator action of mucus-covered cilia (*b*) carries them up to the pharynx for expulsion. Particles smaller than two microns in diameter reach the lung's air sacs (*c*). Some are engulfed by scavenger cells; others are carried to the nearest lymph node. Any that remain may cause fibrous tissue to form.

needles into thousands of alveoli in order to determine what mixture of gas each of them contains, but the nitrogen meter easily samples and analyzes the gas leaving the alveoli as the subject exhales. The first part of the exhaled gas comes from the outermost part of the tracheobronchial tree; it contains pure oxygen that has not traveled far enough down the conducting air path to mix with any alveolar gas. The second part of the exhalation shows a rapidly rising concentration of nitrogen; it represents alveolar gas that has washed some pure oxygen out of the conducting air path during exhalation and has mixed with it in the process. Once the conducting air path has been washed clear of oxygen, the remainder of the exhalation will be entirely alveolar gas. Analysis of the amount of nitrogen in the third part of the exhalation quickly shows whether or not the oxygen is distributed uniformly. If the distribution is uniform, the nitrogen-meter record for this part of the exhalation will be a horizontal line: from beginning to end the alveolar gas will be 40 percent nitrogen. If, on the other hand, the distribution is uneven, the nitrogen-meter record for the third part of the exhalation will rise continuously because the first part of the alveolar gas will come from well-ventilated areas of the lung and the last from poorly ventilated ones. In less than a minute the nitrogen meter can separate individuals with uneven ventilation from individuals whose ventilation is normal.

Having discovered by means of the nitrogen meter that a subject suffers from uneven ventilation somewhere in his lungs, the pulmonary physiologist can now use radioactive gases to determine exactly   here the unevenness lies. The subject inhales a small amount of a relatively insoluble radioactive gas such as xenon 133 and holds his breath. A battery of three radiation counters on each side of the thorax measures the amount of radioactivity in the alveolar gas contained in the upper, middle and lower portions of each lung. Well-ventilated lung areas show a high level of radioactivity; poorly ventilated areas, a low level.

Radiation-counter readings can also be used to measure the uniformity of blood flow through the pulmonary capillaries. In order to do this the radioactive xenon is dissolved in a saline solution and the solution is administered intravenously. As it flows through the pulmonary capillaries the xenon comes out of solution and enters the alveolar gas. A high local concentration of xenon is an indication of a good flow of blood in that area; a low concentration indicates the contrary.

Another use of radioactive tracers to check on blood circulation involves the deliberate clogging of some fine pulmonary blood vessels. Radioactive albumin is treated so that it forms clumps that are about 30 microns in diameter—a size somewhat larger than the pulmonary capillaries or the vessels that lead into them. When the clumps are administered intravenously, they cannot enter blood vessels that are obstructed by disease; instead they collect in the parts of the lung with good circulation, where they block some of the fine vessels for a few hours. The whole thorax can now be scanned for radioactivity; in 10 or 20 minutes the activity of the albumin produces a clear image in which the regions of the lung with good pulmonary blood flow are clearly delineated.

Still another test, which measures the rate of gas exchange across the alveolar membrane, is made possible by the fact that hemoglobin has an extraordinary capacity for combining with carbon monoxide. The subject inhales a very low concentration of this potentially toxic gas. The carbon monoxide molecules diffuse across the capillary membranes and combine with the hemoglobin in the red blood cells. Assuming a normal amount of blood in the pulmonary capillaries, the rate at which the carbon monoxide disappears from the alveolar gas is directly proportional to its rate of diffusion. Unlike the somewhat similar test involving nitrous oxide, the carbon monoxide test measures only the rate of gas diffusion, not the rate of capillary blood flow. The affinity between the gas and the hemoglobin is so great that, even if the circulation of the blood were briefly halted, the stagnant red blood cells could still absorb carbon monoxide. A slow rate of carbon monoxide diffusion therefore indicates that the alveolar membranes have become thickened or that some abnormal fluid or tissue is separating many of the alveoli from the pulmonary capillaries.

The Regulation of Ventilation

The blood and air pumps that feed the lung's gas-exchange apparatus must be able to vary their performance to suit environments that range from sea level to high altitudes and activities that range from complete rest to violent exercise. Whatever the circumstances, exactly the right amount of oxygenated blood must be provided to meet the body's needs; to achieve this result responsive decision centers in the body, controlling both respiration and circulation, must not only be supplied with the necessary information but also possess the capacity to enforce decisions.

The first of these respiratory control centers was discovered by César Legallois of France in 1811. He found that if the cerebrum, the cerebellum and part of the upper brainstem were removed from a rabbit, the animal's breathing remained rhythmic, but that if a small region of the lower medulla was damaged or removed, breathing ceased. In the century and a half since Legallois' time physiologists have continued to accord this region of the brain —a group of interconnected nerve cells in the lower medulla—the paramount role in the control of respiration. This is not, however, the only region of the brain concerned with the regulation of breathing; there are chemically sensitive regions near the lateral surfaces of the upper medulla that call for an increase in ventilation when their carbon dioxide pressure or their acidity increases. Some other parts of the medulla, the cerebral cortex and the part of the brain called the pons can also influence respiration.

In addition to these areas in the brain a variety of respiratory receptors, interconnecting links, pathways and reflexes are found elsewhere in the body. Chemically sensitive cells in the regions of the carotid artery and the aorta initiate reflexes that increase respiration when their oxygen supply is not sufficient to maintain their metabolic needs or when the local carbon dioxide pressure or acidity increases. Stretch-sensitive receptors in the major arteries act through reflexes to increase or decrease respiration in response to low or high arterial blood pressure. Other receptors in the circulatory system, sensitive both to chemical stimuli and to mechanical deformation, can set off reflexes that slow or stop breathing. Respiration can also be regulated by the degree of inflation or deflation of the lungs, by the individual's state of wakefulness or awareness, by the concentration of certain hormones in the blood and by the discharge of the special sensory receptors known as spindles in skeletal muscles (including the respiratory muscles themselves).

We still do not know how some or all of these central and peripheral components interact to achieve the most important (and the most frequent) change in ventilation: the change that takes place when the body's metabolic activity increases. We know that during exercise both the body's oxygen consumption and its carbon dioxide production increase, and that so does the rate of ventilation. It is therefore logical to assume that ventilation is regulated by receptors somewhere in the body that are sensitive to oxygen or to carbon dioxide or to both. A puzzling fact remains: mild and even moderate exercise simply does not decrease the amount of oxygen or increase the amount of carbon dioxide in the arterial blood, yet the respiration rate rises. What causes this increase, which is enough to satisfy both the ordinary needs of the body and the extraordinary needs of exercising muscles? No one knows.

The Upper Respiratory Tract

In the course of taking from four to 10 million breaths a year each individual draws into the alveoli of his lungs air that may be hot or cold, dry or moist, possibly clean but more probably dirty. Each liter of urban air, for example, contains several million particles of foreign matter; in a day a city-dwelling adult inhales perhaps 20 billion such particles. What protects the lungs and the air ducts leading to them from air with undesirable physical characteristics or chemical composition? Sensory receptors in the air ducts and the lungs can initiate protective reflexes when they are suitably stimulated; specialized cells can also engulf foreign particles that have penetrated far into the lung. The main task of protecting the lungs, however, is left to the upper respiratory tract.

The nose, the mouth, the oropharynx, the nasopharynx, the larynx, the trachea and those bronchi that are outside the lung itself together constitute the upper respiratory tract. Although the obvious function of this series of passages is to conduct air to and from the lung, the tract is also a sophisticated air conditioner and filter. It contains built-in warning devices to signal the presence of most pollutants and is carpeted with a remarkable escalator membrane that moves foreign bodies upward and out of the tract at the rate of nearly an inch a minute. Within quite broad limits the initial state of the air a man breathes is of little consequence; thanks to the mediation of the upper respiratory tract the air will be warm, moist and almost free of particles by the time it reaches the alveoli.

The first role in the conditioning process is played by the mucous membrane of the nose, the mouth and the pharynx; this large surface has a rich blood supply that warms cold air, cools hot air and otherwise protects the alveoli under a wide range of conditions. Experimental animals have been exposed to air heated to 500 degrees centigrade and air cooled to −100 degrees C.; in both instances the trip through the respiratory tract had cooled or warmed the air almost to body temperature by the time it had reached the lower trachea.

The upper respiratory tract also filters air. The hairs in the nose block the passage of large particles; beyond these hairs the involuted contours of the nasal turbinate bones force the air to move in numerous narrow streams, so that suspended particles tend to approach either the dividing septum of the nose or the moist mucous membranes of the turbinates. Here many particles either impinge directly on the mucous membranes or settle there in response to gravity.

The filter system of the nose almost completely removes from the air particles with a diameter larger than 10 microns. Particles ranging in diameter from two to 10 microns usually settle on the walls of the trachea, the bronchi and the bronchioles. Only particles between .3 micron and two microns in diameter are likely to reach the alveolar ducts and the alveoli. Particles smaller than .3 micron, if they are not taken up by the blood, are likely to remain in suspension as aerosols and so are washed out of the lungs along with the exhaled air.

Foreign bodies that settle on the walls of the nose, the pharynx, the trachea, the bronchi and the bronchioles may be expelled by the explosive blast of air that is generated by a sneeze or a cough, but more often they are removed by the action of the cilia. These are very primitive structures that are found in many forms of life, from one-celled organisms to man. Resembling hairs, they are powered by a contractile mechanism; in action each cilium makes a fast, forceful forward stroke that is followed by a slower, less forceful return stroke that brings the cilium into starting position again. The strokes of a row of cilia are precisely coordinated so that the hairs move together as a wave. The cilia of the human respiratory tract do not beat in the open air; they operate within a protective sheet of the mucus that is secreted by glands in the trachea and the bronchi. The effect of their wavelike motion is to move the entire mucus sheet—and anything trapped on it—up the respiratory tract to the pharynx, where it can be swallowed or spat out.

The ciliary escalator is in constant operation; it provides a quiet, unobtrusive, round-the-clock mechanism for the removal of foreign matter from the upper respiratory tract. The speed of this upward movement depends on the length of the cilia and the frequency of their motion. Calculations show that a cilium that is 10 microns long and that beats 20 times per second can move the mucus sheet 320 microns per second, or 19.2 millimeters per minute. Speeds of 16 millimeters per minute have actually been observed in experiments.

In spite of all such preventive measures some inhaled particles—particularly those suspended in fluid droplets—manage to pass through the alveolar ducts and reach the alveoli. How do these deeper surfaces, which have no cilia or mucous glands, cleanse themselves? The amoeba-like lymphocytes of the bloodstream and their larger relatives the macrophages engulf and digest some particles of foreign matter. They can also surround the particles in the air ducts and then ride the mucus escalator up to the nasopharynx. Other particles may pass into lymphatic vessels and come to rest in the nearest lymph nodes. Some remain permanently attached to the lung tissue, as the darkened lungs of coal miners demonstrate. Many such intrusions are essentially harmless, but some, for example particles of silica, can result in the formation of tough fibrous tissue that causes serious pulmonary disease.

The filtration mechanism of the upper respiratory tract can thus be credited with several important achievements. It is responsible for the interception and removal of foreign particles. It can remove bacteria suspended in the air and also dispose of bacteria, viruses and even irritant or carcinogenic gases when they are adsorbed onto larger particles. Unless the filter system is overloaded, it keeps the alveoli practically sterile. This, however, is not the only protection the lungs possess. Among the reflex responses to chemical or mechanical irritation of the nose are cessation of breathing, closure of the larynx, constriction of the bronchi and even slowing of the heart. These responses are aimed at preventing potentially harmful gases from reaching the alveoli and, through the alveoli, the pulmonary circulation.

In many animals, for instance, the act of swallowing results in reflex closure of the glottis and the inhibition of respiration. Because the pharynx is a passageway both for air and for food and water, this reflex prevents food or water from entering the respiratory passages during the journey from the mouth to the esophagus. Because the reflex does not operate during unconsciousness it is dangerous to try to arouse an unconscious person by pouring liquids such as alcohol into his mouth.

When specific chemical irritants penetrate beyond the larynx, the reflex response is usually a cough combined with bronchial constriction. Like the swallowing reflex, the cough reflex is depressed or absent during unconsciousness. It also is less active in older people; this is why they are more likely to draw foreign bodies into their lungs.

Bronchial constriction is a response to irritation of the air paths that is less obvious than a cough. When the concentration of dust, smoke or irritant gas is too low to elicit the cough reflex, this constrictive increase in air-path resistance is frequently evident. Smoking a cigarette, for example, induces an immediate twofold or threefold rise in air-path resistance that continues for 10 to 30 minutes. The inhalation of cigarette smoke produces the same effect in smokers and nonsmokers alike. It does not cause shortness of breath, as asthma does; the air-path resistance must increase fourfold or fivefold to produce that effect. Nor has the reflex anything to do with nicotine; no increase in air-path resistance is caused by the inhalation of nicotine aerosols, whereas exactly the same degree of resistance is induced by smoking cigarettes with a normal (2 percent) or a minimal (.5 percent) nicotine content. The reflex is evidently triggered by the settling of particles less than a micron in diameter on the sensory receptors in the air path.

Other air pollutants—irritant gases, vapors, fumes, smokes, aerosols or small particles—may give rise to a similar bronchial constriction. It is one of the ironies of man's urban way of life that exposure to the pollutants that produce severe and repeated bronchial constriction results in excessive secretion of mucus, a reduction in ciliary activity, obstruction of the fine air paths and finally cell damage. These circumstances enable bacteria to penetrate to the alveoli and remain there long enough to initiate infectious lung disease. They are also probably a factor in the development of such tracheobronchial diseases as chronic bronchitis and lung cancer. Thus man's advances in material culture increasingly threaten the air pump that helped to make his evolutionary success possible.

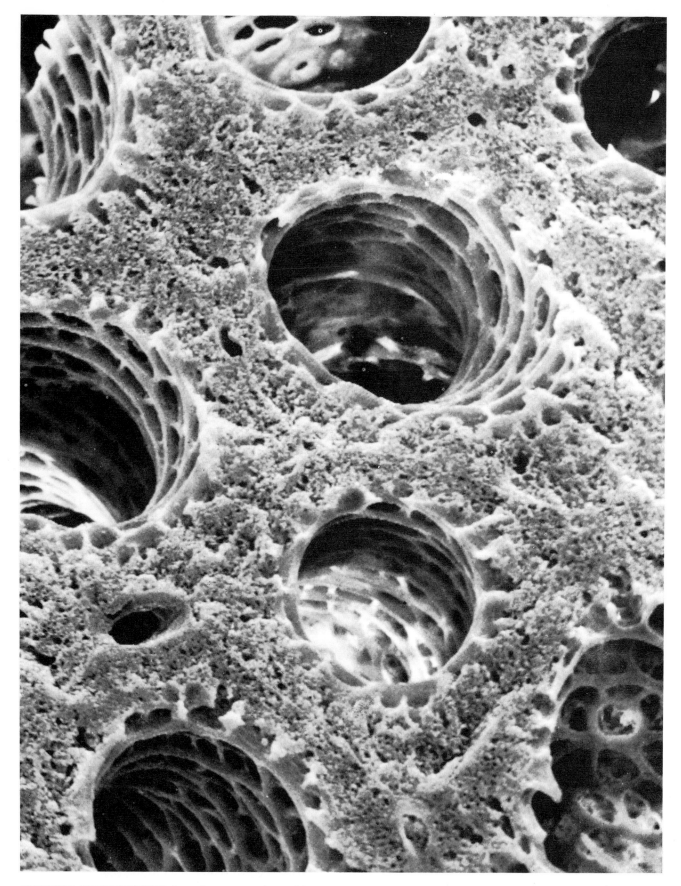

INTERIOR OF BIRD'S LUNG shows the structure that enables air to flow through the lung instead of in and out of it, as in the mammalian lung. This scanning electron micrograph was made by H. R. Duncker of Justus Liebig University at Giessen in West Germany. The circular structures are parabronchi, which are fine branches of the bronchial system, and the surrounding material is lung tissue. Equivalent structures in the mammalian lung are the saclike alveoli. The parabronchi in this micrograph, which are enlarged 180 diameters, are in the lung of a domestic fowl that was 14 days old. The micrograph shows the parabronchi in transverse section.

15

HOW BIRDS BREATHE

KNUT SCHMIDT-NIELSEN

December 1971

The avian respiratory system is different from the mammalian one. The lungs do not simply take air in and then expel it; the air also flows through a series of large sacs and even hollow bones

A bird in flight expends more energy, weight for weight, than a mammal walking or running on the ground. Moreover, the bird's respiratory system can deliver enough oxygen for the animal to fly at altitudes where a mammal can barely function. How do the birds do it? It turns out that the avian respiratory system is quite different from the mammalian one. The remarkable anatomical details of the avian system have been elucidated over a period of three centuries, but precisely how the system operates has been worked out only recently.

One of the first clues to the distinctive nature of the avian respiratory system was the discovery that a bird with a blocked windpipe can still breathe, provided that a connection has been made between one of its bones and the outside air. This phenomenon was demonstrated in 1758 by John Hunter, a fellow of the Royal Society, who wrote: "I next cut the wing through the *os humeri* [the wing bone] in another fowl, and tying up the trachea, as in the cock, found that the air passed to and from the lungs by the canal in this bone. The same experiment was made with the *os femoris* [the leg bone] of a young hawk, and was attended with a similar result."

The bones of birds contain air, not marrow. This is true not only of the larger bones but also often of the smaller ones and of the skull bones, particularly in birds that are good fliers. As Hunter's experiments showed, the air spaces are connected to the respiratory system.

Like mammals, birds have two lungs. They are connected to the outside by the trachea, much as in mammals, but in addition they are connected to several large, thin-walled air sacs that fill much of the chest and the abdominal cavity [*see top illustration on next page*]. The sacs are connected to the air spaces in

the bones. The continuation of the air passages into large, membranous air sacs was discovered in 1653 by William Harvey, the British anatomist who became famous for discovering the circulation of blood in mammals.

The presence in birds of these large air spaces, much larger in volume than the lungs, has given rise to considerable speculation. It has often been said that the air sacs make a bird lighter and are therefore an adaptation to flight. Certainly a bone filled with air weighs less than a bone filled with marrow. The large air sacs, however, do not in themselves make a bird any lighter. As a student I heard a professor of zoology assert that the sacs did make a bird lighter and therefore better suited to flight. Somewhat undiplomatically I suggested that if I were to take a poor flier such as a chicken and pump it up with a bicycle pump, the chicken would be neither any lighter nor better able to fly. The simple logic of the argument must have convinced the professor, because we did not hear any more about the function of the air sacs.

In order to understand the function of the sacs, how air flows in them, how oxygen is taken up by the blood and carbon dioxide is given off and so on it is necessary to consider the main structural features of the system. In this context it is helpful to compare birds with mammals. Birds as a group are much more alike than mammals. In size they range from the hummingbird weighing some three grams to the ostrich weighing about 100 kilograms. In terms of weight the largest bird is roughly 30,000 times bigger than the smallest one.

All birds have two legs and two wings, although the ostrich cannot use its wings for flying and penguins have flipper-like wings modified for swimming. All birds

have feathers, and all birds have a similar respiratory system, with lungs, air sacs and pneumatized bones. (Even the ostrich has the larger leg bones filled with air. Air sacs and pneumatized bones are therefore not restricted to birds that can fly.)

Mammals, on the other hand, range in size from the shrew, which weighs about as much as a hummingbird, to the 100-ton blue whale. The largest mammal is therefore 30 million times bigger than the smallest one. Mammals can be four-legged, two-legged or no-legged (whales). They can even have wings, as bats do. Most mammals have fur, but many do not.

It was once argued that birds needed a respiratory system particularly adapted for flight because of the high requirement for energy and oxygen during flight. At rest birds and mammals of similar size have similar rates of oxygen consumption, although in both birds and mammals the oxygen consumption per unit of body weight increases with decreasing size. In recent years the oxygen consumption of birds in flight has been determined in wind-tunnel experiments [see "The Energetics of Bird Flight," by Vance A. Tucker; SCIENTIFIC AMERICAN, Offprint 1141]. The results show that the oxygen consumption in flight is some 10 to 15 times higher than it is in the bird at rest. This performance is not much different from that of a well-trained human athlete, who can sustain a similar increase in oxygen consumption.

Small mammals such as rats or mice, however, seem unable to increase their oxygen consumption as much as tenfold. Since birds and mammals of the same body size show similar oxygen consumption when they are at rest, is it the special design of the bird's respiratory system that allows the high rates of oxygen consumption during flight?

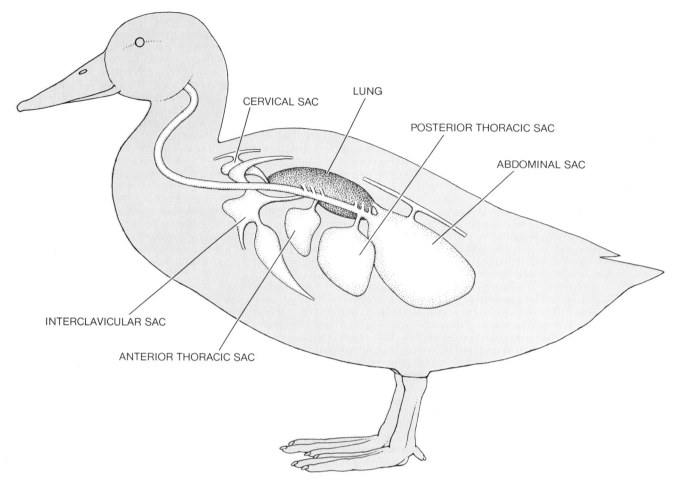

CERVICAL SAC

LUNG

POSTERIOR THORACIC SAC

ABDOMINAL SAC

INTERCLAVICULAR SAC

ANTERIOR THORACIC SAC

AVIAN RESPIRATORY SYSTEM, here represented by the system of a mallard duck, has as a distinctive feature a number of air sacs that are connected to the bronchial passages and the lungs. Most of the air inhaled on a given breath goes directly into the posterior sacs. As the respiratory cycle continues the air passes through the lungs and into the anterior sacs, from which it is exhaled to the outside in the next cycle. The mechanism provides a continuous flow of air through the lung and also, by means of the holding-chamber function of the anterior sacs, keeps the carbon dioxide content of the air at an appropriate physiological level.

The best argument against considering the unusual features of the avian respiratory system as being necessary for flight is provided by bats. They have typical mammalian lungs and do not have air sacs or pneumatized bones, and yet they are excellent fliers. Moreover, it has recently been shown by Steven Thomas and Roderick A. Suthers at Indiana University that bats in flight consume oxygen at a rate comparable to the rate in flying birds. Clearly an avian respiratory system is not necessary for a high rate of oxygen uptake or for flight.

One highly significant phenomenon is that many birds can fly at high altitudes where mammals suffer seriously from lack of oxygen. This fact points to what is perhaps the most important difference between the avian system and the mammalian one.

PARABRONCHI

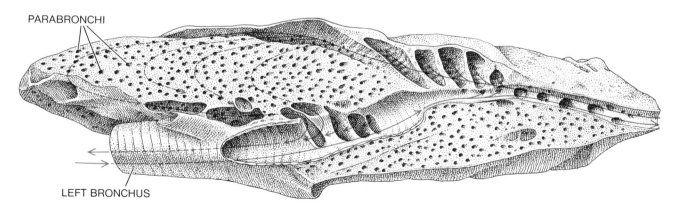

LEFT BRONCHUS

AVIAN LUNG is shown in longitudinal section from the left side of the bird. Since the bird has two lungs, one sees here half of the lung system. The orientation is as it would be with the bird's head at left. Air enters the bronchus and for the most part passes through to the posterior sacs. On its return through the lungs, assisted by a bellows-like action of the posterior sacs, it flows into the many parabronchi, where gases are exchanged with the blood. The flow has some similarity to the flow of water through a sponge.

Let us look more closely at the main features of the avian respiratory system. The trachea of a bird branches into two main bronchi, each leading to one of the lungs. So far the system is similar to the mammalian system. In birds, however, each main bronchus continues through the lung and connects with the most posterior (and usually the largest) air sacs: the abdominal air sacs. On its way through the lung the main bronchus connects to the anterior air sacs and also to the lung [see bottom illustration on opposite page]. In the posterior part the main bronchus has another set of openings that connect to the posterior air sacs as well as to the lung. The air sacs also have direct connections to the lung in addition to the connection through the bronchus.

The lung itself has a most peculiar characteristic: it allows air to pass completely through it. In contrast, the mammalian lung terminates in small saclike structures, the alveoli; air can flow only in and out of it. The bird lung is perforated by the finest branches of the bronchial system, which are called parabronchi. Air flows through the lung somewhat the way water can flow through a sponge.

This feature of the bird lung has led to the suggestion that the air sacs act as bellows helping to push air through the lung, which thus could be supplied with air more effectively than the mammalian lung is. Before accepting this hypothesis one must be sure that the air sacs do not have a lunglike function, that is, that they do not serve as places where oxygen is taken up by the blood. Since the air sacs are thin-walled, they could perhaps be important in the exchange of gases between the air and the blood.

The fact is that the sacs are poorly supplied with blood. Moreover, they have smooth walls, which do not provide the immensely enlarged surface that the finely subdivided lung has. A crucial experiment was performed some 80 years ago by the French investigator J. M. Soum, who admitted carbon monoxide into the air sacs of birds in which he had blocked the connections to the rest of the respiratory system. If the air sacs had played any major role in gas exchange, the birds would of course have been rapidly poisoned by the carbon monoxide. They remained completely unaffected. We can therefore conclude that the air sacs have no direct function in gas exchange. Since the volume of the sacs changes considerably during the respiratory cycle, one can accept the hy-

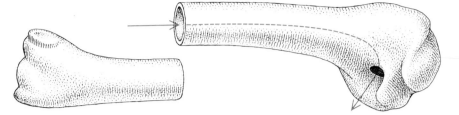

HOLLOW BONE filled with air is characteristic of bird skeleton. Such a structure makes the bird lighter than a bird would be with mammalian bones and so is an aid to flying. The bird's bones are connected to the respiratory system. A bird with a blocked trachea can still breathe if a connection has been made between the wing bone and the outside air.

pothesis that they serve as a bellows.

A suggestion made long ago is that the large sacs could be filled with fresh air and, by alternate contraction of the anterior and posterior sets of sacs, air could be passed back and forth through the lung. The hypothesis has proved to be wrong, however, for the reason that the sacs do not contract in alternation. The pressure changes in the anterior and posterior sacs are similar: on inhalation the pressure drops in both sets of sacs, and all the sacs are filled with air; on exhalation the pressure increases simultaneously in the anterior and posterior sacs, and air passes out of both sets of sacs.

It has even been suggested that birds, by filling their air sacs, could take with them a supply of air to last them during a flight. This adventurous suggestion was supported by the speculation that the chest of a flying bird is so rigidly constrained by muscular contractions that breathing is impossible. The reasoning disregards the most elementary considerations of the amount of oxygen needed for flight.

The question of how air flows in the avian lung can be studied in a number of ways. One useful approach is to introduce a foreign gas as a marker. The flow of the gas and its time of arrival at various points in the respiratory system yield much information. Another approach is to use small probes that are sensitive to airflow and to place them in various parts of the elaborate passageways. In this way the flow directions can be determined directly during the phases of the respiratory cycle. My colleagues and I at Duke University have used both of these approaches, and we believe we now know with reasonable certainty the main features of avian respiration.

The use of a tracer gas has been quite successful in clarifying the flow of air. Our first experiments were with ostriches, which have the advantage of rather slow respiration. An ostrich breathes

about six times per minute, and changes in the composition of gas in its air sacs can therefore be followed rather easily. If an ostrich is given a single breath of pure oxygen and is then returned to breathing normal air, which has an oxygen content of 21 percent, an increased concentration of oxygen in the respiratory system will indicate how the single marked inhalation is distributed.

We used an oxygen electrode to follow changes in oxygen concentration. In the posterior air sacs we picked up a rapid increase in oxygen near the conclusion of the inhalation that carried pure oxygen. In other words, the marker gas flowed directly to the posterior sacs. In contrast, in the anterior sacs the oxygen did not appear until a full cycle later; the rise was noted as the second inhalation was ending [see illustration on page 165]. This finding must mean that the anterior air sacs do not receive inhaled air directly from the outside and that the marker gas that arrived on the second cycle or later must meanwhile have been in some other part of the respiratory system. We concluded that the posterior sacs are filled with air coming from the outside and that air entering the anterior sacs must come from elsewhere, presumably the lungs. Outside air thus enters the anterior sacs only indirectly, through other parts of the respiratory system, and it is delayed by at least one cycle.

It would be tempting to conclude from this experiment that the posterior sacs are well ventilated and that the anterior sacs do not receive much air but contain a rather inert and stagnant mass of air. The composition of the gas in the sacs might seem to support such a conclusion. The posterior air sacs usually contain about 3 or 4 percent carbon dioxide and the anterior sacs 6 or 7 percent, which is comparable to the carbon dioxide content of an air mass that is in equilibrium with venous blood. The conclusion would be wrong.

Whether or not an air sac is well ventilated can be ascertained by introducing a marker gas directly into the sac and determining how fast the marker disappears on being washed out by other air. In the ostrich we injected 100 milliliters of pure oxygen directly into an air sac and measured the time required to reduce by half the increase in oxygen concentration thereby achieved. We found that all the air sacs in the ostrich are highly ventilated and that they wash out rapidly.

The results showed that none of the air sacs contained a stagnant or relatively inert air mass. Since the anterior sacs have about the same washout time as the posterior sacs, they must be equally well ventilated. Why, then, since the renewal rate of air is high in the anterior sacs, do they contain a high concentration of carbon dioxide? This phenomenon can best be explained by postulating that the anterior sacs receive air that has passed through the lungs, where during passage it has exchanged gases with the blood, taking up carbon dioxide and delivering oxygen.

When we had arrived at this stage, it became essential for us to obtain unequivocal information about the flow of air in the bird lung. For this purpose W. L. Bretz in our laboratory designed and built a small probe that could record the direction of airflow at strategic points in the respiratory system of ducks. The information obtained in these experiments can best be summarized by going through the events of inhalation and then through the events of exhalation.

On inhalation air flows directly to the posterior sacs, which therefore initially receive first the air that remained in the trachea from the previous exhalation and then, immediately afterward, fresh outside air. The posterior sacs thus become filled with a mixture of exhalation air and outside air. Experiments with marker gas showed just this sequence; the marker arrived in the posterior sacs as the first inhalation was ending. The flow probe did not show any flow in the connections to the anterior sacs during inhalation, which was what we had expected from the fact that marker gas never arrived directly in the anterior sacs. Since the anterior sacs do expand during inhalation, the air that fills them can come only from the lung. Another finding is that air flows in the connection from the main bronchus to the posterior part of the lung, indicating that some of the inhaled air goes directly to the lung.

During exhalation the posterior sacs decrease in volume. Since the flow probe shows little or no flow in the main bronchus, the air must flow into the lung. The anterior sacs also decrease in volume. A probe placed in their connection to the main bronchus shows a high flow, consistent with direct emptying of these sacs to the outside.

The most interesting conclusion to be drawn from these patterns of flow is that air flows continuously in the same direction through the avian lung during both inhalation and exhalation. This suggestion is not new, but once we are certain that it is correct we can better examine its consequences. The air flowing through the lungs comes mostly from the posterior sacs, where the combination of dead-space air and outside air supplies a mixture that is high in oxygen but also contains a significant amount of carbon dioxide. Here we encounter one of the most elegant features of the system. If completely fresh outside air, which contains only .03 percent carbon dioxide, were passed through the lung, the blood would lose too much carbon dioxide, with serious consequences for the acid-base regulation of the bird's body. Another consequence of excessive loss of carbon dioxide arises from the fact that breathing is regulated primarily by the concentration of carbon dioxide in the blood. An increase in carbon dioxide stimulates breathing; a decrease causes breathing to slow down or even stop for a time.

Hence we see that the avian lung is continuously supplied with a mixture of air that is high in oxygen without being too low in carbon dioxide. The anterior sacs serve as holding chambers for the air coming from the lungs. This air is later discharged to the outside on exhalation, but enough of it remains in the trachea to ensure the right concentration of carbon dioxide in the posterior sacs after the next inhalation.

A few disturbing questions remain. One is why, since the pressure in the anterior air sacs falls during inhalation, air from the outside does not enter these sacs. The system has no valves that can open and close to help direct the flow. The answer is probably that the openings from the main bronchus have an aerodynamic shape that tends to lead the air past the openings. The avian respiratory tract is a low-pressure, high-velocity system in which gas flow may be governed by the principles of fluidics without the need for anatomical values [see "Fluid Control Devices," by Stanley W. Angrist; SCIENTIFIC AMERICAN, December, 1964].

Another conceptual difficulty is why air moves from the posterior sacs to the lungs during exhalation and from the lungs to the anterior sacs during inhalation. These movements require both suitable pressure gradients and a change in the volume of the lungs. It has been said that the bird lung changes little in volume because it is much firmer and less distensible than the mammalian lung. A bird's lung removed from the body retains its shape instead of collapsing to a small fraction of its normal volume as a mammal's lung does.

Another anatomical feature that has been misinterpreted is the bird's diaphragm. Birds have no muscular diaphragm, which is a most important feature in mammalian respiration. In its place they have a thin membrane of connective tissue. The membrane is con-

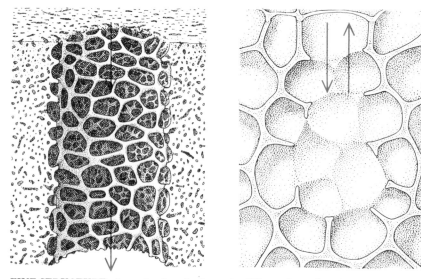

FINE STRUCTURE of an avian lung is also quite different from that of a mammalian lung. In the bird's lung the parabronchi (*left*) enable air to pass through the lung, entering from one side and leaving from the other. In the mammalian lung the baglike alveoli (*right*) are terminals, so that air necessarily flows into and out of the lung rather than through it.

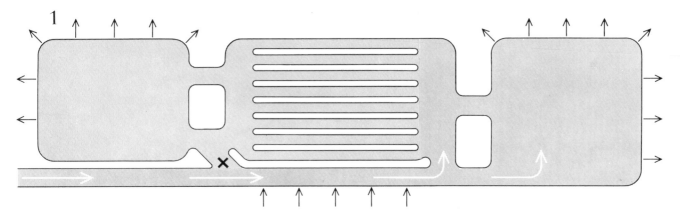

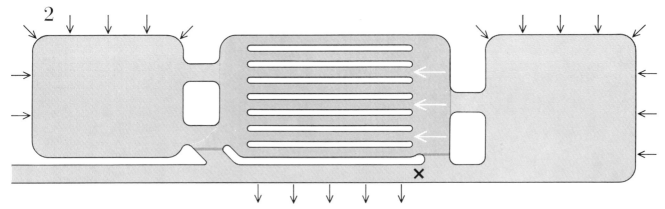

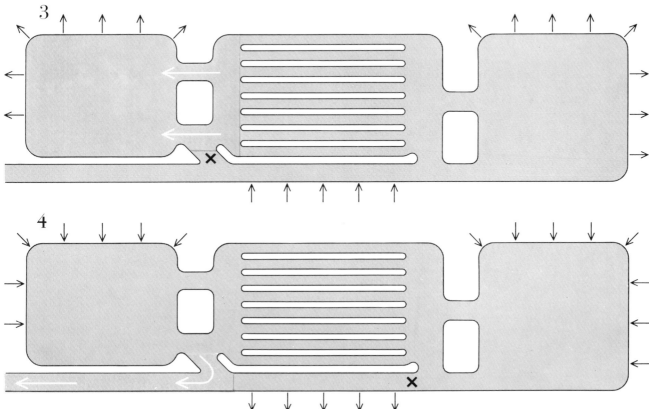

RESPIRATORY CYCLE in a bird is depicted schematically, following a single slug of air through two breaths. On inhalation (1) air flows through the bronchus and mainly into the posterior sacs, represented here by a single chamber. Some air also goes into the lung. The first air to reach the posterior sacs is air that was left in the trachea after the previous exhalation, so that it contains more carbon dioxide than fresh air does. The anterior sacs are bypassed, apparently under fluid-dynamical influences since there are no valves. The air sacs expand. On exhalation (2) the sacs decrease in volume, and air from the posterior sacs flows into the lung. On the next inhalation (3) the slug of air moves from the lungs into the anterior sacs. On the next exhalation (4) it is discharged from them into the trachea and thence to the outside. The system thus provides a continuous, unidirectional flow through the bird's lungs.

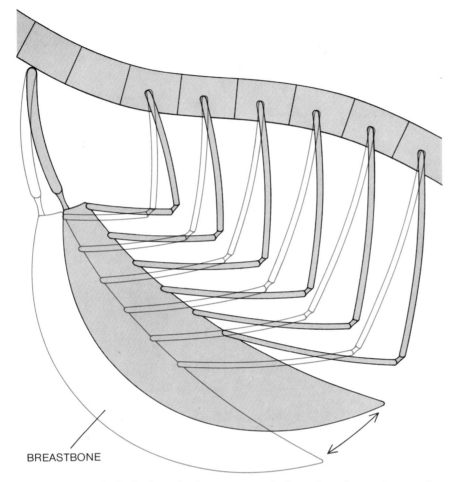

RIB STRUCTURE of a bird is related to respiration by being hinged in such a way that on inhalation the breastbone is lowered. The chest expands, as do the air sacs, but the lung diminishes in volume. On exhalation the process is reversed. Because the lungs expand on exhalation, air flows into them from the posterior sacs. Similarly, as the lungs decrease in volume on the next inhalation, air flows out of them and into the anterior-sac system.

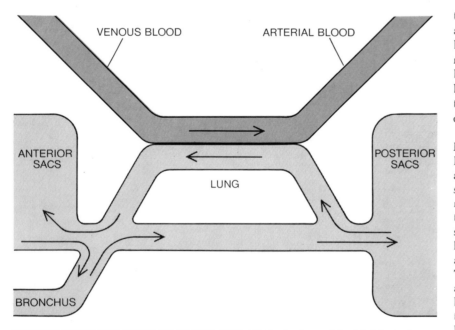

COUNTERCURRENT FLOW OF BLOOD AND AIR in the lung of the bird is the key to the bird's efficient extraction of oxygen and so to its ability to fly at high altitudes. Air flowing through the lung from the posterior sacs gives up more and more oxygen to the blood, and the blood can continuously take up more and more oxygen. Even as blood enters the lung it can take up oxygen because blood at that point has a low oxygen concentration.

nected to muscles that are attached to the body wall. When the muscles contract, they flatten out the membranous diaphragm, thus pulling on the ventral surface of the lung in a manner that is mechanically similar to the pull of the mammalian diaphragm. The avian diaphragm, however, works on a cycle opposite to that of the mammalian diaphragm: it tends to make the lungs expand on exhalation and the volume of the lungs to diminish on inhalation.

This paradoxical cycle provides the necessary mechanism for the movement of air into the lungs. As the lungs expand on exhalation, air flows from the posterior sacs to fill the lungs. As the lungs diminish in volume on inhalation, air flows from the lungs to the anterior sacs.

Earlier in this article I remarked that the complex lung and air sac system of birds is not a prerequisite for flight, but I suggested that it confers a considerable advantage at high altitude. Man and other mammals begin to show marked symptoms of oxygen deficiency at an altitude of 3,000 to 4,000 meters (10,000 to 13,000 feet). A man moving to such an altitude from sea level finds it difficult to exert himself in physical work, although he gradually acclimatizes and is able to perform normally. At higher altitudes work and acclimatization become increasingly difficult; the limit for moderately active functioning of a man, even after long acclimatization, is about 6,000 meters.

Birds, in contrast, have been observed to move about freely and fly at altitudes above 6,000 meters. Airplanes have collided with flying birds as high as 7,000 meters. Birds might reach these altitudes by riding on strong upcurrents of wind, but this would not explain the fact that they fly actively and without apparent difficulty once they are there.

A few years ago Vance A. Tucker of Duke University simultaneously exposed house sparrows and mice to a simulated altitude of 6,100 meters, which represents somewhat less than half the atmospheric pressure at sea level and therefore less than half the partial pressure of oxygen at sea level. At this low level of oxygen the sparrows were still able to fly, but the mice were comatose. The blood of the sparrow does not have any higher affinity for oxygen than the blood of the mouse; otherwise the ability of the sparrows to take up oxygen at low pressure could be explained as a difference in the blood. What can explain the difference between birds and mammals under these conditions is the unidirectional flow of air in the bird's lungs.

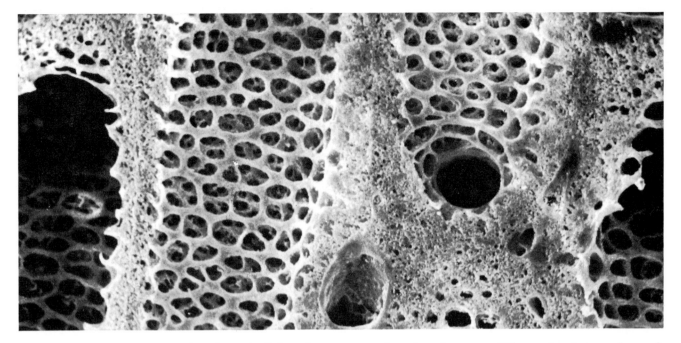

LONGITUDINAL SECTION of parabronchi in a bird's lung shows the spongy structure that enables air to flow through the lung as water flows through a sponge. This scanning electron micrograph was also made by Duncker; the enlargement is 90 diameters.

One can depict the flow of air and blood in the bird's lungs with a simple diagram [*see bottom illustration on opposite page*]. In the diagram the airflow through the lungs is shown as a single stream and the flow of blood as another single stream. The salient point is that the two flows are in opposite directions.

In this way it becomes apparent that the blood, as it is about to leave the lungs, can take up oxygen from air that has the highest oxygen concentration available anywhere in the system. As the air flows through the lungs it gives up more and more oxygen to the blood before it enters the anterior sacs, where it is held until it is exhaled. It will be noted that the air, just before it leaves the lungs, encounters venous blood that is low in oxygen. This blood is therefore able to take up some oxygen, even though much of the oxygen in the air has already been removed. As the blood passes through the lungs it meets air of increasing oxygen concentration and therefore can continuously take up more oxygen until, just before it leaves the lungs, it meets the maximally oxygen-rich air coming from the posterior sacs.

The end result of this countercurrent flow is that more oxygen can be extracted from the air than would otherwise be possible. The system is similar to the flow through the gill of the fish, where the blood and the water flow in opposite directions. The blood just as it leaves the gill therefore encounters water with the highest possible oxygen content. Because of this type of flow, fish can extract from 80 to 90 percent of the oxygen in the water. The oxygen extraction normally reached in mammals under normal conditions is about 20 to 25 percent of the oxygen present in the air.

We are still trying to obtain better evidence that the flow of air and blood in the bird's lungs is as proposed in the scheme I have described, but the performance of birds at high altitude could hardly be explained in any other way. Examining the exchange of carbon dioxide rather than oxygen, we found several years ago that the air in the anterior sacs has a content of carbon dioxide that is much higher than the concentration in the arterial blood. This relation too can only be explained if the air coming from the lungs to the anterior sacs has received carbon dioxide from venous blood instead of being in equilibrium with arterialized blood as exhaled air in mammals is.

To what I have said so far, which I regard as hypotheses well supported by physical evidence, I should like to add a wild speculation. It is well known that some large birds, notably cranes and swans, have an extremely elongated trachea. This long trachea would seem to be a disadvantage, since at the end of an exhalation it would represent dead space filled with exhaled air that would have to be reinhaled at the beginning of the next breath, thus diluting the fresh outside air that follows.

The usual interpretation of the long trachea of swans and cranes is that it aids in vocalization. Such a luxury could not be allowed, however, if the large increase in dead space were physiologically detrimental. In fact, the increase in dead space may be an advantage. For aerodynamic reasons large birds have a slow wingbeat. For anatomical reasons the wingbeat and breathing in flying birds may be synchronized, since the large muscles that provide the downstroke of the wing are inserted at the keel of the breastbone and pull on it. It would therefore seem simple to attain simultaneous movements of wing and chest; indeed, it may be difficult to avoid.

The reasoning now goes as follows. If a slow wingbeat is determined by the size of the bird, and if respiration is synchronized with wingbeat, enough air can be taken in only by making each breath deeper. If each breath is deeper, and it is necessary (as I pointed out earlier) to achieve a certain level of carbon dioxide in the posterior air sacs, the amount of exhaled air reinhaled with each breath must be increased. In other words, to achieve the necessary concentration of carbon dioxide it is necessary to increase the volume of dead space.

Perhaps this speculation will have to be modified as more evidence becomes available. At present, we do not have adequate information about the synchronization of wingbeat and respiration in any of the larger birds. In fact, the respiration of birds during flight remains an interesting and almost uncharted field of physiology.

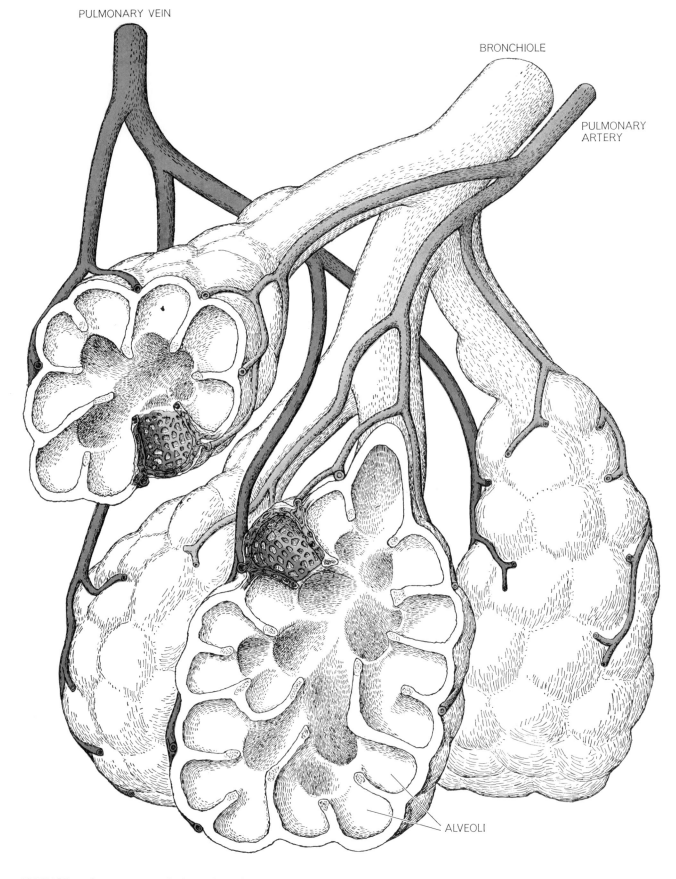

PULMONARY VEIN

BRONCHIOLE

PULMONARY
ARTERY

ALVEOLI

ALVEOLI are the air spaces in the lungs through which oxygen enters the blood and carbon dioxide leaves. A surface-active agent coats the moist alveoli and regulates the elasticity and tension of the lungs as a whole. In this schematic diagram nearly all the smaller blood vessels have been omitted except where capillary networks embedded in the alveolar walls show through from the backs of two alveoli seen in cross section. The average alveolus expands and contracts more than 15,000 times a day during breathing.

16

SURFACE TENSION IN THE LUNGS

JOHN A. CLEMENTS

December 1962

Recent investigations have shown that the air spaces of the lung are coated with a complex substance that lowers surface tension. It now appears that this substance keeps the lungs from collapsing

By far the most extensive surface of the human body in contact with the environment is the moist interior surface of the lungs. To carry on the exchange of carbon dioxide and oxygen between the circulating blood and the atmosphere in sufficient volume to sustain life processes requires approximately one square meter of lung surface for each kilogram of body weight. In the normal adult this amounts to the area of a tennis court. Such an area is encompassed in the comparatively small volume of the chest by the compartmentation of the lungs into hundreds of millions of tiny air spaces called alveoli. These air spaces are connected by confluent passages through the bronchial tree and the trachea to the atmosphere and are thus, topologically speaking, outside the body. Within the walls of the alveoli the blood is spread out in a thin sheet, separated from the air by a membrane about one micron (.001 millimeter) thick.

Since the primary function of the lungs is to present the inner surface of the alveoli to the air, it is not surprising to learn that the vital process of respiration is critically dependent on the physical properties of this surface. There is, of course, much more to the anatomy and physiology of the lungs. In recent years, however, the attention of investigators has been drawn increasingly to the role that is played by surface tension: the manifestation of the universal intermolecular forces that is observed in the surfaces of all fluids. The surface tension in the outermost single layer of molecules in the film of tissue fluid that moistens the surface of the lungs has been found to account for one-half to three-quarters of the elasticity with which the air spaces expand and contract in the course of the 15,000 breaths that are drawn into the

lungs of the average individual each day.

As this knowledge suggests, it has also been found that the body has a way of regulating the surface tension of the lungs. Certain cells in the walls of the alveoli secrete a sort of detergent or wetting agent. This "surface-active" sub-

stance tends to weaken the surface tension. Its presence in the monomolecular layer on the surface of the film of moisture coating the air spaces serves to stabilize the dynamic activity of the lungs. It equalizes the tension in the air spaces as they expand and contract; it

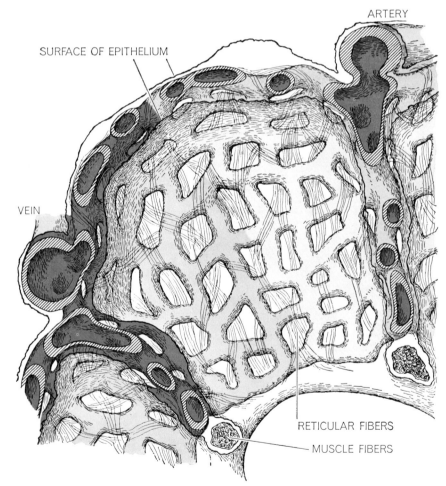

SINGLE ALVEOLUS is actually microscopic in size. The alveolar wall has been rendered transparent in this schematic cross section so that the rich network of blood capillaries and fibers that support the alveolus can be seen. The surface-active substance that plays a key role in stabilizing lung function normally coats the epithelium of every healthy alveolus.

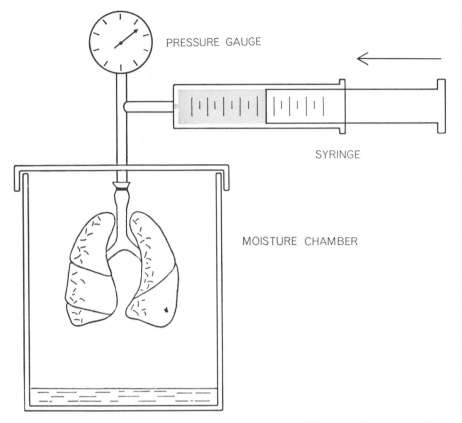

LUNG ELASTICITY is measured by using large syringe to fill lungs with a gas or a fluid (*color*). Lungs taken at autopsy are placed in a moist chamber and attached to a manometer, or pressure gauge. Karl von Neergaard of Zurich made the first such measurement in 1929.

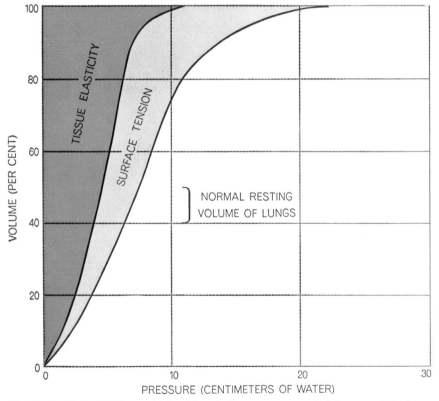

ELASTICITY CURVES are obtained by plotting pressure needed to expand the lungs against volume of lungs. The lungs show much less elasticity when filled with fluid than they do when filled with gas. Surface tension plays a significant role only when lungs are filled with gas. The amount of elasticity contributed by surface tension shows up as the difference between the curves at any particular volume of the lungs. It rises as the lungs expand.

brings about an even distribution of pressure between large and small alveoli, and by decreasing the over-all pressure it reduces the muscular effort required for respiration. This substance has the further function of assisting the osmotic forces acting across the surface of the lungs and so keeping the film of moisture on the surface from drawing fluid into the air spaces.

The critical importance of the surface-active agent becomes apparent when it is not there to do its work. Its absence explains some of the symptoms in the complex organic disease of the newborn recorded variously as fatal respiratory distress, hyaline-membrane disease or atelectasis. The collapse of the lungs and the filling of the air spaces with fluid observed in this disease is promoted by abnormally high surface tension in the alveoli. Some 25,000 newborn infants die of the disease in the U.S. each year. A similar syndrome, although it is not always fatal, has recently appeared as a complication attending heart surgery in some patients whose lungs have been temporarily disengaged from the respiratory function by diversion of the blood through a "heart-lung" machine.

The contribution of surface tension to the elasticity of the lungs was first demonstrated in 1929 by Karl von Neergaard of the University Clinic in Zurich. He distended lung preparations alternately with air and with saline solution and compared the pressures required to do so with each. This experiment, since repeated many times by other workers, showed that it takes a higher pressure to distend the lungs with air. The interpretation of this experiment calls for a more precise definition of a surface: it is an interface between two substances and it is established by the relative cohesion of their constituent molecules. Thus when the fluid on the surface of the lungs forms an interface with air, it exhibits a stronger surface tension than it does at an interface with saline solution. In fact, the tissue fluid forms essentially no interface at all with saline solution of the right concentration and the surface tension is reduced to almost zero. Distention of the lungs with saline solution can therefore be used to measure the elastic properties of the tissue alone, uncomplicated by the effects of surface tension. Since inflation with air yields a measure of both the tissue and the surface-tension components, the effect of surface tension can be derived by subtracting the pressure required to distend the lungs with saline solution from the pressure required to inflate them to the same volume with air.

The technique has recently provided conclusive evidence that surface tension is abnormally high in the fatal respiratory distress of the newborn. After autopsy the lungs of such infants can be expanded at almost normal pressures with saline solution but require three to four times the normal pressure for air inflation. Moreover, the alveoli collapse at abnormally high air pressures during deflation.

In everyday experience with surface tension there is little to suggest that it has such formidable power. It has barely measurable effects on the properties of solids, showing up for example in the measurement of the elasticity of fine-drawn wires. Its action is more prominent in the behavior of liquids, as in the shaping of raindrops or the providing of a platform for certain aquatic insects. But it seems no more than an incidental effect of the geometry that accounts for it. Whereas the molecules in the bulk of a liquid experience forces of mutual attraction that are balanced in all directions, the molecules at the surface are attracted more strongly to their neighbors below the surface and are attracted only weakly to the sparser population of molecules in the air above the surface.

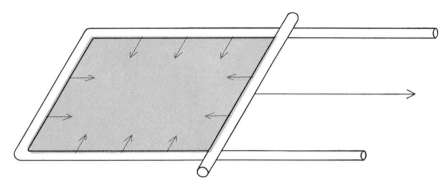

MAXWELL FRAME, used to demonstrate surface tension, consists of U-shaped wire with separate wire across open end. Liquid film (*color*) pulls at cross wire. Force needed to prevent cross wire from moving to bottom of U (*black arrow*) is proportional to the tension.

Because the net pull is downward, the surface particles tend to dive and the surface shrinks to the least possible area.

The resultant force of cohesion at the interface between a liquid and the air can be demonstrated with the help of a Maxwell frame, named for the 19th-century physicist James Clerk Maxwell. This is a U-shaped wire, with the open end of the U closed by a cross wire that can slide along the legs of the U. A film of liquid stretched out on the frame tends to pull the cross wire to the bottom of

the U. The force necessary to resist this pull—to maintain a constant area of film —provides a measure of the surface tension. Since the film in this experiment has two surfaces, the measured force must be divided by twice the width of the U; the result is usually expressed in dynes per centimeter. (A dyne is the force required to accelerate a one-gram mass one centimeter per second.) The surface tension of pure water at body temperature is equal to about 70 dynes per centimeter; that of blood plasma and

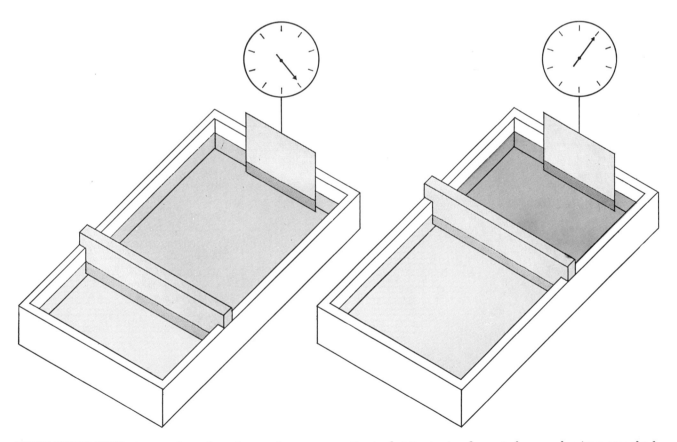

SURFACE BALANCE measures change in surface tension as area of film of surface-active agent on water increases and decreases. Surface tension pulls down on platinum strip (*attached to gauge*). Water alone produces pull of about 70 dynes per centimeter. A detergent in the water makes surface tension about 30 dynes per centimeter but the tension does not change as barrier moves slowly back and forth. Surface-active agent from lungs forms a film on the water and makes the surface tension about 40 dynes per centimeter (*left*). As barrier moves toward strip, compressing the agent, tension drops (*right*). Surface tension rises as the barrier moves back.

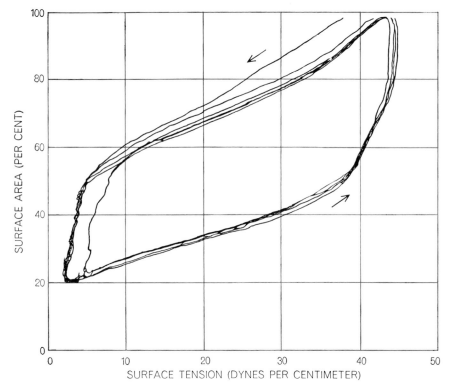

CHANGE IN SURFACE TENSION with area, as measured by surface balance, is large when surface-active agent from normal lungs covers the water. Moving barrier made several trips back and forth during test. The tension at first drops rapidly as barrier moves in (*arrow pointing to left*), and it rises rapidly as barrier moves back (*arrow pointing to right*).

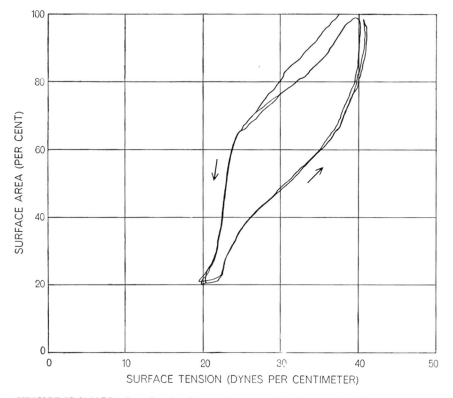

CHANGE IS SMALL when the alveolar coating comes from lungs of newborn infant who succumbed to acute respiratory distress. In such a case the surface tension is about 18 dynes per centimeter. Relative lack of surface activity plays a key role in the fatal disease.

tissue fluid, to about 50 dynes per centimeter.

The results conform with the observation that surface tension does not exert very strong forces. At this point, however, geometry enters the picture again. According to a formula of Pierre Simon de Laplace, the 18th-century astronomer and mathematician, the force exerted in a given surface is equal to twice the tension divided by the radius of the surface. In a flat surface with, so to speak, infinite radius the force is zero. Given the tiny dimensions of the average-sized alveolus, calculation shows that the surface tension of tissue fluid should exert a considerable force. At 50 dynes per centimeter in a surface with a radius of only .05 millimeter, it would produce a force of 20,000 dynes per square centimeter. Expressed as pressure this is equal to 20 centimeters of water.

This computation explains why surface tension influences the elasticity of the lungs so greatly. It does not, however, agree with the actual values for the surface tension in normal lungs obtained by comparison of the pressures required to distend the lungs with liquid and to inflate them with air. At functional or intermediate lung volume, in fact, the calculated effect of surface tension turns out to be from five to 10 times too large. In other words, the surface tension of the tissue fluid would have to be closer to five or 10 dynes per centimeter instead of 50 dynes per centimeter. At larger lung volumes the measured pressure comes into closer agreement with the calculated pressure, indicating a surface tension for the lung tissue of about 40 dynes per centimeter. In short, the surface tension of the tissue is unexpectedly low in the lungs, and it varies with the inflation and deflation of the air spaces.

With these suggestive clues in hand, investigators began to look into the tissue fluid of the lungs for the presence of a surface-active agent. Soaps or detergents are familiar examples of substances of this kind. Their molecules have weaker forces of mutual attraction for one another and for molecules of other species. They tend to accumulate in excess at surfaces and interfaces when mixed in solutions. Acting as bridges between dissimilar substances such as oil and water or water and air, they wet, penetrate and disperse oily substances and stabilize emulsions and foams. The concentration of their weaker attractive forces at an interface reduces the surface tension. At a number of laboratories the presence of a surface-active agent was soon demonstrated in the tissue fluid extracted from the lungs. The extraction can be accomplished in a number of ways: by rinsing the alveoli with saline solution via the air passages; by generating a foam in the alveoli; and by filtration from minced whole tissue. Each of these procedures yields an extract that contains a powerful surface-active agent on which accurate measurements can be made.

The laboratory technique for detecting the presence of this agent and measuring its effect on the surface tension of the tissue fluid provides a nice demonstration of its mode of operation in the alveoli. The extract is placed in a shallow tray and a .001-inch-thick platinum strip is suspended in it from the arm of a sensitive electrobalance or strain gauge. The pull of the surface on the strip provides a measure of the surface tension. A motor-driven barrier slowly sweeps the surface from the far end of the tray, reducing the area of liquid surface in which the platinum strip is hanging to 10 or 20 per cent of its initial size. Since the surface-active agent in the extract spontaneously forms a film at the surface, it is concentrated in the area in front of the barrier. As the concentration builds up, the surface tension falls to low values.

Extracts from normal lungs show a change in surface tension, when measured this way, from about 40 dynes per square centimeter to two dynes per square centimeter—in excellent agreement with the surface tension as estimated in the lung itself. In contrast, the surface tension does not fall below 18 dynes per square centimeter in extracts from the lungs of newborn infants that have succumbed to hyaline-membrane disease.

The change in surface tension with the change in surface area is the key to the action of the surface-active agent in the lung. In an expanded air space the layer of surface-active agent is attenuated, and surface tension is increased accordingly. The increase in tension is partly offset, however, by the increase in the radius of the air space, and the increase in force or counterpressure exerted by the surface tension is diminished. As the air space contracts to perhaps half its expanded size, the increasing concentration of the surface-active agent reduces the surface tension, balancing the Laplace equation in the other direction and again decreasing the pressure in the air space. Similarly, between

LUNGS OF FROGS contain quite large air spaces that are not threatened by pressure of surface tension created by a liquid that coats the inner lining. Therefore a surface-active agent is not necessary, and none has been found. One lung is shown in longitudinal section at right. Color indicates blood vessels. The diagrams on this page are highly schematic.

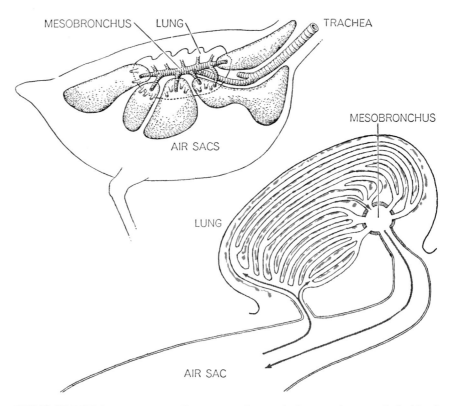

LUNGS OF BIRDS contain air capillaries, tiny tubes in which gas exchange with the blood takes place. Air passes through lungs into and out of large air sacs. Since there is little change in lung volume, surface tension does not need to be adjusted. Surface-active agent has not been found in the lungs of birds. A cross section of a bird lung is seen at right.

the smallest alveolus and the largest, which may be three or four times larger, differences in the concentration of the surface-active agent in the interface of tissue fluid and air bring about a homogeneous distribution of pressure. The performance of hundreds of millions of alveoli, of random size, is thereby smoothed and co-ordinated.

The action of the alveolar surface-active agent in balancing the forces that

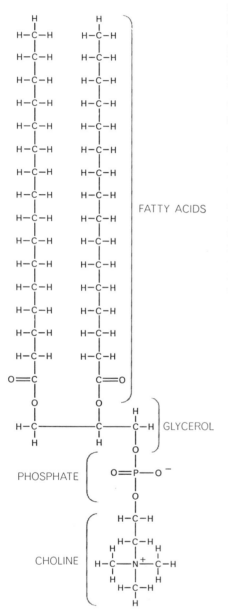

FATTY ACIDS

GLYCEROL

PHOSPHATE

CHOLINE

LECITHIN MOLECULE consists of long-chain fatty-acid groups that are not strongly attracted to water, as well as glycerol and electrically charged, or polar, phosphate and choline groups that are attracted to polar molecules of water. The fatty acid is thought to stand up out of the water when the molecule is part of a surface layer. Lecithin is strongly surface-active; it is a constituent of the substance that coats the alveoli.

otherwise tend to draw fluid out of the capillaries into the alveolar air spaces is also important. The blood pressure in the capillaries, the osmotic pull of the tissue fluid in and on the alveolar membrane, and the surface tension of this fluid all work to move fluids outward. One force, the osmotic pull of the blood plasma, opposes this combination of forces. The maintenance of a favorable balance of forces is assisted by the reduction in surface tension through the action of the surface-active agent. In hyaline-membrane disease, with surface tension sustained at as much as 45 dynes per centimeter, the leakage of fluid from the capillaries blocks the exchange of oxygen and carbon dioxide between blood and air; the process is limited in the end only by engorgement of the air spaces or by their collapse.

Chemical analysis has shown that the surface-active agent of the lungs is a lipoprotein, that is, a compound molecule made up of protein and fatty constituents. The latter are of an appropriately soapy kind, lecithin being the predominant component. A member of the same chemical family that has been made synthetically, a substance called dipalmitoyl lecithin, shows the same surface activity. Dipalmitoyl lecithin has even been isolated from the lung fluid. It is tempting, therefore, to attribute the surface activity of the lung material to dipalmitoyl lecithin. Against this conclusion, however, it can be shown that the active material isolated from the extract of tissue fluid is the intact lipoprotein molecule. Its activity is destroyed by attempts to segregate the lipids from the protein or to isolate any one of the lipid components from the whole. At present the most reasonable opinion is that the native material is a complex of protein and lipids, particularly dipalmitoyl lecithin, and that both are essential to its activity.

The discovery of this remarkable substance in the lung fluid of man has prompted a search for it in other animals. So far it has turned up in all the other mammals that have been tested (the mouse, rat, guinea pig, rabbit, cat, dog and cow), but not in any amphibian (frog and toad), reptile (snake and crocodile) or bird (pigeon and chicken). There appears to be some rationale for this distribution among species. Amphibians and reptiles depend on their environment to supply a major portion of the heat that sustains their metabolism; weight for weight they do not require as much exchange of respiratory gases as

mammals and therefore do not need as much lung surface. Accordingly they have relatively large air spaces in their lungs, and their lung function is not seriously threatened by the action of surface tension. Birds, on the other hand, have small air spaces, more comparable in size to those in the lungs of mammals. But the bird lung is ventilated in a peculiar way. Instead of the tidal ventilation, which alternately inflates and deflates the air spaces of mammals, the exchange of respiratory gases is accomplished by drawing air through the lungs into large air sacs that are separate from the lungs. In this way the change of volume in the air spaces is minimized. The air spaces can remain at or near their maximum volume, and the lungs are stabilized by the elasticity of the lung tissue itself.

In mammals the lung tissues apparently begin to secrete the critical surface-active material late in embryonic development. This is true, at least, of the two species in which the question has been investigated. In the mouse, which has a gestation period of 20 days, surface activity in the lungs appears suddenly at 17 or 18 days. The lungs of the human fetus develop the activity somewhat more gradually, during the fifth to the seventh month of gestation. This is the interval during which prematurely born infants become increasingly viable.

Since it is now reasonably certain that the secretion of a surface-active material is an adaptation peculiar to mammals, investigators are finding new significance in an observation made in 1954 by Charles Clifford Macklin of the University of Western Ontario. He showed that the walls of the alveoli in mammalian lungs contain special cells that he called granular pneumonocytes. He even suggested that the "granules" discharged by these cells "regulated the surface tension" of the alveoli, but he did not enlarge on this idea further. Under the electron microscope it now appears that the granules of Macklin are cellular particles called mitochondria and possibly the products of mitochondria from certain cells in the alveolar membrane. Mitochondria are associated with the metabolic and synthetic activities of all cells; they appear in high concentration in those tissue cells that have specialized secretory functions [see "Energy Transformation in the Cell," by Albert Lehninger; SCIENTIFIC AMERICAN Offprint 69]. Some of the granules can be identified as true mitochondria, with the fine structure that characterizes them in other cells. Others appear to be mitochondria-like bodies involved in a process of transformation by which they lose their fine structure and become relatively featureless. Most remarkable of all, the electron-microscope pictures show these same forms passing through the cell membrane from the cytoplasm into the air space. This process could be the means by which the surface-active substance is secreted into the tissue fluid that coats the surface of the alveoli.

Various stages of the process have been observed in the cells of a half-dozen species of mammals but never in the amphibians or birds in which it has been looked for. In mammals, moreover, it has been found that this peculiar transformation of the mitochondria appears in the lung tissues along with surface activity at the same stage of fetal development.

With this background of evidence established by classical physiology and the most modern techniques of cell biology, the way seems to be cleared for investigation of the hormonal, neural, nutritional, environmental and genetic factors that may influence the production, function and elimination of the alveolar surface-active agent. Diabetes, for example, is associated with derangement of lipid, or fat, metabolism. In view of the importance of the lipid fraction of the surface-active agent one wonders if diabetes in the mother may not be a factor predisposing the fetus to respiratory distress at birth, particularly since the syndrome occurs more frequently among infants born to diabetic mothers. The experience with patients in heart surgery suggests another line of investigation. From the rapid decrease in surface activity and the collapse of the lung that sometimes follows the bypassing of the pulmonary circulation, it can be surmised that the production of the agent depends on blood flow and that distribution of the flow affects the distribution of air in the lungs. The question of neural control is raised by experiments with small animals, in which cutting of the vagus nerve is followed by decline in surface activity, the accumulation of fluid and finally collapse of the air spaces. Pure oxygen, atmospheric pollutants and some industrial chemicals have been shown to affect the alveolar surfaces. Animals in which hormonal activity is high—young animals, females in estrus and animals that have been treated with cortisone—are particularly subject to the toxic effects of pure oxygen on the lungs. Occasionally massive collapse of the lungs follows general anesthesia, with no indication of obstruction to the air passages. It is not too much to hope that problems of this kind can be brought within the reach of effective treatment by the next advances in the understanding of the mechanism that regulates the surface tension in the lungs.

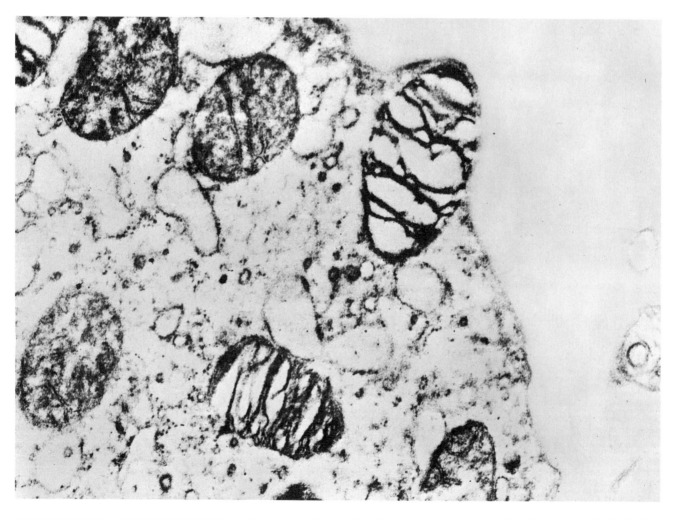

MITOCHONDRIAL TRANSFORMATION can be seen in this electron micrograph by Robert E. Brooks of the University of Oregon. It shows part of an alveolar epithelial cell enlarged approximately 45,000 diameters. Toward upper right a transformed mitochondrion seems to be emerging into the air space. To its left is a normal mitochondrion, and to the left of that, one is beginning to change, losing fine structure. Mitochondria probably produce lung agent; transformed mitochondria may carry it to the surface.

17

THE MASTER SWITCH OF LIFE

P. F. SCHOLANDER
December 1963

*Studies of diving have led to the identification of the
vertebrate animal's ultimate defense against asphyxia: a gross
redistribution of the circulation that concentrates oxygen in
the brain and heart*

In the higher animals breathing and the beating of the heart seem synonymous with life. They implement the central process of animal metabolism: the respiratory gas exchange that brings oxygen to the tissues and removes carbon dioxide. Few events are more dangerous to life than an interruption of breathing or circulation that interferes with this exchange. It is not that all the tissues of an animal need to be continuously supplied with fresh oxygen; most parts of the human body display a considerable tolerance for asphyxia. The tissues of an arm or a leg can be isolated by a tight tourniquet for more than an hour without damage; the kidney can survive without circulation for a similar period and a corneal transplant for many hours. The heart and the brain, however, are exquisitely sensitive to asphyxia. Suffocation or heart failure kills a human being within a few minutes, and the brain suffers irreversible damage if its circulation ceases for more than five minutes.

One might expect that the body would respond with heroic measures to the threat of asphyxia. It does indeed. The defense is a striking circulatory adaptation: a gross redistribution of the blood supply to concentrate the available oxygen in the tissues that need it most. The identification of this defense mechanism has resulted from studies, extending over a number of years, of animals that are specialized to go for an unusual length of time without breathing: the diving mammals and diving birds. Only recently has it become clear that this "master switch" of life is the generalized response of vertebrate animals to the threat of asphyxia from any one of a number of quite different circumstances.

A cat or a dog or a rabbit—or a human being—dies by drowning in a few minutes. A duck, however, can endure submersion for 10 to 20 minutes, a seal for 20 minutes or more and some species of whales for an hour or even two hours. How do they do it? The simplest explanation would be that diving animals have a capacity for oxygen storage that is sufficient for them to remain on normal aerobic, or oxygen-consuming, metabolism throughout their dives. As long ago as the turn of the century the physiologists Charles R. Richet and Christian Bohr realized that this could not be the full story. Many diving species do have a large blood volume and a good supply of oxygen-binding pigments: hemoglobin in the blood and myoglobin in the muscles. Their lungs, however, are not unusually large. Their total store of oxygen is seldom even twice that of comparable nondiving animals and could not, it was clear, account for their much greater ability to remain submerged.

At the University of Oslo during the 1930's I undertook a series of experiments to find out just what goes on when an animal dives. For this purpose it was necessary to bring diving animals into the laboratory, where they could be connected to the proper instruments for recording in detail the physiological events that take place before, during and after submergence. Over the years my colleagues—Laurence Irving in particular—and I have worked with many mammals and birds. We have found seals to be ideal experimental animals: they tame easily and submit readily to a number of diving exercises. At first we confined them to a board that could be lowered and raised in a bathtub full of water. Lately my colleague Robert W. Elsner at the Scripps Institution of Oceanography has trained seals to "dive" voluntarily, keeping their noses under water for as long as seven minutes.

Our first experiments at Oslo confirmed the earlier discovery, by Richet and others, of diving bradycardia, or slowing of the heart action. When the nose of a seal submerges, the animal's heartbeat usually falls to a tenth or so of the normal rate. This happens quickly, indicating that it occurs by reflex action before it can be triggered by any metabolic change. The initiation of bradycardia is affected by psychological factors. It can be induced by many stimuli other than diving, such as a sharp handclap or a threatening movement on the part of the investigator when the seal is completely out of the water. Conversely, bradycardia sometimes fails to develop in a submerged seal if the animal knows it is free to raise its head and breathe whenever it likes. In long dives, however, the slowing down is always pronounced. It is significant that the impulse is so strong it ordinarily continues for the duration of the dive, even when the animal works hard—a situation that would normally cause a rise in the heart rate.

Bradycardia occurs in every diving animal that has been studied. It has been reported in such diverse species as the seal, porpoise, hippopotamus, dugong, beaver, duck, penguin, auk, crocodile and turtle. The same thing happens in fishes when they are taken out of the water. And when such nondivers as cats, dogs and men submerge, bradycardia develops too, although it is often less pronounced than in the specialized divers.

When the heart of a seal beats only five or six times a minute, what happens to the blood pressure? We found that the central blood pressure—in the main artery of a hind flipper, for instance—stays at a normal level. The shape of the pressure trace, however, reveals that

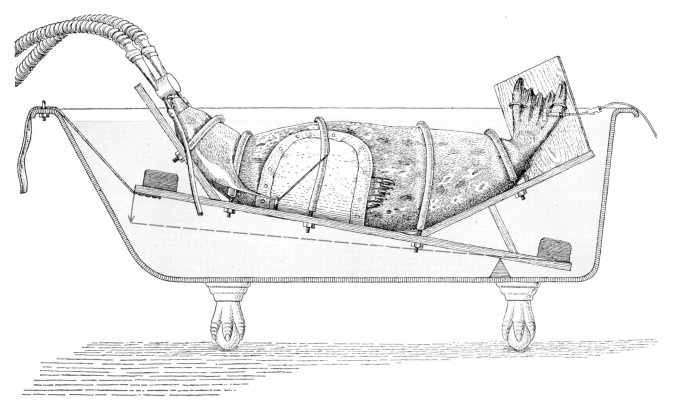

SEAL DIVES IN LABORATORY by being ducked in a bathtub full of water. The animal is strapped loosely to a weighted board. Its head is covered by a mask connected to a device for recording respiration. When the board is tilted down (*broken line*), the mask fills with water and the seal's nose is submerged. An artery in a hind flipper is shown cannulated for removal of blood samples.

VOLUNTARY DIVING eliminates any possibility that restraint affects the seal's responses. This harbor seal (*Phoca vitulina*) is being trained to keep its nose under water until the experimenter lowers his warning finger and instead displays the reward, a fish.

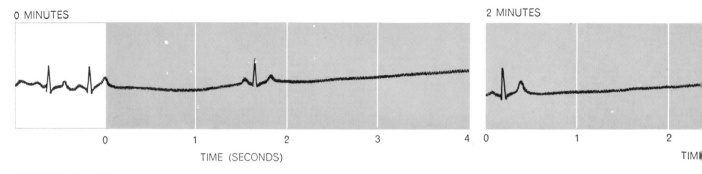

0 MINUTES

2 MINUTES

TIME (SECONDS)

TIM

DIVING BRADYCARDIA, the slowing of the heart rate that occurs in vertebrates when they submerge, is quite apparent in this electrocardiogram of a diving seal. Three segments of the record are shown, made at the beginning of, during and at the end of an

whereas the pressure rise with each beat is normal, the subsequent drop in pressure is gradual and prolonged. This indicates that, although the systolic phase of the heartbeat is almost normal, the diastolic phase, during which the blood is forced through the aorta, encounters resistance: the peripheral blood vessels are constricted. Measurements in a small toe artery in the seal's flipper show that the pressure there drops when the dive begins, falling rapidly to the much lower level maintained in the veins. In other words, we found that the circulation in the flippers shuts down to practically nothing during a dive [*see bottom illustration on opposite page*].

For another clue to circulation we measured the level of lactic acid in the muscles and blood of a diving seal. Lactic acid is the end product of the anaerobic metabolic process from which muscles derive energy in the absence of oxygen. The concentration of this metabolite in muscle tissue rises sharply during a dive but the concentration in the blood does not; then, when the seal begins to breathe again, lactic acid floods into the bloodstream. The same sequence of

events has been found to occur in most other animals, showing that the muscle circulation remains closed down as long as the dive continues. Similarly, oxygen disappears from muscle tissue a few minutes after a seal submerges, whereas the arterial blood still contains plenty of oxygen—enough to keep the myoglobin saturated if the muscles are being supplied with blood [*see illustrations on page 180*]. Other experiments revealed that in the seal both the mesenteric and the renal arteries, supplying the intestines and kidneys respectively, close down during diving. All these findings made it apparent that a major portion of the peripheral circulation shuts off promptly on submergence. This was evidently the reason the heart slows down.

At this point our results tied in nicely with some conclusions reached by Irving, who was then at Swarthmore College. His efforts had been stimulated by pioneering studies of circulatory control conducted in the 1920's by Detlev W. Bronk, then at Swarthmore, and the late Robert Gesell of the University of

Michigan. Bronk and Gesell had discovered in 1927 that in a dog rendered asphyxic by an excess of carbon dioxide and a lack of oxygen the muscle circulation slowed down as the blood pressure remained normal and the brain circulation increased. Irving noted in 1934 that this phenomenon might explain a diving animal's resistance to asphyxia, and he proceeded to measure blood flow in a variety of animals by introducing heated wire probes into various tissues and recording the rate at which their heat was dissipated. His data indicated that during a dive the flow in muscle tissue is reduced but the brain blood flow remains constant or even increases. He decided that the essence of the defense against asphyxia in animals would prove to be some mechanism for the selective redistribution of the circulation, with preferential delivery of the decreasing oxygen store to those organs that can least endure anoxia: the brain and the heart.

When the blood flow closes down in most tissues during a dive, what happens to energy metabolism? This is best studied during a quiet dive, with a seal or

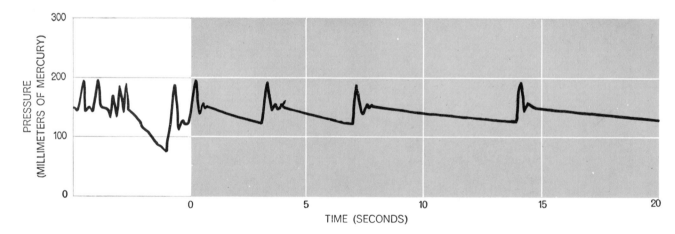

CENTRAL BLOOD PRESSURE stays at about a normal level during a seal's dive (*color*); the rate of increase in pressure during a contraction is also normal. The slow pressure drop between contractions, however, suggests constriction of peripheral blood vessels.

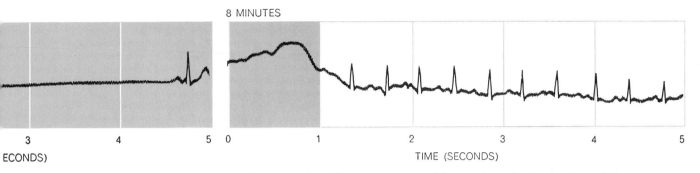

8 MINUTES

3   4   5 0   1   2   3   4   5

ECONDS)   TIME (SECONDS)

eight-minute dive, the duration of which is shown in color. The heart slows down at the start of the dive. The rate remains as low as seven or eight per minute during the dive and then returns to a normal 80 or so per minute as soon as the seal breathes again.

duck trained to remain inactive while under water. The oxygen stores are large enough to provide only a quarter of the energy expended in a predive resting period of the same length. The next question was: Do anaerobic processes, including lactic acid production, substitute fully for the lack of oxygen? Muscle on anaerobic metabolism incurs an "oxygen debt" that must be paid off when oxygen becomes available. The excess oxygen intake on recovery from a dive is a measure of that debt. If an animal consumed energy at the same rate during a dive as before it, this excess intake would be enough to equal a normal oxygen-consumption rate during the dive. We found that it was characteristic in quiet dives, however, for the seal or duck to exhibit an oxygen debt much smaller than this. In the case of the sloth, a tree-living animal that is curiously tolerant of submersion, there was no apparent oxygen debt at all [see illustration on page 181]. The implication was that metabolism must slow down.

We could not settle this definitely by studying the oxygen debt alone; it was conceivable that the debt was being paid off so slowly it eluded us. Temperature measurements, however, confirmed the impression of decreased metabolism. We often noticed that after long dives (20 minutes or so) the seal would be shivering during the recovery period. We found that the animal lost body temperature at a rapid rate while submerged. Now, this could not be because of increased heat loss, since there was no substantial change in the thermal contact between the seal and the water; only the nostrils were submerged for the dive. Moreover, the reduction of circulation meant that heat conductivity was lessened, not increased. The loss in body temperature therefore meant an actual decrease in heat production—a slowing down of metabolism. Apparently the lack of blood in the tissues simply jams the normal metabolic processes by mass action; the flame of metabolism is damped and burns lower. It is quite logical that submergence should bring about a progressive reduction in energy metabolism, considering that the suspension of breathing ultimately terminates in death, or zero metabolic activity.

In most dives under natural conditions, of course, this general metabolic slowing down is masked. The animal is actively gathering food, and its muscles probably expend energy at several times the resting rate for the total animal. After a few minutes the muscles have used up the private store of oxygen in their myoglobin, and then they depend on anaerobic processes resulting in lactic acid formation. After such dives there is a substantial oxygen debt reflecting the amount of exercise; it is therefore impossible to detect the subtle lowering of metabolism that must still occur in the nonactive tissues deprived of circulation.

It has been fascinating, and of particular interest from the point of view of evolution, to discover the very same asphyxial defense in fishes taken out of the water—diving in reverse, as it were. The response is found in a variety of fishes, including many that would never leave the water under normal conditions. It is most striking in the aquatic versions of diving mammals and diving birds: the fishes that routinely make excursions out of the water, such as the fly-

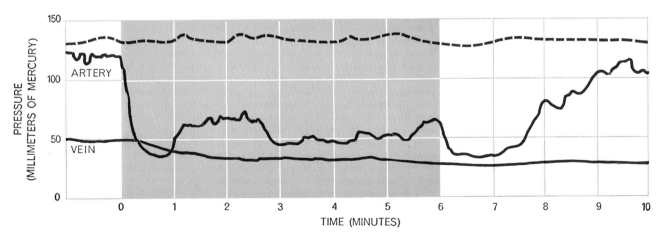

PERIPHERAL PRESSURE, taken in a small toe artery, drops appreciably during a seal's dive (colored area). From near the central blood-pressure level (broken line) it falls almost to the venous level, which indicates a closing down of circulation in the flipper.

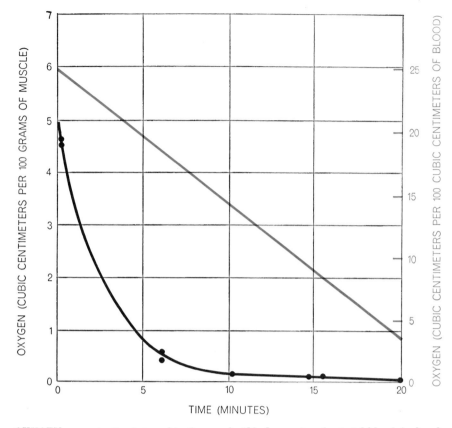

OXYGEN concentration is traced in the muscle (*black curve*) and arterial blood (*colored curve*) of a harbor seal during a dive. The sharp drop in muscle oxygen while the blood is still more than half-saturated suggests that there is no appreciable blood flow in muscle.

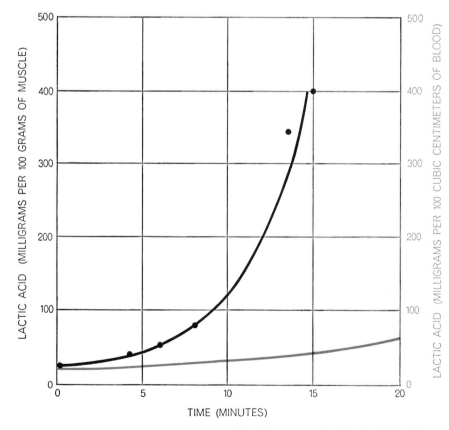

LACTIC ACID concentration confirms lack of blood flow in muscle. Lactic acid builds up in the muscle as the oxygen there is used up, but little enters the blood. The blood lactic acid level rises sharply only after the muscle circulation is restored when breathing resumes.

ing fish. It would be interesting to obtain an electrocardiogram of a flying fish taking off on a natural flight, but this would call for a rather tricky technique. When the leap is simulated, however, by lifting a flying fish out of the water, a profound bradycardia develops immediately.

Another fish that survives on land for some time is the grunion, an amazing little member of the herring family that frequents the coast of California. These fish spawn only on a few nights with maximum tides during the spring. They ride up the beach on a long wave at high tide. As the water recedes the female digs into the sand tail first and deposits her eggs; the male curves around her and fertilizes them. When they have finished, the fish ride out to sea again on another high wave. The spawning procedure can last five or 10 minutes or even longer and is accompanied by much thrashing about; in spite of this activity there is a profound bradycardia during the entire period. Walter F. Garey and Edda D. Bradstreet of the Scripps Institution have studied the lactic acid sequence in grunions caught on the beach and kept overnight in a laboratory tank. The fish are placed in a dish and prodded to keep them wriggling; blood and muscle are sampled during this period and after return to the water. Garey and Miss Bradstreet found that during the anaerobic period lactic acid increases rapidly in the muscles; practically none appears in the circulation until the fish is back in the water. Then, as the peripheral circulation opens up again, lactic acid is flushed out of the muscles and suddenly appears in the blood [*see upper illustration on page 183*].

Whereas fishes such as the grunion dive in reverse, the mudskipper (*Periophthalmus*) performs a double reverse. It spends most of its time out of the water in mangrove swamps at the edge of tropical seas, perching on a mangrove root and slithering, if it is frightened, into a burrow in the mud. These mudholes are frequently devoid of oxygen. By dint of heroic and slippery investigations in northern Australian mangrove swamps, Garey has determined that the heart of a mudskipper in its mudhole develops a pronounced bradycardia. It would seem, then, that the creature has turned evolution around: it is more at home as an air-breathing animal than as a proper fish!

In view of the strikingly similar responses to asphyxia in so many quite different vertebrate animals, it would be strange if human beings did not conform to the common scheme. Indeed, a number of recent studies of human divers, of birth anoxia in babies and of several pathological conditions have turned up exactly the same pattern.

My associates and I obtained valuable information by examining the native pearl divers of northern Australia, who are trained from boyhood to make deep dives. (We found, incidentally, that these experts seldom stay down for longer than a minute; many individual divers can remain submerged for twice as long, but this is evidently too strenuous as a regular practice.) A diver develops bradycardia within 20 to 30 seconds whether he remains quiet or swims about. The arterial blood pressure is normal or even elevated; just as in the seal, the diastolic pressure drop is slowed down, apparently by constriction of the peripheral blood vessels. As we expect-

ed, there is little or no rise in the lactic acid level in the blood during the dive, but there is an acute rise in the recovery period. In all these respects human divers respond like other vertebrates. In one respect, however, human beings may be unique: Pathological arrhythmias, or irregularities of the heartbeat, are alarmingly common in man after only half a minute's dive and such arrhythmias have so far not been observed in animals.

In our laboratory at the Scripps Institution, Elsner has been able to demonstrate ischemia, or lack of blood flow, in the muscles of an extremity simply by having a volunteer submerge his face in a basin of water. An electrocardiograph measures the heart rate, and the flow of blood into the calf is measured by plethysmography. In this technique a cuff placed around the thigh

is inflated just enough to occlude the return of blood through the veins while leaving the arteries open to supply blood to the lower part of the leg. As the calf fills with blood its circumference is measured and traced by a recording device. As soon as the subject immerses his face his heart slows down. At the same time there is a sharp decrease in the extent to which the calf expands when the venous return is obstructed; the constriction of the small arteries diminishes and may virtually stop blood flow into the calf. As soon as the subject lifts his face out of the water and breathes, the arterioles open up again and the calf expands [see illustrations on page 184]. If a subject is merely told to hold his breath without submerging his face, all these effects are less pronounced. As in the case of the seal that is free to breathe at will, psychological

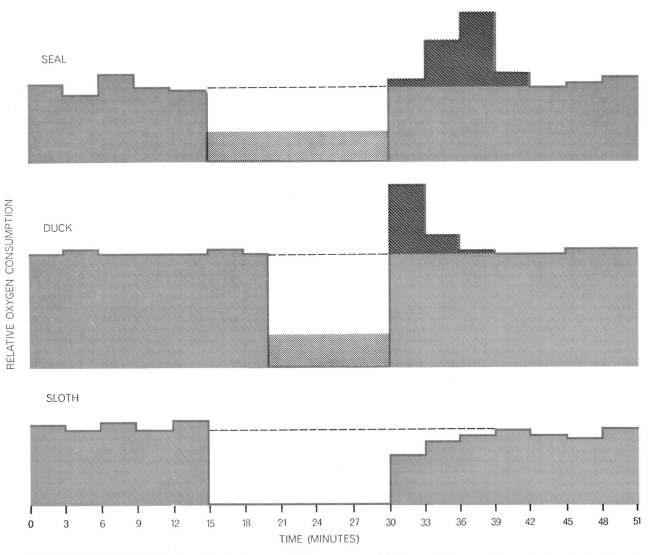

METABOLIC SLOWING DOWN during a dive is demonstrated in three animals by the record of oxygen consumption in successive three-minute periods. In the seal and duck the amount of excess oxygen intake after the dive (hatching on color) represents the oxygen debt incurred by anaerobic metabolism during the dive. This debt (hatching on white) is clearly not enough to have sustained an energy expenditure at a normal rate (broken lines) during the dive. The sloth seems to incur no oxygen debt while diving.

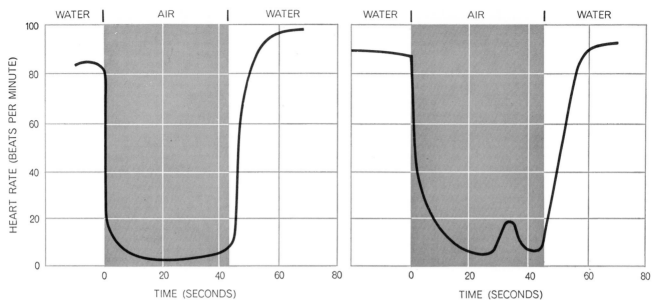

FISH OUT OF WATER develops bradycardia just as a diving animal does when it submerges. These two graphs show the sharp decrease in heart rate that occurs in the grunion (*left*) and the flying fish (*right*) when they are temporarily taken out of water.

factors seem to influence the physiological response to asphyxia.

Physicians have been aware that bradycardia sometimes occurs in babies before, during and immediately after birth and that this can be a sign of asphyxia induced by obstruction or final interruption of the placental blood flow. This concept has been strengthened by lactic acid measurements in newborn infants by Stanley James of the Columbia University College of Physicians and Surgeons. Judging by his data, a normal birth is always followed by a sharp rise in the blood lactic acid. This rise is sharper and higher in babies that have survived a difficult delivery and show clini-

cal symptoms of birth distress; in other words, the longer the period of anoxia, the greater the lactic acid build-up [see *lower illustration on page 183*]. Newborn animals in general have a short period of increased resistance to asphyxia. The sequence of events in babies suggests that selective ischemia is an important asphyxial defense even in newborn infants.

Various pathological conditions that decrease cardiac output, such as arrhythmias and coronary occlusions, are sometimes followed by such apparently unrelated complications as damage to the kidneys or even gangrenous sores in the intestine. Donald D. Van Slyke

and his collaborators at the Rockefeller Institute for Medical Research found in 1944 that severe shock in dogs resulted in decreased kidney function and tissue damage—and that the same symptoms appeared if they simply clamped the renal artery of a healthy dog. Pointing out the analogy to the peripheral vasoconstriction we had reported in diving animals, Van Slyke concluded that under stress the blood supply to the brain is maintained, if necessary, at the cost of restriction of circulation to other areas: the organism, as he said recently, is reduced to "a heart-lung-brain preparation."

More recently Eliot Corday and his

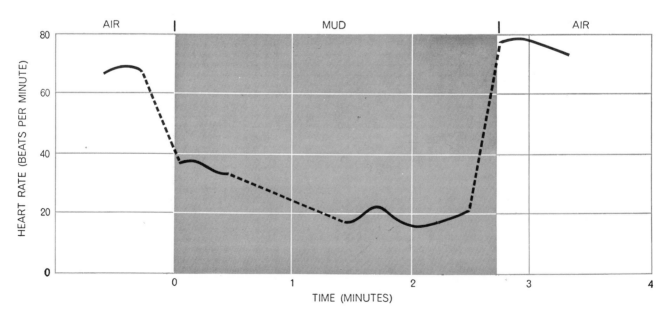

MUDSKIPPER is a curious fish that has become acclimated to breathing air. While it is out of water, its heart rate is normal; when it enters its mud-filled burrow, it develops bradycardia. Broken lines join the various segments of this fragmentary record.

colleagues at the University of California at Los Angeles have found that the same events account for certain gangrenous lesions of the intestine. They impaired the circulation of dogs in various ways, inflicting cardiac arrhythmias by electrical stimulation or decreasing the blood pressure by bleeding the animals. With modern blood-flow-metering techniques and blood-pressure measurements they were able to demonstrate a widespread vasoconstriction that tends to sustain the blood pressure near a normal level but leaves the kidney, the gastrointestinal tract, the muscles and the skin with greatly reduced circulation. These workers again recognized the sequence as a mechanism for maintaining an adequate blood supply to the most sensitive organs.

A quite different physiological event that seems to depend on the same circulatory switch as the prime control is hibernation. In all the relatively few species of mammals and birds that hibernate the body temperature is lowered in the presence of an unfavorable thermal environment. In most animals hibernation is seasonal but in others the temperature drops in a daily cycle. The dormant state is characterized, in any case, by a body temperature only a degree or so warmer than the surroundings; along with this there is a correspondingly low metabolic rate, perhaps a tenth or less of the resting rate in the waking condition. The heart rate is very low—only a few beats per minute—but the central blood pressure remains quite high in relation to this bradycardia. Again the pressure trace shows the slow diastolic emptying of the arteries that suggests a peripheral vasoconstriction. There is good evidence that hibernation is a controlled state; when a decrease in the ambient temperature brings a threat of freezing, the animal increases its heat production and usually emerges from hibernation.

The transition periods during which the animal enters or emerges from hibernation are of particular interest. When a ground squirrel or woodchuck goes into hibernation, the heart rate slows down before the body temperature starts to drop, indicating that the drop in metabolic rate is caused—as in asphyxial defense—by a primary vasoconstriction. Arousal from hibernation is easier to study because it can be precipitated at will by disturbing the animal. This triggers an immediate acceleration of the heartbeat to as much as 100 times the hibernating rate. There follows an in-

tense shivering of the front part of the body, which warms up much more quickly than the rest of the body does as measured by the rectal temperature. Midway through arousal the blood flow in the forelegs of the squirrel is sometimes 10 times greater than in the hind legs. The uneven distribution of metabolic and circulatory activity is apparent-

ly accomplished by a dilation of the blood vessels that begins in the forward parts. When the vessels in the rest of the animal finally dilate, the over-all metabolic rate sometimes rises as high as when the animal exercises. The entire sequence is consistent with the idea that the onset and termination of hibernation are triggered in the first instance by

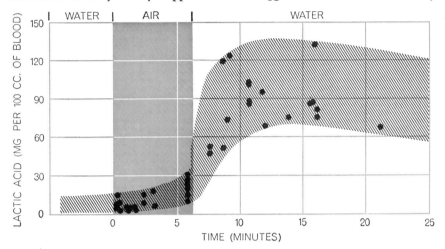

MUSCLE ISCHEMIA, or lack of blood, in grunions results in a lactic acid build-up in muscle while the fish is out of the water. As seen here, the lactic acid does not rise much in the blood until the muscle circulation is restored when the fish re-enters the water.

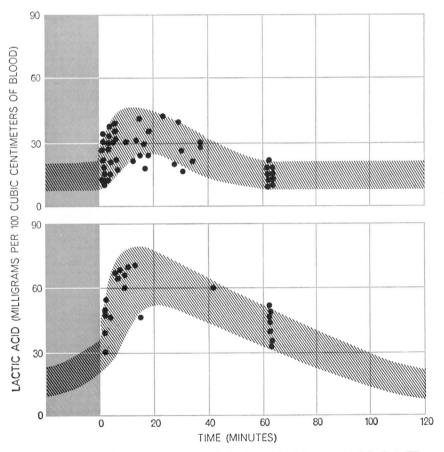

SIMILAR ISCHEMIA apparently protects a baby during the delivery period (color). When breathing begins, the muscle circulation opens up and lactic acid floods the blood. The lactic acid build-up is smaller in a normal delivery (top) than in a long, difficult one (bottom).

vasomotor impulses controlling the size of the small blood vessels. The circulation then throttles metabolism in the tissues to a rate compatible with the blood flow. Going into hibernation seems to call for the same primary vasoconstriction that operates in asphyxial defense.

Any mechanism that operates in many kinds of animals across a wide range of circumstances must be of fundamental physiological significance. In our current work at the Scripps Institution we are trying to learn more about the details of blood flow in animals by implanting ultrasonic measuring devices on arteries and veins. We hope to discover just how the autonomic nervous system responds to environmental changes and the threat of anoxia and what sequence of events actually throws the circulatory switch.

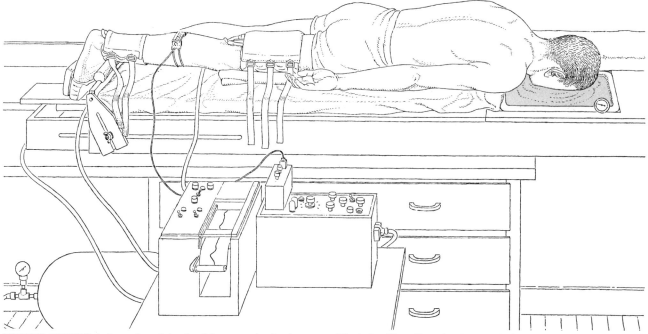

HUMAN DIVING is investigated in the laboratory by having a volunteer immerse his face in a basin of water. In this case the circulation in the lower leg is being measured by plethysmography.

The inflatable cuff on the thigh occludes the veins draining the calf but leaves the arteries open. By measuring the circumference of the calf one can determine the blood flow into the lower leg.

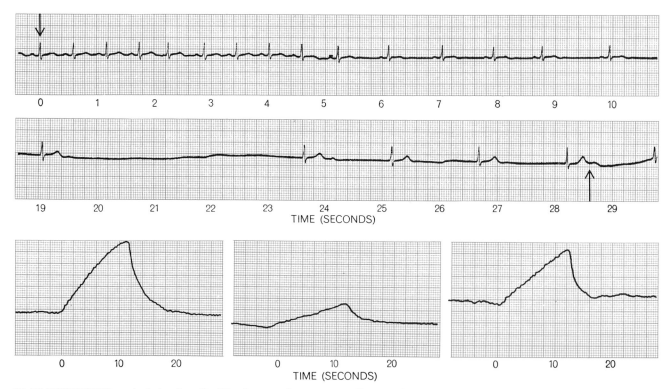

FACE IMMERSION results in bradycardia. The electrocardiogram (*two top strips*) records an extreme case (*arrows mark start and end of dive*). Plethysmographic records (*bottom*) show changes in calf circumference when venous return is occluded for some 12 seconds before (*left*), during (*center*) and after (*right*) face immersion. Blood flow into calf is clearly much reduced during dive.

18

THE DIVING WOMEN OF KOREA AND JAPAN

SUK KI HONG AND HERMANN RAHN
May 1967

Some 30,000 of these breath-holding divers, called ama, are employed in daily foraging for food on the bottom of the sea. Their performance is of particular interest to the physiologist

Off the shores of Korea and southern Japan the ocean bottom is rich in shellfish and edible seaweeds. For at least 1,500 years these crops have been harvested by divers, mostly women, who support their families by daily foraging on the sea bottom. Using no special equipment other than goggles (or glass face masks), these breath-holding divers have become famous the world over for their performances. They sometimes descend to depths of 80 feet and can hold their breath for up to two minutes. Coming up only for brief rests and a few breaths of air, they dive repeatedly, and in warm weather they work four hours a day, with resting intervals of an hour or so away from the water. The Korean women dive even in winter, when the water temperature is 50 degrees Fahrenheit (but only for short periods under such conditions). For those who choose this occupation diving is a lifelong profession; they begin to work in shallow water at the age of 11 or 12 and sometimes continue to 65. Childbearing does not interrupt their work; a pregnant diving woman may work up to the day of delivery and nurse her baby afterward between diving shifts.

The divers are called ama. At present there are some 30,000 of them living and working along the seacoasts of Korea and Japan. About 11,000 ama dwell on the small, rocky island of Cheju off the southern tip of the Korean peninsula, which is believed to be the area where the diving practice originated. Archaeological remains indicate that the practice began before the fourth century. In times past the main objective of the divers may have been pearls, but today it is solely food. Up to the 17th century the ama of Korea included men as well as women; now they are all women. And in Japan, where many of the ama are male, women nevertheless predominate in the occupation. As we shall see, the female is better suited to this work than the male.

In recent years physiologists have found considerable interest in studying the capacities and physiological reactions of the ama, who are probably the most skillful natural divers in the world. What accounts for their remarkable adaptation to the aquatic environment, training or heredity or a combination of both? How do they compare with their nondiving compatriots? The ama themselves have readily cooperated with us in these studies.

We shall begin by describing the dive itself. Basically two different approaches are used. One is a simple system in which the diver operates alone; she is called *cachido* (unassisted diver). The other is a more sophisticated technique; this diver, called a *funado* (assisted diver), has a helper in a boat, usually her husband.

The *cachido* operates from a small float at the surface. She takes several deep breaths, then swims to the bottom, gathers what she can find and swims up to her float again. Because of the oxygen consumption required for her swimming effort she is restricted to comparatively shallow dives and a short time on the bottom. She may on occasion go as deep as 50 or 60 feet, but on the average she limits her foraging to a depth of 15 or 20 feet. Her average dive lasts about 30 seconds, of which 15 seconds is spent working on the bottom. When she surfaces, she hangs on to the float and rests for about 30 seconds, taking deep breaths, and then dives again. Thus the cycle takes about a minute, and the diver averages about 60 dives an hour.

The *funado* dispenses with swimming effort and uses aids that speed her descent and ascent. She carries a counterweight (of about 30 pounds) to pull her to the bottom, and at the end of her dive a helper in a boat above pulls her up with a rope. These aids minimize her oxygen need and hasten her rate of descent and ascent, thereby enabling her to go to greater depths and spend more time on the bottom. The *funado* can work at depths of 60 to 80 feet and average 30 seconds in gathering on the bottom—twice as long as the *cachido*. However, since the total duration of each dive and resting period is twice that of the *cachido*, the *funado* makes only about 30 dives per hour instead of 60. Consequently her bottom time per hour is about the same as the *cachido*'s. Her advantage is that she can harvest deeper bottoms. In economic terms this advantage is partly offset by the fact that the *funado* requires a boat and an assistant.

There are variations, of course, on the two basic diving styles, almost as many variations as there are diving locations. Some divers use assistance to ascend but not to descend; some use only light weights to help in the descent, and so on.

By and large the divers wear minimal clothing, often only a loincloth, during their work in the water. Even in winter the Korean divers wear only cotton bathing suits. In Japan some ama have recently adopted foam-rubber suits, but most of the diving women cannot afford this luxury.

The use of goggles or face masks to improve vision in the water is a comparatively recent development—hardly a century old. It must have revolutionized the diving industry and greatly increased the number of divers and the size of the harvest. The unprotected human eye suffers a basic loss of visual acuity in water because the light passing through water undergoes relatively little refraction when it enters the tissue of the cornea, so that the focal point of the image is considerably behind the retina [*see top*

JAPANESE DIVING WOMAN was photographed by the Italian writer Fosco Maraini near the island of Hekura off the western coast of Japan. The ama's descent is assisted by a string of lead weights tied around her waist. At the time she was diving for aba- lone at a depth of about 30 feet. At the end of each dive a helper in a boat at the surface pulls the ama up by means of the long rope attached to her waist. The other rope belongs to another diver. The ama in this region wear only loincloths during their dives.

illustration on page 191]. Our sharp vision in air is due to the difference in the refractive index between air and the corneal tissue; this difference bends light sharply as it enters the eye and thereby helps to focus images on the retinal surface. (The lens serves for fine adjustments.) Goggles sharpen vision in the water by providing a layer of air at the interface with the eyeball.

Goggles create a hazard, however, when the diver descends below 10 feet in the water. The hydrostatic pressure on the body then increases the internal body pressures, including that of the blood, to a level substantially higher than the air pressure behind the goggles. As a result the blood vessels in the eyelid lining may burst. This conjunctival bleeding is well known to divers who have ventured too deep in the water with only simple goggles. When the Korean and Japanese divers began to use goggles, they soon learned that they must compensate for the pressure factor. Their solution was to attach air-filled, compressible bags (of rubber or thin animal hide) to the goggles. As the diver descends in the water the increasing water pressure compresses the bags, forcing more air into the goggle space and thus raising the air pressure there in proportion to the increase in hydrostatic pressure on the body. Nowadays, in place of goggles, most divers use a face mask covering the nose, so that air from the lungs instead of from external bags can serve to boost the air pressure in front of the eyes.

The ama evolved another technique that may or may not have biological value. During hyperventilation before their dives they purse their lips and emit a loud whistle with each expiration of breath. These whistles, which can be heard for long distances, have become the trademark of the ama. The basic reason for the whistling is quite mysterious. The ama say it makes them "feel better" and "protects the lungs." Various observers have suggested that it may prevent excessive hyperventilation (which can produce unconsciousness in a long dive) or may help by increasing the residual lung volume, but no evidence has been found to verify these hypotheses. Many of the Japanese divers, male and female, do not whistle before they dive.

Preparing for a dive, the ama hyperventilates for five to 10 seconds, takes a final deep breath and then makes the plunge. The hyperventilation serves to

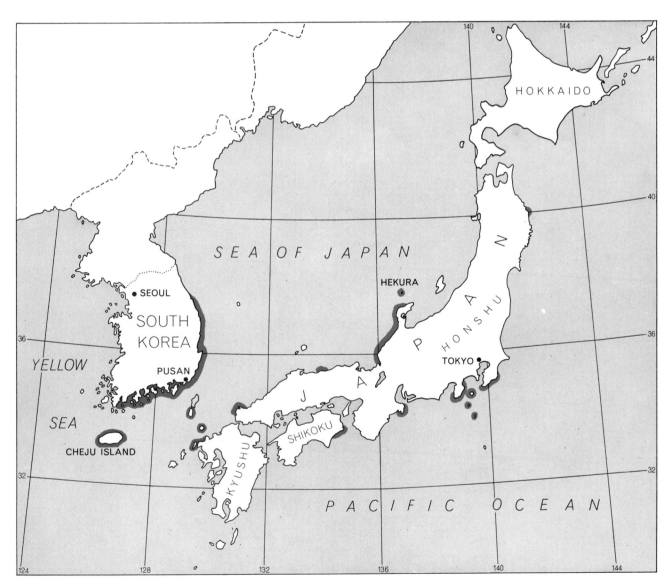

GEOGRAPHIC DISTRIBUTION of the ama divers along the seacoasts of South Korea and southern Japan is indicated by the colored areas. The diving practice is believed to have originated on the small island of Cheju off the southern tip of the Korean peninsula.

UNASSISTED DIVER, called a *cachido,* employs one of the two basic techniques of ama diving. The *cachido* operates from a small float at the surface. On an average dive she swims to a depth of about 15 to 20 feet; the dive lasts about 25 to 30 seconds, of which 15 seconds is spent working on the bottom. The entire diving cycle takes about a minute, and the diver averages 60 dives per hour.

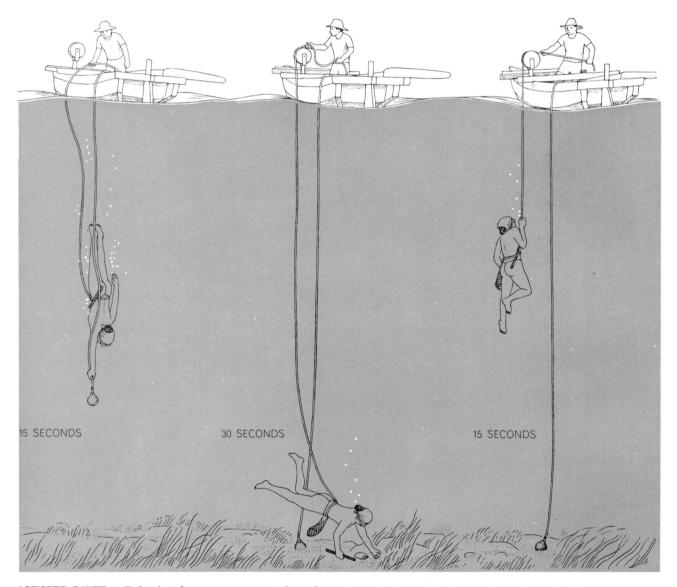

ASSISTED DIVER, called a *funado,* uses a counterweight to descend passively to a depth of 60 to 80 feet. She averages 30 seconds in gathering on the bottom but makes only about 30 dives per hour. At the end of each dive a helper in the boat pulls her up.

remove a considerable amount of carbon dioxide from the blood. The final breath, however, is not a full one but only about 85 percent of what the lungs can hold. Just why the ama limits this breath is not clear; perhaps she does so to avoid uncomfortable pressure in the lungs or to restrict the body's buoyancy in the water.

As the diver descends the water pressure compresses her chest and consequently her lung volume. The depth to which she can go is limited, of course, by the amount of lung compression she can tolerate. If she dives deeper than the level of maximum lung compression (her "residual lung volume"), she becomes subject to a painful lung squeeze; moreover, because the hydrostatic pressure in her blood vessels then exceeds the air pressure in her lungs, the pulmonary blood vessels may burst.

The diver, as we have noted, starts her dive with a lungful of air that is comparatively rich in oxygen and comparatively poor in carbon dioxide. What happens to the composition of this air in the lungs, and to the exchange with the blood, during the dive? In order to investigate this question we needed a means of obtaining samples of the diver's lung air under water without risk to the diver. Edward H. Lanphier and Richard A. Morin of our group (from the State University of New York at Buffalo) devised a simple apparatus into which the diver could blow her lung air and then reinhale most of it, leaving a small sample of air in the device. The divers were understandably reluctant at first to try this device, because it meant giving up their precious lung air deep under water with the possibility that they might not recover it, but they were eventually reassured by tests of the apparatus.

We took four samples of the diver's lung air: one before she entered the water, a second when she had hyperventilated her lungs at the surface and was about to dive, a third when she reached the bottom at a depth of 40 feet and a fourth after she had returned to the surface. In each sample we measured the concentrations and calculated the partial pressures of the principal gases: oxygen, carbon dioxide and nitrogen.

Normally, in a resting person out of the water, the air in the alveoli of the lungs is 14.3 percent oxygen, 5.2 percent carbon dioxide and 80.5 percent nitrogen (disregarding the rare gases and water vapor). We found that after hyperventilation the divers' alveolar air con-

KOREAN DIVING WOMAN from Cheju Island cooperated with the authors in their study of the physiological reactions to breath-hold diving. The large ball slung over her left shoulder is a float that is left at the surface during the dive; attached to the float is a net for collecting the catch. The black belt was provided by the authors to carry a pressure-sensitive bottle and electrocardiograph wires for recording the heart rate. The ama holds an alveolar, or lung, gas sampler in her right hand. The Korean ama wear only light cotton bathing suits even in the winter, when the water temperature can be as low as 50 degrees Fahrenheit.

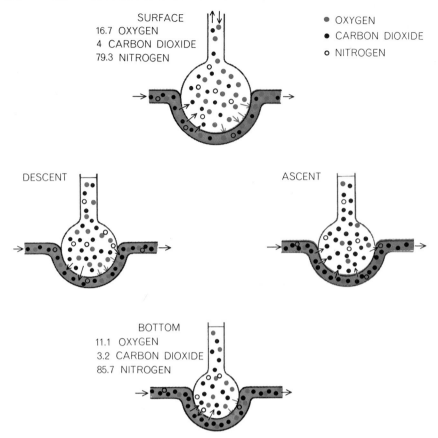

GASES EXCHANGED between a single alveolus, or lung sac, and the bloodstream are shown for four stages of a typical ama dive. The concentrations of three principal gases in the lung at the surface and at the bottom are given in percent. During descent water pressure on the lungs causes all gases to enter the blood. During ascent this situation is reversed.

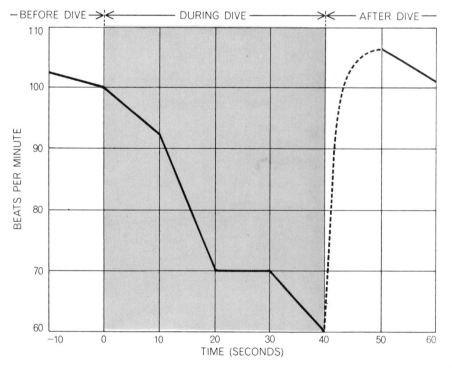

AVERAGE HEART RATE for a group of Korean ama was measured before, during and after their dives. All the dives were to a depth of about 15 feet. The average pattern shown here was substantially the same in the summer, when the water temperature was about 80 degrees Fahrenheit, as it was in winter, when water temperature was about 50 degrees F.

sists of 16.7 percent oxygen, 4 percent carbon dioxide and 79.3 percent nitrogen; translating these figures into partial pressures (in millimeters of mercury), the respective proportions are 120 millimeters for oxygen, 29 for carbon dioxide and 567 for nitrogen.

By the time the *cachido* (unassisted diver) reaches the bottom at a depth of 40 feet the oxygen concentration in her lungs is reduced to 11.1 percent, because of the uptake of oxygen by the blood. However, since at that depth the water pressure has compressed the lungs to somewhat more than half of their pre-dive volume, the oxygen pressure amounts to 149 millimeters of mercury— a greater pressure than before the dive. Consequently oxygen is still being transmitted to the blood at a substantial rate.

For the same reason the blood also takes up carbon dioxide during the dive. The carbon dioxide concentration in the lungs drops from 4 percent at the beginning of the dive to 3.2 percent at the bottom. This is somewhat paradoxical; when a person out of the water holds his breath, the carbon dioxide in his lungs increases. At a depth of 40 feet, however, the compression of the lung volume raises the carbon dioxide pressure to 42 millimeters of mercury, and this is greater than the carbon dioxide pressure in the venous blood. As a result the blood and tissues retain carbon dioxide and even absorb some from the lungs.

As the diver ascends from the bottom, the expansion of the lungs drastically reverses the situation. With the reduction of pressure in the lungs, carbon dioxide comes out of the blood rapidly. Much more important is the precipitous drop of the oxygen partial pressure in the lungs: within 30 seconds it falls from 149 to 41 millimeters of mercury. This is no greater than the partial pressure of oxygen in the venous blood; hence the blood cannot pick up oxygen, and Lanphier has shown that it may actually lose oxygen to the lungs. In all probability that fact explains many of the deaths that have occurred among sports divers returning to the surface after deep, lengthy dives. The cumulative oxygen deficiency in the tissues is sharply accentuated during the ascent.

Our research has also yielded a measure of the nitrogen danger in a long dive. We found that at a depth of 40 feet the nitrogen partial pressure in the compressed lungs is doubled (to 1,134 millimeters of mercury), and throughout the dive the nitrogen tension is sufficient to drive the gas into the blood. Lanphier has calculated that repeated dives to depths of 120 feet, such as are performed by male pearl divers in the Tuamotu Archipelago of the South Pacific, can result in enough accumulation of nitrogen in the blood to cause the bends on ascent. When these divers come to the surface they are sometimes stricken by fatal attacks, which they call *taravana*.

The ama of the Korean area are not so reckless. Long experience has taught them the limits of safety, and, although they undoubtedly have some slight anoxia at the end of each dive, they quickly recover from it. The diving women content themselves with comparatively short dives that they can perform again and again for extended periods without serious danger. They avoid excessive depletion of oxygen and excessive accumulation of nitrogen in their blood.

As far as we have been able to determine, the diving women possess no particular constitutional aptitudes of a hereditary kind. The daughters of Korean farmers can be trained to become just as capable divers as the daughters of divers. The training, however, is important. The most significant adaptation the trained diving women show is an unusually large "vital capacity," measured as the volume of air that can be drawn into the lungs in a single inspiration after a complete expiration. In this attribute the ama are substantially superior to nondiving Korean women. It appears that the divers acquire this capacity through development of the muscles involved in inspiration, which also serve to resist compression of the chest and lung volume in the water.

A large lung capacity, or oxygen intake, is one way to fortify the body for diving; another is conservation of the oxygen stored in the blood. It is now well known, thanks to the researches of P. F. Scholander of the Scripps Institution of Oceanography and other investigators, that certain diving mammals and birds have a built-in mechanism that minimizes their need for oxygen while they are under water [see "The Master Switch of Life," by P. F. Scholander, which begins on page 176 in this book]. This mechanism constricts the blood vessels supplying the kidneys and most of the body muscles so that the blood flow to these organs is drastically reduced; meanwhile a normal flow is maintained to the heart, brain and other organs that require an undiminished supply of oxygen. Thus the heart can slow down, the rate of removal of oxygen from the blood by tissues is reduced,

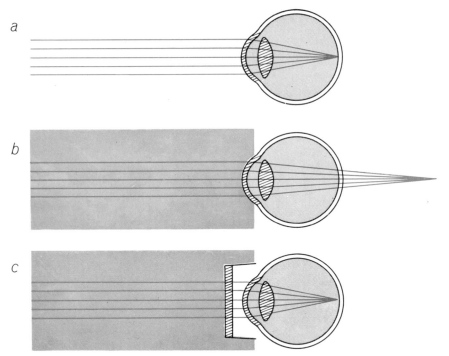

GOGGLES SHARPEN VISION under water by providing a layer of air at the interface with the eyeball (*c*). Vision is normally sharp in air because the difference in refractive index between air and the tissue of the cornea helps to focus images on the retinal surface (*a*). The small difference in the refractive index between water and corneal tissue causes the focal point to move considerably beyond the retina (*b*), reducing visual acuity under water.

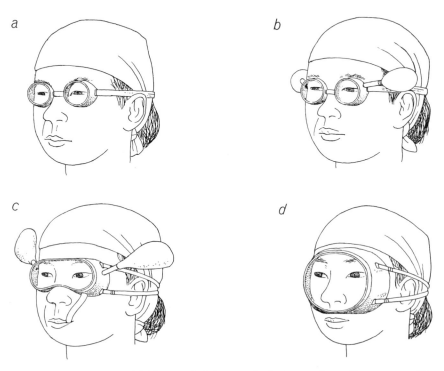

EVOLUTION OF GOGGLES has resulted in several solutions to the problem presented by the increase in hydrostatic pressure on the body during a dive. The earliest goggles (*a*) were uncompensated, and the difference in pressure between the blood vessels in the eyelid and the air behind the goggles could result in conjunctival bleeding. The problem was first solved by attaching air-filled, compressible bags to the goggles (*b*). During a dive the increasing water pressure compresses the bags, raising the air pressure behind the goggles in proportion to the increase in hydrostatic pressure on the body. In some cases (*c*) the lungs were used as an additional compensating gas chamber. With a modern face mask that covers the nose (*d*) the lungs provide the only source of compensating air pressure during a dive.

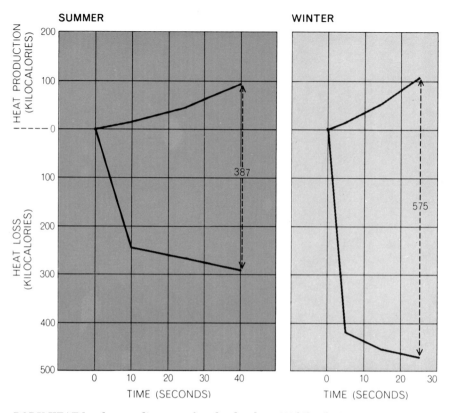

BODY HEAT lost by ama divers was found to be about 400 kilocalories in a summer shift (*left*) and about 600 kilocalories in a winter shift (*right*). The curves above the abscissa at zero kilocalories represent heat generated by swimming and shivering and were estimated by the rate of oxygen consumption. The curves below abscissa represent heat lost by the body to the water and were estimated by changes in rectal temperature and skin temperature.

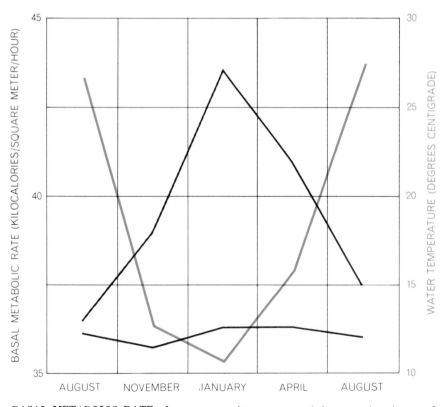

BASAL METABOLIC RATE of ama women (*top gray curve*) increases in winter and decreases in summer. In nondiving Korean women (*bottom gray curve*) basal metabolic rate is constant throughout the year. The colored curve shows the mean seawater temperatures in the diving area of Pusan harbor for the same period covered by the other measurements.

and the animal can prolong its dive.

Several investigators have found recently that human subjects lying under water also slow their heart rate, although not as much as the diving animals do. We made a study of this matter in the ama during their dives. We attached electrodes (sealed from contact with the seawater) to the chests of the divers, and while they dived to the bottom, at the end of a 100-foot cable, an electrocardiograph in our boat recorded their heart rhythms. During their hyperventilation preparatory to diving the divers' heart rate averaged about 100 beats a minute. During the dive the rate fell until, at 20 seconds after submersion, it had dropped to 70 beats; after 30 seconds it dropped further to some 60 beats a minute [*see bottom illustration on page 190*]. When the divers returned to the surface, the heart rate jumped to slightly above normal and then rapidly recovered its usual beat.

Curiously, human subjects who hold their breath out of the water, even in an air pressure chamber, do not show the same degree of slowing of the heart. It was also noteworthy that in about 50 percent of the dives the ama showed some irregularity of heartbeat. These and other findings raise a number of puzzling questions. Nevertheless, one thing is quite clear: the automatic slowing of the heart is an important factor in the ability of human divers to extend their time under water.

In the last analysis the amount of time one can spend in the water, even without holding one's breath, is limited by the loss of body heat. For the working ama this is a critical factor, affecting the length of their working day both in summer and in winter. (They warm themselves at open fires after each long diving shift.) We investigated the effects of their cold exposure from several points of view, including measurements of the heat losses at various water temperatures and analysis of the defensive mechanisms brought into play.

For measuring the amount of the body's heat loss in the water there are two convenient indexes: (1) the increase of heat production by the body (through the exercise of swimming and shivering) and (2) the drop in the body's internal temperature. The body's heat production can be measured by examining its consumption of oxygen; this can be gauged from the oxygen content of the lungs at the end of a dive and during recovery. Our measurements were made on Korean diving women in Pusan harbor at two seasons of the year: in August, when the water temperature was 80.6 degrees F., and in January with the water temperature at 50 degrees.

In both seasons at the end of a single diving shift (40 minutes in the summer, 25 minutes in winter) the deep-body temperature was found to be reduced from the normal 98.6 degrees F. to 95 degrees or less. Combining this information with the measurements of oxygen consumption, we estimated that the ama's body-heat loss was about 400 kilocalories in a summer shift and about 600 kilocalories in a winter shift. On a daily basis, taking into consideration that the ama works in the water for three long shifts each day in summer and only one or two short shifts in winter, the day's total heat loss is estimated to be about the same in all seasons: approximately 1,000 kilocalories per day.

To compensate for this loss the Korean diving woman eats considerably more than her nondiving sisters. The ama's daily food consumption amounts to about 3,000 kilocalories, whereas the average for nondiving Korean women of comparable age is on the order of 2,000 kilocalories per day. Our various items of evidence suggest that the Korean diving woman subjects herself to a daily cold stress greater than that of any other group of human beings yet studied. Her extra food consumption goes entirely into coping with this stress. The Korean diving women are not heavy; on the contrary, they are unusually lean.

It is interesting now to examine whether or not the diving women have developed any special bodily defenses against cold. One such defense would be an elevated rate of basal metabolism, that is, an above-average basic rate of heat production. There was little reason, however, to expect to find the Korean women particularly well endowed in this respect. In the first place, populations of mankind the world over, in cold climates or warm, have been found to differ little in basal metabolism. In the second place, any elevation of the basal rate that might exist in the diving women would be too small to have much effect in offsetting the large heat losses in water.

Yet we found to our surprise that the diving women did show a significant elevation of the basal metabolic rate—but only in the winter months! In that season their basal rate is about 25 percent higher than that of nondiving women of the same community and the same economic background (who show no seasonal change in basal metabolism). Only one other population in the world has been found to have a basal metabolic rate as high as that of the Korean diving women in winter: the Alaskan Eskimos. The available evidence indicates that the warmly clothed Eskimos do not, however, experience consistently severe cold stresses; their elevated basal rate is believed to arise from an exceptionally large amount of protein in their diet. We found that the protein intake of Korean diving women is not particularly high. It therefore seems probable that their elevated basal metabolic rate in winter is a direct reflection of their severe exposure to cold in that season, and that this in turn indicates a latent human mechanism of adaptation to cold that is evoked only under extreme cold stresses such as the Korean divers experience. The response is too feeble to give the divers any significant amount of protection in the winter water. It does, however, raise an interesting physiological question that we are pursuing with further studies, namely the possibility that severe exposure to winter cold may, as a general rule, stimulate the human thyroid gland to a seasonal elevation of activity.

The production of body heat is one aspect of the defense against cold; another is the body's insulation for retaining heat. Here the most important factor (but not the only one) is the layer of fat

BETWEEN DIVES the ama were persuaded to expire air into a large plastic gas bag in order to measure the rate at which oxygen is consumed in swimming and diving to produce heat. The water temperature in Pusan harbor at the time (January) was 50 degrees F. One of the authors (Hong) assists. Data obtained in this way were used to construct the graph at the top of the preceding page.

under the skin. The heat conductivity of fatty tissue is only about half that of muscle tissue; in other words, it is twice as good an insulator. Whales and seals owe their ability to live in arctic and antarctic waters to their very thick layers of subcutaneous fat. Similarly, subcutaneous fat explains why women dominate the diving profession of Korea and Japan; they are more generously endowed with this protection than men are.

Donald W. Rennie of the State University of New York at Buffalo collaborated with one of the authors of this article (Hong) in detailed measurements of the body insulation of Korean women, comparing divers with nondivers. The thickness of the subcutaneous fat can easily be determined by measuring the thickness of folds of skin in various parts of the body. This does not, however, tell the whole story of the body's thermal insulation. To measure this insulation in functional terms, we had our subjects lie in a tank of water for three hours with only the face out of the water. From measurements of the reduction in deep-body temperature and the body's heat production we were then able to calculate the degree of the subject's overall thermal insulation. These studies revealed three particularly interesting facts. They showed, for one thing, that with the same thickness of subcutaneous fat, divers had less heat loss than nondivers. This was taken to indicate that the divers' fatty insulation is supplemented by some kind of vascular adaptation that restricts the loss of heat from the blood vessels to the skin, particularly in the arms and legs. Secondly, the observations disclosed that in winter the diving women lose about half of their subcutaneous fat (although nondivers do not). Presumably this means that during the winter the divers' heat loss is so great that their food intake does not compensate for it sufficiently; in any case, their vascular adaptation helps them to maintain insulation. Thirdly, we found that diving women could tolerate lower water temperatures than nondiving women without shivering. The divers did not shiver when they lay for three hours in water at 82.8 degrees F.; nondivers began to shiver at a temperature of 86 degrees. (Male nondivers shivered at 88 degrees.) It appears that the diving women's resistance to shivering arises from some hardening aspect of their training that inhibits shiver-triggering impulses from the skin. The inhibition of shivering is an advantage because shivering speeds up the emission of body heat. L. G. Pugh, a British physiologist

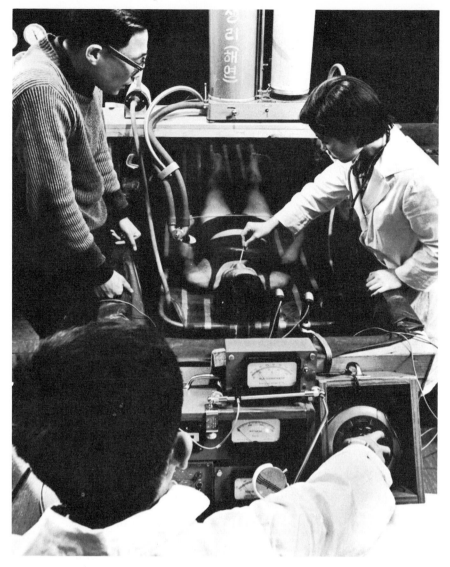

AMA'S THERMAL INSULATION (mainly fat) was measured by having the subjects lie in a tank of water for three hours. From measurements of the reduction of deep-body temperature and the body's heat production the authors were able to calculate the degree of the subject's overall thermal insulation. Once again Hong (*left*) keeps a close eye on the operation.

who has studied long-distance swimmers, discovered the interesting additional fact that swimmers, whether fat or thin, lose heat more rapidly while swimming than while lying motionless in the water. The whole subject of the body's thermal insulation is obviously a rather complicated one that will not be easy to unravel. As a general conclusion, however, it is very clearly established that women are far better insulated than men against cold.

As a concluding observation we should note that the 1,500-year-old diving occupation in Korea and Japan is now declining. The number of divers has dwindled during the past few decades, and by the end of this century the profession may disappear altogether, chiefly because more remunerative and less

arduous ways of making a living are arising. Nonetheless, for the 30,000 practitioners still active in the diving profession (at least in summer) diving remains a proud calling and necessary livelihood. By adopting scuba gear and other modern underwater equipment the divers could greatly increase their production; the present harvest could be obtained by not much more than a tenth of the present number of divers. This would raise havoc, however, with employment and the economy in the hundreds of small villages whose women daily go forth to seek their families' existence on the sea bottom. For that reason innovations are fiercely resisted. Indeed, many villages in Japan have outlawed the foam-rubber suit for divers to prevent too easy and too rapid harvesting of the local waters.

19

THE PHYSIOLOGY OF HIGH ALTITUDE

RAYMOND J. HOCK
February 1970

To meet the stress of life at high altitude, notably lack of oxygen, a number of changes in body processes are required. What are these adaptations, and are they inborn or the result of acclimatization?

At altitudes above 6,000 feet the human organism leaves its accustomed environment and begins to feel the stresses imposed by an insufficiency of oxygen. Yet 25 million people manage to live and work in the high Andes of South America and the Himalayan ranges of Asia. More than 10 million of them live at altitudes above 12,000 feet, and there are mountain dwellers in Peru who daily go to work in a mine at an elevation of 19,000 feet. How does the human physiology contrive to acclimatize itself to such conditions? Over the past half-century, ever since the British physiologist Joseph Barcroft led an expedition to study the physiology of the mountain natives of Peru in the early 1920's, a small host of fascinated investigators has been exploring this puzzle. We now know many of the details of the body's remarkable ability to accommodate itself to life in an oxygen-poor environment.

The study has an intrinsic lure, akin to the challenge of an Everest for a mountain climber, and in these days of man's travels beyond the earth's atmosphere the subject of oxygen's relation to life has taken on added interest. The problem also has its practical aspects on our own planet. Already more and more people each year have recourse to mountain heights for recreations such as camping and skiing; for example, in 1968 there were five million visitor-days in the Inyo National Forest of California, nearly all at elevations of 7,000 feet or higher. From studies of the physiology of adjustment to oxygen deprivation we can expect some beneficial dividends, not only for the problems of living or vacationing in the high mountains but also for medical problems in diseases involving hypoxia.

The native mountain dwellers of Peru and the Sherpa people of Tibet in the Himalayas have served as very helpful subjects for the investigation of acclimatization to high-altitude life. British expeditions led by L. G. C. E. Pugh have conducted several important studies of the Sherpas. In the Andes the research has been centered principally in the Institute of Andean Biology, a permanent station founded in 1928 as a division of the University of San Marcos by Carlos Monge with Alberto Hurtado as director of research [see "Life at High Altitudes," by George W. Gray; SCIENTIFIC AMERICAN, December, 1955]. Recently Pennsylvania State University established another station in Peru at Nuñoa under the supervision of Paul T. Baker and Elsworth R. Buskirk. In the U.S. the University of California has a major center for high-altitude research on White Mountain, with Nello Pace as director. I was associated with the White Mountain Research Station for several years as resident physiologist, working primarily with experimental animals. There are four laboratories in the complex: the Barcroft Laboratory at 12,500 feet and others at 14,250 feet, 10,150 and 4,000 feet. Investigators from the U.S. Army Laboratory of the Fitzsimons Hospital have also been active in human high-altitude research in the Rockies, principally on Pikes Peak.

The High-Altitude Environment

One may wonder why people choose to live in the hostile environment of mountain heights, as the Quechua Indians of Peru, for example, have done for centuries. Life on the mountains is made rigorous not only by hypoxia but also by cold. Even in the equatorial Andes the air temperature decreases by one degree Celsius with each 640 feet of altitude. The winters are long, snowy and windy; summers are short and cool. At the 12,500-foot station on White Mountain in California temperature records over a period of 10 years showed that the mean temperature is below freezing during eight months of the year, and even in some of the summer months the nighttime minimum averages below freezing. Plant life at high altitudes has an extremely short growing season, and few animals can breed successfully at these altitudes. The relatively strong ultraviolet radiation, ionization of the air and other harsh factors no doubt affect life there. There are a few compensating factors. The intense sunlight in the thin atmosphere heats the rocks and provides warm niches for life. The heavy winter snowfall lays down a greater store of moisture than may be available in the surrounding lowlands. For a few months the highlands offer grazing for herdsmen's flocks (which are brought down to lower altitudes when the summer season ends).

Most of the people in the highlands live on herding or agriculture, raising short-season crops such as potatoes or some grains. In the Andes mining also is an important factor in the economy. The highest inhabited settlement in the world is a mining camp at 17,500 feet in Peru. The residents there work in the mine at 19,000 feet that I have already mentioned. The miners daily climb the 1,500 feet from their camp to the mine. Significantly, they rebelled against living in a camp that was built for them at 18,500 feet, complaining that they had no appetite, lost weight and could not sleep. It seems, therefore, that 17,500 feet is the highest altitude at which even acclimatized man can live permanently.

Notwithstanding the rigors of life on the heights, the Peruvian Indians of the "altiplano" have thrived in their environment. It is said that the Incas had

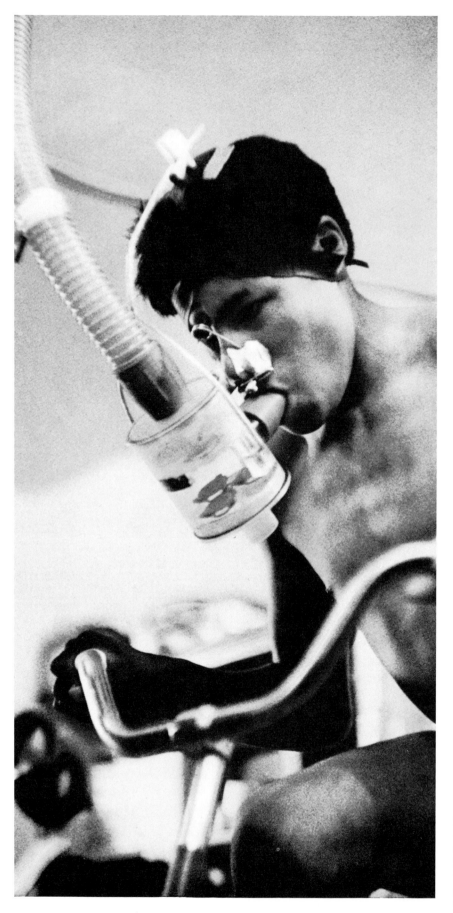

QUECHUA INDIAN breathes from a device that records oxygen intake and carbon dioxide output while he exercises on a bicycle ergometer, which measures the work performed.

two separate armies, one for the lowlands and one for high-mountain duty. Monge and Hurtado believe the high-altitude natives have become a distinctive breed ("Andean man") superbly fitted for life on the heights but probably incapable of surviving long in the lowlands. This is open to doubt. It may be that at sea level the highlanders would succumb to diseases to which they have not been exposed in the mountains, but it has not yet been demonstrated that they would be unable to adjust physiologically to the conditions at low altitudes.

Nevertheless, it is incontestable that high-altitude man, in the Andes and the Himalayas, does indeed possess unusual physiological capabilities. They are evidenced in his responses to hypoxia (defined here as a deficient supply of oxygen in the air). To see his special attributes in perspective, let us first examine the usual reactions of an unacclimatized person to hypoxia.

The Physiological Responses

The proportion of oxygen in the air is not reduced at high altitudes (it is constant at 21 percent throughout the atmosphere), but as the barometric pressure of the air as a whole declines with increasing altitude the partial pressure of the oxygen also declines correspondingly. Thus at 12,500 feet the barometric pressure drops to 480 millimeters of mercury (from 760 millimeters at sea level) and the partial pressure of oxygen is only 100 millimeters, as against 159 millimeters at sea level. That is to say, the number of oxygen and other molecules per cubic foot of air is reduced.

This decrease in oxygen tension, reducing the transfer of oxygen from inspired air to the blood in the lungs, calls forth several immediate reactions by the body. The breathing rate increases, in order to bring more air into the lungs. The heart rate and cardiac output increase, in order to enhance the flow of blood through the lung capillaries and the delivery of arterial blood to the body tissues. The body steps up its production of red blood cells and of hemoglobin to improve the blood's oxygen-carrying capacity. The hemoglobin molecule itself has a physicochemical property that enables it to take in and unload oxygen more readily when necessary at high altitudes. In a person who remains at high altitude these acclimatizing changes take place over a period of time. Investigators who measured them during a Himalayan expedition found that the hemoglobin content of the blood continued to increase for two or three months and then leveled off. As the climbers moved up from 13,000 feet to 19,000 feet and beyond, the number of red cells in the blood increased continuously for as long as 38 weeks.

The adjustments I have just recounted are not sufficient to enable a newcomer to high altitude to expend normal physical effort. Because of the interest stimulated by the holding of the 1968 Olympic Games in Mexico City (at an altitude of 7,500 feet) much study has recently been given to the effects of high altitude on the capacity for exercise. It has been found that at 18,000 feet, for instance, a man's capacity for performing exercise without incurring an oxygen debt is only about 50 percent of that at sea level. The tolerance of such a debt, and of the accumulation of lactic acid in the muscles, also is reduced. This accounts for the fact that mountain climbers at extreme altitudes can take only a few tortured steps at a time and must rest for a considerable period before going on. The limits on the capacity for work are set, of course, by the limits of the body's possible physiological adjustments to the high-altitude conditions. These limits affect the rate of ventilation of the lungs, the heart rate, the cardiac output and the blood flow to the exercising muscles. The limit for hyperventilation, for example, is a flow of 120 liters of air per minute through the lungs. This maximum, invoked at an altitude of about 16,400 feet, supplies two liters of oxygen per minute to the blood. At extreme altitudes the heart can speed up its beat during moderate exercise, but under the stress of maximum exercise the limit for both the

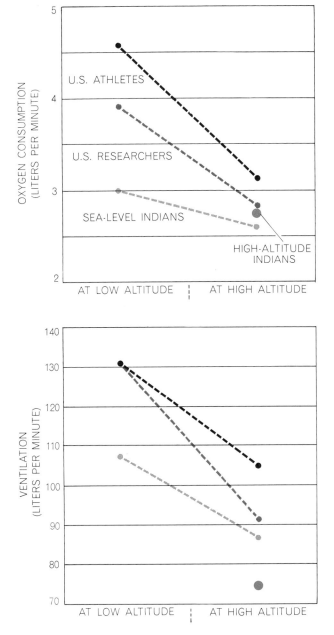

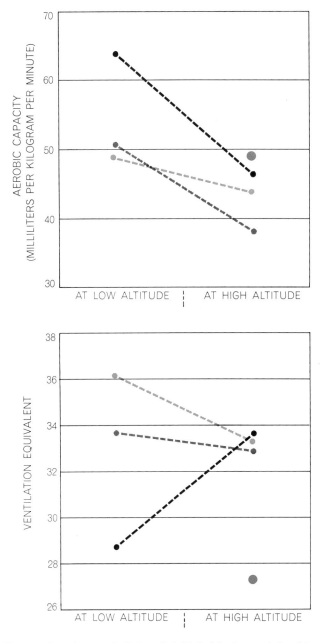

OXYGEN CONSUMPTION of Quechua Indians who were natives of Nuñoa, at 13,000 feet in the Andes (*solid color*), was compared with that of sea-level Quechua (*light color*), U.S. athletes (*black*) and U.S. research workers (*gray*) in a study made by Paul T. Baker of Pennsylvania State University. The data were collected during a bicycle exercise test. (The low-altitude subjects had been at Nuñoa at least four weeks before their high-altitude tests.) Aerobic capacity relates oxygen consumption to an individual's weight and therefore measures the success of his oxygen-transport system. Ventilation is the total volume of air breathed by the individual. Ventilation equivalent is ventilation divided by oxygen consumption; the lower the value, the greater the oxygen-extracting efficiency.

heart rate and the cardiac output is lower than at sea level.

Mountain Natives' Physiology

Let us now turn to the mountain natives' extraordinary adjustments for living at high altitudes. To begin with, the Quechua Indians of the Andes and the Sherpas of the Himalayas have developed an exceptionally large chest and lung volume, enabling them to take in a greater volume of air with each breath. Their breathing rate also is higher than that of dwellers at sea level, but they do not need to hyperventilate as much as lowlanders do when the latter go to high altitudes. The mountain natives also have a high concentration of red cells and hemoglobin in their blood, and their hemoglobin is geared to unload oxygen readily to the tissues.

In the high-altitude native the lung capillaries are dilated, so that the pulmonary circulation carries an unusually large percentage of the body's total blood volume. Moreover, the blood pressure in the lungs is higher than in the rest of the circulatory system. The heart is unusually large, apparently because of the heightened pressure in the pulmonary arteries. The heartbeat is slower than in sea-level dwellers. The mountain dwellers' metabolism also appears to be affected by the hypoxic conditions. Their basal metabolic rate is slightly higher than it is in lowlanders, and when this is considered in terms of body mass, it turns out that the rate of oxygen consumption per unit of metabolizable tissue is unusually high. That is, the hypoxic conditions exact a cost in lowered efficiency in the use of oxygen.

The mountain natives show their superior acclimatization most markedly in their capacity for exercise at high altitude. Sherpas show a smaller increase in ventilation, similar oxygen consumption and a greater heart-rate increase when performing the same exercise as low-altitude subjects who have become thoroughly acclimatized to a high altitude. The ability of mountain natives to perform physical labor daily at altitudes where even acclimatized visitors are quickly exhausted by exercise is itself obvious evidence of the mountaineers' extraordinary physiology.

In general, the physiological adjustments of the permanent mountain dwellers are similar in kind to those developed by sojourners in the mountains after a year of residence there. Furthermore, even mountain natives sometimes lose their acclimatization to high altitude and incur *soroche* (chronic mountain sickness), which is characterized by extreme elevation of the relative number and mass of red cells in the blood, pulmonary hypertension, low peripheral blood pressure, enlargement of the right lobe of the heart and ultimately congestive heart failure if the victim remains at high altitude. In general, the differences between mountain natives and sea-level natives are most apparent in the mountaineers' superior capacity for exercise at high altitude and their ability to produce children in that habitat; newcomers to the mountains, even after extended acclimatization, are much less successful in reproduction. The Spanish conquistadors who settled in the high Andes, for example, found themselves afflicted with relative infertility and a high rate of infant mortality.

Does the special physiology of the mountain people arise from genetic adaptation or is it acquired during their lifelong exposure (from the womb onward) to high altitude? One approach to answering that question has been through investigations with experimental animals. The laboratory studies of animals have also explored the physiological aspects of acclimatization much more exhaustively than is possible in man. Much of this animal work has been done at the White Mountain Research Station.

Hypoxia and Rats

Pace and an associate, Paola S. Timiras, carried out a series of investigations on rats at the White Mountain Research Station. Rats that had been bred at sea level were brought to the Barcroft Laboratory (at 12,500 feet), and the investigators examined the responses to hypoxia in these animals and in the second generation of offspring produced at the high altitude. The development of the animals exposed to the high altitude was compared with that of a control group of rats kept at sea level.

The rats at the Barcroft Laboratory exhibited acclimatizing reactions like those of human newcomers to high altitude. There was a marked increase, for example, in their red cells and hemoglobin: in the imported animals the red-cell concentration rose to 54.6 percent of the blood volume, and in their second-generation offspring it was 66.7 percent, as against 47.5 percent in the control rats of the same age at sea level. The rats also developed an enlargement of the heart like that of human mountain natives. After 10 months at the high altitude they had a 20 percent higher ratio of heart weight to body weight than the sea-level controls did, and in the second-generation rats born at the Barcroft station the increase in heart-weight ratio was 90 percent. An increase in the relative weight of the adrenal glands was also observed in the rats exposed to the high altitude.

The exposure to hypoxia stunted the rats' growth. Up to the age of about 120 days the rats brought to Barcroft (at age 30 days) gained weight at the same rate as the rats left at sea level, but thereafter their growth slowed, and their maximum weight (at about 300 days) was significantly lower than that of the sea-level controls. The growth rate of the second-generation rats at high altitude was lower still: at 130 days they weighed only 250 grams, whereas their parents and the sea-level controls attained this weight in 84 days.

Fenton Kelley at the Barcroft Laboratory investigated the rats' reproduction and high-altitude effects on their young. The hypoxic conditions did not impair the ability to conceive in young, healthy rats: more than 85 percent of the females that were mated after 30 days of acclimatization became pregnant. Their fetuses, however, suffered considerable attrition: by the 15th day of pregnancy 25 percent of the females had abnormally stunted fetuses. Whether this is due to inadequacy of the oxygen supply to the fetus, disturbance of hormone production, neurological anomalies or metabolic disturbances has not yet been determined. At all events, the females bred at high altitude bore substantially smaller litters of live young than those bred at sea level.

The offspring were generally normal in weight at birth, but by the age of 10 days their weight was 30 percent less than the sea-level norm. Their mortality rate in the first 10 days was about 20 percent, 10 times higher than in the sea-level control group. This was not attributable to lack of nursing ability in their mothers; in fact, the high-altitude infant rats had more milk in their stomachs than the sea-level young of the same age did. There were indications that the high-altitude young had metabolic defects that may account for their high mortality and for the slow rate of growth in those offspring that survived the postpartum period. The high-altitude young had a subnormal content of glycogen in the liver, apparently reflecting a defect in carbohydrate metabolism, and there is reason to believe the metabolism of fats and proteins also is affected by hypoxia. The dry atmosphere of high altitude may be another hazard for the newborn, affecting the

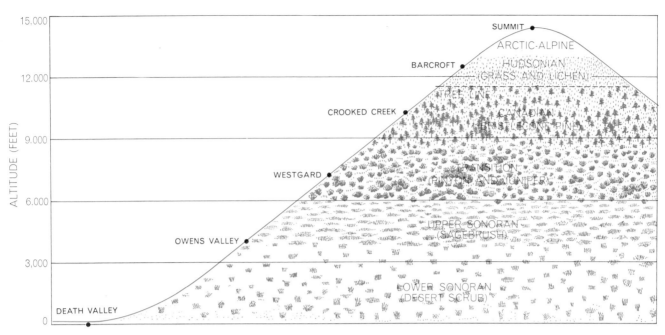

ALTITUDE RANGE of the deer mouse *Peromyscus maniculatus,* the animal the author studied, is remarkably large, extending from sea level to the summit of White Mountain in California. It en-compasses six of the seven "life zones" described some years ago by the American naturalist C. Hart Merriam, which are indicated, with their characteristic vegetation, on this schematic diagram.

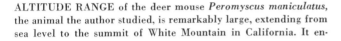
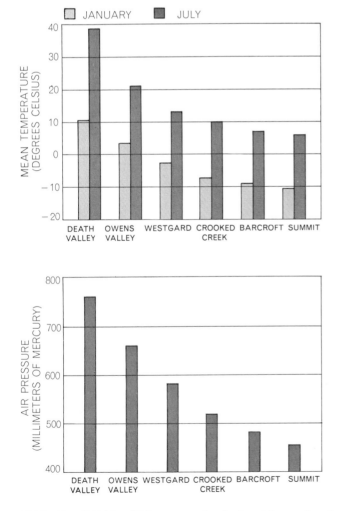

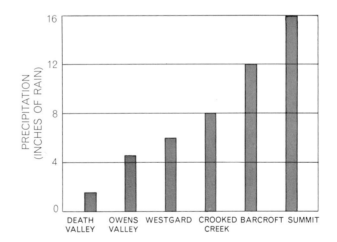

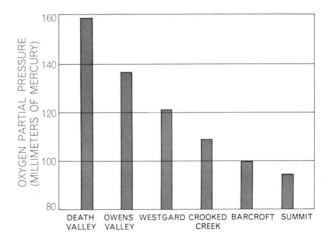

ENVIRONMENTAL DATA are given for the four laboratories of the White Mountain Research Station and two other places where the author worked, Death Valley and Westgard. Altitudes are in-dicated at the top of the page. (Summit precipitation is estimated.)

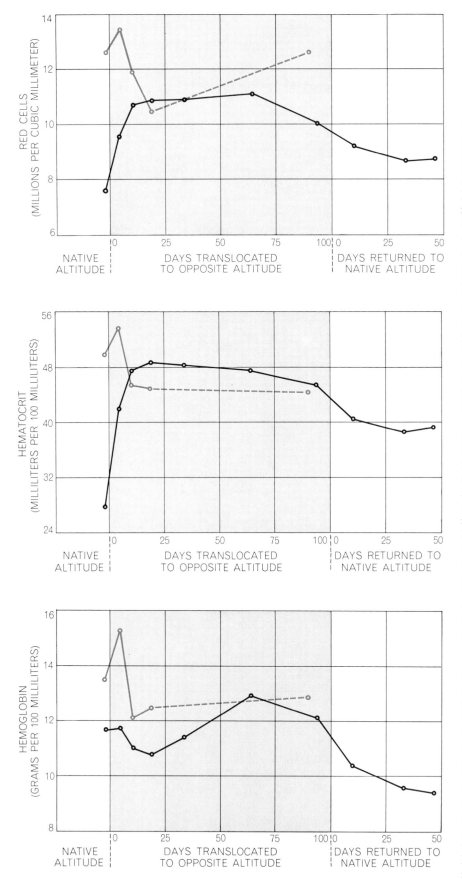

DEER MICE trapped at 12,500 feet (*color*) have more red blood cells, a higher hematocrit (red cells as a proportion of total blood volume) and more hemoglobin than mice trapped at sea level (*black*). As the curves indicate, these values increase in sea-level mice that are transported to the high altitude and decline again with their return to sea level. There is less long-term change, however, in high-altitude mice that are brought down to sea level.

body's water balance, temperature regulation, respiration and vulnerability to infection.

The Deer Mouse

My own investigations focused on the deer mouse (*Peromyscus maniculatus*), a small, white-footed species that is noted for its ubiquitous presence throughout North America and its ability to live in all climatic zones except extreme desert. Deer mice of various species are found inhabiting all altitudes from below sea level to about 15,000 feet. This one species is ideal for our studies not only because of its great variety of physiological responses to different conditions but also because a mouse spends its entire life in the same locality. A tiny deer mouse on White Mountain probably does not range more than a quarter of a mile from its birthplace during its lifetime; consequently we could be sure that a mouse trapped at 12,500 feet on White Mountain had been born at about that altitude (within 500 feet) and had been exposed to it throughout its life.

I trapped deer mice at seven different altitudes, ranging from below sea level to the 14,250-foot White Mountain summit, and for comparison of their native differences each population was examined at the altitude at which it was caught. It was apparent at once that the mice followed only in part the well-known rule that body size increases with exposure to cold: their body weight increased with increasing altitude up to a point. The heaviest weight was found at 10,150 feet; beyond that the oxygen scarcity apparently limits growth in deer mice as it was found to do in the experiments on rats.

There was also a clear progression, with increasing altitude, in the ratio of the heart size to the body weight. In the deer mice caught at 14,250 feet the relative heart weight was one and a half times greater than it was in mice living at an altitude of 4,000 feet. The relative number and mass of red cells and the amount of hemoglobin in the blood also increased with altitude in the deer mice as it does in rats and men. Contrary to what had been observed in rats, however, the mice's adrenal glands shrank with increasing altitude, both in absolute weight and in relation to body weight. Presumably this phenomenon in the mice reflects a different physiological response to the stresses of high altitude from that shown by the rats.

In an effort to determine whether the differences among the natives of various altitudes were genetic or simply the

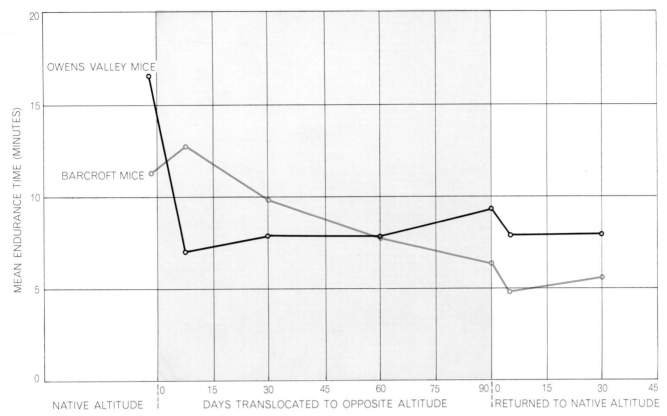

ENDURANCE of mice trapped at 4,000 feet (*black*) was greater than that of mice trapped at 12,500 feet (*color*) when both groups were in their home environments (*left*). The low-altitude mice showed a decided decrease in endurance when they were first taken to 12,500 feet but then improved somewhat. Surprisingly, the high-altitude natives did less well at 4,000 feet than at 12,500.

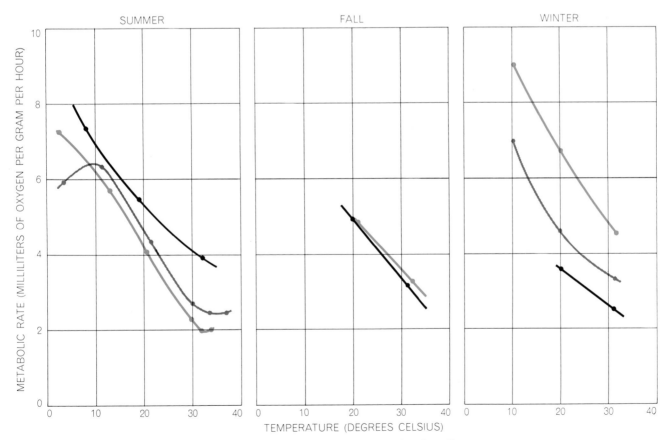

BASAL METABOLIC RATE (oxygen consumption at rest) of mice native to sea level (*black*), 4,000 feet (*gray*) and 12,500 feet (*color*) was measured at their native altitudes in the summer and winter over a wide range of ambient temperatures; two of the three groups were tested in the fall over a narrower temperature range. In the summer the sea-level mice had the highest rates and the 12,500-foot animals had the lowest rates; in the fall they were about equal, and in the winter the summertime findings were reversed (*see text*).

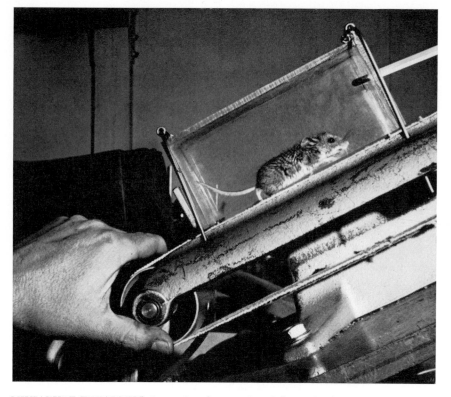

MINIATURE TREADMILL for testing the capacity of deer mice for strenuous exercise was improvised by mounting a linen belt on an old belt sander. The belt was moved at a speed of a mile and a half an hour. The mouse ran on it, restrained by the plastic barriers, until it was exhausted, and the length of time it ran was the measure of its endurance.

decreased with altitude. By the age of maturity (120 days) the average weight of mice born at 4,000 feet was 22.5 grams; of those born at 14,250 feet, 18.8 grams.·

Exercise and Metabolism

In order to test the capacity of the mice for strenuous exercise I developed a miniature treadmill that could be run at various speeds. As was to be expected, the natives at low altitudes showed more endurance than those at higher levels. The mean time before exhaustion on the treadmill in one series of tests, for example, was 16.7 minutes for natives at 4,000 feet and only 11.2 minutes for natives at 12,500 feet. Was the difference in performance due to the handicap of hypoxia at the higher altitude or to some innate physical or physiological difference in the mice themselves? I examined this question by switching the environment for the mice, testing the low-altitude natives at the high station and the high-altitude natives at the lower station.

The 4,000-foot natives, when taken to the 12,500-foot level, at first showed a drop in endurance on the treadmill. Their performance slowly improved, however, as they became acclimatized over a period of 90 days. The transfer of 12,500-foot natives to tests at the 4,000-foot level, on the other hand, yielded a major surprise. Although they were now performing in an atmosphere richer in oxygen, their endurance on the treadmill declined instead of improving. At the end of 90 days of "acclimatization" they were able to run on the treadmill only half as long, before exhaustion, as they had done in their oxygen-poor native environment! The explanation may lie in abnormalities of the heart and certain other functions in the high-altitude mice and in the change to a new climate at the lower altitude.

In one study I compared low-altitude and high-altitude mice with regard to their consumption of oxygen during exercise. When sea-level natives were transferred to high altitude, they did not increase their oxygen consumption more while exercising than they had at sea level. High-altitude natives, on the other hand, showed a considerably greater increase in oxygen consumption during exercise than the low-altitude mice did, both at high and at low altitudes. In short, under both conditions the high-altitude mice paid a higher cost (that is, were less efficient) in the use of oxygen during exercise.

I measured the basal metabolic rate

result of acclimatization from birth, I then began to transfer mice from one altitude to another for study of their responses to the change. When mice from the sea-level colony were transferred to our 12,500-foot laboratory, their relative heart weight did not increase. They showed definite signs of acclimatization, however, in other responses. The adrenal glands diminished in size. The weight of the spleen decreased and the lung weight increased, indicating that circulatory and respiratory adjustments were taking place. The red-cell mass and the hemoglobin content of the cells increased to about the same values as in native high-altitude mice. When the surviving mice were later returned to sea level, they soon reverted to their original sea-level condition: the adrenal glands gained in weight and the red-cell mass and hemoglobin fell back to sea-level norms. It appears, therefore, that most of the adaptive mechanisms found in native high-altitude mice are actually adjustments acquired in the course of their exposure to the conditions of their environment.

The reverse experiment—transferring high-altitude mice to sea level—produced a mixed picture. In these mice the relative heart weight decreased and the spleen weight increased, but the adre-

nals and the lungs showed no change in relative weight. After 90 days of acclimatization to the low altitude the number of red cells in the mice's blood remained unchanged from what it had been at high altitude. This seems to suggest that the high red-cell count in high-altitude mice may represent a genetic adaptation, but it might be explainable on the basis that the translocated animals simply retained their original high red-cell count because there was nothing in the change to low-altitude conditions that would foster destruction of the cells. The total mass of the red cells in proportion to the blood volume did decrease to sea-level values; this may have been due to an increase in the amount of plasma. The concentration of hemoglobin, however, changed only slightly if at all.

The high-altitude natives showed no inferiority to low-altitude mice in fertility; in fact, the average litter size at 10,000 feet or above was six, as against five for females living at 4,000 feet. The high-altitude young, however, had a poor survival rate: mortality among them by the 30th day after birth was 23 percent, whereas all the low-altitude young survived beyond that age. I found that the rate of growth for the young was about the same at all levels during the first 45 days of life, but thereafter it

NUÑOA, at an altitude of 13,000 feet in the Andes Mountains of Peru, is the site of an experimental station operated by Pennsylvania State University for the study of human adaptation to high altitude. The native Indians herd llamas and alpacas, as shown here.

BARCROFT LABORATORY of the University of California is at an altitude of 12,500 feet in the White Mountains, a range just east of the Sierra Nevada in California. The environment appears uninhabitable but is the home of a number of small animal species.

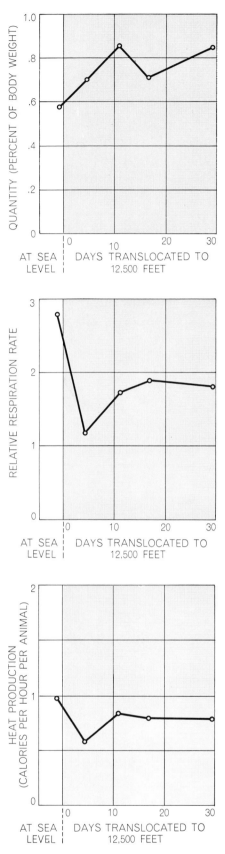

BROWN FAT, a tissue that generates heat, increased in quantity in sea-level deer mice transported to 12,500 feet (*top*). The rate of respiration of the tissue decreased, however, because of the relative lack of oxygen (*middle*). As a result the animal's total heat production was somewhat reduced (*bottom*).

(as indicated by oxygen consumption at rest) of the mice at various altitudes under various temperature conditions. The stations used for comparison were at sea level, at 4,000 feet in the Owens Valley (which is hot in summer) and at 12,500 feet in the Barcroft Laboratory. The determinations were made during three different seasons and at temperatures ranging from near freezing to a maximum of about 99 degrees Fahrenheit. I found that in summer the high-altitude mice had the lowest metabolic rate, the 4,000-foot mice an intermediate rate and the sea-level mice the highest rate. (The highest temperature proved to be lethal for mice from the high altitude, and conversely lower-altitude mice tested at near-freezing temperatures became severely chilled.) In the fall (at temperatures between 68 and 90 degrees F.) the high-altitude mice and sea-level mice had about the same rate of metabolism. In winter (February) the situation was the reverse of the summer picture: now the high-altitude mice had the highest rate at all temperatures, the sea-level mice the lowest.

These observations could be interpreted as follows. At the high altitude the comparatively mild temperatures of summer enable the native mouse to adjust to the hypoxic environment by reducing oxygen consumption to the minimum required for nourishing the body tissues. Because of the mouse's small size it cannot grow a thick enough hair covering to insulate it effectively against the winter cold. Consequently the high-altitude mouse is forced to increase its metabolism in winter to maintain its body temperature. The sea-level mouse, on the other hand, is not subjected to extreme cold in winter. Hence it is adequately protected from the drop in the ambient temperature by a small increase in its furry insulation and by an adjustment in the form of "physiological insulation," that is, reduction of its body temperature, which cuts down heat loss by reducing the temperature difference between the body and the ambient air. (I found that the deep-body temperature of the sea-level mouse does indeed decrease in winter.) Moreover, the sea-level mouse in winter can afford to reduce its metabolic rate from the high rate associated with summer activities without stinting its tissues' needs for oxygen.

With two associates, Robert E. Smith and Jane C. Roberts, I looked into the metabolic response of mice at the cell and tissue levels. We found several marked differences in cell activities at low altitudes and at high altitudes. In most cases it was difficult to tell whether

the observed differences were due to exposure to cold or to hypoxia. Studies of the brown fat in these animals, however, produced some significant findings.

Brown fat, found mainly between an animal's shoulder blades, is a heat-generating tissue [see "The Production of Heat by Fat," by Michael J. R. Dawkins and David Hull, beginning on page 269]. In response to exposure to cold there is an increase in both the mass of brown fat and its heat production, which may multiply severalfold in a few weeks. We found that the mass of brown fat could be increased in deer mice by exposure to cold or to hypoxia (which also lowers the body temperature). It turned out, however, that when sea-level mice were transferred to high altitude, the respiration of their brown fat decreased, so that its heat production was reduced in spite of a clear increase in its mass. On the other hand, when high-altitude natives were brought down to sea level, both the mass and the respiration of their brown fat increased, and its heat production apparently equaled that of sea-level mice that had been acclimated to cold. Evidently the transfer to the oxygen abundance at sea level had improved the tissue's respiration so that it increased its production of heat. From these findings we concluded that hypoxia, although it gives rise to the growth of brown fat, may suppress heat production by limiting the tissue's respiration.

Heredity or Environment?

The extensive investigations leave unsettled the question of whether men native to high altitude are a race apart or merely human beings with a normal heredity who have adjusted to the conditions over a lifetime of habituation beginning in the uterus. In some respects the mountain natives, both animals and men, do seem to show innate physiological differences from their kindred species at sea level. There is a serious objection to considering them a separate strain, however, namely the lack of genetic isolation. There has been no barrier in this case to the pooling of genes, either for the deer mouse or for man. We know that Andean man has intermarried freely with lowlanders. Indeed, many of the mountain miners came from the lowlands and many highlanders have come down to live in the lowlands. It seems likely that the highlanders have derived their special qualities from acclimatization—in short, that their response to their environment is phenotypic rather than genotypic.

IV

WATER BALANCE
AND ITS CONTROL

In Africa the greater part of the wild animals do not drink at all in summer, owing to lack of rains for which reason Libyan mice in captivity die if given drink. The perpetually dry parts of Africa produce the antelope, which owing to the nature of the region goes without drink in quite a remarkable fashion, for the assistance of thirsty people, as the Gaetulian brigands rely on their help to keep going, bladders containing extremely healthy liquid being found in their body.

Pliny
NATURAL HISTORY, V, XCiV.

IV

WATER BALANCE
AND ITS CONTROL

INTRODUCTION

The body of higher vertebrates is about 60 percent water. Somewhat less than half of the water is extracellular, and in it are dissolved a variety of organic and inorganic molecules. Because animals are open systems in that they take in and give off water and many other types of molecules each day, a regulating system that maintains the constancy of the body water and its constituents has evolved. As summarized by A. V. Wolf ("Body Water," *Scientific American*, November 1958), the relation among the quantity of water an animal drinks, the volume of urine it produces, and evaporative loss is so controlled that the total amount of body water is kept within a few percent of its average value. Control of salt content and the balance between different salt ions is equally strict. The core of the system is, of course, the kidney and its physiology.

The functional units of the kidney—the nephrons—appeared early in vertebrate evolution. Each nephron was originally a duct that drained the body cavity (the coelom) to the exterior. In later forms, a cluster of blood capillaries that adjoined a portion of the duct evolved, and the connection between the duct and coelom was closed off. Thereafter, the ducts drained fluid and molecules that filtered from the blood capillaries into the cavity of the ducts. The most primitive kidney we know today is in lamprey larvae, in which a glomus (a tangle of blood vessels) is located in the wall of the coelom near the opening of each nephron. Similar glomi and nephrons are present in adult hagfish (another cyclostome), whose body fluids have the same osmotic pressure as sea water. Some regulation within the hagfish occurs, however, because the ratio between various ions in the body is different from the ratio between those ions in sea water. In adult lampreys and all higher vertebrates the glomus is inserted into the end, or funnel, of the nephron, and in these organisms it is called the glomerulus. The adult lampreys and all fresh-water fishes require an active regulatory system because their blood is hypertonic (i.e., has a higher osmotic pressure) to the fresh water in which they live. This system must eliminate the water that tends to enter the body because of the osmotic gradient, and yet retain the salts, which tend to flow out from the body (because they tend to flow from a region of high concentration—the body fluids—to one of low concentration—the watery habitat).

The glomerular portion of the nephron is the site at which water, salts, and some organic molecules (those having a molecular weight of 68,000 or less) are forced through the wall of the glomerular capillary and nephron capsule. This filtration is directly dependent upon blood pressure in the glomerulus. As the filtrate flows from the capsule and through the tubular portion of the nephron, water, glucose, small proteins, chloride ions and some other molecules are reabsorbed to varying degrees, while other substances—such as urea, magnesium ions, and potassium ions—are secreted into the filtrate by tubule cells. The end product, of course, is the urine. The system functions at a remarkable rate in man: about 175 quarts of fluid per day pass from the blood into the capsules of the kidney nephrons and all but one or two quarts are resorbed by the tubules. Homer Smith (*Scientific American*, January, 1953; Offprint 37) points out that 2.5 pounds of salt enter the capsules with the water, but only a third of an ounce is lost in urine.

As outlined by Scholander in "The Wonderful Net" (Part II), the ability of the mammalian kidney to produce urine hypertonic to blood is dependent upon a countercurrent flow system that transfers sodium ions (Na^+) from one portion of the kidney tubule to the other; consequently, high sodium concentration can be maintained in the medullary (central) part of the kidney. Refer to Figure 1 for help in visualizing the following relationships. First, for clarity, we must trace the route of urine flow: nephron capsule, proximal tubule, descending loop, ascending loop, distal tubule, and, finally, collecting duct. Most of the active transport of sodium ions takes place across the thick-walled ascending loops. The sodium ions removed from these loops are added to the intercellular fluid and then transferred back to the descending loops; thus, the sodium ions tend to move in a circle—descending loop to ascending loop to intercellular fluid to descending loop. The result is to create a very high concentration of sodium ions within and between these loops, that is, in this whole "medullary" portion of the kidney. Consider the effects on water flowing down the kidney tubule: because of the high sodium ion concentration, water tends to move through the walls of both the descending and ascending loops; the result is to reduce the *volume* of urine that ultimately leaves the top of the ascending loop and enters the distal tubule and collecting duct. Furthermore, since sodium ions have also left the urine as it passes upward through the ascending loop, the concentration of the urine is lowered; that is, it becomes dilute. Next, the urine flows down through the thin-walled collecting ducts that actually pass among the medullary descending and ascending loops; recall that the extracellular sodium ion concentration is very high there (four times that of blood, in man). The result is to remove more water from the urine in the collecting ducts. This, of course, raises the concentration of the urine (until it is hypertonic to the blood). In the end, a small volume of highly concentrated urine is produced and sent on to the bladder.

The control system for kidney function is marvelously complex and depends upon activity of both nerve and endocrine cells. Primary control resides in the brain, where osmoreceptor cells of the hypothalamus control the release of antidiuretic hormone (vasopressin). These nerve cells respond to alterations in the osmotic pressure of blood plasma flowing through the brain and make compensatory adjustments in quantities of vasopressin released in the posterior pituitary.

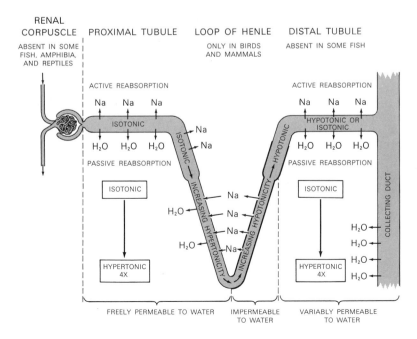

Figure 1. A diagrammatic representation of a nephron unit, showing the manner in which hypertonic urine is produced. [After Pitts, Robert F., *The Physiological Basis of Diuretic Therapy*, Charles C. Thomas, 1959.]

The hormone affects mammalian kidney nephrons (particularly distal tubules and collecting ducts) and the bladder wall to increase resorption of water. In other organisms, it also decreases glomerular filtration rate. Interestingly, this peptide hormone can work physiologically only if it is applied to the connective tissue or basal side of the tissue. The biochemical mechanism of action appears to be a stimulated production of cyclic-AMP (adenosine-3′,5′-phosphate), which in turn alters properties of cell permeability. In amphibians vasopressin also reduces the permeability barrier to sodium ions and so causes increased retention of salts as well as conserving water.

An important element of chemical regulation of physiological processes is the short half-life of chemical mediators in the body; if the half-life were not short, the control machinery would be relatively insensitive to altering conditions in the body. Vasopressin of various vertebrates has a half-life that varies between 1 and 24 minutes; under most circumstances the shorter times (1 to 4 minutes) apply. The half-life is governed by destruction of the peptide; two-thirds is removed from the blood and inactivated by the kidney and the remaining one-third is destroyed in the liver.

Several other coordinating agents control urine content. Aldosterone produced by the cortex of the adrenal gland, controls resorption of sodium ions by the distal tubules of the nephrons. Aldosterone secretion is regulated by angiotensin II, a substance produced in the kidney when an enzyme called renin is released from cells of the juxtaglomerular apparatus. Renin acts on α-2-globulin, one of the proteins in serum, to change it into angiotensin I. Then another blood protein, the "converting enzyme," splits two amino acids from this decapeptide to produce the active octapeptide, angiotensin II. Besides affecting aldosterone release, this agent causes contraction of smooth muscle cells in the walls of blood capillaries, both in the kidney and elsewhere in the body. It seems significant that the posterior pituitary hormones vasopressin and oxytocin, both of which produce contraction of smooth muscles in various other organs of the body, also are composed of eight amino acids.

Angiotensin II has a vital function in a feedback-control loop of the mammalian nephron. The apparatus of this loop is composed of the first portion of the distal tubule (see Figure 1), the cells of which are called the macula densa cells, and the renin-containing cells of the blood vessels approaching the glomerulus. Although in the figure the two kinds of cells appear far apart, in an intact kidney the various parts of the nephron are in fact twisted about one another, and, specifically, the macula densa cells of the tubule and the renin-containing cells of the blood vessels really interdigitate with one another; thus, they can interact easily. The sodium concentration, either of the tubular fluid (i.e., the urine) near the macula densa cells or of the resorbed fluid in the same regions, is somehow measured. This sodium concentration effects release of renin by the blood vessel cells and thus, indirectly, the production of angiotensin I and II. It is angiotensin II that controls the blood pressure (or rate of flow) in the glomerulus. And, as we said before, the glomerular filtration rate is dependent upon blood pressure. Therefore, the loop is complete: the quantity of sodium entering the distal tubular portion of the nephron ultimately determines the filtration rate back in the glomerulus. Thus, a sodium-sensitive feedback loop coordinates glomerular and tubular function, and operates through proteins produced by the kidney (renin) and by the liver (globulin and converting enzyme).

It is often said that the length of the loop of Henle (comprising the ascending and descending loops described above) is the unique feature of mammalian and avian nephrons that permits production of urine that is hypertonic to blood. Recent observations have disproved that generalization. A number of terrestrial birds (such as the Savannah sparrow and the house finch) that lack functional salt glands (see 'Salt Glands" by Knut Schmidt-Nielsen) have been investigated, and they possess extraordinarily large numbers of relatively short loops of Henle. Yet hypertonic urine is produced by these kidneys. Thus the same physiological ability can be derived from either anatomical arrangement—the long loops or the large numbers of short loops.

Those birds and reptiles that cannot produce a hypertonic urine possess a salt-secreting gland near the eyes. More recent work by Schmidt-Nielsen and his collaborators shows that this gland is necessary because of an indirect effect of the final water resorption that occurs in the cloaca (the final chamber through which kidney and intestinal products pass). Water is taken back into the body through the walls of the cloaca, thus drying out the feces and uric acid. Cations tend to move with the water and they must be excreted elsewhere; this excretion is the function of the salt glands of the head. Physiologically the salt-secreting glands act like kidney tubules in response to vasopressin and hypothalamic control.

The biochemical basis for salt gland function is beginning to be understood. A Na^+,K^+,-dependent ATPase appears to be functional at the luminal (outer) end of salt-gland cells and as a result Na^+ ions are pumped out of the cell and the body. What is surprising about salt-gland function is that the *continuous* presence of the neurotransmitter acetylcholine is required for active pumping. The system works like this: receptors in the heart detect elevated blood tonicity due to the drinking of sea water; that information is sent to the brain whence nerve discharge at the salt gland liberates acetylcholine, and Na^+ pumping starts. It is not yet known whether the acetylcholine acts directly on the ATPase enzyme or via cyclic AMP, as is often the case in other situations.

With these comments on salt glands as background, it is worth turning to the closely related case of fish adapted for life in the sea. Salts tend to enter their bodies and must also be excreted, just as in marine birds. Interestingly the same sort of Na^+,K^+,-dependent ATPase appears in the gill "chloride" cells and functions to pump Na^+ ions from the blood, against the concentration gradient, into the sea.

Fish in fresh water have the opposite problem, one that is solved by a shift in the type of enzyme activities present. Thus, the Na^+,K^+,-dependent ATPase is lost; a new Na^+,-dependent ATPase activity appears, and the direction of pumping is altered. Na^+ ions are pumped from fresh water, once again opposite to the concentration gradient, into the blood. In a steelhead trout which migrates back and forth between fresh and salt water, or in brackish-water fishes, both enzyme activities apparently are present.

We see in these cases important supplementary adaptations that complement the kidneys. Another interesting adaptation of various vertebrates is the ability to retain large quantities of urea in the body fluids. Sharks, their relatives, and various amphibia that live in saline environments build up high concentrations of urea in order to counteract the osmotic gradient resulting from higher salt concentrations in the sea outside the body. The urea molecules act as osmotic particles that raise internal osmotic pressure and reduce the tendency for water to leave the body.

In the crab-eating frog (*Rana cancrivora*), immersion in brackish sea water leads to an actual change in metabolism, so that the urea-cycle enzymes are activated, and the urea produced is not excreted. The high levels of urea in these amphibia and in the elasmobranchs (200 to 500 mM) are sufficient to cause marked alterations in the secondary and tertiary structure of proteins characteristic of organisms with "normal," low urea levels. But, high urea has no effect on enzymes, on heart or skeletal muscle, and on other organs in creatures adapted to using high levels of urea as an osmotic aid. The means by which proteins and cells become tolerant of high urea concentrations is unknown.

210

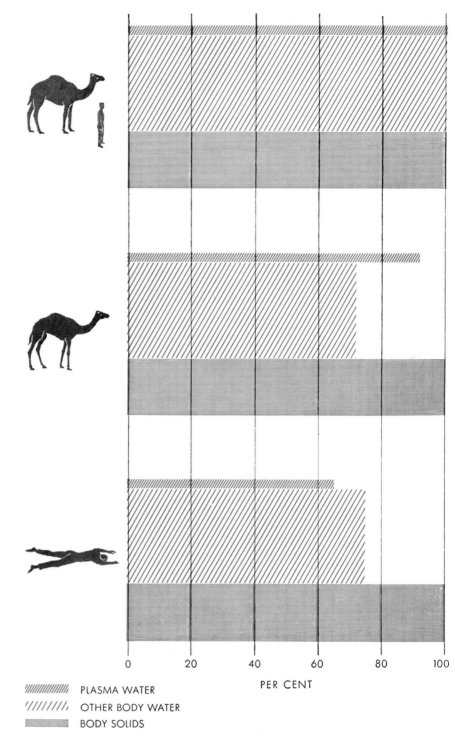

PLASMA WATER

OTHER BODY WATER

BODY SOLIDS

PER CENT

Figure 2. Camel survives dehydration partly by maintaining the volume of its blood. Under normal conditions (*top*) plasma water in both man and camel accounts for about a 12th of total body water. In a camel that has lost about a fourth of its body water (*center*) the blood volume will drop by less than a 10th. Under the same conditions a man's blood volume will drop by a third (*bottom*). The viscous blood circulates too slowly to carry the man's body heat outward to the skin, so that his temperature soon rises to a fatal level. From "The Physiology of the Camel" by Knut Schmidt-Nielsen. Copyright © 1959 by Scientific American, Inc. All rights reserved.

Though these introductory remarks have concentrated on properties of the kidney, there is much more than the kidney involved in permitting survival of vertebrates under extreme environmental conditions. Studies of desert animals (see, for instance, Schmidt-Nielsen, "The Physiology of the Camel," *Scientific American*; Offprint 1096) have revealed a variety of adaptations to extreme heat and low humidity. An intuitively confusing adaptation relates to insulation. Animals such as camels have thick coats of hair, particularly on their backs. Even during the hottest months of the year in the desert, when the thinner, "summer" coat is found over most of the body, the back retains its thick insulating fur. The reason for this is that the insulation is helping to keep heat *out* of the body; consequently the camel can sweat less to maintain a given body temperature. For this same reason, people inhabiting the deserts of the world commonly clothe themselves in several layers of loose-fitting, thick garments. Experience has shown that baring the skin is not a cool move under the desert sun!

Schmidt-Nielsen and his collaborators also made an important discovery concerning blood volume under dehydrating conditions. A human being exposed to the desert sun sweats copiously and becomes severely dehydrated. When about 12 percent of the body water is lost, the individual most likely undergoes "explosive heat death." This may result in large part from the fact that the blood volume—total plasma water—decreases even more precipitously, thus causing a drastic rise in blood viscosity. Even when a camel has lost nearly one-quarter of total body water, its plasma water content falls less than 10 percent; the blood viscosity is not changed appreciably, and so heat death is avoided. How this maintenance of water is achieved in one of the body's compartments is another mystery of vertebrate biology.

C. R. Taylor, in "The Eland and the Oryx," discusses other adaptations for desert survival. In the oryx, we see an interesting variation of the blood vascular system which functions in an opposite sense to the heat-retaining capillary beds of sharks and tuna (see "Fishes with Warm Bodies," by Francis G. Carey). Blood is cooled in the walls of the large nasal cavities of the oryx and then returned past the carotid arteries as they approach the brain. Heat is transferred to the cool nasal blood so that the carotid blood is at a lower temperature than it is when leaving the heart. The result is a lower brain temperature and an increased opportunity to survive under desert conditions.

ELAND is the largest of all African antelopes. An average adult bull weighs more than half a ton and may measure six feet at the shoulder. A gregarious and docile member of the family Bovidae, the eland can thrive in drought-ridden rangeland unfit for cattle.

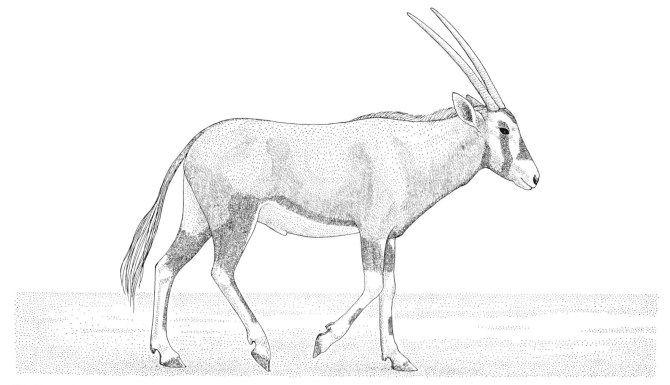

ORYX, another large African antelope, is four feet high at the shoulder. It is even better adapted to arid lands than the eland and is found in barren desert. The oryx, however, is far from docile. It wields its long horns readily and has been known to kill lions.

20

THE ELAND AND THE ORYX

C. R. TAYLOR

January 1969

These large African antelopes can survive indefinitely without drinking. Their feat is made possible by stratagems of physiology that minimize the amount of water they lose through evaporation

When travelers' accounts of snow on the Equator first reached 19th-century London, learned members of the Royal Geographical Society ridiculed the reports. To many zoologists today the existence of antelopes in the deserts of Africa may seem as much of a surprise as equatorial snow was to 19th-century geographers. Such an environment for any member of the family Bovidae is simply unreasonable. If the reports were accompanied by the statement that the animals survive without drinking, some zoologists would replace "unreasonable" with an emphatic "impossible." Yet it is now well known that snow covers the peaks of many equatorial mountains, and it is consistently reported by naturalists, hunters and local people that certain desert antelopes can survive indefinitely without drinking.

Why should desert survival without drinking seem impossible? It is certainly impossible for humans; in the course of a hot day in the desert a man can lose as much as three gallons of water as a result of sweating and evaporation. Without drinking he could not survive one such day. On the other hand, the kangaroo rat and other desert rodents thrive without drinking at all, even when eating dry food. Their water requirements are met by the very small amounts of free water in their food and by the additional water the food yields when it is oxidized in the process of metabolism. Rodents, however, are small; they can escape the high temperatures of the desert day by burrowing underground. No such shelter is available to the larger mammals. In order to regulate their body temperature during the heat of the day they must evaporate substantial quantities of water. Even the camel, probably the best-known desert animal, has the same problem. As the physiologist Knut

Schmidt-Nielsen and his collaborators discovered, the camel has an unusual ability to limit its loss of water by evaporation, but it still must drink in order to survive [see "The Physiology of the Camel," by Knut Schmidt-Nielsen; SCIENTIFIC AMERICAN Offprint 1096]. If the antelopes of the African desert did not have similar abilities, survival would be impossible.

Numerous eyewitness accounts identify two antelopes—the eland and the oryx—as animals that do not need drinking water. The eland (*Taurotragus*) is a large, tractable animal that occupies a variety of East African habitats, including the edge of the Sahara. It is easily domesticated, and it has often been proposed as a means of utilizing rangeland that during droughts is too arid for cattle.

Although the eland does not require water, it does not penetrate the most barren deserts. The oryx, on the other hand, is truly a desert species. Unlike the eland, it does not seek shelter during the midday heat but remains exposed to the hot sun throughout the day. The oryx is

as aggressive as the eland is tractable. It wields its rapier-like horns with great facility; those who study its physiology get physical as well as mental exercise.

Some years ago, with the help of Charles P. Lyman of Harvard University, I set out to see if these two antelopes really did live in the African deserts without drinking and, if so, how. I had three simple questions. First, do the eland and the oryx possess any unusual mechanisms for conserving water? Second, if they do, how much water do they require when the mechanisms are operating? Third, can they get this amount of water in some way other than drinking? I had the opportunity to investigate these problems at the East African Veterinary Research Organization at Muguga in Kenya, where the directors (initially Howard R. Binns and later Marcel Burdin) generously provided the needed laboratory space and equipment.

The first step was to establish a laboratory environment that would simulate a hot desert and make it possible to find out how much water the antelopes lost.

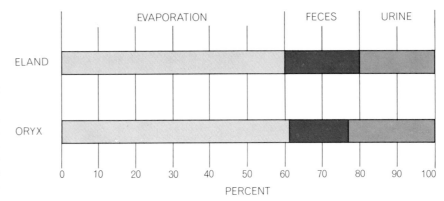

EVAPORATION from the skin and respiratory tract proved to be the major avenue of water loss for antelopes allowed to drink freely in a simulated desert environment. The only means of saving significant amounts of water in the desert heat is to reduce evaporation.

I did not attempt to duplicate the desert's day and night temperature extremes, but I maintained the average daytime temperature for 12 hours and the average nighttime temperature for the same period. With heating and air-conditioning equipment I was able to rapidly raise the temperature of the animals' room to 40 degrees Celsius (104 degrees Fahrenheit) or lower it to 22 degrees C. (72 degrees F.), thus simulating the desert's average day and night temperatures.

In this environment I measured the various ways the animals lost water. I found that loss by evaporation was the most important [*see illustration on preceding page*]. The function served by evaporation is to prevent the animal from overheating. Most mammals maintain their body temperature at a nearly constant level, usually about 37 degrees C. (98.6 degrees F., the "normal" point on a clinical thermometer). If their body temperature rises too high, they die; for most mammals a body temperature of 43 degrees C. (109 degrees F.) for a few hours would be fatal.

When the temperature of the environment is lower than the animal's temperature, which is usually the case, heat flows from the animal to the surroundings (by conduction and radiation). The maintenance of a constant body temperature in these circumstances requires

that the animal's metabolic machinery generate an amount of heat equal to the net outward heat flow. When the environmental temperature is higher than the animal's, however, the direction of heat flow reverses and heat is transferred from the environment to the animal. If under these circumstances the body temperature is to remain constant, the heat gained from the environment, as well as the heat generated by metabolism, must be dissipated by evaporation. Each gram of water that is evaporated carries away .58 kilocalorie of heat, and the combined heat load on a man may be so great that he is obliged to evaporate more than a liter (about a quart) of water an hour.

One way the body can reduce evaporation under heat stress is to abandon the maintenance of a constant body temperature. Schmidt-Nielsen and his collaborators found that when the camel is confronted with a shortage of water, its body temperature rises during the course of the day by as much as seven degrees C. To see if the eland or the oryx had the same ability, I recorded their rectal temperature during the laboratory's hot 12-hour day. The animals had all the water they could drink, so that nothing prevented the maintenance of a constant body temperature by evaporation. Nonetheless, during 12 hours at 40 degrees C. the eland's temperature on occasion rose by more than seven degrees (from 33.9

to 41.2 degrees C.) and the oryx's by more than six degrees (from 35.7 to 42.1 degrees) before increased evaporation prevented any further rise (although usually the temperature rise was less extreme). Thus instead of spending water to maintain a constant body temperature, the animals "stored" heat in their bodies.

In an eland weighing 500 kilograms a 7.3-degree rise in temperature means that the animal has managed to store some 3,000 kilocalories. To dissipate the same amount of heat by evaporation would cost it more than five liters of water. In the wild, as the lower night temperature allows a reversal of heat flow from the animal back to the environment, this stored heat is dissipated by conduction and radiation rather than by evaporation.

I found that when the eland and the oryx were exposed to the high experimental temperature and had water available to drink, their body temperature usually increased by three or four degrees during a 12-hour day. After three or four hours' exposure the animals' evaporative processes accelerated sufficiently to prevent a further rise in their temperature, which remained below the temperature of the laboratory even at the end of the full 12 hours. In the wild such a pattern of gradual warming before an increase in evaporation means that the animals might get past the hottest hours

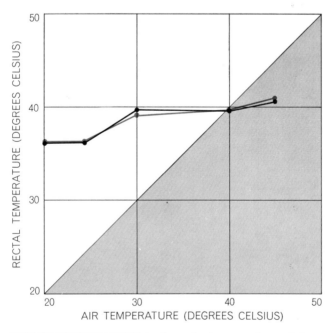 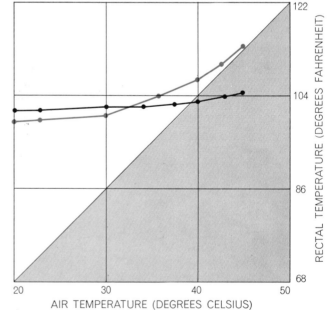

RISE IN TEMPERATURE in the author's test enclosure brings much the same physiological response from the eland (*left*) whether abundantly watered (*black*) or dehydrated (*color*) as from the oryx (*right*) when abundantly watered (*black*). At air temperatures over 40 degrees Celsius, the animals' rectal temperatures no longer remain higher than the temperature of the environment (*di-*

agonal connects points where air and rectal temperature are equal). Instead evaporative cooling keeps rectal temperature below air temperature, so that heat flows from the environment to the animal. The temperature of the dehydrated oryx (*color*), however, continues to rise, and heat flows from animal to environment. The dehydrated oryx therefore gains no heat from its surroundings.

of the day before expending precious water for cooling.

These first observations showed that in desert-adapted antelopes, as in the camel, water loss was reduced by the increase in body temperature. So far, however, I had measured the responses of the antelopes only under circumstances where they had been free to drink as much water as they wanted; this was a long way from testing their reported ability to get along without drinking at all. Before making my next series of measurements, therefore, I restricted the animals' water intake until they became dehydrated and lost weight. They were given just enough water to keep them at the point where they maintained their body weight at 85 percent of their original weight. I then exposed them to the same experimental 12-hour hot days as before. The body temperature of the dehydrated eland still remained below the temperature of the environment, even after 12 hours at 40 degrees C., and of course this can only be achieved by maintaining a high rate of evaporation. The temperature of the dehydrated oryx, in contrast, routinely exceeded that of the environment by a wide enough margin for metabolic heat to be lost by conduction and radiation; evaporation did not increase during the entire 12-hour exposure.

Although I had selected 40 degrees C. as the temperature of the hot periods, I knew that the desert air temperature is higher for a few hours every day, and that the solar radiation at midday near the Equator is literally searing. During the worst heat of the day the eland moves into the shade but the oryx appears to be oblivious of the intense heat. Oryx species with the lightest-colored coat are the ones that penetrate farthest into the desert; the coat's reflectivity must help to reduce the radiant heat gain. Unless the coat is perfectly reflective, however, an oryx standing in the hot desert sun should absorb far more heat than one in the laboratory at 40 degrees C.

I decided to test the animals' reactions to an even more severe heat load by raising the laboratory temperature to 45 degrees C. (113 degrees F.). Dehydrated or not, the eland managed to maintain its temperature some five degrees below the new high. So did the oryx, when it was supplied with water. The dehydrated oryx's physiological response was quite different. Its temperature rose until it exceeded the temperature of the laboratory and remained above 45 degrees C. for as long as eight hours without evident

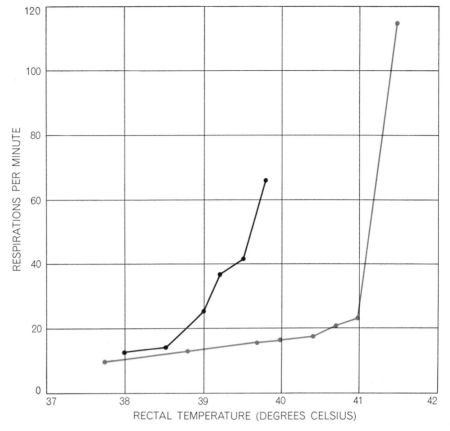

EVAPORATIVE COOLING through panting rather than through sweating is characteristic of the dehydrated oryx. When abundantly watered (*black*), an oryx not only sweats but also starts panting as its temperature reaches 39 degrees C. Although dehydrated oryx (*color*) does not sweat, it pants vigorously once its temperature exceeds 41 degrees C.

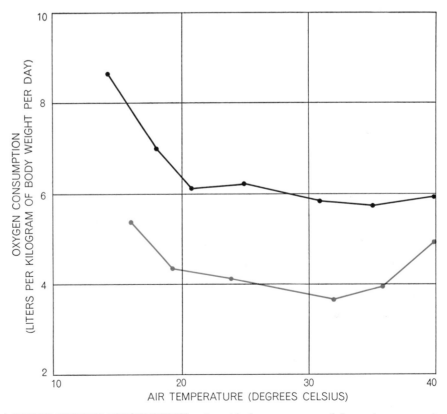

ORYX'S OXYGEN CONSUMPTION varies with the temperature of the environment and is drastically reduced by dehydration (*color*). As a result the dehydrated oryx generates much less metabolic heat and the quantity of heat to be lost by evaporation is reduced.

ill effect. I believe such a high continuous temperature has been observed in only one other mammal: the small desert gazelle *Gazella granti*. This hyperthermia in both the oryx and the gazelle enables them to save large amounts of water even under severe heat loads, and this is probably the critical factor in their survival under desert conditions.

How do the oryx and the gazelle survive these high internal temperatures? The brain, with its complex integrative functions, is probably the part of the body most sensitive to high temperatures. It is possible that in both animals the brain remains substantially cooler than the rest of the body. As the external carotid artery, which supplies most of the blood to the brain in these animals, passes through the region called the cavernous sinus, it divides into hundreds of small parallel arteries. Cool venous blood from the nasal passages drains into the sinus, presumably reducing the temperature of the arterial blood on its way to the brain. Evidence for this view is that, when temperature readings are taken during exercise, the brain of a gazelle proves to be cooler than the arterial blood leaving the heart by as much as 2.9 degrees C. Mary A. Baker and James N. Hayward of the University of California at Los Angeles have demonstrated that such a mechanism also operates in the sheep.

The manner in which an animal increases evaporation to keep cool can make a difference in the amount of heat it gains from a hot environment. Some animals pant, some sweat and some spread saliva on the body. An animal that depends on sweating or salivation for loss of body heat will necessarily have a skin temperature that is lower than its internal temperature. The blood must flow rapidly to the skin, carrying the internal heat to the evaporative surface. Conversely, a high skin temperature and a low flow of blood to the skin reduces the rate of heat flow from the hot environment to the animal. If the animal can pant rather than sweat, it can dispose of body heat by respiratory evaporation and at the same time have a higher skin temperature that minimizes its accumulation of environmental heat. Accordingly I wanted to find out whether the oryx under heat load increased its evaporation by sweating, by panting or by both. I also wondered what effect dehydration might have on the relative importance of the two evaporative routes. When I measured sweating and panting in an oryx freely supplied with water, I found that the evaporation rate increased in both routes but that evaporation from the skin accounted for more than 75 percent of the total. When the animal was deprived of water, it did not sweat at all in response to heat, but it began to pant when its body temperature exceeded 41 degrees C.

David Robertshaw of the Hannah Dairy Research Institute in Scotland and I had previously found that the sweat

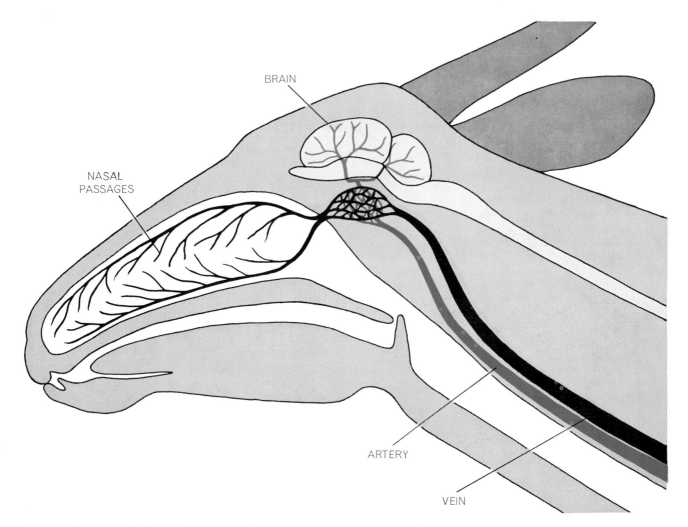

COUNTERCURRENT COOLING of arterial blood on its way from heart to brain occurs in the cavernous sinus, where the carotid artery ramifies into hundreds of smaller vessels (*color*). There venous blood (*black*) from the oryx's nasal passages, cooled by respiratory evaporation, lowers the arterial blood temperature. A brain cooler than the body temperature may be vital to desert survival.

glands of the oryx are controlled by nerve cells that release adrenalin. For example, an oryx can be made to sweat by means of a small intravenous injection of adrenalin. When we gave the same doses of adrenalin to a dehydrated oryx, the animal also sweated. The response makes it clear that the sweat glands of a dehydrated oryx can still function, but that the animal's nervous system has simply stopped stimulating them.

Increases in the temperature of the body and of the skin are not the only stratagems that minimize evaporation. A lowering of the animal's usual metabolic rate would reduce the amount contributed to the total heat load by the animal itself. To see whether or not this potential saving was being accomplished, I measured the metabolic rate of both the eland and the oryx over a broad range of temperatures. The metabolic rate of the dehydrated eland was somewhat reduced, but in the oryx the reduction was much greater. At 40 degrees C. the observed reduction in the metabolic rate of the oryx was sufficiently low to reduce its evaporation by 17 percent as compared to evaporation from animals freely supplied with water.

Taken together, these findings indicate that when water is scarce, the eland and the oryx reduce their rate of evaporation during the hot desert day in various ways. Both animals store heat; both decrease the heat flow from the environment, the eland by seeking shade and the oryx by accommodating to an extreme body temperature; both decrease the amount of metabolic heat they produce.

What about water loss during the night? When the sun goes down, of course, the antelopes are no longer under a heat load from the environment. The heat generated by their own metabolic processes is easily lost to the cooler surroundings by means other than evaporative cooling. Nonetheless, some evaporation, both from the respiratory tract and from the skin, continues at night.

The water lost through the skin at night is not lost through sweating. It is probably lost through simple diffusion. Skin is slightly permeable to water; even apparently dry-skinned reptiles lose appreciable amounts of water through the skin. So do the eland and the oryx. During the 12-hour cool night in the laboratory I found that both animals, when they had free access to water, lost about half a liter of water per 100 kilograms of body weight through the skin. The water loss was reduced when the animals were dehydrated; their skin seemed drier and

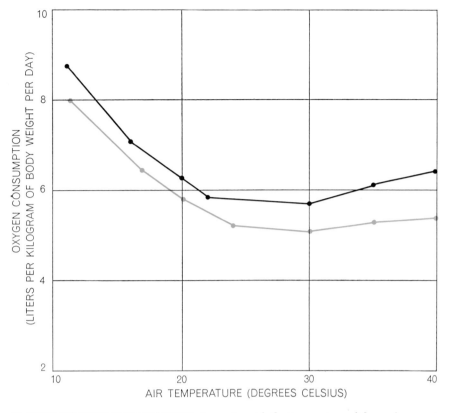

ELAND'S OXYGEN CONSUMPTION also varies with the temperature of the environment but is only slightly reduced by dehydration (*color*). Like the oryx, the eland increases its oxygen consumption at the low temperatures of the desert night. The increase threatens both animals with a net loss of water if night feeding produces less than 10 percent water.

less permeable. The water loss from the dehydrated eland was 30 percent less than when it could drink freely. In the dehydrated oryx the water loss from the skin was reduced by nearly 60 percent.

I wondered if the loss of water from the respiratory tract could also be reduced. As a mammal breathes, the inhaled air is warmed to body temperature and is saturated with water vapor in the respiratory tract before it reaches the lungs. Normally most mammals then exhale saturated air that is still at body temperature. Donald C. Jackson of the University of Pennsylvania School of Medicine and Schmidt-Nielsen have observed, however, that two species of small rodents manage to exhale air that is much cooler than their body temperature. The rodents apparently recondense some water vapor within the respiratory tract. This is one way water loss through respiratory evaporation could be minimized by antelopes. Two other possible means of water economy are related to oxygen requirements. First, if more oxygen can be extracted from each breath, an animal does not need to move as great a volume of air through its respiratory tract, thus reducing the loss of water to the respiratory air. Second, if the ani-

mal's oxygen consumption is lowered, the volume of inhaled air (and therefore the loss of water) is also reduced.

It is known that domesticated cattle exhale saturated air at body temperature. It is probable that the eland and the oryx, with their large nasal passages and relatively slow rate of respiration, do the same, so that water economy by recondensation is not available to them. Both animals have a lower body temperature at night than during the day, however, and the difference is enough to significantly reduce the amount of water needed to saturate the respiratory air. Air saturated at 39 degrees C., a typical daytime temperature for the eland, contains some 48 milligrams of water per liter. Air at the typical night temperature of 33.8 degrees contains some 25 percent less water.

When I investigated the animals at nighttime temperatures, I found that both the eland and the oryx extracted more oxygen from the air and breathed more slowly when they had a low body temperature. When the eland's temperature is at a nighttime low of 33.8 degrees C., about twice as much oxygen is extracted from each liter of air that it breathes. As oxygen extraction increases, the amount of air inspired with each breath also increases. Only part of the air an animal breathes actually reaches the lung, where oxygen and carbon dioxide are exchanged. The rest fills the respiratory passages, where the air is warmed and saturated with water vapor but where no exchange of gas takes place—the "dead-space volume." As the eland breathes more deeply, the dead-space volume remains constant but a greater proportion of the total inspired air reaches the lungs: the same amount of oxygen is extracted from each volume of air within the lungs but the volume has been increased. Any animal that breathes more slowly and more deeply will extract more oxygen from the inspired air and lose less water (and heat) with its expired air.

Is water economy also aided by a lower oxygen consumption? I had already found that in both the eland and the oryx

the rate of metabolism is reduced when the animals are dehydrated. During the cool nighttime period the metabolic rate of the dehydrated eland was about 5 percent lower than the rate of the freely watered eland, and the metabolic rate of the dehydrated oryx was more than 30 percent lower than the rate of the freely watered oryx. Other things being equal, then, when water is scarce in the wild, the respiratory water loss of both antelopes would be reduced by an amount equal to the reduction in their metabolic rate. Although some loss of water through respiration remains unavoidable, it is minimized by the combination of lowered body temperature, increased oxygen extraction and reduced metabolism.

When I measured oxygen consumption at various temperatures, I found that metabolism increased at temperatures below 20 degrees C. This seemed odd, because both antelopes frequently en-

counter temperatures below 20 degrees C. at night. The increase, which only serves to keep the animals warm, would not be necessary if the animals had a slightly thicker fur, which would also reduce the heat gain during the hot day. It seems possible that such an adaptation has not appeared because the increased metabolism at night means the difference between a net loss or gain of water. If an animal increases its metabolism at night, it also eats more, takes in more free water with its food and generates more oxidation water. At the same time additional water is lost by breathing more air to get the necessary additional oxygen. When one calculates the amount of water needed in the food to offset a net loss of water through increased metabolism at night, it works out at about 10 percent water content in the food. Thus if the eland and the oryx feed at night on plants with a water content higher than 10 percent, they achieve a net gain of water. As we shall see, they favor plants that have a water content considerably above this level.

Having found that the two animals do indeed possess unusual mechanisms for conserving water, I next undertook to find out how much water—or rather how little—each required. To do this I kept dehydrated animals in two contrasting laboratory environments. One was the usual alternation of 12-hour hot days and cool nights; the other was constantly cool. In the cool environment the dehydrated eland managed to stay at an even 85 percent of its original weight when its total water intake (free water in food, oxidation water and drinking water) was slightly in excess of 3.5 liters per 100 kilograms of body weight per day. The dehydrated oryx got by on scarcely half that amount: a little less than two liters per 100 kilograms. On a regime of cool nights and hot days the eland's water requirement increased to nearly 5.5 liters and the oryx's to three liters. These findings brought me to a final question: Could the two animals obtain this minimum amount of water without drinking?

The eland not only finds shade beneath acacia trees; acacia leaves are one of its favored foods. I collected acacia leaves and measured their moisture content. Even during a severe drought they contained an average of 58 percent water. Calculating the weight of acacia leaves that an eland would ingest in a day to meet its normal metabolic requirements, I found that the leaves would provide about 5.3 liters of water per 100 kilograms of body weight, or almost ex-

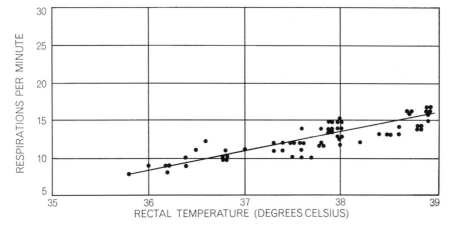

RESPIRATION RATE of an eland is controlled by the animal's temperature, decreasing as the temperature falls. When it breathes more slowly, the eland also breathes more deeply, so that a greater part of each breath reaches the lung areas where gas exchange occurs.

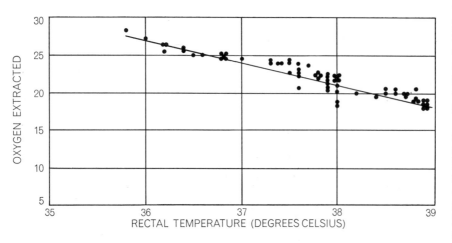

OXYGEN EXTRACTION from each breath of air increases as the eland's temperature falls and the animal breathes more slowly and deeply (see upper illustration). Because the eland's oxygen needs are met by a lesser volume of air, respiratory water loss is reduced.

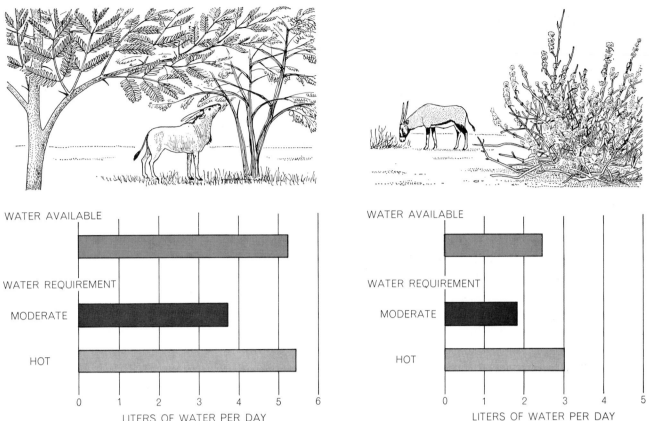

SURVIVAL WITHOUT DRINKING is possible for the oryx and the eland because their food contains almost all the water they need. Even in droughts the leaves of the acacia (*left*), the eland's preferred fodder, are 58 percent water. The leaves of a shrub, *Disperma*, and other fodder preferred by the oryx (*right*) contain little water by day but may average 30 percent water at night. Thus the amount of water each animal can obtain by feeding (*color*) is more than the animal needs for survival in a moderate environment (*black*) when dehydrated and closely approaches the quantities necessary for both antelopes' survival under desert conditions (*gray*).

actly the amount needed by the dehydrated eland for survival. An eland that obtains this much water by browsing only can probably live indefinitely without drinking.

The oryx favors grasses and shrubs, particularly a shrub of the genus *Disperma*. In the daytime these plants are so dry their leaves fall apart when they are touched; my measurements showed that they contain as little as 1 percent water. At first this seemed an impossible contradiction. Nonetheless, there is a way for the oryx to get all the water it needs by grazing. As long ago as 1930 the British naturalist Patrick A. Buxton observed that at night dry grass collects moisture from the desert air, even when there is no dew. The reason is that the drop in nighttime temperatures raises the relative humidity of the desert air and that the dry plant material can absorb moisture. To determine whether or not this mechanism was of importance to the oryx, I exposed some of the plants to laboratory air of the same average temperature and humidity as desert night air.

Within 10 hours the formerly parched plants had acquired a water content of 42 percent.

In the wild, of course, the plants would not always contain this much water. At sunset their water content would be less, but later at night the plants could be substantially cooler than the surrounding air because of radiation to the night sky and thus might collect more water than the plants in the laboratory experiment. It seems entirely possible that by eating mainly at night the oryx could take in food containing an average of 30 percent water. If this is the case, the oryx, which needs only half as much water as the eland, would also be independent of drinking water as it roams the desert.

Hence we see that both eland and oryx have unusual physiological and behavioral adaptations for life in an arid environment. It is therefore tempting to conclude that ranching eland and oryx in arid regions would be an excellent way to expand Africa's meat supply. The conservation of beautiful and interesting species would be an additional benefit. Serious problems, however, exist in getting antelope protein off the hoof and to the market at a price competitive with beef (equivalent to about 28 cents per pound in most of East Africa). The success of the antelopes in arid regions with sparse vegetation depends on their low density per square mile; this makes it difficult to locate them in the vast areas where they live. To harvest them economically would require the development of inexpensive ways to find, kill, butcher and transport them from the isolated deserts to the cities and towns of Africa. The alternative to wild ranching —domestication, fencing and concentrated feeding—dissipates the physiological and behavioral advantages of antelopes over cattle. In fact, there is every reason to believe man's intensive breeding of cattle for meat production has produced an animal superior to antelopes under these conditions. Tapping the potential of antelope meat awaits economists and agriculturists who can solve these seemingly insurmountable problems.

21

SALT GLANDS

KNUT SCHMIDT-NIELSEN

January 1959

A special organ which eliminates salt with great efficiency enables marine birds to meet their fluid needs by drinking sea water. Similar organs have been found in marine reptiles

As the writers of stories about castaways are apt to point out, a man who drinks sea water will only intensify his thirst. He must excrete the salt contained in the water through his kidneys, and this process requires additional water which is taken from the fluids of his body. The dehydration is aggravated by the fact that sea water, in addition to common salt or sodium chloride, also contains magnesium sulfate, which causes diarrhea. Most air-breathing vertebrates are similarly unable to tolerate the drinking of sea water, but some are not so restricted. Many birds, mammals and reptiles whose ancestors dwelt on land now live on or in the sea, often hundreds of miles from any source of fresh water. Some, like the sea turtles, seals and albatrosses, return to the land only to reproduce. Whales, sea cows and some sea snakes, which bear living young in the water, have given up the land entirely.

Yet all these animals, like man, must limit the concentration of salt in their blood and body fluids to about 1 per cent—less than a third of the salt concentration in sea water. If they drink sea water, they must somehow get rid of the excess salt. Our castaway can do so only at the price of dehydrating his tissues. Since his kidneys can at best se-

PETREL EJECTS DROPLETS of solution produced by its salt gland through a pair of tubes atop its beak, as shown in this high-speed photograph. The salt-gland secretions of most birds drip from the tip of the beak. The petrel, however, remains in the air almost continuously and has apparently evolved this "water pistol" mechanism as a means of eliminating the fluid while in flight.

crete a 2-per-cent salt solution, he must eliminate up to a quart and a half of urine for every quart of sea water he drinks, with his body fluids making up the difference. If other animals drink sea water, how do they escape dehydration? If they do not drink sea water, where do they obtain the water which their bodies require?

The elimination of salt by sea birds and marine reptiles poses these questions in particularly troublesome form. Their kidneys are far less efficient than our own: a gull would have to produce more than two quarts of urine to dispose of the salt in a quart of sea water. Yet many observers have seen marine birds drinking from the ocean. Physiologists have held that the appearance of drinking is no proof that the birds actually swallow water, and that the low efficiency of their kidneys proves that they do not. Our experiments during the past two years have shown that while the physiologists are right about the kidneys, the observations of drinking are also correct. Marine birds do drink sea water. Their main salt-eliminating organ is not the kidney, however, but a special gland in the head which disposes of salt more rapidly than any kidney does. Our studies indicate that all marine birds and probably all marine reptiles possess this gland.

The obvious way to find out whether birds can tolerate sea water is to make them drink it. If gulls in captivity are given only sea water, they will drink it without ill effects. To measure the exact amount of sea water ingested we administered it through a stomach tube, and found that the birds could tolerate large quantities. Their output of urine increased sharply but accounted for only a small part of the salt they had ingested. Most of the salt showed up in a clear, colorless fluid which dripped from the tip of the beak. In seeking the source of this fluid our attention was drawn to the so-called nasal glands, paired structures of hitherto unknown function found in the heads of all birds. Anatomists described these organs more than a century ago, and noted that they are much larger in sea birds than in land birds. The difference in size suggested that the glands must perform some special function in marine species. Some investigators proposed that the organs produce a secretion akin to tears which serves to rinse sea water from the birds' sensitive nasal membranes.

We were able to collect samples of the secretion from the gland by inserting a thin tube into its duct. The fluid turned out to be an almost pure 5-per-cent solution of sodium chloride—many times saltier than tears and nearly twice as salty as sea water. The gland, it was plain, had nothing to do with rinsing the nasal membranes but a great deal to do with eliminating salt. By sampling the output of other glands in the bird's head, we established that the nasal gland was the only one that produces this concentrated solution.

The nasal glands can handle relatively enormous quantities of salt. In one experiment we gave a gull 134 cubic centimeters of sea water—equal to about a tenth of the gull's body weight. In man this would correspond to about two gallons. No man could tolerate this much sea water; he would sicken after drinking a small fraction of it. The gull, however, seemed unaffected; within three hours it had excreted nearly all the salt. Its salt glands had produced only about two thirds as much fluid as its kidneys, but had excreted more than 90 per cent of the salt.

The fluid produced by the salt gland is about five times as salty as the bird's blood and other body fluids. How does the organ manage to produce so concentrated a solution? Microscopic examination of the gland reveals that it consists of many parallel cylindrical lobes, each composed of several thousand branching tubules radiating from a central duct like bristles from a bottle brush. These tubules, about a thousandth of an inch in diameter, secrete the salty fluid.

A network of capillaries carries the blood parallel to the flow of salt solution in the tubules, but in the opposite direction [see illustration on opposite page]. This arrangement brings into play the principle of counter-current flow, which seems to amplify the transfer of salt from the blood in the capillaries to the fluid in the tubules. A similar arrangement in the kidneys of mammals appears to account for their efficiency in the concentration of urine [see "'The Wonderful Net," by P. F. Scholander, beginning on page 125]. No such provision for counter-current flow is found in the kidneys of reptiles, and it is only slightly developed in birds.

Counter-current flow, however, does not of itself account for the gland's capacity to concentrate salt. The secret of this process lies in the structure of the tubules and the cells that compose them.

The microscopic structure of a salt-gland tubule resembles a stack of pies with a small hole in the middle. Each "pie" consists of five to seven individual

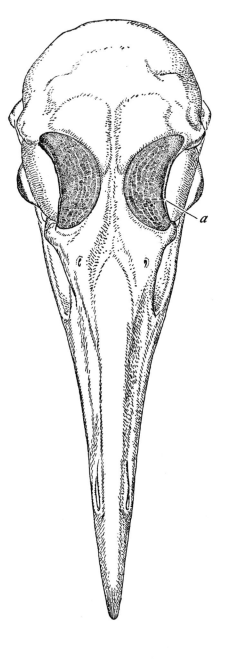

STRUCTURE of salt gland is essentially the same in all sea birds. In the gull the glands lie above the bird's eyes, as shown at left. Cross section of a gland (a) shows that it consists of many lobes (b). Each of these

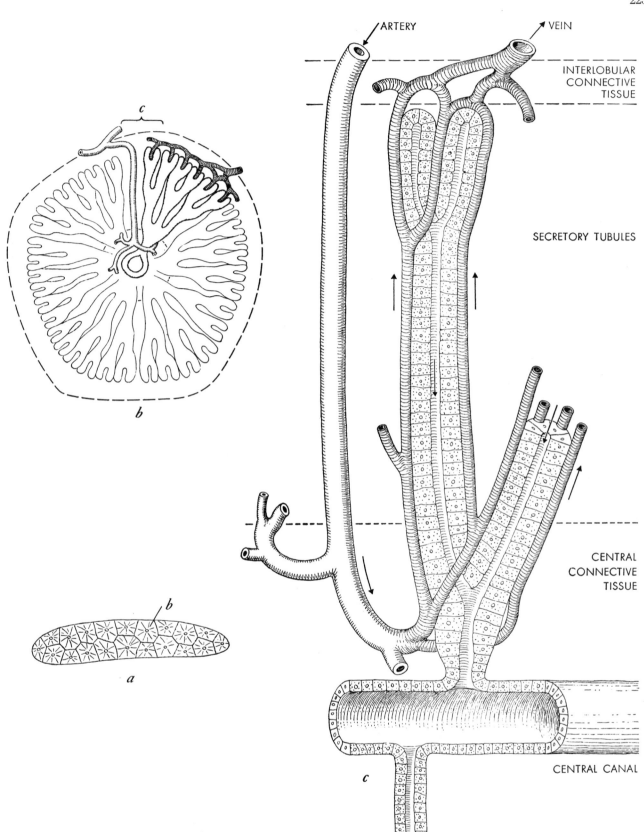

ARTERY

VEIN

INTERLOBULAR CONNECTIVE TISSUE

SECRETORY TUBULES

CENTRAL CONNECTIVE TISSUE

CENTRAL CANAL

c

b

b

a

c

lobes contains several thousand branching tubules which radiate from a central duct like the hairs of a bottle brush. Enlargement of a single tubule (*c*) reveals that it is surrounded by capillaries in which blood flows counter to the flow of salt secretion in the tubule. This counter-current flow, which also occurs in the kidneys of mammals, facilitates the transfer of salt from the blood to the tubule. The tubule wall, only one cell thick, consists of rings of five to seven wedge-shaped cells. These rings, stacked one on top of another, encircle a small hole, or lumen, through which the salty secretion flows from the tubule into the central canal of the lobe.

cells arranged like wedges. The hole, or lumen, funnels the secretion into the central duct. When we inject dye into the lumen, colored fluid seeps out into a system of irregular crevices in the walls of the tubule. More detailed examination with the electron microscope reveals a similar, interlocking system of deep folds which extend inward from the outer surface of the tubule. This structure may be important in that it greatly multiplies the surface area of the cell. It is worth noting that cells with similar, though shallower, folds are found in the tubules of the mammalian kidney.

Evidently some physiological mechanism in the cell "pumps" sodium and chloride ions against the osmotic gradient, from the dilute salt solution of the blood to the more concentrated solution in the lumen. Nerve cells similarly "pump" out the sodium which they absorb when stimulated [see "The Nerve Impulse and the Squid," by Richard D. Keynes; SCIENTIFIC AMERICAN Offprint 58]. Of course the mechanisms in the two processes may be quite different. In the tubule cells the transport of sodium and chloride ions seems to involve the mitochondria, the intracellular particles in which carbohydrates are oxidized to produce energy.

The similarities between the salt gland and the mammalian kidney should not obscure their important differences. For one thing, the salt gland is essentially a much simpler organ. The composition of its secretions, which apart from a trace of potassium contain only sodium chloride and water, indicates that its sole function is to eliminate salt. In contrast, the kidney performs a variety of regulatory and eliminative tasks and produces a fluid of complex and variable composition, depending on the animal's physiological needs at a particular time.

The salt gland's distinctive structure, elegantly specialized to a single end, enables it to perform an almost unbelievable amount of osmotic work in a short time. In one minute it can produce up to half its own weight of concentrated salt solution. The human kidney can produce at most about a twentieth of its weight in urine per minute, and its normal output is much less.

Another major difference between the two glands is that the salt gland functions only intermittently, in response to the need to eliminate salt. The kidney, on the other hand, secretes continuously, though at a varying rate. The salt gland's activity depends on the concentration of salt in the blood. The injection of salt solutions into a bird's bloodstream causes

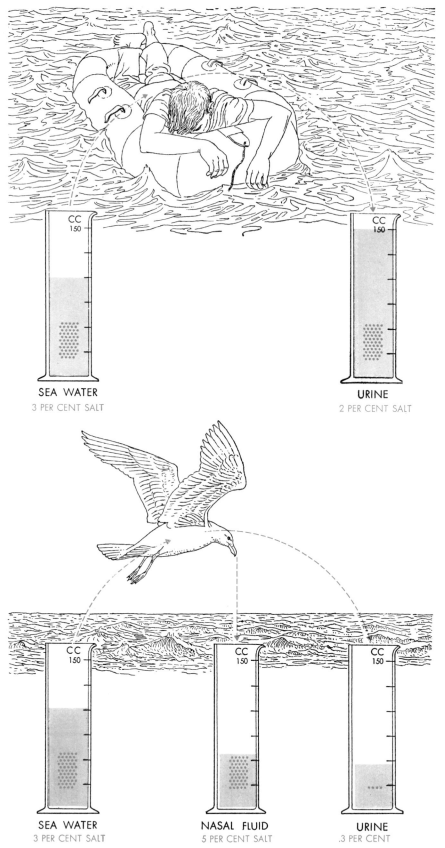

SALT EXCRETION IN MEN AND BIRDS is compared in these drawings. Castaway at top cannot drink sea water because in eliminating the salt it contains (*colored dots*) he will lose more water than he has drunk. His kidney secretions have a salt content lower than that of sea water. Gull (*below*) can drink sea water even though its kidneys are far less efficient than a man's. It eliminates salt mainly through its salt, or "nasal," glands. These organs, more efficient than any kidney, secrete a fluid which is nearly twice as salty as sea water.

the gland to secrete, indicating that some center, probably in the brain, responds to the salt concentration. The gland responds to impulses in a branch of the facial nerve, for electric stimulation of this nerve causes the gland to secrete.

While the structure and function of the salt gland is essentially the same in all sea birds, its location varies. In the gull and many other birds the glands are located on top of the head above the eye sockets [see illustrations on this page]; in the cormorant and the gannet they lie between the eye and the nasal cavity. The duct of the gland in either case opens into the nasal cavity. The salty fluid flows out through the nostrils of most species and drips from the tip of the beak, but there are some interesting variations on this general scheme. The pelican, for example, has a pair of grooves in its long upper beak which lead the fluid down to the tip; the solution would otherwise trickle into the pouch of the lower beak and be reingested. In the cormorant and the gannet the nostrils are nonfunctional and covered with skin; the fluid makes its exit through the internal nostrils in the roof of the mouth and flows to the tip of the beak.

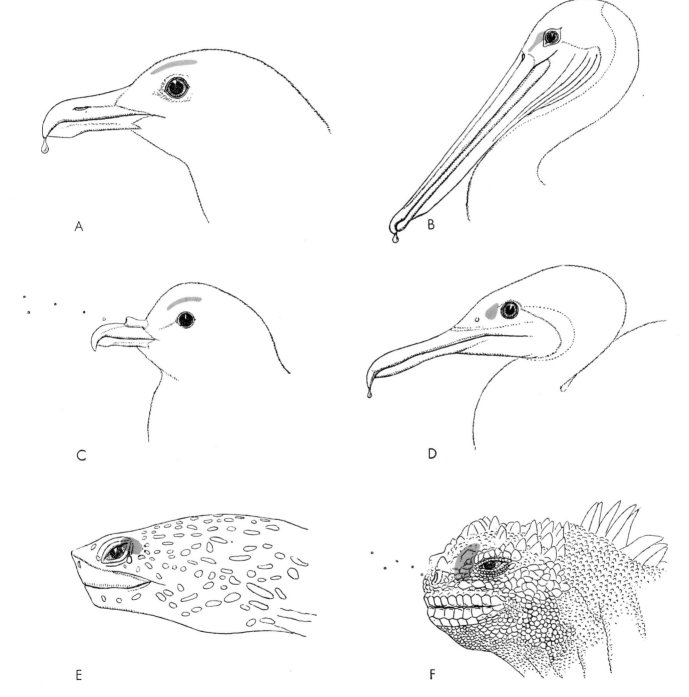

LOCATION OF SALT GLAND (*color*) varies in different species of marine birds and reptiles. In the gull (*A*) the gland's secretions emerge from the nostril and drip from the beak; in the cormorant (*D*) the fluid flows along the roof of the mouth. The pelican (*B*) has grooves along its upper beak which keep the fluid from dripping into its pouch; the petrel (*C*) ejects the fluid through tubular nostrils. In the turtle (*E*) the gland opens at the back corner of the eye; in the marine iguana (*F*) it opens into the nasal cavity.

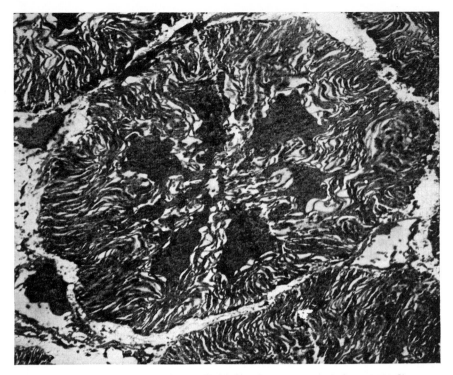

CROSS SECTION OF SALT-GLAND TUBULE is shown magnified about 5,700 diameters in this electron micrograph made by William L. Doyle of the University of Chicago. To emphasize the cell-structure the specimen was kept in a solution which shrank and distorted the cells and their nuclei. Most of the material of the cells lies in folded, leaflike layers; cells with a somewhat similar structure are found in the kidney tubules of mammals.

The petrel displays an especially interesting mechanism for getting rid of the fluid. Its nostrils are extended in two short tubes along the top of its beak. When its salt glands are working, the bird shoots droplets of the fluid outward through the tubes [see illustration on page 221]. This curious design may reflect a special adaptation to the petrel's mode of life. Though the bird remains at sea for months at a time, it rarely settles down on the water to rest. Presumably the airstream from its almost continuous flight would hamper the elimination of fluid from the bird's nostrils, were it not for the water-pistol function of the tubes.

Our studies so far have demonstrated the existence of the salt gland in the herring gull, black-backed gull, common tern, black skimmer, guillemot, Louisiana heron, little blue heron, double-crested cormorant, brown pelican, gannet, petrel, albatross, eider duck and Humboldt penguin. These species, from a wide variety of geographical locations, represent all the major orders of marine birds. There is little doubt that this remarkable organ makes it possible for all sea birds to eliminate salt and live without fresh water.

The discovery of the salt gland in sea birds prompted us to look for a similar organ in other air-breathing sea animals.

In *Alice's Adventures in Wonderland* the Mock Turtle weeps perpetual tears because he is not a real turtle; real turtles, at least the marine species, also weep after a fashion. A. F. Carr, Jr., a distinguished specialist in marine turtles, gives us a vivid account of a Pacific Ridley turtle that came ashore to lay its eggs. The animal "began secreting copious tears shortly after she left the water, and these continued to flow after the nest was dug. By the time she had begun to lay, her eyes were closed and plastered over with tear-soaked sand and the effect was doleful in the extreme." Thus Carr makes it clear that the turtle's tears do not serve to wash its eyes free of sand, an explanation that otherwise might seem reasonable. The suggestion that the turtle weeps from the pangs of egg-laying is even wider of the mark.

With the loggerhead turtle as our subject, we have found that the sea turtle's tears come from a large gland behind its eyeball. The tears have much the same composition as that of the salt-gland secretions of the sea bird. Thus it would seem more than likely that the turtle's "weeping" serves to eliminate salt. The salt gland of the turtle has a structure similar to that of the gland in sea birds, with tubules radiating from a central duct, and it seems that this structure is

essential for the elaboration of a fluid with a high salt concentration. The similarity is the more striking because the location of the gland in the turtle indicates that it has a different evolutionary origin. Still a third independent line of evolution may be represented by the salt gland in the Galápagos marine iguana, the only true marine lizard.

Anatomical studies of the other marine reptiles—the sea snakes and the marine crocodiles—have established that their heads contain large glands whose function may be similar to that of the salt gland. When we succeed in obtaining living specimens of these creatures, we expect to determine whether their glands have the same function.

Investigations of marine mammals thus far indicate that these animals handle the elimination of salt from their systems in a more conventional manner. The seal and some whales apparently satisfy their need for water with the fluids of the fish on which they feed. The elimination of such salt as these fluids contain requires kidneys of no more than human efficiency. But other whales, and walruses, whose diet of squid, plankton or shellfish is no less salty than sea water, must surely eliminate large quantities of excess salt even if they do not drink from the ocean itself. Our knowledge of their physiology suggests that their kidneys, which are more powerful than ours, can eliminate all the salts in their food. Some mammalian kidneys do function at this high level. The kangaroo rat, whose desert habitat compels it to conserve water to the utmost, can produce urine twice as salty as the ocean, and thrives in the laboratory on a diet of sea water and dried soybeans [see "The Desert Rat," by Knut and Bodil Schmidt-Nielsen; Scientific American Offprint 1050].

We should like to study salt excretion in whales, but these animals are obviously not easy to work with. We have undertaken, however, some pilot studies on seals. When we injected them with salt solutions that stimulate the salt glands of birds and reptiles, they merely increased their output of urine. Methacholine, a drug which also stimulates the salt gland, gave equally negative results. Whatever the seal's need to eliminate salt, its kidneys are evidently adequate to the task. We must therefore assume that the salt gland has evolved only in the birds and reptiles, animals whose kidneys cannot produce concentrated salt solutions.

V

TEMPERATURE ADAPTATIONS

. . . the males [bears] lie in hiding for periods of forty days and the females four months. If they have not got caves, they build rain-proof dens by heaping up branches and brushwood, with a carpet of soft foliage on the floor. For the first fortnight they sleep so soundly that they cannot be aroused even by wounds; at this period they get fat with sloth to a remarkable degree. . . . As a result of these days of sleep they shrink in bulk and they live by sucking their fore-paws. . . . No evidence of food and only the smallest amount of water is found in the belly at this stage, and . . . there are only a few drops of blood in the neighborhood of the heart and none in the rest of the body.

Pliny
NATURAL HISTORY, VIII, liv.

V

TEMPERATURE ADAPTATIONS

INTRODUCTION

Fishes and amphibians are poikilotherms—animals whose temperature varies with that of the environment. They are adapted in a variety of ways to their normal environmental temperature range, and, as we will see, they show ability to become acclimated to different temperatures by a number of biochemical modifications. However, as is so often the case, the evolutionary versatility of the vertebrate stock provides us with exceptions to the generality that fishes are "cold blooded." As explained by Francis Carey in "Fishes with Warm Bodies," heat conservation mechanisms have evolved quite independently in certain large bony fish (the tuna, for example) and sharks (the mako). The result is internal body temperatures that can be maintained for long periods at levels considerably above those of the surrounding sea water. An interesting problem with respect to the "bends" arises in the heat exchangers of these warm bodied fishes. One might anticipate that, as oxygen-loaded hemoglobin goes from the cool gills to the warm heat exchanger, the typical vertebrate response would occur: affinity for oxygen would drop markedly, and oxygen would be released, even enough to form bends-producing bubbles! But, again "nature's forethought," as Pliny called it, has operated: blue-fin tuna hemoglobin shows virtually no change in affinity as a function of temperature; hence it does not give up its oxygen precipitously, and the bends are avoided. This is still another example of a situation in which evolutionary modification at the anatomical level in one system (blood vessels) is accompanied by changes at the biochemical level in another system (the red cells).

Reptiles, too, are poikilotherms, but, as Charles M. Bogert points out in "How Reptiles Regulate Their Body Temperature," they behave in special ways in order to raise and regulate their body temperature at least part of the time. The hormonal and neuronal mechanisms contributing to this capacity in reptiles are still present in birds and mammals and are part of the more complex control machinery used for true homeothermy. (Homeotherms are animals whose temperature is high and is held constant despite fluctuations in environmental temperature.) Most important of the reptilian properties is a group of cells, apparently located in the hypothalamus, that responds to temperature by altering blood pressure and the rate of heart beat. In fact, the vascular system occupies a critical position among those adaptations that permit temperature regulation. Reptiles are sometimes called "ectotherms" because they derive most of their heat from outside the body; the blood fluid transports heat from the surface to the internal regions of the body. Because of its ubiquitous flow, moving blood fluid is also the best means of providing the organism with a continuous, rapidly responding source of information about overall body temperature. In birds and mammals, control of vasoconstriction of peripheral blood vessels has become the primary means of regulating the degree of heat conservation (the insulation derived from feathers or fur is not considered a regulator). The foregoing, then, emphasizes the central role of the vascular system in temperature relations and homeothermy.

Since temperature control originated relatively late in the history of vertebrates, most other physical properties of the internal environment were already under regulatory control. Development of the capacity to govern temperature seems to have involved utilizing a number of existing organ systems for the secondary purpose of participating in thermal control. A lizard, for example, when it is frightened, shows a brief shivering tremor as a means of increasing heat production. It never shivers when exposed to chilling conditions, but the same sort of shivering is used by birds and mammals to generate heat in cold environments. The biologist V. H. Hutchison has shown that the Indian python goes one step further. The female coils about a cluster of eggs and incubates them at elevated temperature by continually contracting her body musculature. The heat generated is sufficient to maintain a body and egg temperature of as much as 7 degrees centigrade above ambient. To perform this feat the female functions like a homeotherm as environmental temperature falls: it increases its consumption of oxygen and so is able to produce the extra heat needed for constancy of internal temperature. Nonincubating pythons and all other poikilotherms show rather strict proportionality of oxygen utilization to temperature, so that the quantities of oxygen used fall rapidly as the environment cools.

Still another example of reptiles employing a temperature-regulatory activity has been described by George A. Bartholomew. Several types of lizards breathe faster and faster as their body warms (thus meeting the need for more oxygen). In addition, some large lizards take special deep, rapid breaths, utilizing in particular the floor of the mouth and the neck, as temperature rises above 38 degrees centigrade. This increases the volume of air pumped and the amount of evaporation from the respiratory system. Obviously, here is an equivalent to panting, one of the emergency measures used by birds and mammals for cooling the body. There are many such neuronal or endocrine-controlled processes, which originated in reptiles but are now included in the temperature-regulating arsenal of the homeotherm.

From a study of the monotremes (the spiny anteater *Echidna*, or the duckbill platypus *Ornithorhynchus*), the most primitive living mammals, we attain further insight into the development of homeothermy. These "prototherian" animals diverged from the early mammals long before the marsupial and placental insectivores appeared. *Echidna* can maintain its body temperature between 20 and 30 degrees centigrade as the ambient environmental temperature falls toward 0 degrees. It is able to do this by increasing its rate of heat production; the metabolic rate is raised as the environment chills, and the heat produced is conserved by fur, which serves as an insulator. The result is quite different if temperature rises much above 30 to 35 degrees centigrade: regulation breaks down, and heat death can result. One reason is that spiny anteaters lack the panting reflex and, at best, have only a few sweat glands; they are essentially unable to combat rising body temperature except by decreasing heat production. Obviously animals with such primitive control abilities are limited in their activity; thus, monotremes commonly hibernate during cool weather and remain inactive and out of the sun in hot weather.

Marsupials (the Metatheria) are more like the placentals (the Eutheria) than the monotremes are because they have a higher normal body temperature (about 33 degrees centigrade) and are able to control it over a wider range of ambient temperature. Sweat glands, panting, and peripheral vasoconstriction are all present, so that control in both upper and lower danger zones is feasible.

Properties of the eutherian temperature-regulatory system are described by Laurence Irving in "Adaptations to Cold" and by T. H. Benzinger in "The Human Thermostat." The evolution of the system was the result of the ability certain neurons acquired to respond to altered temperature. The output from the medulla of the brain that elicits a coordinated response to compensate for such a temperature change is dependent upon

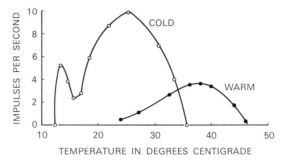

Figure 1. The rates of discharge of mammalian hot and cold temperature-sensing cells. Note that the cold receptor is virtually quiet when 37° C is approached.
[After Zotterman, Y., *Ann. Rev. Physiol.* **15**, 1953.]

sensory information from the body periphery, from deep body temperature sensors, and from the hypothalamic sensor cells themselves.

Recent experiments on the thermostat of mammals provide an interesting illustration of the scientific method, in addition to suggesting how the thermostat may be regulated. One hypothesis states that the crucial feature of the microenvironment near the sensor cells is the ratio Na^+/Ca^{++}. If a solution with elevated Na^+ levels is pipetted into the region, a mammal responds by shivering, erecting its fur, and showing vasoconstriction—all responses designed to conserve heat. If, on the other hand, elevated Ca^{++} levels are applied, the respiratory rate falls, blood vessels vasodilate, and the animal is quiet—all phenomena associated with overheating. Interestingly, if prolonged perfusion with a solution of abnormal Na^+/Ca^{++} ratio is carried out, the animal gradually comes to regulate about a new "set-point." If a bacterial toxin, typhoid vaccine, is injected into a cat, Ca^{++} falls relative to Na^+, a new ratio of the ions is established, and a fever results. A fever then is regulation about a new, abnormally high set point, perhaps because of these effects upon the metal ions in the hypothalamus. Despite all these consistencies in evidence, new questions have been raised by the finding that elevated calcium in the hypothalamus also depresses breathing rate, heart rate, and oxygen consumption. Thus, the altered ion ratios may lead to general depressant effects and only indirectly to altered set points. This is the way that science works: data generates an hypothesis; new experiments force its modification. Ultimately a satisfactory hypothesis is formulated that meets each experimental test.

The activity of temperature-sensing nerves is typified by the following data on the responses of *cold* receptors in the skin of monkeys. A characteristic discharge rate of nerve impulses from the cold receptor is found for each temperature. Under conditions of constant ambient temperature this rate is nearly maximal when surface temperature of the skin is between 20 and 35 degrees centigrade. Discharge falls to zero as temperature rises from 35 to 40 degrees centigrade. When the temperature drops from about 40 degrees centigrade to 20 degrees, a peak discharge of 150 impulses per second is attained. A change as small as 0.5 degrees centigrade alters the discharge rates of these nerves, and even smaller increments affect the rates of some other temperature sensors. It is not clear how changing temperature evokes nerve discharge in such cells as these, or in the temperature sensors of the central nervous system.

A variety of biochemical adaptations to temperature are seen in various vertebrates. The amazing heat-generating brown fat of immature or hibernating mammals is described in "The Production of Heat by Fat" by Michael Dawkins and David Hull. This type of heat production is dependent upon the activity of different kinds of differentiated cells (brown or white fat cells) of one organism. In work performed since the appearance of the article on brown fat, it has become clear that the release of epinephrine from nerve endings in the brown-fat tissue results in activation of adenylcyclase, production of cyclic AMP, and action of that "second messenger" on the lipase enzyme (see the diagram on page 274).

A great deal is being learned about the biochemical control of heat production. Perhaps one of the most significant achievements is the demonstration that the crucial vertebrate hormone thyroxine (see "An Essay on Vertebrates") acts on muscle, liver, and kidney cells to activate a Na^+,K^+-dependent ATPase enzyme associated with the cell membranes. The enzyme splits ATP, and heat is generated as a byproduct. This process may well be the basis for nonshivering thermiogenesis—that is, the basic heat production under nonemergency conditions. As is emphasized by Peter Hochachka and George Somero (*Strategies of Biochemical Adaptation*), this is a superb example of a case where a truly ancient enzyme has been used for a completely new purpose by relative newcomers to the biological world, the birds and mammals.

Some biochemical adaptations to temperature may vary between closely related species. An example of this type is seen in lizards that "prefer" different normal body temperatures ranging from 30 to 38.8 degrees centigrade. Differences are seen in myosin ATPase, an important enzyme in ATP metabolism during muscle contraction. Paul Licht of the University of California has shown that the temperature optimum for the enzyme differs between the species and correlates nicely with preferred body temperatures. Interestingly, lizards preferring high normal temperatures tend to have narrow preferred temperature ranges and the optima for their enzyme is similarly narrow; in contrast, animals preferring cooler conditions have more widely ranging preferred body temperatures and myosin ATPases with broad optima. It seems reasonable to assume that these functional differences in the enzymes are probably due to minor variations in the structure of the protein chains (i.e., they reflect the sorts of mutations discussed already for vertebrate hemoglobins).

Another type of temperature adaptation is the alteration of proteins within single cells. It can be seen in animals that are placed at new temperatures in the laboratory so that they become acclimated. In rainbow trout, for instance, electrophoretically distinct forms of brain acetyl cholinesterase are present at 2 degrees and 17 degrees centigrade. Peter Hochochka of the University of British Columbia has shown that at 12 degrees centigrade, on the other hand, both forms of the enzyme are present. These differences are seen in nature during seasonal temperature fluctuations. Besides this sort of variation in different enzyme forms, other strategies have appeared during evolution: for instance, in certain goldfish the activity of succinic dehydrogenase is markedly altered by the type of lipid present. Thus, changes in membrane lipids, which might occur as part of acclimatizing to different environmental temperatures, may be capable of affecting the basic enzymatic machinery of cells.

In addition to these specific sorts of biochemical adaptations, there is a general difference between the basic catalytic efficiency of enzymes obtained from ectotherms (poikilotherms) and endotherms: enzymes from the former animals are the most efficient in terms of turnover number of substrate or other physico-chemical indices. Why this should be so—why the enzymes of mammals should have become *less* efficient as mammals evolved from ectothermal reptiles—is a mystery for the future to solve.

A well-known biochemical adaptation for life in extreme cold is that of fats found at different levels in the long legs of caribou, animals that often stand in the snow. As one proceeds from the thigh to the foot the degree of saturation of fatty acids decreases; this means that the melting point of the fat is lowered and the fat remains soft and pliable despite low temperature. If normal fats, with melting points as much as 30 centigrade degrees higher, were present in the foot, the tissues would be quite stiff and immobile. In some vertebrates this type of property also varies with season of the year, with unsaturated fats being manufactured during winter months and not in the seasons of warm weather.

Other differences are seen in the extremities of homeotherms that are exposed to temperature extremes, and, in particular, to cold extremes. In birds and mammals, for instance, most neurons stop conducting impulses at temperatures of about 15 to 20 degrees centigrade. In contrast, in cold-adapted legs, the peripheral nerves continue to function at 3 degrees centigrade. This is truly amazing when one realizes that such differences may be found within *single* long nerve cells. In a cold-acclimated herring gull, for instance, a nerve in the tibial portion of the leg showed blockage of conduction at temperatures of 11.7 to 14 degrees centigrade, while the metatarsal region remained active until temperature fell to between 2.8 and 3.9 degrees centigrade. Although it is not known yet what temperature-dependent biochemical property is responsible for these differences, this adaptation of the gull implies a remarkable distributional control within one cytoplasmic mass.

An area of research not treated in Laurence Irving's "Adaptations to Cold" concerns vertebrate fishes that live in water at subfreezing temperatures. One strategy is used by marine fishes that live in shallow waters in Arctic regions. During the coldest months, such fish live in a "supercooled" condition by staying near the bottom of the sea and avoiding conditions that will act to "seed" ice-crystal formation in their bodies. A quite different tactic is used by other fishes that synthetize special "antifreeze" glycoprotein molecules. Among different species of one genus, the quantity of these antifreeze molecules is directly related to the depth at which the fish live; most molecules are present in pelagic fishes where micro-sized ice crystals are encountered frequently; progressively fewer molecules are found with increasing depth and decreased likelihood of encountering ice seeding. Though the structure of the glycoprotein is known, it is still a mystery how the glycoprotein is rendered an antifreeze by the regularly spaced hydroxyl groups.

All of these adaptations may ultimately be found to result from changed enzymatic activities (i.e., the enzymes may be altered, different enzymes may be produced, and so on). As an example, hemoglobin provides a clear demonstration of how the activity of a protein is affected by amino acid substitutions. The hemoglobin of each vertebrate shows optimal oxygen-carrying capacity near normal body temperatures; differences in hemoglobins between animals result from different primary protein structures of globins. All hemoglobins respond in the same basic way to altered temperature: a lowered affinity for oxygen when temperature rises, and an increased affinity as it falls (this response is an adaptation, since temperature can fall in the lung capillaries, where oxygen loading takes place). The changing affinity is probably caused by altered folding of globin in response to temperature; the effect is great if one compares hemoglobin from vertebrates well separated by evolution. For example, if a frog's body temperature has been raised to mammalian body temperature, the hemoglobin has such low affinity for oxygen that it cannot load; conversely, human hemoglobin chilled to temperatures common for amphibia cannot unload its oxygen. In a given species the effect may provide a physiological limitation on behavior. Thus, as a lizard's body and blood temperature rise above 40 degrees centigrade, hemoglobin can no longer load adequately and metabolism necessarily falls; clearly, the possibility of death is strong under such circumstances because effective behavior to counteract the high temperature might be precluded by the relative lack of oxygen.

The gross anatomy of homeotherms exhibits a number of adaptations that can be traced to temperature regulation. Body size of terrestrial mammals tends to be larger, and the limbs shorter, in temperate or cool latitudes than in tropical ones. Both properties obviously result from the relationship of the body mass, which is the source of heat, to the surface area, from which heat is lost. Although there are many exceptions to generalizations like these, it is correct to say that the small animal in a cold habitat must have extraordinary physiological or behavioral adaptations to survive. Two other generalizations that are more difficult to dispute are these: (1) the thickness and quality of insulation is greater in colder climates; (2) extremities show modifications designed to permit tolerance of low temperatures and continuing physiological functioning under conditions in which central body organs would be inoperative (an example of this type of modification is the nerve conduction discussed above).

We have mentioned several specializations of the extremities that permit survival in cold environments. In warm environments, the extremities, as sites of controlled heat loss, are equally important. The hairless tail of a muskrat, the horns of a goat, the tail of a beaver, the flippers of seals, the legs of birds, and the huge ears of a jackrabbit all function in this way. In these organs, insulation tends to be sparse, vasoconstriction and rete mirabile nets are well controlled, and biochemical specializations are present. A bird's body lacks sweat glands, and little water leaks through the skin. Hyperventilation and air sac evaporation, and radiation from the featherless legs are the ways in which birds lower their body heat. Interestingly, an overheated wood stork will urinate on its long legs as often as once a minute, using evaporation from the warm legs as a means of lowering body heat. A final example of interest is the dog, which of course lacks sweat glands. The tongue hanging from the mouth of an overheated dog is one obvious means that evaporative cooling can occur. But in addition special nasal glands that open just within the nostrils copiously secrete fluid when a dog pants and, in fact, may account for up to 36 percent of the water that is evaporated for cooling purposes. These glands are in a sense the sweat glands of a dog.

The final article in this section provides a good view of how extreme temperatures affect the lives of mammals. After generalizing about physiology or morphology, it is often most instructive to return to specific examples of wild animals and to examine as many parameters of their biology as possible, for only in this way can perspective on laboratory results be gained. "Desert Ground Squirrels" by George Bartholomew and Jack Hudson is concerned with two kinds of ground squirrels living in a desert environment. The details of the temperature regulation and the water-balance physiology of these creatures are a remarkable lesson about mam-

Figure 2. Responses of a stork to altering environmental temperature. As room temperature (indicated by solid circles) rises, the respiratory rate (indicated by open circles) rises to increase evaporative cooling. In addition, the stork begins to urinate on its legs (see the bars at the top) so that evaporation from these uninsulated regions can aid the cooling process.
[After Kahl, M. P., Jr., *Physiol. Zool.* 36, 1963. The University of Chicago Press.]

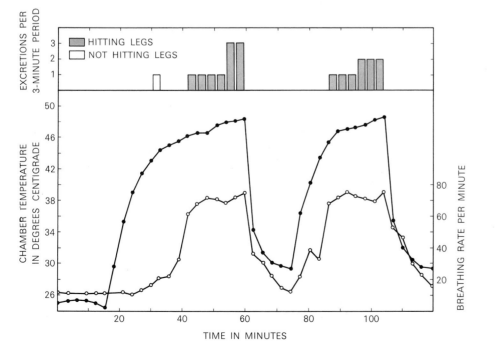

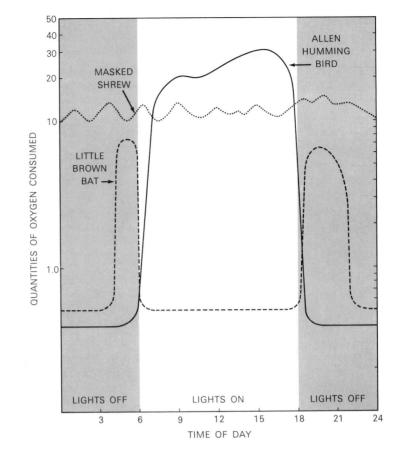

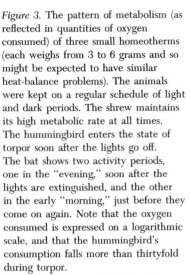

Figure 3. The pattern of metabolism (as reflected in quantities of oxygen consumed) of three small homeotherms (each weighs from 3 to 6 grams and so might be expected to have similar heat-balance problems). The animals were kept on a regular schedule of light and dark periods. The shrew maintains its high metabolic rate at all times. The hummingbird enters the state of torpor soon after the lights go off. The bat shows two activity periods, one in the "evening," soon after the lights are extinguished, and the other in the early "morning," just before they come on again. Note that the oxygen consumed is expressed on a logarithmic scale, and that the hummingbird's consumption falls more than thirtyfold during torpor.
[After Bartholomew, G.; in M. S. Gordon, ed., *Animal Function: Principles and Adaptations*, Macmillan, 1968.]

malian evolution, and they show how a given habitat can be shared by species that do not really compete with each other.

Despite the many vertebrate adaptations to life in cold and hot environments, some animals simply avoid extremes of temperature by performing complex behavior. The best-known examples are those in response to incipient cold weather: namely, migration to warmer climates, or hibernation.

Hibernation of mammals, large or small, is the suspension of maintenance of high body temperature. It necessitates a long preparative period during which stores of body fat are built up. The endocrine system has a central function in coordinating the process. During hibernation, temperature regulation occurs about a new low "set point," usually near 5 degrees centigrade, and as little as 2 percent of normal body heat is produced. The basal metabolic rate may drop to 1 percent of the normal rate, and, as a result, physiological adjustments commensurate with low temperature and low metabolism operate. Although heart rate slows (in a ground squirrel, for example, from 200 beats per minute to 10 beats per minute), the organ continues to function normally as a pump. Clotting could result from the slow blood flow, but it does not, because of the altered clotting capacity of the plasma. Severe peripheral vasoconstriction is maintained, and it keeps central blood pressure high despite the low rate of pumping by the heart. Nerves continue to conduct impulses when the body temperature falls far below the normal blockage temperature for impulses. Neither the heart nor the brain of a nonhibernator can function at such low temperatures. Low rates of metabolism demand much less oxygen: this permits breathing in the ground squirrel to slow from 100 breaths per minute to 4 breaths. Awakening and rewarming at the normal termination of hibernation is a complex process in itself, and necessitates great energy expenditure; the squirrel can increase its body heat from 4 degrees centigrade to 35 degrees in only 4 hours.

Emergency control mechanisms operate during hibernation to awaken a mammal if the body temperature falls toward lethal cold. Shivering starts, the body warms, and behavior to avoid the danger can be initiated. This kind of response consumes large quantities of the animal's stored fat; this may make survival through the long winter period impossible, but, since stored food would be of little use to a frozen animal, the hibernator awakens!

A parallel to hibernation is seen in many small birds and mammals that suspend high temperature maintenance each night and become torpid. This torpor is necessitated by the large relative surface areas of these animals and the resultant heat loss during periods of inactivity. Bats and humming-birds become poikilotherms, and they show the same basic regulatory phenomena and awakening sequence, including repeated muscle flexion, as is seen in hibernators. The amount of energy saved by torpor is significant: for a hummingbird, 10.3 kilocalories per 24 hours would be lost if sleeping took place at normal body temperatures, but only 7.6 kilocalories are lost in 24 hours if the torpid state is entered at night. If hummingbirds did not lower their nighttime temperature they would be unable to survive unless they awoke and fed during the night. This, of course, would require an additional set of special adaptations for navigation and food gathering in the dark. Such additional adaptations and continual activity at high metabolic rates would seem to be of little value for survival to reproduce; instead, small birds and mammals utilize torpidity and limited times of activity.

22

FISHES WITH WARM BODIES

FRANCIS G. CAREY
February 1973

*Not all fishes are cold-blooded. In some fast-swimming
species of tuna and mackerel shark a "wonderful net" of
arteries and veins conserves the heat of metabolism to
increase the power of the swimming muscles*

Fishes are regarded as cold-blooded animals, that is, animals whose body temperature is the same as the temperature of their surroundings. The fact is that not all fishes are cold-blooded. The first man to write about a warm fish was a British physician named John Davy. He was voyaging in the Tropics in 1835, and the ship's company was supplementing its rations by fishing. Davy was intrigued by the copious blood and mammal-like red flesh of one of the fishes brought aboard, a species of tuna known as the skipjack. He took the temperature of several skipjack and discovered that they were warmer than the water they swam in by 10 degrees Celsius (18 degrees Fahrenheit). In the years since Davy's observation most other species of tuna have proved to be warm-blooded and so has one family of sharks: the mackerel sharks. At the Woods Hole Oceanographic Institution my colleagues and I have long been fascinated by these warm-blooded fishes.

What is it that prevents most fishes from maintaining a body temperature that is more than negligibly higher than the temperature of the surrounding water? It is not, as might be supposed, the loss of surface heat that comes from being immersed in a cool medium. Rather it is the heat loss implicit in the fishes' mode of respiration. All land animals and some aquatic ones obtain the oxygen required to support their metabolism by breathing air, which is rich in oxygen and has a low capacity for absorbing heat. Water, from which fish must obtain their oxygen, contains scarcely 2.5 percent of the oxygen found in an equal volume of air but has 3,000 times the capacity of air for absorbing heat. As a result the heat lost in extracting oxygen from water can be 100,000 times greater than the loss in extracting the same amount of oxygen from air.

The oxygen in the water flowing through the gills of a fish diffuses into venous blood pumped to the gills by the heart. The blood may reach the gills at a somewhat elevated temperature as a result of having accumulated metabolic heat. Heat, however, diffuses 10 times faster than oxygen molecules do, so that by the time the blood in the gills is saturated with oxygen its temperature has dropped to the temperature of the water.

The oxygenated blood now flows back through the fish's circulatory system, and the oxygen is utilized in the body tissues. It is this process that generates the metabolic heat; the rise in temperature is proportional to the amount of oxygen extracted from the blood. In any given volume of oxygenated blood the volume of extractable oxygen is less than 20 percent. Taking into account the usual caloric yield of a fish's foodstuffs, that limited supply of oxygen will support only enough metabolism to raise the temperature of the blood one degree C. Moreover, even that slight temperature increase is lost to the water when the blood next passes through the gills, so that no heat accumulates to keep the fish warmer than the surrounding water.

Changes in the fish's rate of metabolism do not affect the cycle, because any exercise or other activity that produces more metabolic heat also demands more oxygen. An increased demand of oxygen means a greater loss of heat through the gills, so that whether the fish is at rest or exercising violently it will always be cold. Indeed, for a fish to raise its body temperature by metabolic processes alone would seem to be an impossibility.

Tunas and mackerel sharks manage to stay warm in apparent defiance of nature because they possess a special structure in their circulatory system. It is the *rete mirabile*, or "wonderful net," a tissue composed of closely intermingled veins and arteries that was first noted in vertebrates by students of anatomy in the 19th century. The French naturalist Georges Cuvier observed the presence of the *rete* in the tuna in 1831, the last year of his life. Four years later, the same year that Davy measured the temperature of the skipjack tuna, the German anatomist Johannes Müller formally described the structures in the viscera of the bluefin.

The *rete* provides a thermal barrier against the loss of metabolic heat. The mass of fine veins and arteries permits the free flow of blood for transport of oxygen and other molecules, but at the same time it short-circuits the flow of heat from the body tissues to the gills and shunts the accumulated heat back to the tissues. In the *rete* an artery, carrying cool, oxygen-rich blood from the gills to the body tissues, branches to form a mass of small vessels. The arterial network runs parallel to and is intermingled with a similar mass of fine veins carrying warm, oxygen-depleted blood from the tissues back to the gills. The system of closely associated arteries and

RETE MIRABILE, or "wonderful net," of the bluefin tuna is seen in cross section. The *rete* is a system of parallel small arteries and veins that supplies and drains the band of dark-colored muscle that is used for sustained swimming effort. The system constitutes a countercurrent heat exchanger: venous blood warmed by metabolism gives up its heat to cold, newly oxygenated arterial blood fresh from the fish's gills. The effect is to increase the temperature and thus the power of the muscle. In this bluefin *rete* sample the vessels have been visualized by the injection of latex: red latex in the arteries and blue latex in the veins.

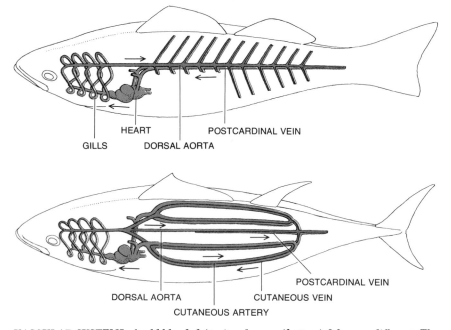

HEART

GILLS   DORSAL AORTA   POSTCARDINAL VEIN

POSTCARDINAL VEIN

DORSAL AORTA   CUTANEOUS VEIN

CUTANEOUS ARTERY

VASCULAR SYSTEMS of cold-blooded (*top*) and warm (*bottom*) fishes are different. The main vessels in most fishes, the cold-blooded ones, run along the backbone and radiate outward to the small vessels (*not shown*) that supply the muscle. In warm fishes the central vessels are smaller; most of the blood flows through cutaneous vessels under the skin and thence through countercurrent nets (*not shown*) that supply the muscle. This arrangement puts the cold end of the exchanger near the skin and the warm end in warm tissues.

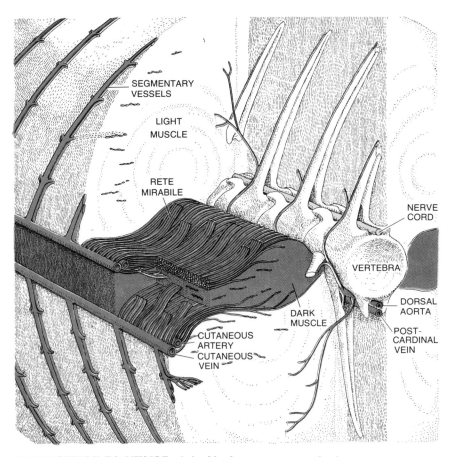

SEGMENTARY VESSELS

LIGHT MUSCLE

RETE MIRABILE

NERVE CORD

VERTEBRA

DORSAL AORTA

DARK MUSCLE

POST-CARDINAL VEIN

CUTANEOUS ARTERY
CUTANEOUS VEIN

BLOOD SUPPLY TO MUSCLE of the bluefin tuna is primarily through four pairs of cutaneous vessels. A multitude of smaller vessels branch from them, forming the thick vascular slab above and below the dark muscle, the *rete,* from which the dark muscle is supplied. The light muscle is supplied by bands of alternating arteries and veins that slant through the muscle from the segmental vessels, thus serving as two-dimensional heat exchangers. Other, non-heat-exchanging arteries and veins run out from the central vessels.

veins acts as a countercurrent heat exchanger. There are large areas of contact between the two kinds of fine vessel, and heat is readily transferred through the vessels' thin walls. Therefore the warm venous blood is cooled as it flows through the *rete,* so that little or no heat is lost when the venous blood at last reaches the gills. At the same time the cold arterial blood is warmed, so that when it reaches the interior of the fish, it has nearly reached the temperature of the warm body tissues.

A *rete* of this type serves exclusively as a countercurrent heat exchanger. Its vessels, although small, are too large and their walls are too thick to allow any significant diffusion of oxygen molecules from the arterial blood to the venous blood. Other *retia* that do serve as gas exchangers are found in many species of fish [see "'The Wonderful Net,'" by P. F. Scholander, beginning on page 125]. The tiny vessels that comprise them are capillaries, and so oxygen molecules readily pass from one to another. Such *retia* are found in the swim bladder and the eye.

A major adaptive advantage of an elevated body temperature is greatly enhanced muscle power. If the difference in temperature between two otherwise equivalent muscles is 10 degrees C., the warmer muscle is able to contract and relax three times more rapidly, with no reduction in the force applied at each contraction; thus it can generate three times as much power. This kind of increase in power is evidently essential in flying animals such as birds and bats; their body temperatures are characteristically high. Even some large insects with a heavy wing loading elevate their body temperature when they fly. For example, cicadas and locusts are often unable to take off until they have warmed up by shivering or by basking in the sun.

Like flying, high-speed swimming is a demanding form of locomotion. Water is a dense and viscous medium, and to move through it rapidly requires not only fine streamlining but also substantial muscle power. The high body temperature of tunas and mackerel sharks apparently helps to provide the extra power needed for high-speed swimming: the most important of the vascular heat exchangers are those that serve the dark-colored swimming muscles.

Vertebrate muscle is generally a mixture of dark-colored fibers and light-colored ones. The dark fibers contain high concentrations of the oxygen-transporting pigments and enzymes associated with oxidative metabolism; they are also richly supplied with blood. The light-colored fibers usually have a less well-developed blood supply. They can function without oxygen during activity by metabolizing foodstuff through fermentation, recovering later when they are at rest. Muscles that are continuously active, for example the heart, are made up largely or entirely of dark fibers. Muscles that alternate between periods of rest and short intervals of activity contain few, if any, dark fibers.

As anyone who has seen a fish steak knows, the two kinds of muscle fiber are sharply segregated in the axial musculature of fishes. For example, in many tunas a large mass of very dark muscle, well supplied with blood, lies in a broad band between the backbone and the skin along the midplane of the body. This dark muscle can operate continuously; its contractions propel the fish as it swims at ordinary cruising speeds. The blood that supplies the muscle reaches the muscle fibers after passing through a large heat-exchanging *rete*. As a result the muscle is warmer than any other part of the fish's body.

In order to provide for this large heat exchanger the arrangement of the tuna's circulatory system has been drastically altered. Most fishes have a central distribution system for the blood. The main blood supply to the muscle flows outward through a large artery, the dorsal aorta, and returns through the postcardinal vein; both are deep inside the fish just below the backbone. From these central blood vessels segmental arteries and veins radiate out to the periphery.

In the tuna the normal pattern of circulation to the muscle has been almost entirely reversed. The major blood vessels are no longer the central artery and vein. Instead four artery-vein pairs just under the skin, two on each side of the fish, provide the main blood supply [see *top illustration on opposite page*]. From these cutaneous vessels large numbers of tiny blood vessels, each only a tenth of a millimeter in diameter, branch off. The tiny vessels intermingle to form slabs of vascular tissue, the *retia*, that lie close to the upper and lower surfaces of the dark muscle [see *bottom illustration on opposite page*]. These *retia* are the heat exchangers that ensure the warmth of the dark muscle. The blood to the light muscle also comes from the large cutaneous vessels by way of pairs of segmental arteries and veins that run up and down over the surface of the muscle and send numerous branches into it. These branches are in the form of vascular bands: ribbons of alternating arteries and veins that act as heat exchangers.

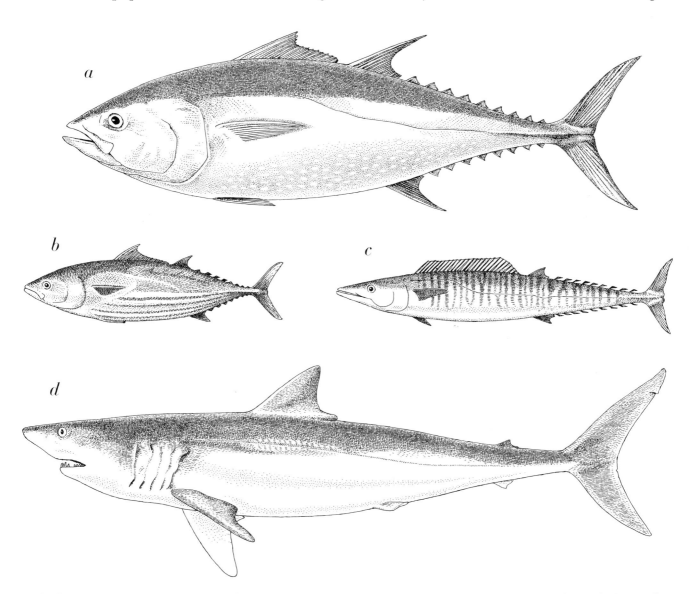

FAST-SWIMMING FISHES with warm bodies have streamlined bodies and heavily muscled tails with crescent-shaped caudal fins. The ones illustrated here are three tunas: the bluefin (*a*), the skipjack (*b*) and the wahoo (*c*), and a mackerel shark, the mako (*d*).

The area of thermal contact between arteries and veins in these two-dimensional structures is smaller than it is in the massive heat exchangers serving the dark muscle, but apparently it is adequate for the typically low rate of blood flow to the light muscle and ensures that much of this tissue too is warm.

One result of the inside-out arrangement of the main blood vessels in a warm fish is that the large arteries and veins that give rise to the *retia* are just under the skin. This puts the cool end of the heat exchanger near the surface of the fish and the warm end deep in the interior, an arrangement that minimizes surface heat loss.

By measuring the temperature of various parts of freshly caught bluefin tuna we have learned that the warmest regions are within the dark muscle [*see illustration on page 242*]. The hot spots are not, as one might expect, in the deepest part of the muscle. The reason is that the deep interior receives some of its blood supply from the dorsal aorta through a system that has no specialized heat exchangers. The highest muscle temperature is found near the center of the dark-muscle band on each side of the fish, but substantial areas of muscle tissue, both dark and light, are considerably warmer than the temperature of the water from which the fish are taken.

The bigeye tuna and the albacore have circulatory systems that closely resemble the bluefin's. Other tuna species, particularly the skipjack and to a lesser extent the yellowfin, elevate the temperature of the muscle in a different way. Although the pairs of cutaneous arteries and veins are still present, the heat exchangers that arise from them may contain no more than a single layer of fine blood vessels. In these fishes the principal heat exchanger is connected to the central artery-vein pair below the backbone. This central *rete*, a rodlike vascular mass that may be larger in diameter than the backbone itself, extends along the vertebrae in the region above the

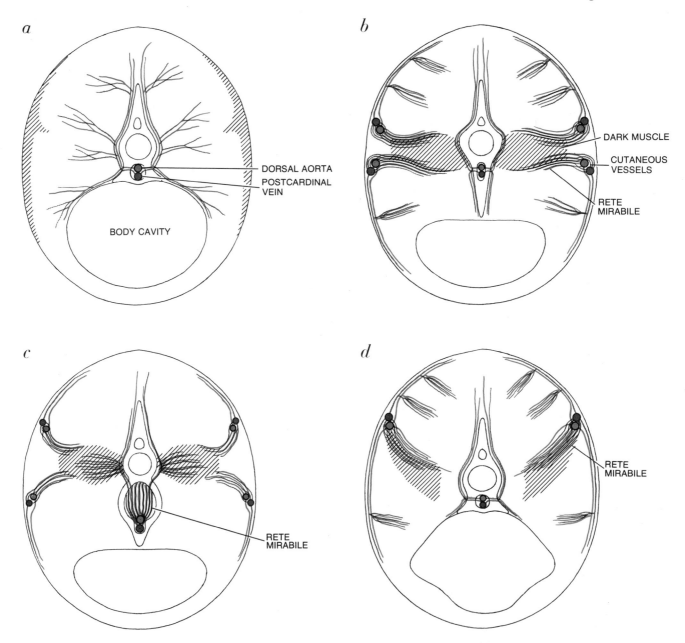

MAJOR BLOOD VESSELS of a typical cold-blooded fish (*a*) radiate from the large central vessels. In warm fishes the vascular system is rearranged so that warmed blood is supplied to the dark muscle. In the bluefin (*b*) the *retia* are slabs of vascular tissue above and below the dark muscle. In the skipjack (*c*), on the other hand, the main *rete* is a net of small vessels running vertically in a cavity below the backbone. The mako shark (*d*) has a single cutaneous blood-vessel pair, supplying a single massive *rete*, on each side.

body cavity [*see illustration on opposite page*]. Body-temperature measurements of skipjack attest to the efficiency of the rod-shaped heat exchanger: these little fish are often 10 degrees C. warmer than the water and their warmest muscle is adjacent to the backbone.

It is an interesting coincidence that the mackerel sharks, a group that is evolutionarily far removed from higher bony fishes such as the tunas, should have developed a heat-exchange system that is so much like the bluefin's. The deep interior blood vessels along a mackerel shark's backbone are small. The major blood supply circulates through a single artery-vein pair running just below the skin along each side of the fish. The paired cutaneous blood vessels give rise to a single massive *rete* that warms the shark's dark muscle. In some mackerel sharks, such as the mako shark, the heat exchanger is a solid slab of blood-vessel tissue resembling the bluefin's [*see illustration on opposite page*]. In others, such as the porbeagle shark and the white shark, the *rete* is diffuse, with many bundles of fine blood vessels extending into the dark muscle. In mackerel sharks the body-temperature distribution closely resembles the distribution in the bluefin; most of the sharks' muscle is warmer than the water they swim in, and the warmest parts of all are twin regions within the dark muscle.

Tunas and mackerel sharks possess heat exchangers that are not associated with muscle tissue but serve the organs of the body cavity. In the albacore, bigeye and bluefin tunas dense bundles of intermingled fine arteries and veins are found on the surface of the liver; in a large bluefin the bundles may be five centimeters in diameter. The arrangement in the mackerel sharks is different because the artery that in other fishes carries the main blood supply to the viscera is either very small or nonexistent. Instead the visceral blood supply is provided by arteries that are insignificant in most other fishes. These arteries penetrate into a much enlarged venous space known as the hepatic sinus. Venous blood flows through this space on its way from the viscera back to the heart. The arteries branch and branch again within the hepatic sinus until the space is virtually filled with a spongelike mass of tiny vessels; the cool blood within them is warmed by the bath of venous blood. Visceral *retia* do not keep the organs of the body cavity warm continuously; the temperature can vary from a level equal to the warmest muscle to one only slightly above the water temperature.

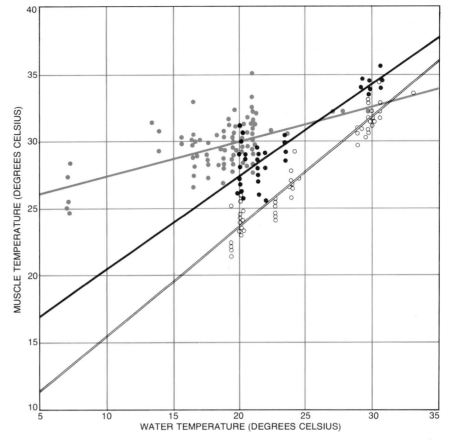

MAXIMUM MUSCLE TEMPERATURES are plotted against water temperature for bluefin (*color*), skipjack (*solid black*) and yellowfin (*open black*). The bluefin regulates its temperature almost independently of water temperature. The skipjack is warmer than the yellowfin but both tend to maintain a fixed temperature difference above water temperature.

The variations may be correlated with the mackerel sharks' and tunas' digestive activity. The tunas in particular have remarkably small body cavities, and the mass of their visceral organs is limited. Raising the temperature of the viscera may speed digestion. Conversely, during times of digestive inactivity the viscera may be allowed to cool.

Since tunas are warm, we wondered whether they could control their temperature at a constant level and thus be independent of water temperature or whether they warmed and cooled as the water temperature changed. Other investigators had noted that yellowfin tuna and skipjack tend to maintain a temperature a fixed number of degrees above that of the water—something that might be accomplished by an unregulated heat exchanger. These two species are found only within a rather narrow range of water temperature, however, and so we turned our attention to the bluefin tuna, which is equally at home in tropical and in subpolar waters. We measured the maximum temperature in the muscle of bluefin taken along the coast from the Bahamas to Newfoundland. Comparison of their muscle temperature with the temperature of the water in which they were captured suggested that they control their temperature quite well. Over a 24-degree C. range of water temperature the average muscle temperature changed only six degrees; bluefin from the 30-degree waters of the Bahamas were only a few degrees warmer than the water, whereas those from seven-degree northern waters were 20 degrees above the water temperature.

There are two ways an animal might achieve such control. One is by means of the swift thermoregulatory adjustment to environmental temperature changes that is characteristic of birds and mammals, which can vary their rate of heat loss in several ways and also generate extra heat by increasing their metabolism. The other is the much slower process of acclimatization, by which cold-blooded animals adjust their activities to changing temperature and which involves cellular and enzyme changes over an extended period of time. In order to discover which of these mechanisms the bluefin were using we decided to try a field experiment in which we would monitor the body tem-

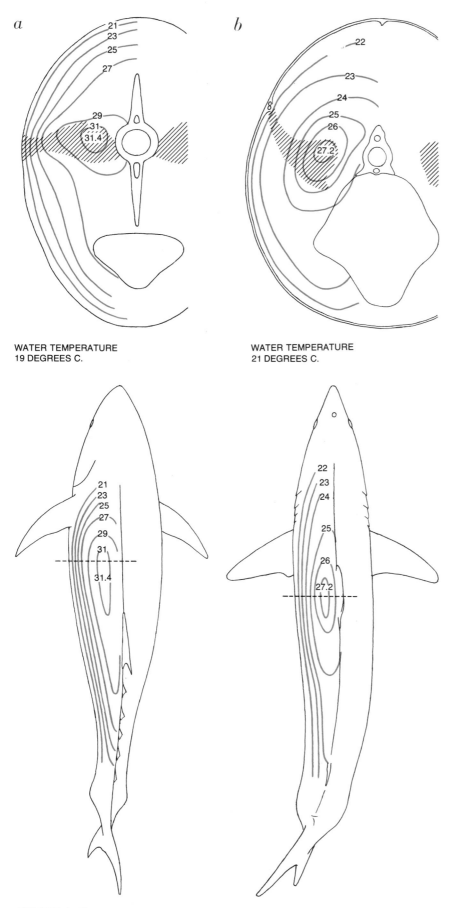

a

WATER TEMPERATURE
19 DEGREES C.

b

WATER TEMPERATURE
21 DEGREES C.

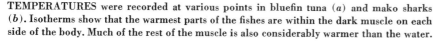

TEMPERATURES were recorded at various points in bluefin tuna (*a*) and mako sharks (*b*). Isotherms show that the warmest parts of the fishes are within the dark muscle on each side of the body. Much of the rest of the muscle is also considerably warmer than the water.

perature of fish as they swam from warm into cold water. If the bluefin were capable of rapid thermoregulation of the kind shown by mammals, their temperature should remain constant during a sudden change in water temperature.

We found that the coast of Nova Scotia was an ideal setting for our project. The Coolen brothers of Fox Point operate pound nets in St. Margaret's Bay, and they frequently catch bluefin tuna in their nets. Moreover, the coastal waters of Nova Scotia are characterized by a marked thermocline (a sharp drop in temperature as the depth increases), so that a free-swimming fish might encounter a wide range of water temperatures.

We used a small harpoon equipped with two thermistors in our first attempts to record bluefin temperatures. One thermistor sensed water temperature and the other muscle temperature; they were connected to an indicator by a 1,000-yard length of field-telephone wire that could unreel as the harpooned fish swam away. We failed on almost every attempt: the fish died or the harpoon was pulled out when the wire was snagged by kelp or by lobster-pot lines. It was at this point that my colleague John Kanwisher suggested that we telemeter the information from the fish.

The telemetering devices we selected were battery-powered acoustical transmitters. Their service life was from one day to three days and their range was as much as five miles. The instrument package was mounted on a small harpoon that could be driven through the fish's thick skin with a minimum of injury to the muscle. The thermistor that sensed the muscle temperature was in the tip of the harpoon and the one that sensed the water temperature was attached to the transmitter, outside the fish's body on the harpoon shaft. For other experiments, in which we measured the stomach temperature, an instrument package was pushed down the bluefin's throat and into its stomach; the thermistor that measured the water temperature was at the end of a wire we led out through the tuna's gill slit. The transmitters broadcast the temperature of the fish and the temperature of the water during alternate one-minute periods.

Our tracking vessel was equipped with a directional hydrophone that enabled us to follow the tuna. We rotated the hydrophone until the telemetered signal was at a maximum and then steered the vessel on that heading. At hourly intervals we lowered a bathythermograph to measure the temperature of the water at various depths. By comparing these readings with the telemetered temperature of the water surrounding the fish we could estimate the bluefin's swimming depth, and we could approximate the fish's course by plotting our own course as we followed the signals.

We placed transmitters on or in 14 bluefin tuna. The longest we tracked a fish was 54 hours; by then the bluefin had reached a position 130 miles offshore. We soon found that, as commercial fishermen had told us, tuna avoid changes in water temperature if they can. Most of our specimens remained near the surface or at least on the warm side of the steep thermocline that separated the upper waters from the cold depths. Some of the bluefin would dive through the thermocline, but they spent only a few minutes in the colder water before returning to the warm side.

Near the end of our efforts we were lucky enough to get a most satisfactory result. The specimen was a 600-pound bluefin with a transmitter in its stomach. The fish had been handled quite roughly while the instrument was being installed. Perhaps for this reason as soon as it was released from the pound at about 9:00 A.M. it swam down through the thermocline into water 14 degrees C. colder than the surface water in the pound. When the fish was released, its stomach temperature registered 21 degrees C. During its four-hour stay in the five-degree water the temperature of its stomach gradually fell to about 19 degrees. Early in the afternoon the fish returned to the warm side of the thermocline and remained in water that registered between 13 and 14 degrees for the rest of the day. In spite of a change in water temperature of nearly 10 degrees C. the temperature of the fish's stomach remained around 18 degrees. The fact that the deep-body temperature of the fish remained nearly constant over extended periods in both cold water and warm indicates that the bluefin was indeed thermoregulating. Just how the fish do this we do not know. Presumably it is by somehow varying the efficiency of the heat-exchanging *retia*.

We were not fortunate enough in our muscle-temperature experiments to have a bluefin stay on the cold side of the thermocline for any substantial length of time. One of the muscle-tagged fish, however, did swim for hours in water that was gradually decreasing in temperature. The water temperature dropped four degrees C. in one 90-minute period. During this interval the muscle temperature of the bluefin slowly rose from a

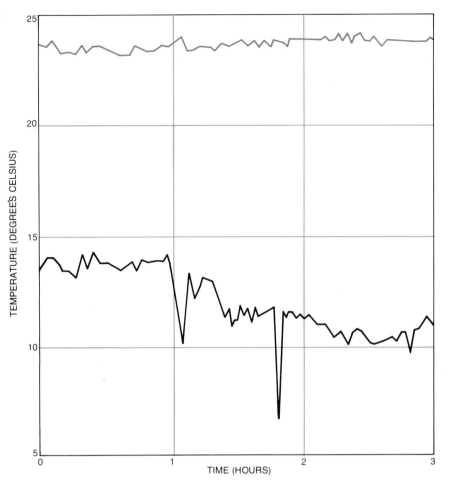

TELEMETRY RECORD compares the temperature in the muscle of a bluefin (*color*) with that of the water (*black*) through which it swam for three hours. The muscle temperature was held constant as the water temperature declined gradually from about 14 to 10 degrees.

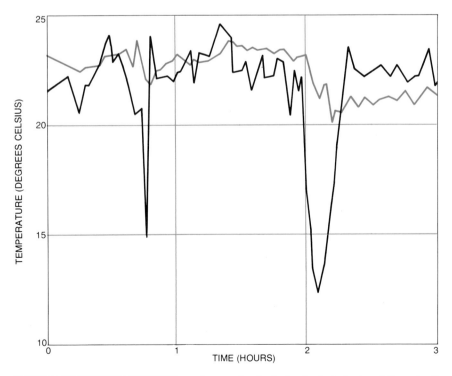

RECORD OF TEMPERATURES from the muscle of a dusky shark (*color*) and the water it swam in (*black*) shows a different relation. The dusky shark is a cold-blooded fish and its muscle temperature stayed close to that of the water, dipping during even short dives.

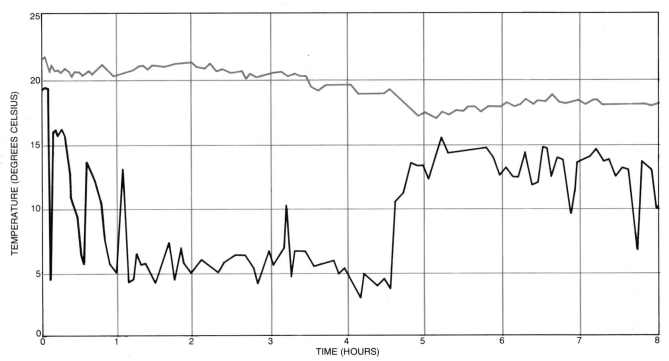

RECORD FROM TUNA'S STOMACH shows the effect of the visceral, as opposed to the muscular, temperature-control system. The temperature was telemetered from thermistors in the tuna's stomach (*color*) and outside the fish in the water (*black*). When the tuna was released, it swam down into cold water and stayed there for four hours, but its stomach temperature decreased only a little.

little more than 23 degrees C. to above 24 degrees.

Our experiments showed that the bluefin does not maintain its body temperature at a constant level as, for example, human beings do. The temperatures of fish from the same school can fluctuate over a range of five degrees or so. The greatest fluctuation we recorded was provided by a second tuna that carried a stomach temperature transmitter. At the time of release the temperature of the fish's stomach was 19 degrees C. During the first day of tracking its stomach temperature slowly began to rise. By the second day, although the water temperature had remained constant, the fish's stomach was registering 26 degrees C. Perhaps the fish was trying to digest the transmitter!

In other readings we found that muscle temperature tended to cycle over a narrow range that is unrelated to the temperature of the water. All these variations tell us that whereas the bluefin's various organs are being kept much warmer than the surrounding water, precise temperature control is either not necessary or not possible. In effect, the fish has a sloppy thermostat. Nonetheless, these fish can be said to thermoregulate in the same sense that birds and mammals do.

How did the tunas and mackerel sharks come to develop these remarkably efficient heat exchangers? Although the *rete* is certainly a complex vascular array, its mode of evolution was probably quite simple. For one thing, arteries and veins generally follow the same path as they travel between organs and within tissues. Every such pairing of an artery and a vein is a rudimentary heat exchanger; if at one end of the system there is either heating or cooling, then there will be some heat transfer between the paired blood vessels. This rudimentary exchange is probably what accounts for our observation that fishes that ought not to be able to warm themselves at all by metabolic means nevertheless may have muscle temperatures from one to two degrees C. higher than the water temperature.

Suppose that such a slight rise in tissue temperature offers a genuine selective advantage to an evolving species. In such an event there already exists (as part of the normal embryonic development of the circulatory system) a basis for the development of more advanced heat exchangers. The circulatory system of the developing embryo first takes the form not of discrete blood vessels but of beds of interconnecting spaces and channels. Later, as certain channels become important, these routes become larger and other channels atrophy and disappear. Moreover, there is a tendency toward the development of multiple channels. Several such channels may form within a single bed, with the result that the vascular system of the adult will have some duplicate components. Therefore in order to form the mass of parallel blood vessels that is an efficient heat exchanger, only a modification of the normal embryonic pattern is required, not a radical genetic change.

A fast-swimming predatory fish finds an abundance of food awaiting it in the form of fast-swimming squid, herring and mackerel that slower predators cannot capture. In terms both of streamlining and of muscle power the tunas are probably the swiftest predators of the open ocean. Yellowfin tuna and wahoo tuna are noted for their speed: they have been observed to reach speeds in excess of 40 miles per hour during 10- to 20-second sprints. Mackerel sharks (at least the mako and porbeagle) probably also enjoy the benefits of fast swimming. Such a shark, with its bulky, muscular body, its streamlining and its narrow, crescent-shaped tail, looks remarkably like a tuna. That the two unrelated groups of fishes independently evolved the same means of raising their body temperature must surely be connected to the adaptive advantage of increased swimming speed; the extra power available from warm muscle must have been decisive in achieving that speed. It is a classic example of parallel adaptations that evidently gave both groups access to an underexploited source of food.

23

HOW REPTILES REGULATE THEIR BODY TEMPERATURE

CHARLES M. BOGERT

April 1959

Although they lack internal controls, they can maintain a high temperature by their behavior. A lizard, for example, can raise its temperature by changing the position of its body in the sun

More than 50 years ago Sir Charles Martin, a distinguished British physiologist, compared the regulation of body temperature in a number of mammals with that of a lizard. He showed that the mammals were able to maintain their temperatures within a fairly narrow range during wide variations in the temperature of the laboratory environment. The temperature of the lizard, on the other hand, rose and fell almost as rapidly as that of the environment. Observations of this sort long ago established the textbook aphorism to the effect that "reptiles have the temperature of the surrounding atmosphere."

It is true that reptiles are "cold-blooded" animals and have no mechanism of temperature regulation such as that of mammals. The laboratory observations correctly reflect what happens to a lizard's body temperature when the laboratory temperature is changed. But the conclusion drawn from these observations holds true only for the lizard in the laboratory. In their natural habitats during the day, lizards forage, mate, defend territories and flee at body temperatures that may be even higher than our own, and they maintain their temperatures within narrow limits despite wide variation in air temperature. The greater earless lizard (*Holbrookia texana*), an inhabitant of the foothills of the U. S. Southwest, has a mean temperature of 101.3 degrees Fahrenheit (38.5 degrees centigrade), slightly above our own, and while the lizard is active its temperature is within 3.3 de-

grees of this level 75 per cent of the time. At 14,600 feet in the Peruvian Andes, with the temperature of the thin air at the freezing point, Oliver P. Pearson of the University of California found that the lizard *Liolaemus multiformis* had a body temperature of 87.8 degrees F. (31 degrees C.); at temperatures as much as eight degrees below freezing he found other lizards abroad, a trifle sluggish, with body temperatures of 58 degrees (14.4 degrees C.), or 34 degrees above the temperature of the air. In my own studies over the past 14 years I have measured the temperatures of lizards of many different North American species. I have found that members of the same or closely related species show the same high and constant temperature in widely different

CHUCKWALLA (*Sauromalus obesus*) is found in deserts of the southwestern states. When this lizard cannot take shelter from the sun, it pants to cool itself by the evaporation of moisture from its lungs and places itself parallel to the sun's rays in order to expose a minimum of body surface to the radiant heat. This animal was photographed at three o'clock in the afternoon on a clear day.

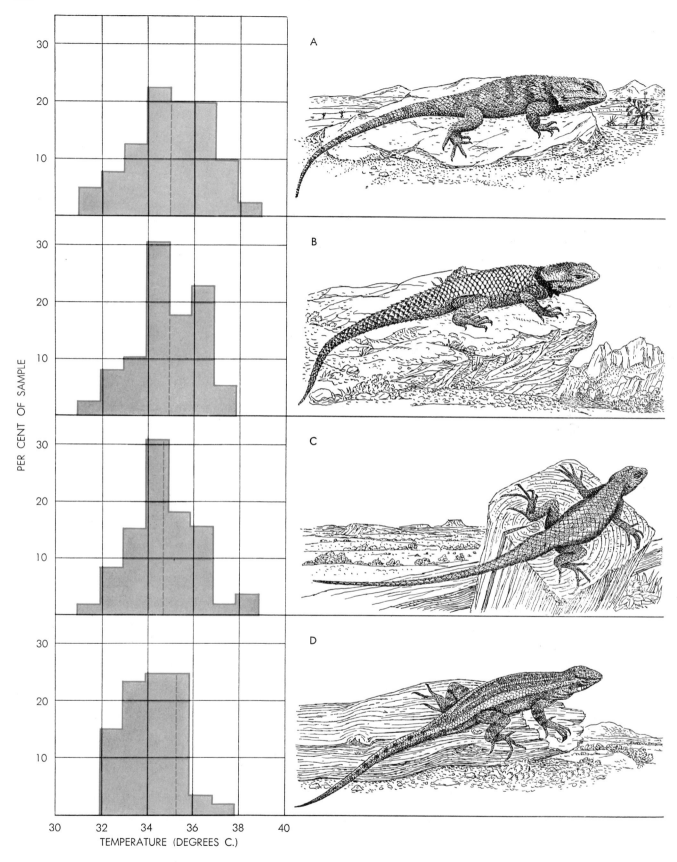

TEMPERATURES OF SPINY LIZARDS of four species were tested by the author, who sampled 39 or more active animals of each kind in their native habitats. In each species the average temperature (*broken line*) fell close to 35 degrees C. (95 degrees F.). The lizards tested were *Sceloporus magister* (A), *Sceloporus jarrovi* (B), *Sceloporus undulatus* (C), *Sceloporus variabilis* (D).

environments, although members of distantly related species maintain distinctly different mean temperatures in the same environment.

Why were the laboratory observations so misleading? For much the same reason that a man with a heavy iron ball chained to one leg cannot demonstrate how fast he can run! The analogy is to the point, because lizards regulate their temperatures to a large degree by their behavior. Many of these creatures are "heliotherms," deriving the heat they need to energize their body chemistry directly from the sun. In consequence of this dependence they have developed basking to a fine art. Lizards do not merely crawl out of their nocturnal shelter and rest in the sunlight. When their temperatures are below the threshold for normal activity, they orient their bodies at right angles to the sun's rays in order to maximize their exposure and even seek inclined surfaces to achieve such orientation with respect to the slanting rays of the morning sun. In the desert, where the ground becomes warmer than the air, lizards often press their bodies close to the surface, shifting slowly from side to side in the loose sand to secure better conduction of heat. On a rocky mountainside that warms up more slowly, they do their basking on mats of dead grass that insulate them from the cold ground. When a lizard's temperature approaches the upper limit of tolerance, on the other hand, it faces the sun, exposing the least possible surface, or returns to cooler temperatures in the shade or underground.

The size and shape of their bodies and the pigmentation of their skins play a part in determining and regulating the rate at which lizards absorb heat from their surroundings. But the decisive factor is behavior. In the artificial situation of the laboratory the lizard could not show what it can do.

When our department at the American Museum of Natural History and R. B. Cowles of the University of California at Los Angeles undertook to study heat regulation in reptiles, we had a general understanding of the factors involved. We expected to find, however, that temperature tolerances played an important part in determining the distribution of species and that the various species would show different optimal body temperatures in their various environments. In framing these assumptions, it turned out, we underestimated the efficiency of our subjects' heat-regulating behavior and equipment.

We chose the spiny lizards as our subjects, because their 50 species are abundant all over North America, from coast to coast and from southern Canada to Panama. Few groups of lizards have penetrated more environments. As many as five species may occur in a single locality, each in its own ecological niche. Some live on the ground, others on trees or shrubs; they variously frequent rocky hillsides, canyon walls, sand dunes, grassy plains and even human habitations. We sought them out in coastal areas, foothills, plateaus and mountains; in arid regions with little vegetation; in pine barrens, short-tree forests, pine forests and high-altitude cloud forests. We made our measurements in habitats ranging from sea level to near the timber line at 12,500 feet on the Nevado de Toluca in Mexico.

In all of this diversity of habitat, to our surprise, measurements indicate that spiny lizards go about their active lives at a mean body temperature of about 93 degrees (34 degrees C.). This approximates the average for all species, but does not imply that spiny lizards

have no leeway in temperature. They function at apparently full efficiency with body temperatures between 86 and 104 degrees (30 and 40 degrees C.). In their natural environments, however, these extremes are exceptional. Once their basking has brought them to the temperature threshold at which activity begins, spiny lizards maintain their body temperatures within 4.5 degrees of the 93-degree mean for about 80 per cent of the time and over the entire range of environmental temperature to which they voluntarily subject themselves during their daily routine.

These lizards regulate their heat intake largely by exposing themselves to direct sunlight, prolonging their forays by suitably orienting their bodies to the sun much as they do when basking, or by retreating from the sunlight when their temperatures run high. Comparisons of the temperature curves of various spiny lizards reveal only minor differences between species. Most of the peaks fall near the 93-degree average for the group. For species living in the tropical lowlands of Mexico and Honduras,

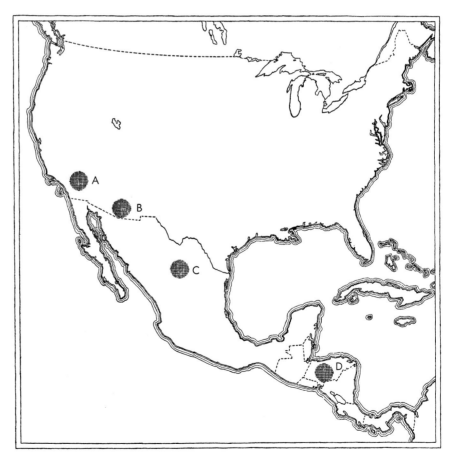

HABITATS of the lizards shown on the opposite page are marked with the corresponding letters on this map. They are (A) the Mojave Desert of California, elevation 3,000 to 4,000 feet; (B) the Chiricahua Mountains of Arizona, elevation 5,000 to 9,000 feet; (C) La Goma, Durango, Mexico, elevation 4,000 feet; (D) El Zamorano, Honduras, 2,600 to 3,000 feet.

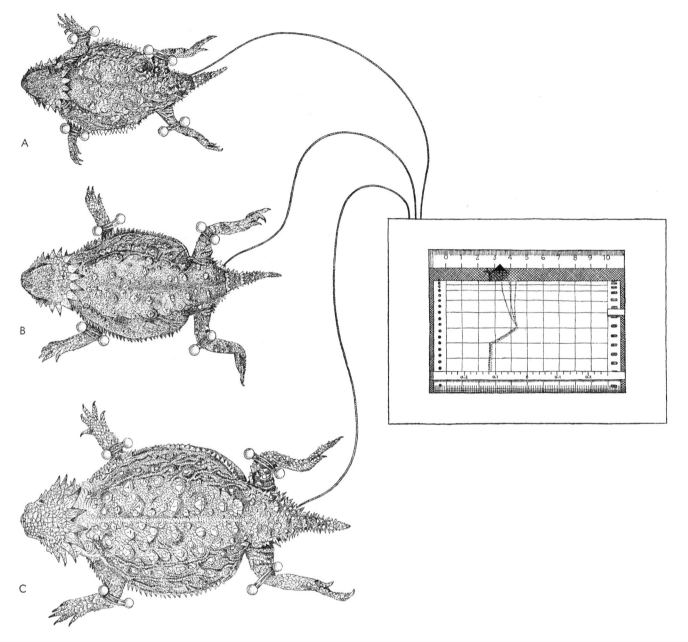

REGAL HORNED LIZARDS (*Phrynosoma solare*) weighing respectively 12.4 grams (A), 29.4 grams (B) and 85.5 grams (C) were exposed to the midday sun while their body temperatures were continuously measured by means of wires from the cloaca of each to a recording potentiometer. The colors of the lizards were nearly the same at start but then changed, smallest lizard becoming palest.

the curves shift scarcely three or four degrees toward the warmer end of the scale. For those in higher mountains, the shift is two or three degrees toward the lower end of the scale.

We do not know whether such slight differences result from variations in the physiological adaptation of the animals or from the limitations of behavioral regulation. We sometimes lose sight of the fact that temperature refers to the amount of heat per unit of mass, or the degree of heat concentration, as measured on one scale or another. Optimum heat concentrations for biological processes, however, lie not at points but within zones on the scale. The "normal temperature" of human beings is actually an average for a zone between limits set roughly at 98 to 99.5 degrees (37 to 37.5 degrees C.). Hence the differences observed between the mean temperatures of the spiny lizard in different environments may still lie within the zone of optimum temperature for their biological activity.

In the Arizona desert the air temperatures recorded at sites of capture averaged 90 degrees (32.2 degrees C.); the temperature of the spiny lizards averaged 95 degrees (35 degrees C.). In contrast, the average body temperature of cloud-forest lizards is 91 degrees (32.8 degrees C.) when the air around them averages 66 degrees. The greater differential between air and body temperature in the case of the cloud-forest lizard should accelerate the loss of heat to the air. Nevertheless the lizards absorb enough solar heat to compensate for these losses and thus keep their body temperature within the zone permitting them to be active.

So, we learned, spiny lizards do not have different body temperatures reflecting different physiological adaptation to different environmental temperatures. Instead these reptiles restrict the fluctuation of their body temperature to a relatively narrow zone suitable to the similar physiological needs of all species. Probably the optimum zone for activity in spiny lizards is an ancestral trait. If so, their physiological adaptation became stabilized for a narrow zone centering around 93 degrees early in the evolution of the stock. The subsequent diversification and dispersal of the stock is largely a history of changes in such characteristics as size, shape, pigmentation and basking behavior, as species after species became adapted for survival under various combinations of environmental conditions.

From one standpoint the evolution of the spiny lizard, as a minor current in the broad stream of evolution, followed the course of least resistance. Once the delicate equilibrium of its physiological processes had been established in a particular zone of temperature, any adaptation to lower internal temperatures would have entailed revision of the whole complex system. The lizard's behavior, on the other hand, provided the necessary leeway to permit it to invade different environments successfully. There, under the pressure of selection, the various species developed adaptations in pigmentation and other physical characteristics associated with the regulation of body heat.

Spiny lizards are not the only lizards whose normal physiological activity is restricted to a narrow zone of temperature characteristic of the group as a whole. Other groups exhibit a similar identity. The whip-tailed lizards (*Cnemidophorus*), with species widely distributed in North, Central and South America, are active at body temperatures in a zone higher on the scale. They maintain a mean temperature close to 104 or 106 degrees (40 or 41 degrees C.), with no significant difference between the temperatures of populations in climates and landscapes as widely different as Arizona, Florida and Honduras.

Whip-tailed and spiny lizards often occur side by side in the same habitat. Invariably they show an average difference of 10 or 12 degrees between the means of their respective ranges in temperature, even though the identical sources of external heat are available to both. This alone is evidence of the effectiveness of behavioral control of heat intake and dissipation, augmented by pigmentation and enhanced by adaptation in structure.

This interaction of structure and behavior makes it difficult to design experiments testing one or another of the reptile's temperature-regulating at-

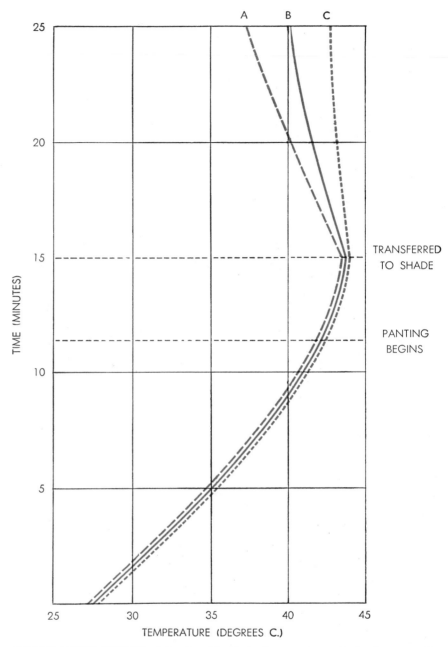

TEMPERATURE CURVES of the three lizards in the experiment depicted on page 108 were virtually identical as the animals absorbed heat. The smaller animals lightened in color, which aided the absorption of heat and compensated for their smaller ratio of surface to mass. Actually the lines on the chart were nearly superimposed. They were separated by transferring the smallest lizard (A) to the shade, while the next largest lizard (B) had only its head in the sun and the largest lizard (C) had its head and half its body in the sun.

tributes. Unquestionably size is subject to adaptive changes. Whenever we compare adequate samples from various portions of the range of a widely distributed reptile, we nearly always find that the average size of adults in one area is larger than that of adults in another area. Often the trend in size parallels the trend in climate, with the larger animals restricted to warmer areas or to regions with longer growing-seasons. Such correlations are suggestive, but without experimental evidence we cannot be certain that heat and size are directly re-

lated. The length or bulk of a reptile may be governed by such other variables as the food supply or gene combinations.

The problem of evaluating the adaptive significance of size is not simplified by the existence of adults and young of the same species in the same environments at the same time. In the desert-dwelling whip-tailed lizard (*Cnemidophorus tigris*) we found that juveniles of the species, weighing two or three grams, keep their bodies at mean temperatures identical with those of

TEXAS HORNED LIZARD, like the regal horned lizard, has the ability to change the color of its skin. When its body temperature is low, the animal is quite dark (*left*), but the same animal becomes paler (*right*) after it has been exposed to higher temperatures.

adults weighing up to 16 grams. However, we found a difference between adults and juveniles when we recorded the daily variation of their temperatures. The smaller lizards restrict their temperature fluctuations to a narrow zone of five or six degrees, while the activity zones in adults range over 10 or 11 degrees, from 99 to 110 degrees (37.2 to 43.3 degrees C.). The juveniles may be more responsive to heat and so may adjust their exposure more sensitively, or their body temperature may adjust more quickly by virtue of their smaller bulk.

Other things being equal, the temperature of a smaller lizard should rise or fall faster because it has more surface in proportion to its mass. In the first rough experiment we performed to test the validity of this generalization from the physics of inanimate objects, however, we discovered how important "other things" can be if they are not equal. Our subjects were adult specimens of two species of spiny lizard. The larger of the two was a green spiny lizard (*Sceloporus formosus*), a species restricted to open areas in moist forests of broad-leaved trees at elevations above 5,000 feet in central Honduras; its green

skin is marked with black pigment sparsely distributed in a reticulated pattern of lines and smaller blotches. The other was a much smaller, slate-colored spiny lizard (*Sceloporus variabilis*) that lives at elevations up to 3,000 feet in the arid valleys below the cloud forests on the mountain summits. In their very different environments the two species keep nearly the same average body temperature.

Though our two specimens had roughly the same bodily proportions, the greenish one, weighing 27.8 grams, had four times the bulk of the other, which weighed only 6.9 grams. The temperature of both lizards was 77 degrees (25 degrees C.) when they were placed in full sunlight with the air temperature at about 90 degrees. Temperatures were recorded at intervals of three minutes. During the first nine minutes the body temperature of the larger lizard lagged less than a degree behind that of the smaller lizard. But after 12 minutes the temperature of the larger lizard rose slightly above that of the smaller. At the conclusion of the experiment, after 18 minutes, the temperature of the larger lizard was 109 degrees (43 degrees C.),

and that of the smaller was 108.7 degrees (42.6 degrees C.). If the two lizards had absorbed heat at rates predicated solely on their weight, the heavier should have required approximately 10 more minutes to reach the temperature attained by the smaller animal in 18 minutes. Though inexact, the results of our simple experiment suggest that the cloud-forest lizard is better equipped, figuratively, "to make hay while the sun shines." Because the pigments in its skin absorb heat so rapidly, it can attain its threshold temperature quickly enough and often enough during the year to permit it to forage and fuel itself.

We suspected that the outcome of this experiment may have been influenced by changes in the pigmentation of one or the other reptile in the course of the experiment. To find out how important such changes are in regulating the absorption of heat we performed an experiment with individuals of different weights but belonging to the same species. This time we used regal horned lizards (*Phrynosoma solare*) weighing respectively 12.4, 29.4 and 85.5 grams. The experiment was conducted in

August on a clear day, with no wind, in the foothills of the Chiricahua Mountains in Arizona. The body temperature of each lizard was 80 degrees (27 degrees C.) at the start of the experiment; within 15 minutes their temperatures simultaneously reached 109 degrees (43 degrees C.), with the curves on the recording instrument indicating that they had risen at a virtually identical rate. About halfway through the experiment the temperature of the smallest lizard ran a degree ahead of the others, but shifted back to the curves being plotted for the other two, as though some mechanism were regulating the rate of heat absorption. This proved to be the case. Although the three lizards were not conspicuously different in color at the beginning of the experiment, we could discern distinct differences at the conclusion. The largest lizard was the darkest of the three, and the smallest lizard the palest.

While their broadly flattened bodies are adapted for the rapid absorption of heat, it is apparent that horned lizards are equally well equipped by pigmenta-tion to regulate the rate at which they absorb heat. The black-pigmented cells, or melanophores, of their skin expand laterally when the animal is cold, thus darkening the body and increasing the rate at which it absorbs radiant energy. When the body is warm, the same cells contract, thereby exposing light pigments in adjacent cells that reflect infra-red radiation. To match such efficiency we would need a mechanism that automatically exchanged our dark winter clothing for white linens with the advent of hot weather.

There appears to be an upper limit of bulk beyond which behavioral and physiological adaptations can no longer secure adequate regulation of a reptile's internal body heat. At the Archbold Biological Station in Florida, E. H. Colbert, R. B. Cowles and I conducted a suggestive experiment with a five-foot, 30-pound alligator as the subject. We found that it took the summer sun 7.5 minutes to increase the animal's temperature by two degrees. To see what this signifies for large reptiles, let us consider the plight of a 10-ton dinosaur under the same circumstances. This creature would have to bask in a blazing sun for more than an hour to elevate its body temperature to the same extent. Suppose our hypothetical dinosaur were active in daytime, and subject to cooling at night, as it would be in any desert region today. If its temperature dropped even four or five degrees below its threshold for activity, the dinosaur would have to bask for a large part of the following day in order to regain the threshold temperature of activity. The odds favor the deduction that dinosaurs, at least the larger ones, lived under fairly constant environmental temperatures. This is an important piece of evidence favoring the conclusion that the earth's climate was once quite uniformly tropical, for the distribution of fossils shows that the large dinosaurs roamed the earth far beyond the borders of the modern tropics.

All the truly large reptiles still abroad in the world today—pythons in Asia as long as 33 feet, anacondas in South America as long as 28 feet, monitor liz-

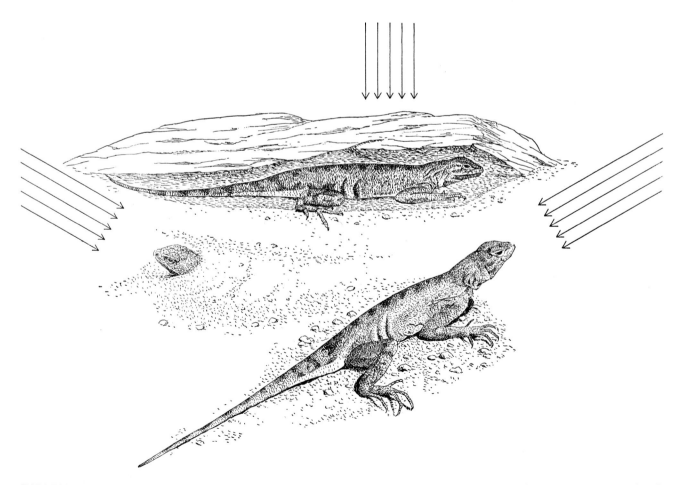

EARLESS LIZARD of the Southwest regulates its temperature within narrow limits by means of its behavior. The morning sun (*arrows at left*) warms the blood in the animal's head while the rest of it remains hidden in the sand until it is warm enough to be active. At noon (*top*) the lizard seeks shelter from the hot sun, but later it emerges and lies parallel to the sun's rays (*bottom*).

ards in Indonesia as long as 10 feet, crocodilians in the Americas as long as 23 feet and tortoises in the Galápagos Islands weighing more than 500 pounds—are residents of the tropics. The only exception is the enormous leatherback turtle *Dermochelys*, which is known to reach and possibly to exceed 1,500 pounds and is perhaps the largest living reptile. But this creature is protected by the constant warmth of its oceanic environment. The occasional specimen that turns up on the coast of Nova Scotia is probably carried there by the Gulf Stream.

In general the tropics afford stable temperatures that fluctuate in the narrow and comfortable range from 68 to 86 degrees year in and year out. The hottest places on earth are found not in the tropics but in the desert regions of the zones so inappropriately called "temperate." In winter these same deserts may be bitterly cold, and the daily fluctuation in temperature throughout the year far exceeds any encountered in the tropics. The night temperature of a tropical forest may drop only a few degrees below that of the day. Thus none of the large reptiles is ever exposed to either the freezing or the high temperatures encountered by reptiles in the middle-latitude deserts.

The green iguana (*Iguana iguana*), the largest lizard found on the American continent, illustrates the dependence of large reptiles on the constancy of the tropical temperature-environment. This animal, widely distributed in the lowlands of Latin America, reaches a length of at least 5.5 feet and a weight of 13 pounds. In a population we studied on the west coast of Mexico we found that the body temperature of the green iguana fluctuates much more than smaller lizards living in the same area or in the deserts 700 miles to the north. Green iguanas spend virtually all of their time high up in the crowns of large trees, feeding on leaves, buds or fruits during the day, and sleeping at night with only branches or foliage to conceal them from predators and to protect them from heat loss. They evidently do some basking; we found that the temperatures of nearly 50 iguanas were 10 to 15 degrees above the level of the air. But it is doubtful that there has ever been sufficient stress from heat fluctuations in the environment of the green iguana to induce adaptive changes in its behavior.

We had an opportunity to test this supposition by exposing several green iguanas to summer conditions in the desert. Instead of fleeing from the heat of direct sunlight, as any reptile native to the desert would have done, they literally sat in the sun until they died. The increase in their body temperatures was slowed a little by respiratory cooling when their breathing turned to panting, but this merely prolonged their discomfiture before breathing stopped. Green iguanas are equally unprepared for exposure to low temperatures. One specimen at the San Diego Zoo had to be moved indoors overnight because it showed no disposition to seek shelter of its own accord. However, one evening preceding a cold snap the creature chanced to be overlooked. Next morning it was found in a state of cold narcosis,

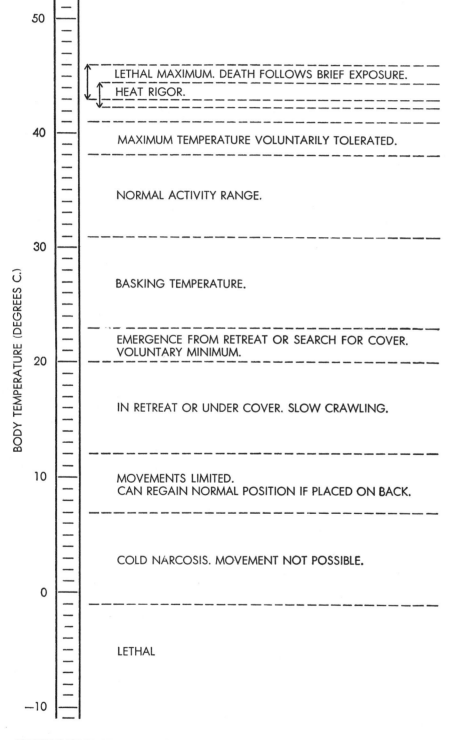

SIGNIFICANCE OF BODY TEMPERATURE to the behavior of reptiles is indicated by this chart giving the approximate temperatures for various activities of spiny lizards. The effects of exposure to heat levels near the extremes depend on duration. Even temperatures near the upper limit for the lizards' normal activity become lethal if exposure is prolonged.

suspended by a single claw accidentally hooked to a twig.

Considering that all reptiles would be just about equally vulnerable to cold and heat were it not for their behavioral adaptations, it is remarkable that so many lizard species are established in the temperate zone. It should be noted that snakes and turtles also have adapted themselves by much the same means to survive in rigorous climates. Snakes in particular have gotten around the surface-to-mass limitation by the lengthening of their bodies, which exposes more surface per unit of mass to absorb solar energy. In Canada snakes outnumber lizards and turtles combined, and in Europe one viper (*Vipera berus*) has penetrated to regions within the Arctic Circle. However, one European lizard (*Lacerta vivipara*) ranges even farther to the north, farther from the Equator than any other living reptile.

In common with the majority of reptiles near the northern limit of their distribution, both the snake and the lizard give birth to living young. Some snakes and lizards transfer nutrients and oxygen from maternal blood to the developing embryo by means of mammal-like placental structures. Most of these live-bearing reptiles, however, incubate the eggs within the body of the female, where behavioral regulation of temperatures keeps the eggs at heat levels within the optimum zone for development. Such modification of the reproductive pattern, peculiar to lizards and snakes, gives them a leeway in dispersal not open to turtles and crocodilians. Snakes and lizards are accordingly the most widely distributed reptiles, with lizards having a slight edge in number of species. Of the two groups it is the lizard that best exemplifies the complexity of adaptations involving the coordination of structure, physiology and behavior in response to the interplay of selective forces in the environment.

Perhaps the most amazing behavioral adaptation is that of the earless lizard, which is almost never found abroad with its body temperature below 96 degrees. We exposed its secret only by observing what it did in a laboratory cage provided with sources of radiant heat. From its overnight retreat, submerged in the sand, the earless lizard first thrusts its inconspicuous head above the surface; there it waits until the blood coursing through a large sinus in its head has absorbed enough heat from the sunlight to raise the temperature of its entire body. When its temperature is well above the threshold for efficient activity, this wary reptile emerges from the sand, preheated and ready to take off at top speed.

24

ADAPTATIONS TO COLD

LAURENCE IRVING
January 1966

One mechanism is increased generation of heat by a rise in the rate of metabolism, but this process has its limits. The alternatives are insulation and changes in the circulation of heat by the blood

All living organisms abhor cold. For many susceptible forms of life a temperature difference of a few degrees means the difference between life and death. Everyone knows how critical temperature is for the growth of plants. Insects and fishes are similarly sensitive; a drop of two degrees in temperature when the sun goes behind a cloud, for instance, can convert a fly from a swift flier to a slow walker. In view of the general hostility of cold to life and activity, the ability of mammals and birds to survive and flourish in all climates is altogether remarkable.

It is not that these animals are basically more tolerant of cold. We know from our own reactions how sensitive the human body is to chilling. A naked, inactive human being soon becomes miserable in air colder than 28 degrees centigrade (about 82 degrees Fahrenheit), only 10 degrees C. below his body temperature. Even in the Tropics the coolness of night can make a person uncomfortable. The discomfort of cold is one of the most vivid of experiences; it stands out as a persistent memory in a soldier's recollections of the unpleasantness of his episodes in the field. The coming of winter in temperate climates has a profound effect on human well-being and activity. Cold weather, or cold living quarters, compounds the misery of illness or poverty. Over the entire planet a large proportion of man's efforts, culture and economy is devoted to the simple necessity of protection against cold.

Yet strangely enough neither man nor other mammals have consistently avoided cold climates. Indeed, the venturesome human species often goes out of its way to seek a cold environment, for sport or for the adventure of living in a challenging situation. One of the

marvels of man's history is the endurance and stability of the human settlements that have been established in arctic latitudes.

The Norse colonists who settled in Greenland 1,000 years ago found Eskimos already living there. Archaeologists today are finding many sites and relics of earlier ancestors of the Eskimos who occupied arctic North America as long as 6,000 years ago. In the middens left by these ancient inhabitants are bones and hunting implements that indicate man was accompanied in the cold north by many other warm-blooded animals: caribou, moose, bison, bears, hares, seals, walruses and whales. All the species, including man, seem to have been well adapted to arctic life for thousands of years.

It is therefore a matter of more than idle interest to look closely into how mammals adapt to cold. In all climates and everywhere on the earth mammals maintain a body temperature of about 38 degrees C. It looks as if evolution has settled on this temperature as an optimum for the mammalian class. (In birds the standard body temperature is a few degrees higher.) To keep their internal temperature at a viable level the mammals must be capable of adjusting to a wide range of environmental temperatures. In tropical air at 30 degrees C. (86 degrees F.), for example, the environment is only eight degrees cooler than the body temperature; in arctic air at −50 degrees C. it is 88 degrees colder. A man or other mammal in the Arctic must adjust to both extremes as seasons change.

The mechanisms available for making the adjustments are (1) the generation of body heat by the metabolic burning of food as fuel and (2) the use

of insulation and other devices to retain body heat. The requirements can be expressed quantitatively in a Newtonian formula concerning the cooling of warm bodies. A calculation based on the formula shows that to maintain the necessary warmth of its body a mammal must generate 10 times more heat in the Arctic than in the Tropics or clothe itself in 10 times more effective insulation or employ some intermediate combination of the two mechanisms.

We need not dwell on the metabolic requirement; it is rarely a major factor. An animal can increase its food intake and generation of heat to only a very modest degree. Moreover, even if metabolic capacity and the food supply were unlimited, no animal could spend all its time eating. Like man, nearly all other mammals spend a great deal of time in curious exploration of their surroundings, in play and in family and social activities. In the arctic winter a herd of caribou often rests and ruminates while the young engage in aimless play. I have seen caribou resting calmly with wolves lying asleep in the snow in plain view only a few hundred yards away. There is a common impression that life in the cold climates is more active than in the Tropics, but the fact is that for the natural populations of mammals, including man, life goes on at the same leisurely pace in the Arctic as it does in warmer regions; in all climates there is the same requirement of rest and social activities.

The decisive difference in resisting cold, then, lies in the mechanisms for conserving body heat. In the Institute of Arctic Biology at the University of Alaska we are continuing studies that have been in progress there and elsewhere for 18 years to compare the

ARCTIC ZONE (20 TO −60 DEGREES C.)

TEMPERATE ZONE (20 TO −20 DEGREES C.)

TROPICAL ZONE (35 TO 25 DEGREES C.)

RANGE OF TEMPERATURES to which warm-blooded animals must adapt is indicated. All the animals shown have a body temperature close to 100 degrees Fahrenheit, yet they survive at outside temperatures that, for the arctic animals, can be more than 100 degrees cooler. Insulation by fur is a major means of adaptation to cold. Man is insulated by clothing; some other relatively hairless animals, by fat. Some animals have a mechanism for conserving heat internally so that it is not dissipated at the extremities.

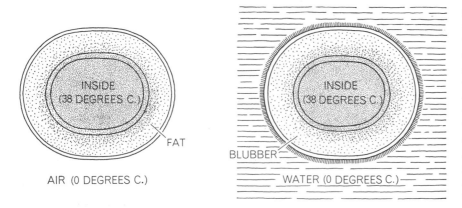

TEMPERATURE GRADIENTS in the outer parts of the body of a pig (*left*) and of a seal (*right*) result from two effects: the insulation provided by fat and the exchange of heat between arterial and venous blood, which produces lower temperatures near the surface.

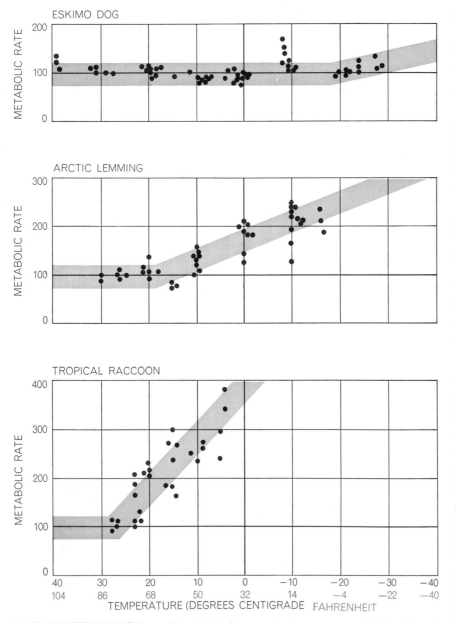

RATE OF METABOLISM provides a limited means of adaptation to cold. The effect of declining temperatures on the metabolic rate is shown for an Eskimo dog (*top*), an arctic lemming (*middle*) and a tropical raccoon (*bottom*). Animals in warmer climates tend to increase metabolism more rapidly than arctic animals do when the temperature declines.

mechanisms for conservation of heat in arctic and tropical animals. The investigations have covered a wide variety of mammals and birds and have yielded conclusions of general physiological interest.

The studies began with an examination of body insulation. The fur of arctic animals is considerably thicker, of course, than that of tropical animals. Actual measurements showed that its insulating power is many times greater. An arctic fox clothed in its winter fur can rest comfortably at a temperature of −50 degrees C. without increasing its resting rate of metabolism. On the other hand, a tropical animal of the same size (a coati, related to the raccoon) must increase its metabolic effort when the temperature drops to 20 degrees C. That is to say, the fox's insulation is so far superior that the animal can withstand air 88 degrees C. colder than its body at resting metabolism, whereas the coati can withstand a difference of only 18 degrees C. Naked man is less well protected by natural insulation than the coati; if unclothed, he begins shivering and raising his metabolic rate when the air temperature falls to 28 degrees C.

Obviously as animals decrease in size they become less able to carry a thick fur. The arctic hare is about the smallest mammal with enough fur to enable it to endure continual exposure to winter cold. The smaller animals take shelter under the snow in winter. Weasels, for example, venture out of their burrows only for short periods; mice spend the winter in nests and sheltered runways under the snow and rarely come to the surface.

No animal, large or small, can cover all of its body with insulating fur. Organs such as the feet, legs and nose must be left unencumbered if they are to be functional. Yet if these extremities allowed the escape of body heat, neither mammals nor birds could survive in cold climates. A gull or duck swimming in icy water would lose heat through its webbed feet faster than the bird could generate it. Warm feet standing on snow or ice would melt it and soon be frozen solidly to the place where they stood. For the unprotected extremities, therefore, nature has evolved a simple but effective mechanism to reduce the loss of heat: the warm outgoing blood in the arteries heats the cool blood returning in the veins from the extremities. This exchange occurs in the *rete mirabile* (wonderful net), a network of small arteries and veins near the junction between the trunk of the animal and the extremity [see "'The Wonderful Net,'" by P. F. Scholander, beginning on page 125]. Hence the extremities can become much colder than the body without either draining off body heat or losing their ability to function.

This mechanism serves a dual purpose. When necessary, the thickly furred animals can use their bare extremities to release excess heat from the body. A heavily insulated animal would soon be overheated by running or other active exercise were it not for these outlets. The generation of heat by exercise turns on the flow of blood to the extremities so that they radiate heat. The large, bare flippers of a resting fur seal are normally cold, but we have found that when these animals on the Pribilof Islands are driven overland at their laborious gait, the flippers become warm. In contrast to the warm flippers, the rest of the fur seal's body surface feels cold, because very little heat escapes through the animal's dense fur. Heat can also be dissipated by evaporation from the mouth and tongue. Thus a dog or a caribou begins to pant, as a means of evaporative cooling, as soon as it starts to run.

In the pig the adaptation to cold by means of a variable circulation of heat in the blood achieves a high degree of refinement. The pig, with its skin only thinly covered with bristles, is as naked as a man. Yet it does well in the Alaskan winter without clothing. We can read the animal's response to cold by its expressions of comfort or discomfort, and we have measured its physiological reactions. In cold air the circulation of heat in the blood of swine is shunted away from the entire body surface, so that the surface becomes an effective insulator against loss of body heat. The pig can withstand considerable cooling of its body surface. Although a man is highly uncomfortable when his skin is cooled to 7 degrees C. below the internal temperature, a pig can be comfortable with its skin 30 degrees C. colder than the interior, that is, at a temperature of 8 degrees C. (about 46 degrees F.). Not until the air temperature drops below the freezing point (0 degrees C.) does the pig increase its rate of metabolism; in contrast a man, as I have mentioned, must do so at an air temperature of 28 degrees C.

With thermocouples in the form of needles we have probed the tissues of pigs below the skin surface. (Some pigs, like some people, will accept a little

TEMPERATURES AT EXTREMITIES of arctic animals are far lower than the internal body temperature of about 38 degrees centigrade, as shown by measurements made on Eskimo dogs, caribou and sea gulls. Some extremities approach the outside temperature.

pain to win a reward.) We found that with the air temperature at −12 degrees C. the cooling of the pig's tissues extended as deep as 100 millimeters (about four inches) into its body. In warmer air the thermal gradient through the tissues was shorter and less steep. In short, the insulating mechanism of the hog involves a considerable depth of the animal's fatty mantle.

Even more striking examples of this kind of mechanism are to be found in whales, walruses and hair seals that dwell in the icy arctic seas. The whale

and the walrus are completely bare; the hair seal is covered only with thin, short hair that provides almost no insulation when it is sleeked down in the water. Yet these animals remain comfortable in water around the freezing point although water, with a much greater heat capacity than air, can extract a great deal more heat from a warm body.

Examining hair seals from cold waters of the North Atlantic, we found that even in ice water these animals did not raise their rate of metabolism. Their skin was only one degree or so warmer

than the water, and the cooling effect extended deep into the tissues—as much as a quarter of the distance through the thick part of the body. Hour after hour the animal's flippers all the way through would remain only a few degrees above freezing without the seals' showing any sign of discomfort. When the seals were moved into warmer water, their outer tissues rapidly warmed up. They would accept a transfer from warm water to ice water with equanimity and with no diminution of their characteristic liveliness.

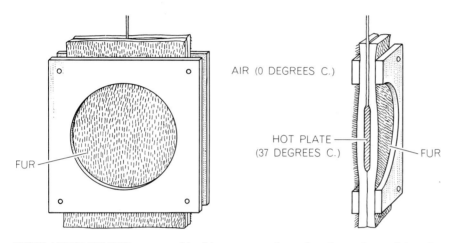

INSULATION BY FUR was tested in this apparatus, shown in a front view at left and a side view at right. The battery-operated heating unit provided the equivalent of body temperature on one side of the fur; outdoor temperatures were approximated on the other side.

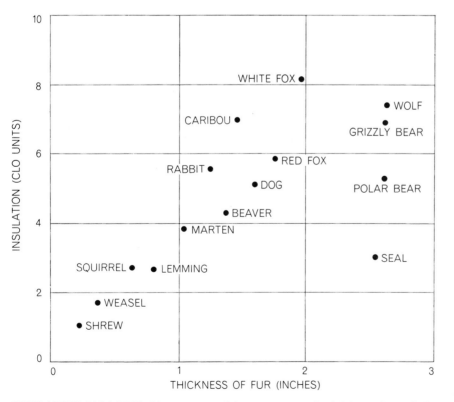

INSULATING CAPACITY of fur is compared for various animals. A "clo unit" equals the amount of insulation provided by the clothing a man usually wears at room temperature.

How are the chilled tissues of all these animals able to function normally at temperatures close to freezing? There is first of all the puzzle of the response of fatty tissue. Animal fat usually becomes hard and brittle when it is cooled to low temperatures. This is true even of the land mammals of the Arctic, as far as their internal fats are concerned. If it were also true of extremities such as their feet, however, in cold weather their feet would become too inflexible to be useful. Actually it turns out that the fats in these organs behave differently from those in the warm internal tissues. Farmers have known for a long time that neat's-foot oil, extracted from the feet of cattle, can be used to keep leather boots and harness flexible in cold weather. By laboratory examination we have found that the fats in the bones of the lower leg and foot of the caribou remain soft even at 0 degrees C. The melting point of the fats in the leg steadily goes up in the higher portions of the leg. Eskimos have long been aware that fat from a caribou's foot will serve as a fluid lubricant in the cold, whereas the marrow fat from the upper leg is a solid food even at room temperature.

About the nonfatty substances in tissues we have little information; I have seen no reports by biochemists on the effects of temperature on their properties. It is known, however, that many of the organic substances of animal tissues are highly sensitive to temperature. We must therefore wonder how the tissues can maintain their serviceability over the very wide range of temperatures that the body surface experiences in the arctic climate.

We have approached this question by studies of the behavior of tissues at various temperatures. Nature offers many illustrations of the slowing of tissue functions by cold. Fishes, frogs and water insects are noticeably slowed down by cool water. Cooling by 10 degrees

C. will immobilize most insects. A grasshopper in the warm noonday sun can be caught only by a swift bird, but in the chill of early morning it is so sluggish that anyone can seize it. I had a vivid demonstration of the temperature effect one summer day when I went hunting on the arctic tundra near Point Barrow for flies to use in experiments. When the sun was behind clouds, I had no trouble picking up the flies as they crawled about in the sparse vegetation, but as soon as the sun came out the flies took off and were uncatchable. Measuring the temperature of flies on the ground, I ascertained that the difference between the flying and the slow-crawling state was a matter of only 2 degrees C.

Sea gulls walking barefoot on the ice in the Arctic are just as nimble as gulls on the warm beaches of California. We know from our own sensations that our fingers and hands are numbed by cold. I have used a simple test to measure the amount of this desensitization. After cooling the skin on my fingertips to about 20 degrees C. (68 degrees F.) by keeping them on ice-filled bags, I tested their sensitivity by dropping a light ball (weighing about one milligram) on them from a measured height. The weight multiplied by the distance of fall gave me a measure of the impact on the skin. I found that the skin at a temperature of 20 degrees C. was only a sixth as sensitive as at 35 degrees C. (95 degrees F.); that is, the impact had to be six times greater to be felt.

We know that even the human body surface has some adaptability to cold. Men who make their living by fishing can handle their nets and fish with wet hands in cold that other people cannot endure. The hands of fishermen, Eskimos and Indians have been found to be capable of maintaining an exceptionally vigorous blood circulation in the cold. This is possible, however, only at the cost of a higher metabolic production of body heat, and the production in any case has a limit. What must arouse our wonder is the extraordinary adaptability of an animal such as the hair seal. It swims in icy waters with its flippers and the skin over its body at close to the freezing temperature, and yet under the ice in the dark arctic sea it remains sensitive enough to capture moving prey and find its way to breathing holes.

Here lies an inviting challenge for all biologists. By what devices is an animal able to preserve nervous sensitivity in tissues cooled to low temperatures? Beyond this is a more universal and more

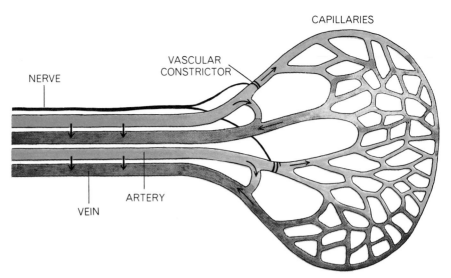

ROLE OF BLOOD in adaptation to cold is depicted schematically. One mechanism, indicated by the vertical arrows, is an exchange of heat between arterial and venous blood. The cold venous blood returning from an extremity acquires heat from an arterial network. The outgoing arterial blood is thus cooled. Hence the exchange helps to keep heat in the body and away from the extremities when the extremities are exposed to low temperatures. The effect is enhanced by the fact that blood vessels near the surface constrict in cold.

interesting question: How do the warm-blooded animals preserve their overall stability in the varying environments to which they are exposed? Adjustment to changes in temperature requires them to make a variety of adaptations in the various tissues of the body. Yet these changes must be harmonized to maintain the integration of the organism as a whole. I predict that further studies of the mechanisms involved in adaptation to cold will yield exciting new insights into the processes that sustain the integrity of warm-blooded animals.

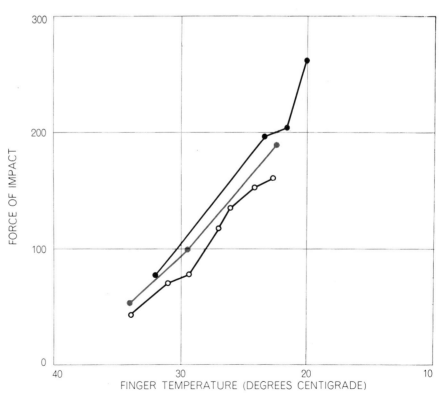

FINGER EXPERIMENT performed by the author showed that the more a finger was chilled, the farther a one-milligram ball had to be dropped for its impact to be felt on the finger. The vertical scale is arbitrary but reflects the relative increase in the force of impact.

25

THE HUMAN THERMOSTAT

T. H. BENZINGER

January 1961

A newly discovered sensory organ in the brain precisely measures the body temperature and trips the heat-dissipating mechanisms that maintain the temperature within a fraction of one degree

Fever is usually the first symptom to arouse concern in illness. The rise in body temperature is not great in the absolute sense. On the contrary, the attention it attracts is a measure of the constancy with which the body temperature is normally maintained. Compared to the daily and seasonal variation in the temperature of "cold-blooded" animals, whose internal temperature depends upon that of the environment, a fever represents a tiny variation in temperature. Yet it is many times greater than the normal variation in the regulated temperature of the healthy body. In spite of large differences in environmental temperature—from the arctic tundra and windswept highlands to fiery deserts and steaming jungles, from season to season and from day to night—the body temperature departs little from the norm of 37 degrees centigrade (98.6 degrees Fahrenheit). Life in the cells continues undisturbed, although the metabolic processes are irrevocably linked to temperature by the laws of thermodynamics and the kinetics of chemical reactions.

Heat is a by-product of these processes. With the body at rest, the heat of basal metabolism easily supplies the necessary interior warmth when external conditions are comfortably cool. Only under extreme conditions does the system fail; as when, in a hot environment, physical effort fans the flame of the metabolic furnace beyond control by the regulatory system; or when, in a very cold environment, the loss of heat by radiation, conduction and convection overbalances the metabolic production of heat and reduces body temperature to a fatal degree. Man of course shares this vital capacity with other mammals and with birds. Favored in consequence with nervous systems maintained at op-

timal working temperatures under all environmental conditions, the "warm-blooded" animals have become masters of the living world on our planet.

The question of how the body keeps its temperature constant within such narrow limits has engaged the efforts of an astonishing number of investigators. It is only recently, however, that one of the two parts of the regulating mechanism—defense against overheating—has been clarified. Max Rubner of Berlin had recognized in 1900 that sweating and the dilation of the peripheral blood vessels constitute the effector mechanisms for the dissipation of excess heat from the body. E. Aronsohn and J. Sachs, two medical students at the University of Berlin, came upon the center of control in the brain as long ago as 1884, when they damaged in animals an area "adjacent to the corpus striatum toward the midline." A few months earlier Charles Richet of Paris had also produced excessive body temperature by puncturing the forebrain. It now seems certain that in both cases the investigators damaged the hypothalamus, an area at the base of the brain stem just above the crossing of the optic nerves.

But how does the body sense and measure its temperature and bring the control center into action? The investigation of this question was confused for a long time by the conspicuous part that the temperature-sensitive nerve endings in the skin play in the feeling of warmth and cold. In recent years, however, a new approach to the problem has been made possible by the development of a new principle of measurement called gradient calorimetry and of the instrumentation to go with it. Experiments employing this instrumentation have now located the sensory end-organ at which the body "takes" its own tempera-

ture when it becomes too warm. The discovery is an unusual one at this late date in the history of physiology. The body's "thermostat" must now be included in the short list of major sensory organs adapted to the primary reception and measurement of physical or chemical quantities. Moreover, it now becomes possible to measure the characteristic responses of the thermostat and perhaps to produce or to suppress those responses artificially. Such investigation will lead to a better understanding not only of the aberration of fever but also of the precise regulation of internal temperature that is so important to the vital function of the body, particularly to the function of the delicate nervous system. With the thermostat identified, it has also become possible to explain

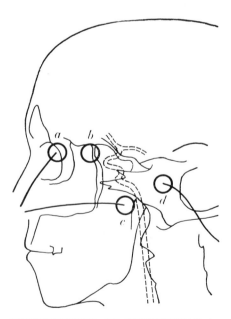

TEMPERATURE AT THERMOSTAT in the brain is measured by thermocouples

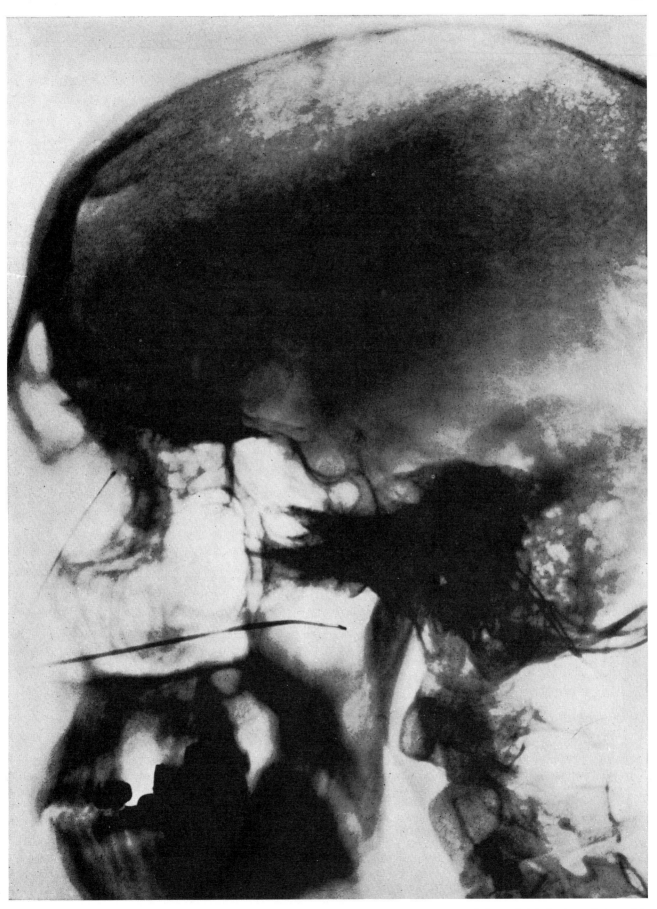

placed at forward wall of ethmoid sinus ("a" in diagram at left), deep in rear wall of nasopharyngeal cavity (c) and at eardrum (d). Thermostat itself is in hypothalamus behind sphenoid sinus (b), at which temperature has also been measured by thermocouple.

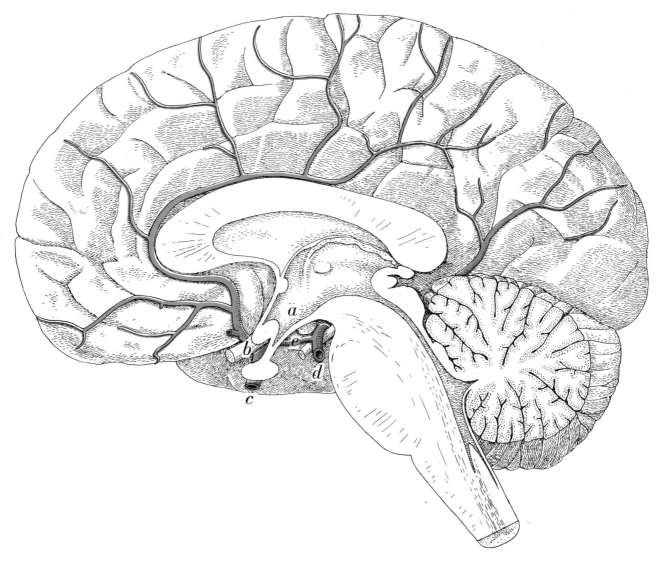

HEAT CONTROL CENTER is located in forward part of hypothalamus (*a*), shown in cross section (*left*) and from below (*right*).

Hypothalamus, centrally located under great hemispheres of brain, rests on the Circle of Willis (*e*), an arterial ring through which

the effects of such mundane factors as a hot meal, a cold drink, a hot bath and a cold shower.

When one encounters a physical or chemical quantity in technology or in a living organism that is maintained at a constant level against disturbances from outside, one looks for a "servomechanism." Pressure, rate of flow, chemical composition or temperature are automatically controlled by such mechanisms in the realm of engineering. The servomechanisms of the body control the same kinds of variable. In manmade devices the chain of control begins with a "sensory" instrument, perhaps a thermometer, which measures the variable in question. The measurement is relayed to a "controller" which compares it with a set point to which the variable is to be held. Whenever the need arises,

the controller sends instruction to an "effector" mechanism, perhaps to the heating system, which brings the temperature into accord with the set point in the controller [see "Feedback," by Arnold Tustin; SCIENTIFIC AMERICAN Offprint 327].

The corresponding elements in biological servomechanisms and the nervous and chemical pathways that interconnect the sites of stimulus and response constitute systems of far greater complexity. They are nonetheless put together in a similar way. A servomechanism in the human body may be considered to be clarified when the sensory organ, the controller and the effector mechanism are known, and when a reproducible, inseparable and quantitative relation has been established between the magnitudes of the stimuli and of the responses they induce. The net

effect of the response must be the restoration of "homeostatic" equilibrium; that is, the variable in question must return to the optimum, stable level essential to the life of the cells.

In the control circuit that prevents overheating of the body, classical physiology had identified the effector and the controller mechanism, but not the sensory organ. In a hot environment, as Rubner showed, the effector is the dilation of the blood vessels in the skin, which increases the transport of heat from the interior of the body to the surface; and sweating, which increases the rate of total heat loss from the surface to the environment as energy is absorbed in evaporation. In a cold environment the corresponding mechanism is increased metabolic heat-production.

The location of the controller in the hypothalamus by Aronsohn and Sachs

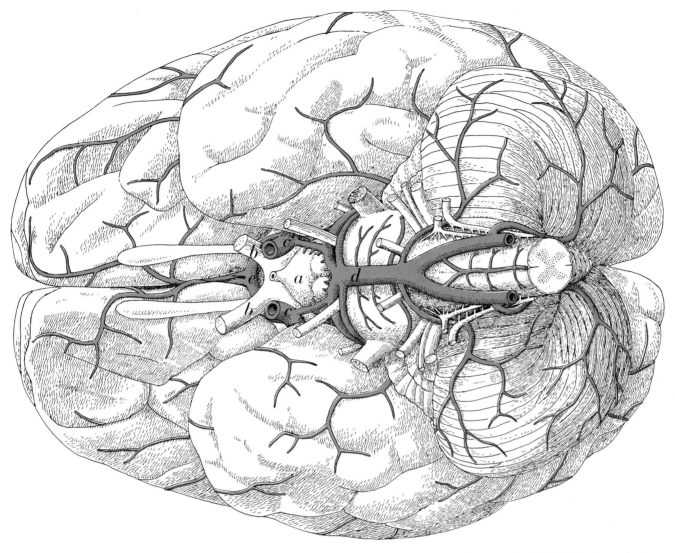

blood supply to the brain flows from carotid arteries (*c and c′*) and basilar artery (*d*). Hypothalamus, optic nerves (*b and b′*) and ret- **ina derive from same tissue matrix. Bulb of pituitary gland attached to hypothalamus appears at left, but is cut away at right.**

was confirmed by later investigators. Some of them applied the stimulus of temperature directly to the site. In 1904 Richard Hans Kahn of the German University in Prague found that heating the head arteries of a dog lowered its body temperature. In 1912 Henry Gray Barbour, a young American physician working in Vienna, carrying out an experiment designed by the pharmacologist H. Meyer, applied warm and cold probes to the general area of the hypothalamus. He observed the expected thermoregulatory responses. In 1938 Horace W. Magoun, now at the University of California at Los Angeles, discovered that this function is mediated by a circumscribed area in the forward part of the hypothalamus. Bengt Andersson of the Royal Swedish Veterinary Institute in 1956 delineated the organ with unprecedented precision in goats. In 1950 Curt

von Euler of the Nobel Neurophysiological Institute in Stockholm even succeeded in recording, in parallel with temperature changes, slow electrical "action potentials" from this area of the hypothalamus of cats.

But the body is also equipped with an elaborate system of millions of tiny sensitive nerve endings, distributed throughout the skin, which produce conscious sensations of warmth. The scientific literature tended to support the view that the skin and not the hypothalamus furnishes the primary temperature measurements to the control center for sweating and the dilation of the arteries. Some investigators held that both systems were involved; a rise in the temperature of the "heat center" in the hypothalamus supposedly made it more responsive to incoming impulses from the temperature-sensing organs of the skin. The

question, in this view, was one of determining the relative importance of the two sites. It was also possible, as some believed, that the body possessed a third area sensitive to temperature or heat flow, and that neither the skin nor the brain was involved.

It was not easy to design a conclusive experiment. In experimental animals one might destroy the nervous pathways from the thermoreceptors in the skin to the heat center. But the results of such an experiment would not exclude the possibility that the temperature of these centers played a role in heat regulation under normal conditions. It would still be necessary to carry out the reverse experiment and destroy the heat-sensitive part of the hypothalamus. Since this structure is intimately involved with the temperature-control center itself, it seems impossible to secure the final evidence by surgical procedures alone.

To observe the operation of sensory receptors in the skin and in the brain independently of each other in the intact organism presented comparable difficulties. No one, apparently, had succeeded in keeping one of the two sites at a constant temperature while observing the effects brought about by a temperature change in the other. This approach called for techniques to measure temperature in the human body at the two sites of presumed temperature-reception—the skin and the hypothalamus—and some way to record, rapidly and continuously, the effector responses of vasodilation and sweating.

The gradient calorimeter has satisfied the second of these two requirements. This rapidly responding and continuously recording successor to the classical calorimeter makes it possible to record for the first time the total output of the effector mechanisms. It measures separately the heat that is carried from the body by radiation and convection and the heat that is dissipated by the evaporation of sweat. From working models made and tested by Charlotte Kitzinger at the Naval Medical Research Institute in Bethesda, Md., the first full-scale human gradient calorimeter was constructed under the direction of Richard G. Huebscher at the laboratory of the American Society of Heating and Ventilating Engineers in Cleveland. Similar units have now been constructed at other laboratories. The gradient calorimeter now operated at Bethesda is a chamber large enough to hold a man stretched out at full length [*see illustration below*]. The subject is suspended in an openweave sling, out of contact with the floor or walls of the chamber, and is free to go through the motions of prescribed exercise when the experiment calls for such exertion.

The new and essential feature of gradient calorimetry is the "gradient layer," a thin foil of material with a uniform resistance to heat flow which lines the entire inner surface of the chamber. Some thousands of thermoelectric junctions interlace the foil in a regular pattern and measure the local difference in temperature (and hence the local heat flow) at as many points across the foil. The junctions are wired in series; their readings are thus recorded in a single potential at the terminals of the circuit. That potential measures the total energetic output from the subject's skin, independent of his position with respect to the surfaces of the gradient layer lining the chamber. The rate of blood flow through the skin can be derived by computing this measurement against the temperature of the outgoing blood (measured internally) and the temperature of the returning blood (measured on the skin), since the observed transfer of heat per unit time at any given difference between internal and external temperature can be effected by only one calculable rate of blood flow. The energy dissipated by evaporation from the subject's skin is also measured by gradient layers which line heat-exchange meters at the inlet and outlet of the air circuit of the calorimeter. Measurements taken for control make it possible to maintain the same temperature and humidity in the air at these two points, so that the air neither gains nor loses energy as it passes through the system. The unbalanced output from the additional gradient layers thus precisely measures the heat loss by evaporation and hence the sweat-gland activity. Heat loss through the lungs is measured separately and subtracted from the total.

With the help of the gradient calorimeter our group at Bethesda set out to establish the correlations obtaining, on the one hand, between the performance of the effector mechanisms and the temperature of the skin and, on the other hand, between the performance of the effector mechanisms and the internal temperature of the body. In these first experiments it was assumed that rectal temperature provided an adequate index of internal temperature as measured at the internal temperature-sensing organ, wherever that might be located. But no correlation could be found, in either resting or "working" subjects, between rectal temperature and the observed rates of sweating. Measurement of skin temperature against the same heat-dissipation variable yielded equally meaningless plots. For a time it seemed that all the effort that had gone into the design of the gradient calorimeter had been wasted. The results made sense only in terms of the classical notion that the thermostat in the interior of the body and the temperature-sensing nerve endings in the skin have indissolubly interlaced effects upon the vasodilation and sweating responses.

Then we found a way to measure the internal temperature of the body at a site near the center of temperature regulation in the brain. We introduced a thermocouple through the outer ear canal and held it against the eardrum membrane under slight pressure. The eardrum is near the hypothalamus and shares a common blood supply with it from the internal carotid artery. At the

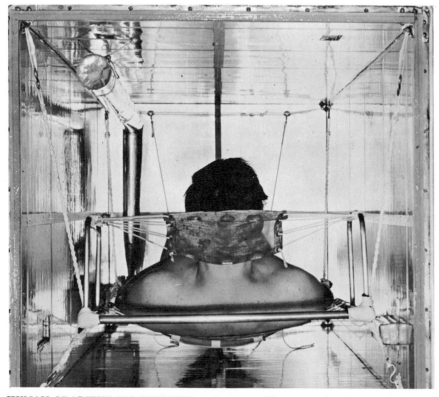

HUMAN GRADIENT CALORIMETER makes it possible to correlate body temperature with dissipation of heat by radiation and convection from skin and by evaporation of sweat. Lining of chamber is interlaced with thermoelectric junctions which measure heat loss from skin; loss by sweating is measured by temperature and humidity control system of calorimeter.

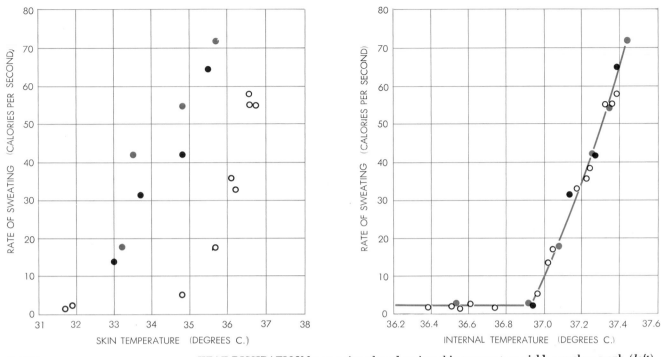

O REST

● WORK (6 CALORIES PER SECOND)

● WORK (12 CALORIES PER SECOND)

HEAT DISSIPATION by sweating plotted against skin temperature yields senseless graph (*left*) when natural correlation between skin and internal temperature is broken by internal heating through exercise. When the same measurements of sweating are plotted instead against internal head temperature (*right*), an inseparable and always reproducible relation appears between the stimulus of temperature and the response of the sweat glands, whether the subject is sweating or not.

very first attempt we observed temperature changes associated with the eating of ice or the drinking of hot fluids, and we soon found we could detect variations caused by immersion of the limbs in warm water. Parallel rectal measurement did not show these variations at all. To make sure that the entire region of the head supplied by the carotid arteries can be expected to show the same temperature variations as the eardrum, we tried other sites. With the help of local anesthesia H. W. Taylor, a surgeon at the Naval Hospital in Bethesda, placed thermocouples in our heads: at the main trunk of the internal carotid in the rear of the nasopharyngeal cavity, in the nasal cavity below the forebrain and at the forward wall of the sphenoid sinus only one inch away from the hypothalamus [*see illustration on pages 260 and 261*]. Continuous measurements of temperature at these points showed large discrepancies with internal temperature as measured at the rectum.

The discrepancies appeared before and after the subject exerted himself by physical exercise, after internal cooling by the eating of ice, after warming the arms or legs in warm water and cooling them in cold water and after immersing the whole body in warm or cold water. These were precisely the situations in which earlier experimenters had found the same absence of correlation between

rectal temperature and the heat-dissipating responses of vasodilation and sweating. It was clear that the temperature at the rectum could under no circumstances be trusted as reflecting the temperature at the internal temperature-sensing organ. The hypothalamus was plainly the place to look for correlation between changes of internal temperature and the responses that regulate it. Since the eardrum is by all odds the most accessible of the four sites thus measured in the head, it was adopted in the experiments that followed. Readings could be taken here with an error of .01 degree centigrade against a standard of temperature maintained with an error of .002 degree C.

The subjects now spent time in the calorimeter on many different days at different environmental temperatures ranging from almost intolerably cold for the nude body at rest to almost intolerably hot for the subject undergoing exertion. Between these two extremes, measurements were made for all the intermediate levels at five-degree temperature intervals and with the subject at rest and at work. Under each set of circumstances the instruments kept a continuous record as the state of homeostasis was reached and maintained for one hour. This arduous series of experiments, extended over two months, made it possi-

ble to plot for the first time the heat-dissipating responses of vasodilation and sweating independently against skin temperature and against internal head temperature. The volunteer subject for this series, Lawrence R. Neff, was observed with a cool skin and cool interior (resting in a cold environment), with a cool skin and warm interior (working in a cold environment), with a hot skin and a relatively cool interior (resting in a hot environment) and with a hot skin and warm interior (working in a hot environment).

The records showed the familiar disordered relationship between the heat-dissipating responses and skin temperature. But the plot of the responses against eardrum temperature showed an almost perfect undisturbed relation. Whatever the temperature of the skin, one certain specific rate of sweating and no other invariably showed up in association with a given internal temperature measured at the eardrum. A reproducible, inseparable and quantitative relation between the stimulus of temperature and one of the heat-dissipating responses had at last been observed. It exhibited a sharply defined breakoff at 36.9 degrees C. (98.4 degrees F.). This was no doubt the set point of the human thermostat in this subject at the time of the experiment. The response proved to be so forceful that a mere .01-degree-C.

rise in temperature was sufficient to increase the dissipation of heat through sweating by one calorie per second and to raise the blood flow through the skin by 15 milliliters per minute.

The success of this series of experiments in distinguishing between the variations in skin and brain temperature was confirmed in many other experiments that subjected the body to quite different sets of extremes. In one of them the skin and the interior of the body were warmed as the subject accommodated himself in the calorimeter to an environmental temperature of 45 degrees C. (113 degrees F.). With homeostasis attained, the subject gulped down large measured helpings of sherbet three times at suitable intervals. On each occasion, as the melting ice withdrew heat from the internal organs and the circulating blood, the brain temperature declined. No less impressively, the skin temperature was observed to rise. The curve

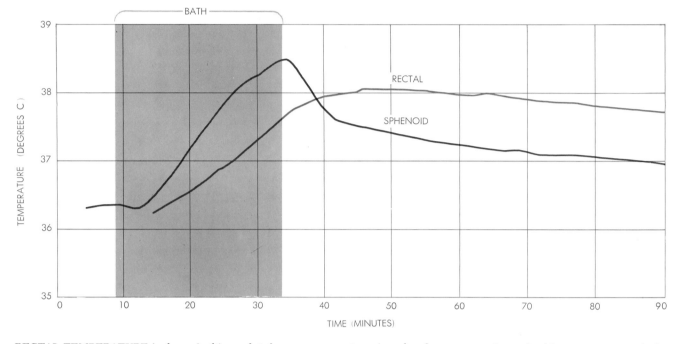

RECTAL TEMPERATURE is shown in this graph to have an uncertain relationship to the hypothalamic temperature measured at the forward wall of the sphenoid sinus. The sphenoid temperature rises sharply as an experimental subject enters a warm bath and falls off sharply when the subject leaves the bath. The rectal temperature reaches a peak only after the subject has left the bath.

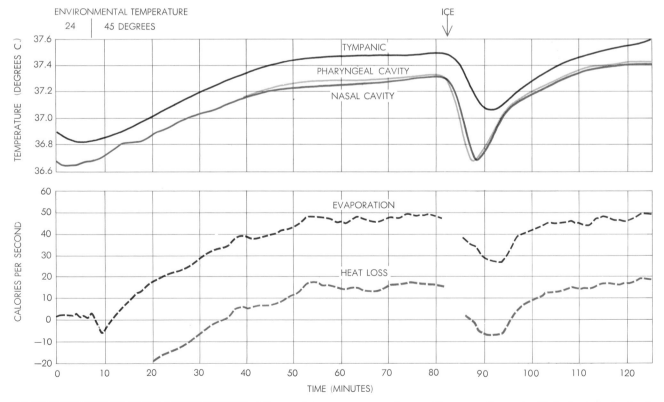

MEASUREMENTS OF HEAD TEMPERATURE at three points near the hypothalamus—at eardrum (*tympanic*), at ethmoid sinus (*nasal cavity*) and in rear wall of pharyngeal cavity—show close correspondence with one another and with the heat dissipation by evaporation of sweat and heat loss by vasodilation. Location of these points is shown in the illustration on pages 260 and 261.

drawn by sweat-gland activity now showed unequivocally which temperature-sensing system controls the heat-dissipating responses: the rate of sweating fell off and rose in perfect parallel with the decline and rise in internal head temperature. It was the consequent drying of the skin that caused the skin to be heated by radiation and conduction in the hot environment. But the sensory reception of heat in the skin brought no response from the heat-dissipating mechanism.

These observations accord well with the familiar constancy of the body temperature. It is difficult to see how it could be maintained within the same narrow range, year in and year out, if the heat-controlling responses were not always triggered at the same set point. As these experiments show, moreover, the responses always closely match the magnitude of the stimulus. Such precise regulation of temperature could not be achieved by measurement of skin temperature. As in all feedback systems, the quantity which is controlled must itself be measured. An architect who wants to control the temperature of a house does not distribute thousands of thermometers over the outside walls. One thermostat in the living room suffices. It responds not only to warming and chilling from out-of-doors but to overheating from within. The thermostat in the hypothalamus similarly monitors the internal temperature of the body from the inside and thereby maintains its constancy.

This is not to say that the warm-sensitive nerve endings of the skin have no function in the regulation of body temperature. They are the sensory organs for another system which operates via the centers of consciousness in the cortex, bypassing the unconscious control center in the hypothalamus. To sensations of heat or cold reported by the skin the body reacts by using the muscles as effector organs. Under the stimulus of discomfort from the extremes of both heat and cold, man seeks a cooler or a warmer environment or takes the measures necessary to make his environment comfortably cool or warm. But for all the mastery of external circumstances that follows from this linkage in the body's temperature-sensing equipment, the skin thermoreceptors cannot regulate internal temperature with any degree of precision. They can contribute directly to the regulation of skin temperature alone. The automatic system of hypothalamic temperature regulation takes over from there and achieves the final adjustment with almost unbelievable sensitivity and precision.

In the regulation of internal temperature, therefore, the hypothalamus can

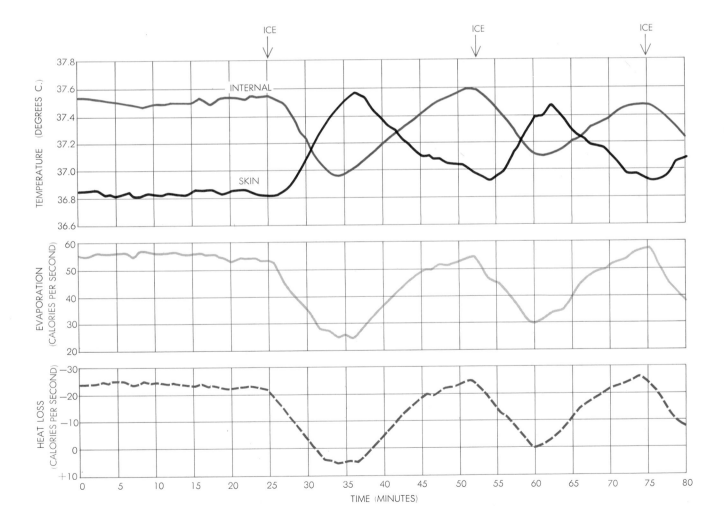

no longer be regarded simply as a controller which converts incoming sensory stimuli into outgoing impulses to the effector system. It is itself the site of a receptor end-organ, an "eye" for temperature comparable to the retina—the receptor organ for light. This analogy between the temperature eye and the optical eye has, in fact, a sound anatomical basis. Both are derived from the same matrix: the bottom of the third ventricle of the brain. These are two parts of the brain that have a proved sensory receptor function. In the course of evolution the optical eye moved outward to connect with a dioptric apparatus partly derived from the skin, and thereby gained a view of the external world. The temperature eye in the hypothalamus is located in the interior of the head, where it properly belongs. It measures as well as regulates the temperature of the blood which bathes its cells and the rest of the brain, the vital function of which requires a closely maintained optimal temperature.

The feedback system that dissipates heat and thus keeps the body from overheating under normal conditions has thus been elucidated. The same cannot be said, however, of the regulatory system that steps up metabolic heat production and keeps the body temperature from falling below the optimum level. It appears that the two systems operate quite differently and that in the metabolic warming-up of the body the temperature-eye performs its task by inhibition of sensory impulses originating elsewhere.

On the other hand, the sure location of the thermostat in experiments on the "warm side" now makes it possible to renew the study of many interesting questions. Temperature measurements at sites that reliably reflect hypothalamic temperature should replace rectal observations of temperature in all these studies and even in some clinical situations. How bacterial toxins produce fever and how drugs act to reduce it can now be redefined in terms of shifts in the set point of the thermostat and may be made the subject of quantitative investigation. The same direct attack may also be made upon individual or group tolerances and the adaptability of human temperature regulation. These are important objectives in connection with hypothermia (the reduction of the body temperature to low levels) in surgical operations, and with the conquest of new spaces for the life of the human species.

26

THE PRODUCTION OF HEAT BY FAT

MICHAEL J. R. DAWKINS AND DAVID HULL

August 1965

*In addition to normal "white" fat, many newborn mammals
and adults of hibernating species have "brown" fat deposits.
It is metabolism in brown fat cells that increases heat output
as a response to cold*

When a warm-blooded animal is exposed to cold, it increases its production of body heat by shivering. Mammals face a cool environment for the first time at birth, however, and many newborn mammals (including human infants) do not shiver. Yet they somehow manage to generate heat in response to a cool environment. The mystery of how this is done has only recently been cleared up. It turns out that the young of many species (and adults of hibernating species as well) are fortified with a special tissue that is exceptionally efficient in producing heat. The tissue in question, long a puzzle to investigators, has become a highly interesting object of physiological and chemical study within the past few years.

Our own interest in it was aroused in 1963 by a chance observation at the Nuffield Institute for Medical Research of the University of Oxford, where we were working in a group studying physiological problems of the newborn. Examining newborn rabbits, we noticed that they had striking pads of brown adipose tissue around the neck and between the shoulder blades. Adipose tissue is a salient feature of all warm-blooded animals. It constitutes the layer of fat underlying the skin over most of the body, and it is known to serve not only as an insulating blanket but also as a storehouse of food and en-

ergy. In the adult animal the adipose tissue is almost entirely of the white variety. The large deposits of adipose tissue we saw in the newborn rabbits were in the brown form, and this was a phenomenon that called for explanation.

Looking back through the literature, we found that the brown adipose tissue had mystified investigators for hundreds of years. It was noted as early as 1551 by the Swiss naturalist Konrad von Gesner, who was impressed by the mass of this tissue he observed between the shoulder blades of a marmot. Some observers confused the tissue with the thymus gland, another mysterious structure that had been found to be particularly prominent in newborn ani-

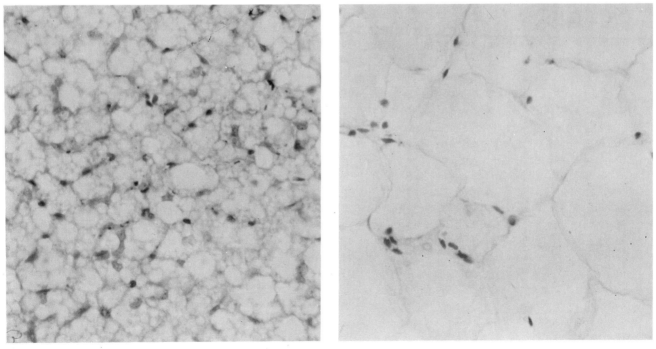

ADIPOSE TISSUE is enlarged 425 diameters in these photomicrographs made by the authors. Brown fat cells (*left*) from between the shoulder blades of a newborn rabbit have numerous small drop-lets of fat suspended in the stained cytoplasm. White fat cells (*right*) from an adult rabbit have large droplets of fat surrounded by narrow rims of cytoplasm. The stain was hematoxylin and eosin.

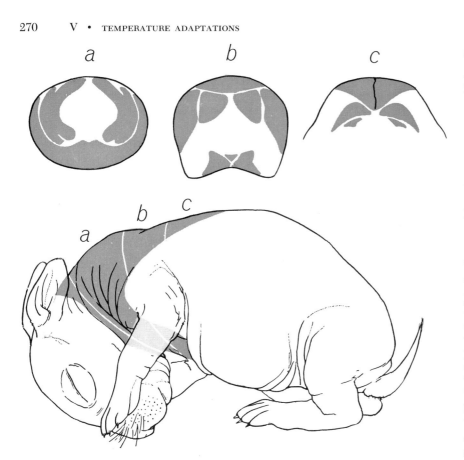

BROWN FAT accounts for 5 or 6 percent of the body weight of the newborn rabbit. It is concentrated, as shown in sections, around the neck and between the shoulder blades.

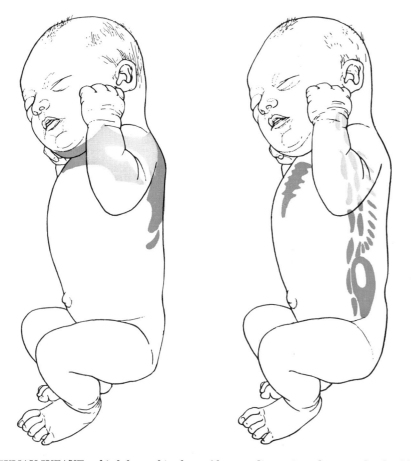

HUMAN INFANT at birth has a thin sheet of brown adipose tissue between the shoulder blades and around the neck, and small deposits behind the breastbone and along the spine.

mals. Other zoologists, noting that a brown adipose mass was typical of hibernating animals, called it the hibernation gland. In this century more modern theories were advanced. Some physiologists suggested that brown adipose tissue had something to do with the formation of blood cells; others, that it was an endocrine gland. It does, in fact, contain hormones similar to those secreted by the adrenal cortex, but experiments in administering extracts from the tissue failed to show any consistent evidence of hormonal effects.

In 1961 two physiologists independently suggested a more plausible hypothesis. George F. Cahill, Jr., of the Harvard Medical School, noting that adipose tissue has an active metabolism that must generate heat as a by-product, proposed that the layer of white fat clothing the body should be regarded "not merely as a simple insulating blanket but perhaps as an electric blanket." And Robert E. Smith of the University of California School of Medicine at Los Angeles specifically called attention to the high heat-producing potentiality of brown adipose tissue, whose oxidative metabolism he had found to be much more active than that of white adipose tissue.

The cells of adipose tissue are characterized by droplets of fat in the cytoplasm (the part of the cell that lies outside the nucleus). In the white adipose cell there is a single large droplet, surrounded by a small amount of cytoplasm. The brown adipose cell, on the other hand, has many small droplets of fat, suspended in a considerably larger amount of cytoplasm. With the electron microscope one can see that the brown fat cells contain many mitochondria, whereas the white fat cells have comparatively few. Mitochondria, the small bodies sometimes called the powerhouses of cells, carry the enzymes needed for oxidative metabolism. What gives the brown fat cells their color is a high concentration of iron-containing cytochrome pigments—an essential part of the oxidizing enzyme apparatus—in the mitochondria.

It is easy to show by experiment that brown fat cells, loaded as they are with mitochondria, have a large capacity for generating energy through oxidation of substrates. Tested, for example, on succinic acid, an intermediate product in the Krebs energy-producing cycle, the brown fat cells of rabbits prove to have a capacity for oxidizing this substance that is 20 times greater than the oxidative capacity of white fat cells

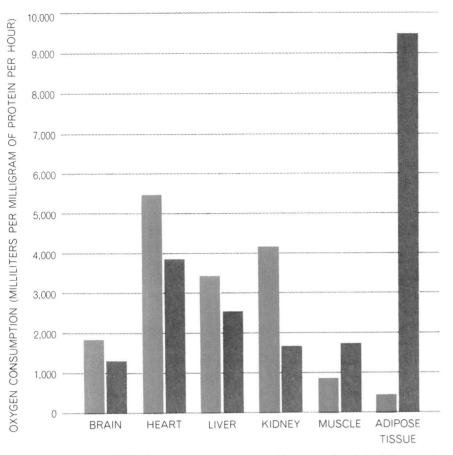

OXIDATIVE CAPACITY of various tissues is compared by measuring their ability to oxidize succinic acid. The colored bars are for adult rabbit tissues, the grey bars for tissues from newborn rabbits. Newborn brown fat (*right*) is the most active oxidizer by far.

and is even greater than that of the hardworking cells of the heart muscle.

To explore the role of the brown adipose tissue in newborn rabbits we began by measuring the animals' total heat production. An indirect measure of this production is the animal's consumption of oxygen: each milliliter of oxygen consumed is equivalent to about five calories of heat in the body. We found that at an environmental temperature of 35 degrees centigrade (95 degrees Fahrenheit) newborn rabbits produced heat at a minimal rate. When their environment was cooled to 25 degrees C. (77 degrees F.), they trebled their heat output. At 20 degrees C. (68 degrees F.) their heat production reached a peak: 400 calories per kilogram of body weight per minute. If newborn rabbits were as fully protected against heat loss and as large as adults are, this rate of production would be sufficient to maintain their internal body temperature at a normal level even in a cold environment as low as 30 degrees below zero C. A newborn rabbit, however, has little or no fur and a large surface area in relation to its body mass;

hence its deep-body temperature falls when the animal is only mildly chilled by the outside air.

The next step was to determine whether or not heat production was concentrated in the brown adipose tissue. In a newborn rabbit this tissue makes up 5 to 6 percent of the total body weight and is localized, as we have noted, around the neck and between the shoulder blades. We inserted a fine thermocouple under the skin next to the brown adipose tissue to measure any change in the temperature of that tissue, and for comparison we inserted a second thermocouple in back-muscle tissue at the same distance from the skin and a third in the colon to record the deep-body temperature. At the neutral environmental temperature of 35 degrees C. the temperatures at all three sites in the body were the same. When the environmental temperature was lowered to 25 degrees C., differences developed: the temperature at the brown adipose tissue then was 2.5 degrees higher than that in muscle tissue in the back and 1.3 degrees higher than the deep-body temperature [*see illustra-*

tion on the next page]. The temperature difference persisted for many hours, until the fat stored in the brown tissue was almost completely exhausted.

This clear indication that the brown adipose tissue produced heat was strengthened by an experiment in which the newborn animals were deprived of oxygen, the oxygen content of the air in the experimental chamber being reduced from the normal 21 percent to 5 percent. Deprived of the oxygen required for oxidative metabolism, the brown adipose tissue promptly cooled to the same low temperature as the muscle tissue. When the oxygen concentration in the air was restored to the normal 21 percent, the brown adipose tissue immediately warmed up again, with the muscle tissue and deep-body temperature trailing after it in recovery.

Is the brown adipose tissue solely responsible for the newborn animal's increase in heat production in response to cold? We examined this question in a series of experiments with Malcolm M. Segall collaborating. The experiments consisted simply in observing the effect of excising most of the brown adipose tissue (amounting to a few grams) from newborn rabbits. When 80 percent of this tissue was removed (by surgery under anesthesia), the animals no longer increased their heat production in response to exposure to cold. In short, removal of the few grams of this specific tissue practically abolished the newborn rabbit's ability to multiply its oxygen consumption threefold and step up its heat production correspondingly. Evidently, then, the brown adipose tissue was entirely, or almost entirely, responsible for this ability.

Our results did not necessarily mean that all the metabolic heat in response to cold was produced within the brown fat cells themselves. Those cells might release fat in some form into the bloodstream for transport to other tissues, where it might be oxidized. Fortunately this question too could be investigated experimentally.

The fat in the droplets in adipose cells is in the form of triglyceride molecules. A triglyceride consists of a glycerol molecule with three long-chain fatty acids attached [*see middle illustration on page 273*]. Before the triglyceride can be oxidized it must be split into smaller, more soluble units—that is, into glycerol and free fatty acids. Glycerol cannot be used for metabolism in a fat cell, because that type of cell does not contain the necessary enzymes. Consequently all the glycerol molecules freed by the splitting of triglycerides in fat cells are discharged into the bloodstream. The glycerol level in the blood therefore provides an index of the rate of breakdown of triglycerides in fat cells. Now, if the level of free fatty acids in the blood corresponds to the glycerol level, we can assume that fatty acids also are released from these cells in substantial amounts for distribution to other tissues.

We examined the blood of newborn rabbits from this point of view. To begin with, at the neutral incubation temperature of 35 degrees C. the level of free fatty acids in the blood was slightly higher than that of glycerol. When the environmental temperature was lowered to 20 degrees, the glycerol level in the blood increased threefold. The concentration of free fatty acids in the blood, however, rose only a little. This showed that most of the fatty acid molecules freed by the splitting of triglycerides in fat cells must have remained in the cells and been metabolized there. Studies of adipose tissue in the test tube indeed demonstrated that less than 10 percent of the freed fatty acid is released from the cell. We can conclude that the brown fat cells are the main site of cold-stimulated heat production.

How is the heat produced? Brown fat cells are admirably suited, by virtue of their abundance of mitochondria, for generating heat by means of the oxidation of fatty acids. In this process a key role is played by adenosine triphosphate (ATP), the packaged chemical energy that powers all forms of biological work, from the contraction of muscle to the light of the firefly. The probable cycle of reactions that turns chemical energy into heat in adipose tissue cells is shown in the illustration on page 274.

Triggered by the stimulus of cold, the brown fat cell splits triglyceride molecules into glycerol and fatty acids. The glycerol and a small proportion of the free fatty acids are released into the bloodstream for metabolism by other tissues (probably liver and muscle). More than 90 percent of the fatty acid molecules remain, however, in the fat cell. They combine with coenzyme A, the energy for this combination being donated by ATP. Since the donation involves the splitting of high-energy bonds in ATP, with its consequent hydrolysis to adenosine monophosphate (AMP), the cell has to regenerate ATP.

This is accomplished by oxidative phosphorylation: the addition of inorganic phosphate to AMP with the simultaneous oxidation of a substrate.

Now, some molecules of the fatty acid–coenzyme A compound formed with the help of ATP are oxidized to provide energy for the regeneration of ATP. But most of this complex is reconverted, by combination with alpha-glycerol phosphate, to the original triglyceride. In short, there is an apparently purposeless cycle that breaks triglyceride down to fatty acids only

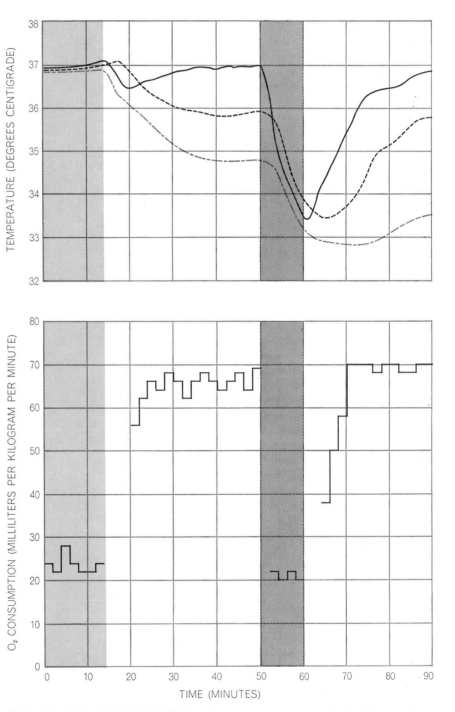

SITE AND OXYGEN DEPENDENCE of heat production are established by data from newborn rabbits subjected to cold and temporarily deprived of oxygen. The top curves are for body temperature measured near brown fat (*solid black line*), in muscle (*broken gray*) and in the colon (*broken black*). The bottom curves trace metabolic activity. In the period covered by the colored band the environmental temperature was 35 degrees centigrade; thereafter it was 25 degrees. The gray band marks a period during which the oxygen concentration was cut from 21 percent to 5 percent. Apparently brown-fat metabolism in the presence of adequate oxygen accounts for a rabbit's ability to respond to a drop in temperature.

to resynthesize the latter back to triglyceride. Although the cycle seems pointless in chemical terms, it is clearly significant in terms of work. The cycle is, in fact, a device for turning the chemical-bond energy of fatty acids into heat. The energy driving the cycle comes fundamentally from the oxidation of the fatty acids, and the fact that the cycle is exceptionally active in brown adipose tissue is demonstrated by that tissue's high consumption of oxygen. Judging from the proportion of free fatty acids retained by the cells of brown adipose tissue, and from the effects of surgical removal, this tissue accounts for more than 80 percent of the increased body heat produced by a newborn rabbit in the cold.

The heat must of course be distributed to the rest of the body by the bloodstream. The newborn animal's brown adipose tissue has an extremely rich blood supply. During exposure to cold the blood flow through this tissue may increase to several times its normal rate, and indirect calculations suggest that as much as a third of the total cardiac output is directed through the tissue.

How does cold stimulate the brown adipose tissue to generate heat? There are two possible means by which the body's sensation of cold may be communicated to the tissue: by nerve impulses and by hormones, the chemical "messengers" of the body. We found that the hormone noradrenaline has a specific stimulating effect on the brown adipose tissue. An intravenous infusion of noradrenaline in a newborn rabbit will bring about a large increase in the animal's oxygen consumption and heat production in its brown fat. If the brown fat is removed, the hormone no longer produces any increase in the body's oxygen consumption. The question remains: Is the hormone delivered to the intact tissue by way of the bloodstream or by release at sympathetic nerve endings, which are known to secrete noradrenaline close to cells? Several clues suggest that the nerve endings, rather than the bloodstream, are the agent of delivery. For one thing, the adipose tissue's rapid response to cold indicates that the message travels via the nerves. Second, experiments show that drugs that block the action of noradrenaline circulating in the blood do not block the tissue's response to cold. Third and conclusively, direct electrical stimulation of the sympathetic nerves going to the tissue

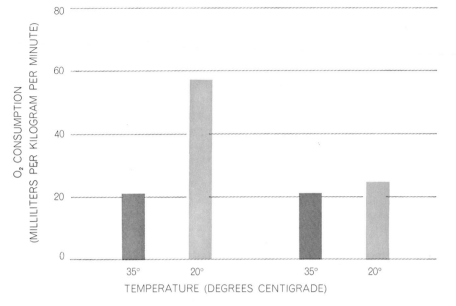

REMOVAL OF BROWN FAT by surgery sharply curtails the response to cold. The two bars at the left show the oxygen consumption in intact newborn rabbits at two temperatures. The bars at the right are for rabbits that have had 80 percent of their brown fat removed: metabolism at 35 degrees is unchanged but there is virtually no increase at 20 degrees.

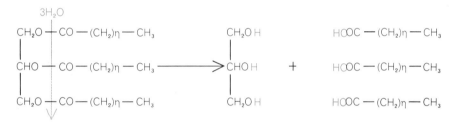

METABOLISM OF BROWN FAT begins with the hydrolysis of the triglyceride molecule (*left*), yielding one molecule of glycerol (*center*) and three of free fatty acid (*right*).

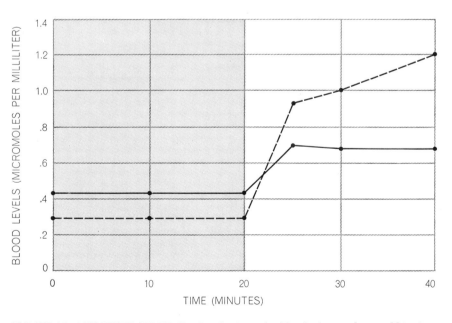

GLYCEROL CONCENTRATION (*broken line*) in the blood of a newborn rabbit rises sharply when the environmental temperature drops from 35 degrees centigrade (*colored area*) to 20 degrees (*white area*). The concentration of fatty acids in the blood rises only slightly, however (*solid line*). Apparently fatty acids are largely metabolized in the cell.

will cause the brown adipose tissue to produce heat, whereas when the sympathetic nerves are cut, the tissue can no longer burn its fat when the animal is exposed to cold.

Various findings indicate that the overall system controlling the production of heat by brown adipose tissue is probably as follows: The temperature receptors in the skin, on sensing cold, send nerve impulses to the brain. The brain's temperature-regulating center then relays impulses along the sympathetic nerves to the brown adipose tissue, where the nerve endings release noradrenaline. The hormone activates an enzyme that splits triglyceride molecules into glycerol and free fatty acids and thereby triggers the heat-producing cycle. Thus the rate of heat production is controlled by the sympathetic nervous system.

Among the animals that have brown adipose tissue at birth, the amount varies considerably from species to species. As in the rabbit, there are large deposits between the shoulder blades in the newborn guinea pig and the coypu (a water rodent). In the cat, dog and sheep at birth there are sheets of brown adipose tissue between the muscles of the trunk and around the kidneys. The human infant has well-marked deposits of such tissue [see bottom illustration on page 270], and recent studies indicate that this tissue is a source of heat for a baby as it is for other newborn animals. When a baby is exposed to cold, its blood shows a small but definite rise in the glycerol level with no significant change in the level of fatty acids; on prolonged exposure to cold the fat in its brown adipose tissue is used up.

In most species of animals born with brown adipose tissue the tissue appears to be largely converted to the white form by the time the animal has reached adulthood. Certain animals retain at least some tissue in the brown form, however. The adult rat, for example, has small amounts of brown adipose tissue in the shoulder blade region and elsewhere. The rat's venous system indicates that there is a rich flow of blood from this region to the plexus of veins around the spinal cord; this suggests that the brown adipose tissue in the adult rat may serve particularly to warm vital structures in the animal's body core during exposure to cold. When a laboratory rat is kept in a cold environment, it develops additional brown adipose tissue and an increased ability to produce heat without shivering.

For hibernating animals brown adipose tissue is an all-important necessity throughout life. These animals possess large amounts of the tissue, and direct studies have now shown that the brown adipose tissue is responsible for the animals' rapid warming and awakening from the torpid hibernating state.

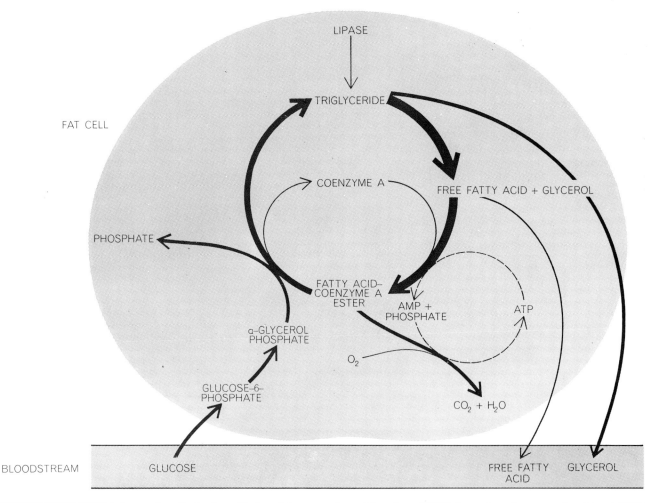

HEAT IS PRODUCED in brown fat cells by the oxidation of fatty acids. In response to cold an enzyme, lipase, splits triglyceride. The glycerol leaves the cell. The fatty acid forms an ester with coenzyme A, the reaction acquiring energy from the splitting of adenosine triphosphate (ATP) into adenosine monophosphate (AMP) and inorganic phosphate. Some of the ester is oxidized to regenerate ATP; the rest goes to resynthesize triglyceride. The effect of the cycle, then, is to turn chemical-bond energy into heat energy.

27

DESERT GROUND SQUIRRELS

GEORGE A. BARTHOLOMEW AND JACK W. HUDSON
November 1961

Two little animals of the Mohave Desert have evolved remarkable adaptations to heat and aridity. Each has adapted in its own way, which apparently enables them to live together without competing

Among the handful of animals that inhabit the hot, dry and sparsely vegetated Mohave Desert of California are two species of ground squirrel: the antelope ground squirrel and the mohave ground squirrel. Both species live in burrows, both are active and aboveground during the day and both feed on the small amount of plant life that is available.

This is an uncommon situation in nature. Species as closely related as these two, and as much alike in their food and habitat requirements, seldom live together even in more favorable environments. In his *Origin of Species* Charles Darwin suggested the reason. "As the species of the same genus usually have ...much similarity in habits and constitution, and always in structure," he wrote, "the struggle will generally be more severe between them, if they come into competition with each other, than between the species of distinct genera." Implicit in Darwin's statement is a concept now fundamental to biology. It is known as the principle of competitive

exclusion and it says, in brief, that two noninterbreeding populations that stand in precisely the same relationship to their environment cannot occupy the same territory indefinitely. They cannot, in other words, live in "sympatry" forever. Sooner or later one will displace the other.

Such a displacement could be under way in the Mohave Desert right now. The mohave ground squirrel may well be a species in the process of extinction. Not only does it have a smaller total population and a narrower geographical distribution than the antelope ground squirrel (which is one of the commonest ground squirrels of the southwestern U.S.); it also appears to be less numerous in the small section of the Mohave Desert to which it is restricted. But since no historical information is available on the population trends of the two animals, there is no way of knowing exactly what the present difference in their number portends.

In any case the mohave ground squirrel is not as yet extinct. This raises a

number of intriguing questions. Do the two species have the same way of life and the same relationship to their environment? If so, the competition between them must be severe. Or are there differences in their adaptation to their common environment? If there are, do these differences reduce competition between them sufficiently to permit them, at least temporarily, a period of peaceful coexistence?

These questions become more intriguing when one considers the nature of the desert environment. Aridity and heat make particularly severe demands on animals, and animals that live in deserts must be equipped with special physiological and behavioral adaptations to meet these demands. The camel, for example, withstands aridity because it can tolerate a high degree of dehydration, can restore its body fluids quickly and can travel long distances in search of water and succulent vegetation. It withstands heat through its tolerance of a wide range of body temperatures and

ANTELOPE GROUND SQUIRREL is found in the Mohave Desert and throughout the southwestern U.S. It is active during the day all year round, in spite of extremes of heat and aridity. Both of these photographs were made in the laboratory by Jack W. Hudson.

MOHAVE GROUND SQUIRREL is found only in one corner of the Mohave Desert. It is active during the day from March to August but remains in its burrow the rest of the year. Before retiring underground it becomes very fat, as this photograph shows.

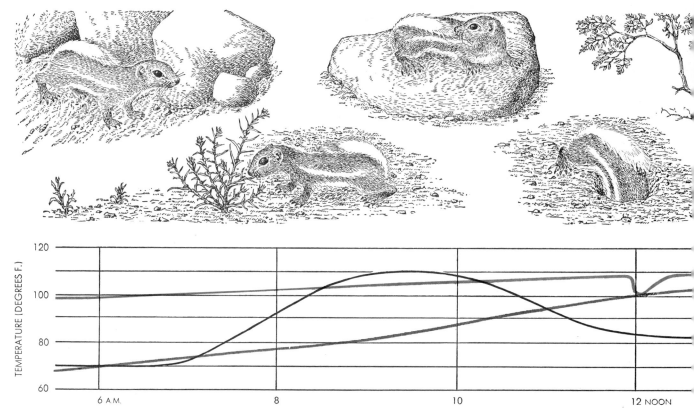

ACTIVITY OF ANTELOPE GROUND SQUIRREL on a typical summer day is charted. On emerging from its burrow animal runs to feeding area. Then it suns and grooms itself. When its body temperature rises too high, it goes to a special retreat burrow to cool

—————— BODY TEMPERATURE
—————— ENVIRONMENTAL TEMPERATURE
—————— LEVEL OF ACTIVITY

through the insulating qualities of its coat [see "The Physiology of the Camel," by Knut Schmidt-Nielsen, beginning on AMERICAN Offprint 1096]. The desert rat of the U.S. Southwest has adapted equally well but in quite different ways. To combat aridity it conserves its body water; the desert rat's kidney is so efficient that it uses only about a fourth of the amount of water that the human kidney requires to excrete the same amount of urea. This adaptation enables the animal to meet a substantial fraction of

its water needs by the oxidation of food-stuff, as opposed to drinking. The desert rat deals with heat by avoiding it: the animal remains in its burrow during the daylight hours, emerging only at night, when the air and soil are cool [see "The Desert Rat," by Knut and Bodil Schmidt-Nielsen; SCIENTIFIC AMERICAN Offprint 1050].

Like the desert rat, most small, burrowing desert rodents are nocturnal. But both the mohave ground squirrel and the antelope ground squirrel are diurnal.

They emerge from their burrows near sunrise and forage outdoors throughout the day. They do so even in summer, when the air temperature may reach 110 degrees Fahrenheit or higher, and when the surface temperature of the soil may rise above 150 degrees F. Since the desert is as arid as it is hot, they must sustain their exposure to heat with a minimum loss of water for evaporative cooling.

In appearance and temperament the antelope ground squirrel resembles the

RETREAT BURROW OF ANTELOPE GROUND SQUIRREL is shown at right in this drawing. It is usually dug in soft soil close

to desert vegetation and is about one foot deep and 12 to 15 feet long. The animal seems to use this burrow to unload body heat

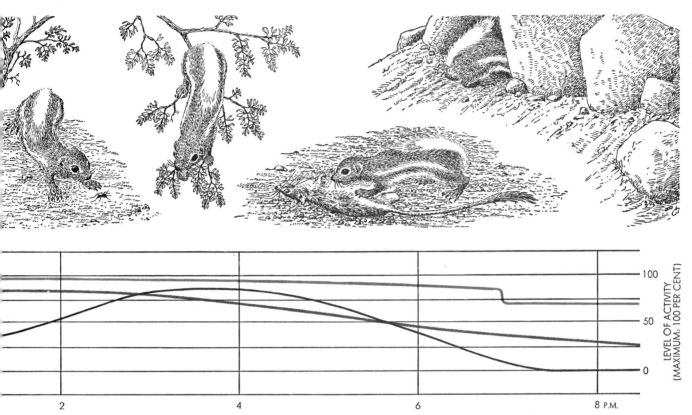

LEVEL OF ACTIVITY
(MAXIMUM: 100 PER CENT)

100

50

0

2   4   6   8 P.M.

off. In early afternoon it stays in the shade. Before retiring to its home burrow it returns to feeding area, and at any time may catch insects or feed on dead animals. On the graph, dip in body tem- **perature is shown only at noon. But dips occur often, whenever animal goes underground to unload heat. At all other times its temperature is a few degrees above the environmental temperature.**

chipmunk. Its body is about six inches long; its weight is about 90 grams. It has two white stripes down its grayish-brown back. It carries its tail high, exposing a white rump; this suggests the appearance of the pronghorn antelope, for which it is named. An extraordinarily active and high-strung animal, the antelope ground squirrel is constantly in motion, dashing from place to place, often traveling hundreds of feet from its home burrow. That it can maintain such hyperactivity even in soaring tempera-

tures is in itself evidence of unusual adaptive mechanisms.

For every animal the ability to adapt to external temperatures depends on two internal factors: the range of body temperatures in which it can function effectively and the rate at which it can produce body heat. Below a lower critical environmental temperature the body loses so much heat that internal temperature can be maintained only if the animal can step up its production of body heat sufficiently. Above an upper critical

environmental temperature the body retains so much metabolic heat that internal temperature can be held within the required range only if the animal can get rid of heat, in most cases by evaporative cooling; that is, by sweating or panting at a sufficient rate. Between the upper and lower critical temperatures—in the thermal neutral zone—an animal can maintain its optimum body temperature without having either to increase its metabolic rate or to lose body water. Such stratagems as contracting or dilat-

and to store food, but not as a living place; the dens in the burrow contain neither nests nor fecal matter. The antelope ground squir- **rel's living burrows have not been excavated. They are probably dug under rocky buttes, like that seen at left side of drawing.**

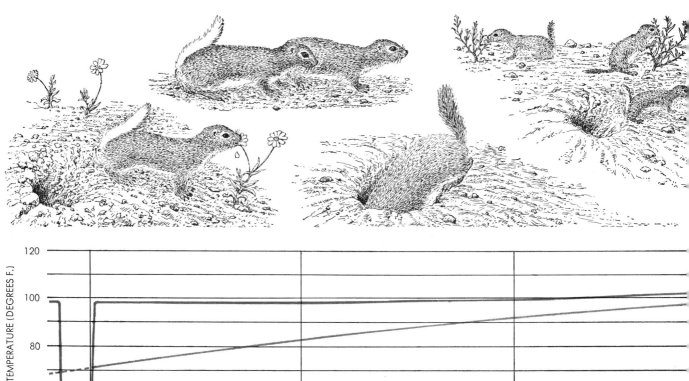

ACTIVITY OF MOHAVE GROUND SQUIRREL for six months is shown here. Animal emerges from burrow in March. In April young are born. From May through July it fattens on desert vegetation and in August returns underground for seven months. Broken line

ing cutaneous blood vessels and depressing or raising hair or feathers allow the animal to function at a minimum cost of energy for temperature maintenance.

In terms of this analysis of the adjustment of body temperature to environmental temperature, the adaptation of the antelope ground squirrel is admirable. It has a broad thermal neutral zone and one that accommodates to high environmental temperatures. Between environmental temperatures of 90 and 107 degrees F. its metabolic rate remains virtually constant. No other nonsweating mammal has a thermal neutral zone extending so high [see illustration on page 280].

Unlike man, the antelope ground squirrel can tolerate a high body temperature; in other words, it can "run a fever" without debility. It can therefore permit its temperature to rise with the temperature of the environment. Like the camel, it can store heat, and it does not have to dispose of heat until its body temperature reaches an extreme point. The antelope ground squirrel shows no serious discomfort even when its body temperature goes above 110 degrees. Throughout the thermal neutral zone it runs a temperature a few degrees above that of the environment [see top illustration on pages 276 and 277]. Instead of expending energy to cool itself and thereby adding to its heat load—as man must do—this animal actually disposes of a portion of its metabolic heat to the lower-temperature environment by

LIVING BURROW OF MOHAVE GROUND SQUIRREL is seen here at various stages and times of year. Burrow is dug in soft sand near the desert plants the animal eats. It is about 18 feet long and three feet deep. First panel shows burrow in early spring,

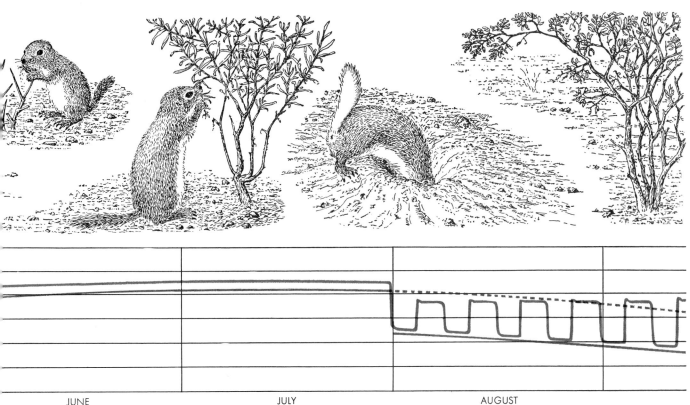

on graph is desert air temperature when animal is in burrow. Gray line under it is burrow temperature then. Rise and fall of animal's temperature in August corresponds to its periods of wakefulness and torpor. Body temperature is always higher than environmental temperature, although when animal is active its temperature fluctuates sharply. A mean body temperature is shown here.

conduction, convection and radiation.

But environmental temperatures in the desert are commonly far higher than tissues can tolerate, and small animals heat up rapidly. The antelope ground squirrel must therefore unload some of its accumulated body heat at intervals during the day. It does this either by flattening itself against the soil in a shaded area or by retreating underground to its burrow. When its body temperature gets dangerously high, it has only to return to the relative coolness of its burrow and remain quiet for a few minutes until its fever has subsided. In our laboratory at the University of California at Los Angeles, antelope ground squirrels have lowered their body temperature from above 107 degrees to about 100 degrees within three minutes after being transferred from an environmental temperature of 104 degrees to one of 77 degrees.

The antelope ground squirrel contends with heat in still another way. Under protracted heat stress it will begin to drool. The animal then systematically spreads the saliva over its cheeks and head with its forepaws as though it were grooming itself. On very hot days, when it has had to tolerate air temperatures of 104 degrees or more for several hours, the antelope ground squirrel may be soaking wet around the head.

Drooling, with its high cost in water losses, is a last resort. But even when the temperature is not extreme, the antelope ground squirrel loses a considerable amount of body water. At 100

when animal emerges. Second shows animal digging new burrow. Third shows it closing burrow in August before retiring underground. In last panel it is winter and animal is torpid. Periods of torpor probably last longer in winter months than in summer.

degrees, long before it has begun to drool, this hyperactive animal gives up water equal to 10 per cent of its body weight in respiration and evaporation through its skin in the course of a day. This is 15 per cent of its total body water. Fortunately the animal withstands dehydration well. Although it gives up three times more water every day than it can extract from its food by oxidation, it can survive from three to five weeks on a completely dry diet. If the antelope ground squirrel is to maintain itself in a healthy state, however, it must find sources of preformed water. It is therefore hardly surprising that the animal is omnivorous, eating insects as well as desert vegetation. When it is seen on the highways, as it often is, it is probably feeding on the corpse of some animal, perhaps another of its species that has been hit by a car.

The antelope ground squirrel is able to stretch its scanty water supply because, like the desert rat, it loses a minimal amount of water in the excretion of nitrogenous wastes. On a dry diet this animal can produce urine with a mean concentration of 3,700 milliosmols (the

maximum concentration of human urine is about 1,300 milliosmols). The urine of the desert rat is somewhat more concentrated. But the antelope ground squirrel's urine is still 10 times more concentrated than its body fluids. Its ability to turn salty water to physiological use is even more impressive. The desert rat can maintain itself on sea water; the antelope ground squirrel can drink water approximately 1.4 times saltier than sea water and still remain in good health. No other mammal can process water of such high salinity. This capacity is important in the desert, where the little surface water that is available is usually highly mineralized.

The structure of the animal's kidney explains its efficient use of water. As in several other desert mammals, the renal papilla—that part of the kidney which contains the ascending and descending kidney tubules—of the antelope ground squirrel is extremely large, extending as far down as the ureter [see illustration on page 281]. In the formation of urine the kidney first extracts a filtrate containing all the constituents of blood except proteins and blood cells. This filtrate

is then converted to urine by the selective reabsorption of water and essential solutes in the kidney tubules. The longer the tubules, the greater the amount of water they can absorb and the greater the amount the body retains. As the antelope ground squirrel's tolerance for high body temperatures constitutes its major physiological adaptation to heat, so the efficiency of its kidney embodies its major adaptation to aridity.

Considering the success with which the antelope ground squirrel occupies its narrow desert niche, how does the mohave ground squirrel manage to find a place beside it? The question cannot be fully answered, because less is known about the life history of the mohave ground squirrel. This in itself is significant, because it appears that the mohave ground squirrel manages to persist largely by staying out of the way of the antelope ground squirrel.

Of the two animals the mohave ground squirrel is the bigger and fatter, and it has the temperament that goes with its more generous proportions. Its body length is about six and a half

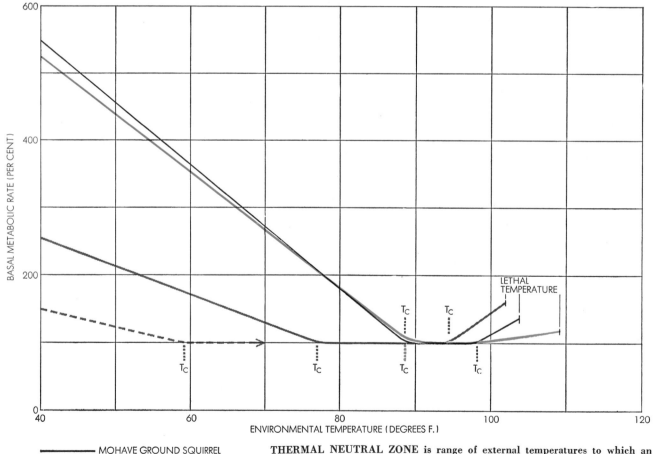

THERMAL NEUTRAL ZONE is range of external temperatures to which an animal is best adjusted. In it body temperature and metabolism can be held at optimum levels. At critical temperatures (T_C) metabolic rate begins to rise. Lemming is an arctic rodent. Kangaroo rat and ground squirrels are desert dwellers.

inches; its weight in its natural habitat, about 150 grams. In temperament it is placid, docile and sedentary. This little brown animal seldom wanders far from its home burrow, which it digs in loose sand, generally in the shade of the desert plants that provide it with its food. For more than half the year, from August to March, it remains in its burrow. During the spring months, when the desert vegetation is at its annual peak, the mohave ground squirrel emerges to reproduce and to fatten itself in preparation for its return underground.

The major proportion of the mohave ground squirrel's life is therefore normally concealed from observation. Fortunately it seems to show a comparable pattern of behavior in the laboratory. Here, as in its natural habitat, the animal is active throughout the day from March to August. During the remainder of the year, however, even at room temperature, and in spite of the continuous availability of food and water, it is intermittently torpid for periods lasting from several hours to several days. If food and water are at hand, it will eat and drink in its periods of wakefulness. If they are not, it does not seem to be disturbed. We do not know whether or not in its natural habitat it stores food in its burrow. We do know that it is usually thin in early spring, when it emerges from its burrow, and that it can add as much as 100 grams to its body weight in the period before its retirement underground. We also know that in the laboratory, where the animal becomes exceedingly fat, it loses an appreciable amount of weight during its period of dormancy only if no food and water have been made available to it.

Since this pattern of intermittent dormancy extends from late summer to early spring, it involves what would normally be considered two separate processes—hibernation and the summer dormancy called estivation. Our studies indicate, however, that in the mohave ground squirrel the two processes are merely aspects of the same physiological phenomenon. From early August to the end of February, whether the temperature in the laboratory is one that would normally be associated with estivation or whether it is one at which hibernation would be expected to occur, the same events take place. As the animal becomes torpid, its oxygen consumption and its body temperature drop sharply. Then both level off, and the body temperature stabilizes at the environmental temperature or very slightly above it. During the time the animal is dormant its torpor is more pronounced than

deep sleep, its breathing is suspended for long periods and its heart rate is profoundly reduced. On arousal it restores its body temperature to normal through increases in breathing movements, acceleration of heartbeat, shivering (which releases heat) and increased oxygen consumption. In the laboratory arousal may come about spontaneously or it may be induced by a touch or a sound. In either event it is extremely rapid. Although the animal can take as long as six hours to enter torpor, it can wake in less than one hour. Oxygen consumption can reach its peak in 15 to 20 minutes, and body temperature can rise from 68 to 86 degrees in 20 to 35 minutes.

Such rapid alterations in temperature do not occur during the five-month period in which the mohave ground squirrel is active. Even then, however, its body temperature is remarkably variable and fluctuates over a broad range. We have measured a deep-body temper-

ature as low as 88 degrees in individual animals engaged in normal activity, and yet the animal does not seem to suffer any ill effects from body temperatures as high as 107 degrees. Its thermal neutral zone does not, however, extend as high as that of the antelope ground squirrel. The metabolism of the mohave ground squirrel begins to rise at an environmental temperature of about 98 degrees. Its tolerance for high body temperatures is of major adaptive value in June and July, when the desert is particularly hot.

Obviously the mohave ground squirrel's dormancy serves the function traditionally associated with hibernation: it conserves energy. At an environmental temperature of 68 degrees the oxygen consumption of a dormant mohave ground squirrel is only about a tenth that of the same animal active at the same temperature. Fat is undoubtedly the major energy source. Since the oxidation

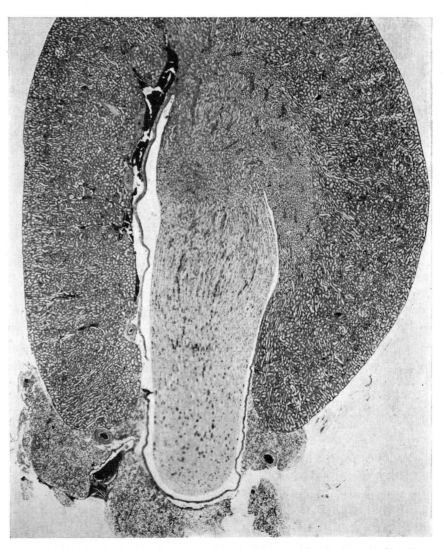

SECTION OF KIDNEY of antelope ground squirrel is magnified 20 times in this micrograph by Hudson. Mass outlined in white is papilla, containing renal tubules. In this species it is very large. This permits reabsorption of much water and production of concentrated urine.

of one gram of fat requires two liters of oxygen, a simple calculation shows that a torpid squirrel, weighing 300 grams and consuming oxygen at a rate of .08 cubic centimeter per gram per hour, will burn .29 gram of fat a day. Some 50 grams of fat would therefore supply it with all its energy requirements for 172 days; this is just half the fat supply it usually accumulates in the active months of the year. According to this calculation a mohave ground squirrel should be able to remain torpid for a whole year if it did not arouse at all. We do not know, of course, how much its energy requirements are increased by periods of arousal during the months of dormancy. Not being able to observe the animals in their burrows, we do not know how long these periods last. But from laboratory evidence we deduce that the cycles of torpor and wakefulness are repeated every week, with three to five days spent in torpor. The mohave ground squirrel should therefore be able to get along on its accumulated fat.

Energy conservation is not the only function that dormancy serves. Like the antelope ground squirrel, the mohave ground squirrel loses considerable body water in evaporative cooling and, like the antelope ground squirrel, it must have preformed water in its food. The sedentary mohave ground squirrel does not go in search of water; it gets its water almost entirely from the desert plants it eats. During the period the animal is underground this vegetation is in decline and the desert is at its driest. Thus dormancy is an important adaptation to seasonal aridity.

The mohave ground squirrel's seven months of estivation and hibernation serves still another function: it minimizes competition with the more active and abundant antelope ground squirrel. During the time the mohave ground squirrel is aboveground the food and water available are probably adequate to sustain both animals. But during its months of dormancy both food and water are in short supply.

The two animals have adapted well to desert life—the antelope ground squirrel by its tolerance for high body temperatures and the efficiency of its kidney, and the mohave ground squirrel by its avoidance underground of the most rigorous months of the year. One may conclude that these adaptations are sufficiently different to permit the two animals to live in sympatry in spite of Darwin's stern injunction.

VI

ORIENTATION
AND NAVIGATION

. . . It is a vast distance, if one calculates it, over which (the cranes) come from the eastern sea. They agree together when to start, and they fly high so as to see their route in front of them; they choose a leader to follow, and have some of their number stationed in turns at the end of the line to shout orders and keep the flock together with their cries. . . . Geese and swans also migrate on a similar principle. . . . They travel in a pointed formation like fast galleys, so cleaving the air more easily than if they drove at it with a straight front; while in the rear the flight stretches out in a gradually widening wedge, and presents a broad surface to the drive of a following breeze.

Pliny
NATURAL HISTORY, X, XXX, XXXII.

VI

ORIENTATION
AND NAVIGATION

INTRODUCTION

In this section we will study the ways in which vertebrates orient to their environment. Homing and electric location by fishes, infrared sensing by snakes, sonar navigation by bats, and sun navigation by birds are the topics of the articles that are included. They involve the senses of smell, electric field and heat measurements, hearing, and sight. In this introduction we will discuss these senses in general and how they are used by some vertebrates other than those described in the accompanying articles.

A convenient means of distinguishing "tasting" from "smelling" in vertebrates and in other animals is as follows: taste is the response to quite high concentrations of chemicals present in the mouth, with the result being acceptance or rejection of material as food; smell is the response to low concentrations of chemicals in the nasal passage, with the result being locomotion or complicated behavioral responses. Smell can be thought of as a long-distance orienting sense; and taste, a short-distance feeding sense. There is some overlap: for example, in human beings odor can help to assign a characteristic "taste" to food; as a consequence, during illnesses that reduce smell sensitivity, food tastes "flat" and abnormal. The reason is that smell complements taste in man. Nevertheless, the basic distinction between the two is generally obvious in terrestrial vertebrates, but less so in aquatic vertebrates. The hypothesis is supported strongly by the fact that sensory cells for the two processes differ morphologically in all vertebrates.

Smell is initiated by excitation of bipolar nerve cells, of which one end is exposed to the environment and the other is connected to the dendrites of an associational neuron that transmits nerve impulses within the central nervous system. The outer receptor process of the nasal bipolar cell extends between epithelial cells of the nasal organ so that its outer end is exposed directly to the nasal cavity. Only a layer of liquid or mucus covers the outer plasma membrane of the cell, where foreign odoriferous substances initially make contact.

The taste sensation is initiated on modified columnar epithelial cells. The outer ends of many of the cells are hair-like processes; the inner ends contact dendrites of nerve cells. In land vertebrates, these taste cells are found only in the mouth, whereas, in fishes, the same kinds of taste receptors may also be scattered over portions of the outer surface. Obviously this fact complicates our semantic separation of taste from smell in fish, for in a sense these "taste" receptors are sampling fluid-borne chemicals outside of the mouth, much as smell receptors function in land vertebrates.

To be complete we must mention the most primitive class of chemoreceptors of all: they are nerve endings that are unspecialized morphologically, but that respond to specific chemical conditions by changing the pattern of their nerve impulses. Such endings in the aortic bodies of vertebrates are stimulated by alterations in pH or in carbon dioxide and so influence the control of breathing. It seems reasonable to guess that smell receptors, at least, are specialized hypersensitive forms of these basic chemoreceptors.

All substances that can be smelled are volatile at normal environmental temperatures, and most are of reasonably large molecular size. Various

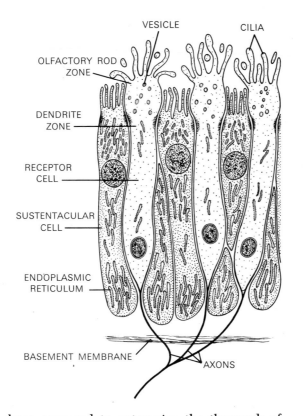

VESICLE   CILIA

OLFACTORY ROD
ZONE

DENDRITE
ZONE

RECEPTOR
CELL

SUSTENTACULAR
CELL

ENDOPLASMIC
RETICULUM

BASEMENT MEMBRANE
AXONS

Figure 1. A typical portion of the nasal epithelium. The sensory cells extend from the outer surface—where their ends are exposed to foreign molecules— through the epithelium, and then continue as nerve axons toward the olfactory bulb of the brain. Note that there is no discrete localization of these sensory cells, as there is of taste cells in "buds." The sustentacular cells are thought to "insulate" receptor cells, that is, to isolate them from each other. [From DeLorenzo, A. J. D.; in Y. Zotterman, *Wenner-Gren Center International Symposium Series*, Vol. 1, Macmillan, 1963.]

"classes" of odors have been proposed to categorize the thousands of chemicals that can be detected by olfaction. It seems clear that certain molecular arrangements evoke the sensation of a "smell"; considerable variation in molecular detail can occur, but as long as a specific grouping is present a given "name" of the smell will usually be assigned. As was pointed out by A. J. Haagen-Smit ("Smell and Taste," *Scientific American*, March, 1952; Offprint 404), however, as small a difference as the replacement of hydrogen atom by a methyl group changes the odor from "mint" to "camphor." It seems likely that there are different kinds of receptor molecules on the outer surface of odor-sensing cells, and that the steric fit between air-borne agent and receptor determines whether association between the molecules will occur. We do not know how the combination of receptor with foreign molecule is transduced into altered cell properties and so to depolarization and impulse initiation. Nor is it clear what happens to the molecular complex so that the ability to continue to respond to a particular substance is possible.

Other phenomena of odor detection are equally inexplicable: we do not understand, for example, why certain substances, in small quantity, will dominate large quantities of others, so that only the small quantities are "smelled;" also confusing is the fact that some substances smell differently when present in large or small amount. Whether phenomena such as these occur in the receptors or in the subsequent neural pathways has not been established.

Neurophysiologists have recorded electrical activity from single smell receptors and have established that different cells are sensitive to different classes of molecules. No unique pattern of response has been observed, however: the rate of impulse initiation in a given cell may go up or down, or it can be turned on or off in response to an agent. What seems to be important in evoking the sensation of a particular smell, is the distribution of many cells responding in different degrees to a foreign molecule: in a sense, then, the organization of the sensory epithelium is a critical feature of the system, because that organization determines which of the thousands of olfactory nerve axons will initiate a new pattern of impulses and so send information to the brain. Other studies show that the terminations of nerve tracts leading from the olfactory epithelium to the olfactory bulb of the brain are distributed nonrandomly. On the olfactory bulb is a topographical representation of the nasal epithelium: that is, the spatial relationship between points on the epithelium is reflected on the bulb. Interestingly, cells that detect water-soluble chemicals tend to be located at one end of the bulb, and those that sense oil-soluble molecules are localized at the other end.

Most substances that can be tasted are soluble in water. They must be present at about 3,000 times the minimal concentration required for smell to operate. Taste is simpler than smell in that four discrete groups—salt, sour, sweet, and bitter—are known. Variation in sensation occurs within some of the groups. In the absence of smell, for instance, all acids (H^+ ions) taste alike—sour; but many substances that taste "sweet" can be distinguished from each other, as indicated by conditioning experiments or verbal description by human subjects.

It has recently be found for the first time that two purified proteins from plants taste sweet, in fact with an intensity about 10^5 times that of sucrose. These substances, as well as "miraculin," a glycoprotein that makes acids taste sweet instead of sour, all may act on the membrane of receptor cells to elicit the series of events (membrane depolarization, etc.) that culminates in the sensation of taste. The resultant implication—that molecules do not have to enter cells to elicit "taste"—agrees with data showing that the hydrogen ions bind to the phosphate groups of membrane lipids on gustatory cells and thereby cause alterations leading to the sour taste. It is possible that a quite distinctive mechanism is involved for bitterness. When quinine, nicotine, or other "bitter" substances are applied to the tongue, the enzyme phosphodiesterase is activated to hydrolyze cyclic AMP. The decrease leads by unknown means to the bitter sensation. Here, the important physiological event is destruction of the cyclic nucleotide; in many situations (particularly during hormone action), it is *production* of cyclic AMP that triggers the physiological response. If it can be proved that bitter substances act at the cell membrane to activate phosphodiesterase, then it will be clear that enzymes catalyzing both synthesis and destruction of the important "secondary messenger" are subject to control by extracellular molecules. In the case of taste, these results are consistent with those cited above for certain proteins and hydrogen ions: the cell surface assumes a central role in the regulatory network of vertebrate organisms.

Like the olfactory cells, individual taste receptor cells respond to specific sets of molecules. *One cell*, for instance, may cause nerve impulse discharge when sodium chloride or potassium chloride ("salt"), or hydrochloric acid ("sour") are present, but not when sucrose ("sweet") or quinine ("bitter") are applied; another will respond only to sour and bitter; others, to different sets. Thus, there is no correspondence between these classes of receptor cells and the four basic types of tastes. Like smell, the ultimate taste sensation recorded in higher centers of the brain may be dependent on the pattern of impulses from many taste-sensitive cells. Again, the spatial distribution of the receptor cells that respond to various combinations of tastable substances and their balance of activity (whether impulse initiation speeds or slows, etc.) seem to be the important features of the ability to taste. This would suggest that the gustatory sensory apparatus may be a simpler version of the olfactory one, a not unexpected conclusion if the terrestrial olfactory system evolved from the taste receptors of the first vertebrate fish.

Chemoreception has many functions in the lives of fishes. In parasitic lampreys, it leads the fish to a host. Even young lampreys fresh from metamorphosis respond to the odor of a potential host fish by oriented swimming behavior. Since these fish had not been exposed to the odor of any other fish during larval life, this is a good example of a behavioral pattern that is built into the nervous system during development and maturation. It is innate behavior, which does not depend upon learning or conditioning. The substances that cause the reaction have been extracted from trout: they are amines. The same lamprey species (*Petromyzon marinus*) that displays this orienting behavior uses a pulsating electric field to locate prey within 15 to 20 centimeters; the chemical sensing is used for long-distance orientation.

Other examples of odor-elicited behavior in fishes are the fright reaction and homing in order to reproduce. Particularly in a number of small schooling fishes, substances that cause an immediate fear reaction of the school are present in the skin cells. When the skin of one fish is injured, the chemical is released into the water, causing rapid escape swimming and dispersion of the clustered fishes. This substance is a warning device that helps others of the same species avoid imminent danger of predation.

Arthur Hasler and James Larsen review the homing of salmon in their article, "The Homing Salmon." A variety of experiments conducted since the article was published confirm and elaborate on the hypotheses presented by these authors. Although it has not been proved, it seems reasonable to suppose that a salmon far out in the Pacific finds its way back to the general area of its river system by use of celestial navigation. This process, which is the orientation in relation to the sun or stars, is described in greater detail toward the end of this introduction. A number of experiments have established that fishes, like birds, show oriented swimming under clear skies but not under cloudy ones. Once the fishes are near the home river system, the senses of smell (and perhaps taste) can be assumed to lead the fish back to the stream where it grew up.

Consistent with this hypothesis is the finding that "home" water perfused into the nasal epithelium of a salmon elicits a burst of nerve impulse firing in the olfactory bulb. Water from nearby regions draining similar geological areas fails to alter the basic electroencephalogram firing pattern. Interestingly, the rate of the electrical activity in the olfactory bulb is high in young fish—at a time when they may be in the process of learning to recognize home water—but it is lower during adult life when the eyes are said to be the dominant sense. Finally, electrical activity increases again at the time of homing when smell apparently leads the fish to its spawning site.

How is detection of an odor translated in directional swimming? Some experiments suggest that, if the head is swung back and forth in the water (a normal occurrence during swimming of fish), it may be possible for first one nasal sac, and then the other, to sample for odorous molecules. Comparison of intensity in the central nervous system could then keep the fish "homed in" toward the spawning-site source of the odors. Moreover, branches of the trigeminal nerve have endings in the olfactory epithelium and connect centrally in such a way that the separate processing of information may be possible.

Young salmon may be trained in the laboratory with a given type of water, and they will return to it from the sea years later. Apparently they are imprinted with certain odorous substances of the home stream, so that these substances alone lead to a successful homing reaction. It is not known yet whether a single organic molecular type functions to mark the home territory, or whether groups of molecules combine to give a characteristic smell to each portion of each breeding stream.

Although fishes can employ smell for long-distance orientation, they can use electric locationing only for short distances. As pointed out by H. W. Lissmann in "Electric Location by Fishes," the sense is used under conditions of low visibility, due either to darkness or to abundant suspended material in the water. The two families of fresh-water fishes in which the ability is found are quite distinct morphologically and only distantly related. Nevertheless, they have arrived at the same solution to problems of orientation in a difficult environment. In fact, we can assume that the only way such fishes can survive today is to be active at night, when nonelectric bony fishes are at rest in the water. This is because the mode of swimming of electric fishes—with a stiff body axis and propulsion by the fins—does not allow the quick movement needed for evasion of predators that is possible in typical sine-wave swimming.

Organs used to detect electric fields are probably modifications of the ordinary lateral line organs found on all vertebrate fishes. Each lateral line organ includes a structure that can be displaced by water currents; as a result of this displacement, sensory hairs are moved, and altered patterns of nerve impulses ensue. Thus, disturbances in the water pattern caused by objects near the fishes are detectable. In acting this way, the lateral line serves as an extension of the sense of touch.

Variations of the lateral line organ are found in sharks and their relatives and in electric fishes. In the electric fishes, the sensory epithelium is beneath the surface of the skin. In some electric receptors the canal leading from the sensory area to the surface is filled with jelly; this jelly has remarkable electrical conductivity (85 to 90 percent of that of sea water, compared to 45 percent in body fluids). In another type of electroreceptor, thin covering cells lie on top of the sensory cells. Neither these nor the cells under the jelly show any hint of ciliary derivation, a common feature of the lateral line neuromasts, as described above. The electric sensing cells are always found in close proximity to nerve endings (presumed to be sensory in nature). Even chemical coupling, like that in the main vertebrate nervous system, seems to exist between some of these sensory cells and nerve endings; for example, vesicles thought to contain a transmitter substance are found in the sensory cell cytoplasm, in the intercellular space, and in the nerve ending cytoplasm.

Electric organs can detect weak potential changes in the water—such as those caused by another fish or even by a bare metal wire—or they may be used to measure alterations in the electric field generated by the same organism. The remarkable feature of the electric organs is their great sensitivity; responses to currents as small as 2×10^{-5} microamperes per square centimeter have been observed. It seems clear that, in the course of the evolution of these various fishes, the basic mechanoreceptor properties of the lateral line have been modified to permit detection of electrical signals in the environment. The absence of any hint of cilia on the electric sensing cells emphasizes the change in sensory modality that has occurred in this evolutionary transition.

Certain snakes are able to detect still another portion of the electromagnetic spectrum in order to hunt prey or sense warm environments. As outlined by Igor Gamow and John Harris in "The Infrared Receptors of Snakes," these animals have evolved an energy-transducing system for heat which is more analogous to a green plant absorbing light rays in chlorophyll than it is to a vertebrate eye where light is used for "information" purposes rather than as an energy source. Unlike the plant however where the energy is employed in synthetic processes, the snakes use the energy-dependent nerve firing as a source of information about the environment.

Another mechanism is used by some marine vertebrates for orientation in the water as well as by terrestrial vertebrates for orientation in the air: it is echolocation. This capacity to use sound pulses and their echoes has evolved several times in the vertebrates. The oil birds of the caves in Venezuela, the dolphins and their relatives, the bats, and some insectivores employ the mechanism. In each group the process is employed under conditions of low visibility, and, in each, rapid movement is made so that appropriate analysis of the environment by means of smell or taste is impossible. Both the bats and cetaceans (dolphins, whales, and their relatives) are reasonably close relatives of the insectivores, the ancestral group of modern mammals. Bats, of course, have retained the nocturnal habits of insectivores and many of them have adopted the echolocation method of orientation.

As described by Donald R. Griffin in "More about Bat 'Radar,'" the process depends upon the emission of pulses of high frequency sound from the mouth or the nose, and the detection of objects by distinguishing sound echoes with a hypersensitive auditory system. Except for a few kinds of bats, the length of time during which an individual pulse is emitted is quite short. Therefore, the pulse of sound energy traveling through the air is short also. This distinguishes such bat sounds from most sounds that we produce when speaking—common words are trains of sound waves in the air that may be hundreds of feet in length. Since the bat may have reduced sensitivity of hearing while it is emitting each high-energy pulse from the throat, it is obviously advantageous to have short sound pulses. Short pulses of sound are particularly important for hunting insects: first, when the bat is searching for insects, it is listening for faint echoes that may come from long distances; second, when the bat is homing-in rapidly, immediately before seizing the prey, echoes return to the ear after very short periods of time. In both situations, the auditory system might not function optimally if echoes were returning from objects while a long sound pulse was still being emitted.

Sound pulses are produced by specialized vibrating sheets of tissue in the larynx (the voice box). These sheets are only 8 microns thick, and they are stretched tautly across shallow depressions in the lateral walls of the larynx. Extraordinarily large cricothyroid muscles stretch the membranes tightly, and vary the tension during pulse emission in order to alter the frequency of emitted sound. It is still not known how the pulses are turned on and off with such rapidity, or how emission of sound is coordinated with muscle contractions in the ear. The actual intensity of each ultrasonic pulse generated by the larynx is amazingly high. Two to three inches in front of the bat's mouth the intensity may be six times the intensity of a subway train roaring through a station (60 versus 10 dynes per square centimeter). One might guess that these sounds are very loud to other organisms whose hearing abilities range from 20 to 80 kilohertz.

A variety of adaptations coordinate activity in the larynx with that in the ear. For instance, the large stapedius muscles in the ears contract and pull on the ear bones to reduce sensitivity of the hearing apparatus. As might be expected, the muscles normally contract just before the larynx begins to produce a pulse of sound; muscle relaxation starts during the pulse emission so the ear is returned to full sensitivity for at least the last part of the interval between sound pulses. But, as a flying insect is approached, the muscles in some bats remain contracted constantly; this apparently permits the higher pulse emission rate and prevents damage to the ear. How the faint echo is detected above the high frequency outgoing pulses when the ear is in a state of reduced sensitivity is a mystery.

A final interesting aspect of echolocation is the method of detecting directionality. Some species of bats emit narrow cones of sound directly in front of their head, whereas others send out a wide arc of sound. In part the structure of the nostrils determines which pattern is used and whether the sounds are amplified after emission from the larynx. The ears also function in directional sensing. Many bats have exceedingly large ears. The rhinolophids can move the ears back and forth as pulses are emitted from the nose. During flight, the ears move a distance of 8 millimeters about 50 times per second. If the movement is stopped by denervating the muscles that move the ears, the bats can no longer echolocate effectively.

Other bats may determine the direction of an object by the difference in intensity of an echo at each ear, or by the difference in the length of time it takes for an echo from one source to arrive at each ear. These processes are analogous to the proposed use by fish of first one and then the other nasal sac for direction orientation to odors.

To add even further to the incredulity that these echolocating adaptations evoke, think of the problem *of discrimination between echoes*. How can the echoes from a moth be distinguished from those rebounding from tree trunks, twigs, leaves, other bats, or even rain drops? With many other bats flying about, the noise level must be intense. One possibility is that this discrimination is based upon a special programming, or preconditioning, that occurs in the auditory system as each sound pulse is emitted. Thus, the neurons receiving information from the ear may be made especially sensitive to sound patterns with the same characteristics as those of the pulse that had just been emitted. In this way an ordered and predictable pattern could be distinguished from the din of echoes and pulses in the environment. Another possibility is that some bats use a Doppler effect to detect moving objects in their environment. Thus, as the bat flies through the air, echoes from a flying insect will be shifted upward or downward in frequency, unlike the echoes from fixed objects. The presence of special neurons in the auditory portion of the brain that fire only in response to FM (frequency modulated) echoes makes it reasonable to assume that during the course of evolution a large variety of nerve cells or circuits can appear in order to accomplish the special sorts of tasks associated with echolocation.

As an addendum to Griffin's article, it is now known how fish-eating bats detect their prey. Roderick A. Suthers, an associate of Griffin, found that the pulses of sound from these bats are reflected off minute waves in the surface of the water or from fins or tiny objects protruding through the surface film (even as small an object as a 1-millimeter length of wire, 0.21 millimeters in diameter, reflected sound). Fish that are swimming or hovering just beneath the surface are thought to set up irregularities in the water surface, and the swooping bat responds by dipping its feet into the water as it skims along in order to snatch up the unsuspecting fish. These bats have been seen occasionally to "trawl" with their hooked claws dragging in the water as they swoop along. How they avoid alligators or gars remains to be explained.

The use of sound by cetaceans is of particular interest, because some of these animals have the most complex brain of any mammal other than man. Dolphins emit two types of sound: high-frequency pulses that sound like "clicks" to us, and whistling noises reminiscent of canary "chirps." The sounds are generated by passing air from one nasal sac to the other through specially constructed valves. No air is lost from the body in the process. The clicks are echolocating pulses, and just as a bat increases the pulse repetition rate when a flying insect is detected, so a dolphin increases its rate of pulsing when a fish is thrown into the water in front of it. Here the rate may rise from five to several hundred pulses per second. The frequency of sound (20 to 100 kilohertz) emitted with each pulse is complex, and it seems likely that low frequencies (sometimes coupled with high intensity and long duration) are used for long-range general assessment of the environment, whereas brief, high-frequency ultrasonic pulses are of use in short-range discrimination.

The purpose of dolphin whistles is not as well defined. Each emission of sound lasts for longer periods (for as much as half a second, versus milliseconds for a click), and the frequency of sound rises from 7,000 to 15,000 hertz as the whistle continues. Winthrop N. Kellogg has proposed that such sounds may be used as a form of frequency-modulated echolocation, in contrast to the pulse-modulation of clicking. The constantly changing emission of the whistle would produce echoes of constantly changing frequency, so that any one echo would be of a frequency different from that of the sound being emitted. Thus the echoes could be heard even though sound was still being generated. Because their frequency is lower than that of clicks, whistles would attenuate less in water and so would be of greater use in long-range assessment of the environment. Whether they are actually employed for such purposes or for communication is yet unknown.

Directionality in echolocation has not yet been studied extensively, but two interesting mechanisms are known. Some dolphins can emit highly directional clicks, with most of the sound energy being directed in a narrow cone directly in front of the animal. The other form of behavior is seen in the bottlenose dolphin as it swims toward a small target in the water: the animal's head moves back and forth in the water once every 2 to 3 seconds (the head moves about 5 degrees on each side of the axis of swimming movement). These head oscillations are obviously a form of auditory scanning, with the ears being moved back and forth to establish the source of the echo. As the echoes reach one ear and then the other, phase differences, intensity differences, and time differences can be detected by moving the head in this manner. Thus, the bottlenose dolphin has arrived at the same basic solution to locating its target in space as have rhinolophid bats.

The ear of cetaceans is truly remarkable. Because the ear is necessarily streamlined into the head, there is no external ear. Studies by Bullock and Ridgway have shown that the lower jaw region of the porpoise body is the most sensitive to sound. By comparison, the external ear region itself is quite insensitive. It is likely that echoes strike the mandible and are transmitted by a fat-filled canal to the middle ear where the normal hearing process can go on.

In order for such an arrangement to be useful in directional hearing, it is essential that sound-induced vibrations do not reach the middle ear by other routes. The secret may lie in the fact that the middle and inner ears are encased in special massive bones called the *tympanopetromastoids*. In contrast to the structures in most terrestrial mammals, these bones are not attached directly to the skull or to other body skeletal parts. Consequently, they do not vibrate when the rest of the skeleton does. In addition, the bones are so massive that they probably do not vibrate in response to sound waves over 150 hertz. Finally, a gelatinous "foam" insulates the internal ear parts from the ear bones themselves; this air-filled foam probably reflects and absorbs most of the sound energy that succeeds in passing through the ear bones. The result of all of these structural adaptations is to insulate the ear effectively from vibrations in all directions but one—from each side of the lower jaw via the fat-lined canals. This means that sound

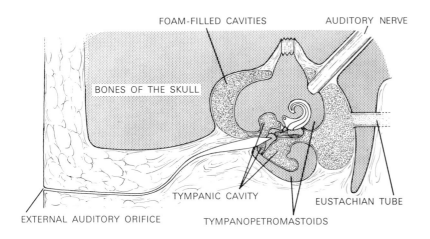

Figure 2. A diagrammatic section through the ear region of a dolphin. The internal ear parts are located in the tympanic cavity enclosed by the ear bones. "Foamy" material insulates the bones from the skull. [After deHaan, Reysenbach, *Acta Oto-Laryngologica*, Supplementum 134, 1957.]

is detected from a circumscribed region in front of and below the animal and that echolocation of objects in the water is feasible.

In cetaceans, the auditory nerve leading from the cochlea of the ear to the brain is very large, and the lateral lobes of the cerebral cortex are huge. This is a major reason why the large complex brain of a dolphin has a width greater than its length. In man, for whom sight is dominant, or in a dog, for whom smell is so important, expansion of the frontal or olfactory centers causes the brain to be lengthened in the anterior-posterior direction. The brain of the bottlenose dolphin is an example of the opposite extreme, for the olfactory lobes and tracts are missing altogether, and the brain is short and wide. The degree of specialization of different portions of the vertebrate brain is emphasized by Bullock, who points out that the inferior colliculus alone of a porpoise (the area where "click" echoes are processed) is larger than the *whole* brain of a bat and is almost as large itself as some small bats. Why so many nerve cells are present in this area of a porpoise brain is a mystery of evolution and physiology.

How else, aside from catching food, do cetaceans use their echolocation abilities? One example is observed in whales during annual migrations. As gray whales migrate from the Arctic Ocean to Baja California each year, they tend to travel fairly close to the shore. Many other whales also remain above continental shelves for portions of their long migrations. Typical echolocation signals have been recorded for most of them, and it is supposed that in regions of shallow water they use their sound system as a fathometer, or depth-gauging device. In regions of deep oceanic trenches, it is common to see whales surfacing repeatedly, coming high out of the water, almost to the level of their fins. The biologist Kenneth Norris has proposed that they use eyesight at such times to guide them across the deep water to areas where depth detection is once more possible. Since in the course of most whale migrations extended periods of deep-water cruising take place, it seems likely that celestial navigation or some other undefined orienting device may be used. In effect, we can assume that

Figure 3. The main migratory routes of humpback whales in the southern hemisphere. Note the prevalence of coastal routes for long portions of the journeys. Many whales spend winter in the tropics where they breed. They migrate to the Antarctic and its abundant food supply for the summer and then return to the tropics where the young are born (often in shallow coastal waters). The young are weaned the summer following the return journey to the Antarctic. [After Slijper, E. J., *Whales,* Basic Books, 1962.]

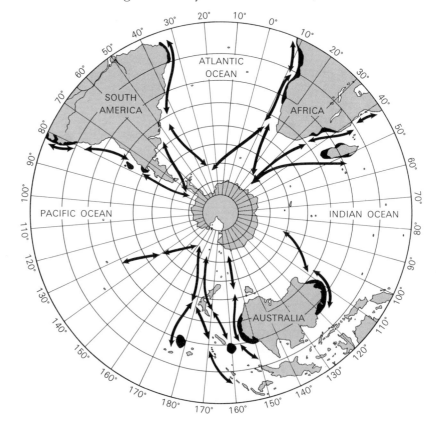

the sensory system best suited for particular circumstances will be employed, and that echolocation is utilized in shallow-water swimming.

Cetaceans also use sound for communication. A captured animal detained next to a ship emits characteristic whistles or click patterns; others of the school respond behaviorally and emit sounds in return. The clearest demonstration of such communication comes from an experiment in which two bottlenose dolphins were placed in separate tanks connected only by an acoustic link. Six types of whistles were exchanged between the two animals, as well as occasional trains of clicks. Evidence implies that certain of the whistles may be a call or recognition signal, whereas several of the others are much less stereotyped and so may have more variable information content. As might be expected from the complexity of the cetacean brain, communication behavior depends on age, sex, and emotional state of the animal. Innumerable historical accounts relate instances in which the observed behavior of the animals seems to have been dependent upon sound communication.

Roger Payne of Rockefeller University has succeeded in recording the "songs" of humpback whales in the vicinity of Bermuda. What is so very astonishing is the stereotyped, repetitious nature of the songs. An individual song may last from 7 to 30 minutes and include a fantastic array of frequencies, modulations, and intervals between notes and "themes." However, when the song is done, it starts over again and is repeated, sound by sound, theme by theme. Thus, despite the undeniable complexity of the sounds, and resultant capacity for great information content, the stereotyped character of the songs raises deep and completely unanswered questions about the significance of whistle-type communication in these marine mammals.

The possibility that whistles and low-frequency sounds are used for higher-order behavior is indirectly supported by studies on the electrical responses of various parts of the brain to different sorts of sound signals. In porpoises, it is the collicular areas of the midbrain that are specialized to fire when "clicks" are heard. Those neurons are relatively inactive when other sounds are aimed at the animal. When low-frequency whistles are used, a portion of the cerebral cortex is active but remains silent for clicks. Thus, the echolocation processing may occur at a lower level in the central nervous system than the cortex, where communication between individuals may be interpreted and controlled.

Another fascinating use of echolocation is found in blind human beings. It has been known for many years that the blind have "facial vision," that is, an ability to detect objects in their vicinity, presumably by sensing differences in air pressure on the face. In fact, this ability is dependent upon the sense of hearing, so that plugging the ears causes disorientation of the subject. When blind people were allowed to select noises that they found most effective at producing echoes (i.e., "awareness" of objects in the environment), all chose forms of "hissing" or tongue "clicking." The skill in detecting objects in the environment varies from person to person; a typical subject identified a 7.1-centimeter target, located 61 centimeters from his face, 100 percent of the time, and a 30.5-centimeter target, 274 centimeters away, 95 percent of the time. The frequency of correct identification diminishes with decreasing target size, no matter what the distance is.

The sounds emitted by these subjects varied in frequency from 170 hertz to 16 kilohertz and even the clicks were of relatively long duration (0.025 seconds). This meant that echoes were returning before the sound production and emission were completed. Clearly, this learned adaptation of atypical human beings does not equal the highly evolved system of a bat.

Do these human subjects have good directional sensing ability? The experiments showed that some people could emit clicks and localize targets with the use of only one ear. With both ears uncovered and sound directed from the mouth straight in front of the face, these subjects could detect targets as much as 90 degrees to the right or the left and point them out with good accuracy. All of these results attest to the adaptability of the human organism in employing substitute sensory mechanisms when normal primary orientation by vision is not possible.

The last type of vertebrate orientation that we will discuss—celestial navigation—is one of the most fascinating aspects of vertebrate biology. Birds, of course, provide the most widely quoted cases of long-range navigation, with flights of thousands of miles often occurring twice each year for many species. As pointed out by the late German behaviorist Gustav Kramer, true navigation requires that the organism have at least three pieces of information: (1) knowledge of its current position on the earth; (2) knowledge of the position (latitude and longitude) of the target of navigation; (3) knowledge of the direction from site one to site two. The basic question is whether birds or other animals are truly capable of such navigation, or whether compass direction alone is the parameter employed during oriented movement. Because it does seem likely that many animals use land marks extensively in initiating and terminating migration, compass direction might be the crucial information for them if navigation normally starts and ends only in known regions.

Sun navigation of birds has been investigated by many research workers since Kramer demonstrated "fluttering" orientation of starlings. Owing to altering day lengths, hormonal changes occur in a bird that bring it into a migratory condition. When caged, such birds become hyperactive and hop and flutter in the direction that they would fly if free to migrate. This oriented behavior occurs *only when the sun is visible,* and the direction of fluttering shifts a predictable number of degrees when the image of the sun is displaced a known amount by mirrors. This finding demonstrates that direction is determined in response to the sun; it does not prove true navigational ability, which demands, in addition, at least assessment of latitude. One of the best cases of measurement of the height of the sun (so that latitude can be determined) comes from fishes that give clear evidence of compensation in orientation when the sun's altitude in the sky is raised with mirrors or by other means.

If the sun or the stars are to be used as guides in navigation, the organism must possess an internal biological "clock." As the sun moves across the sky, the bird is able to maintain a straight migratory course over the surface of the earth only if it can compensate for movement of the sun; thus, with the aid of its clock, the bird continually alters the angle between the sun and the axis of migration. The presence of the clock may be demonstrated by two types of experiments: In the first, a bird is displaced a long distance east or west. Despite the displacement in longitude the bird's clock still runs on the original "local" time, and, as shown in Figure 5, the bird orients as it normally would at home. A bird makes the same error in the second experiment, in which its clock has been "reset" by a gradual altering of the day-night cycle with artificial lighting. Afterward, even when it is exposed to sunlight in its home environment, it acts according to the new time and orients in the wrong direction. As discussed in "The Navigation of Penguins" by John T. Emlen and Richard L. Penney, exposure to sunlight in the new longitude for several weeks will allow the clock to become gradually reset to the new local time.

All of these physiological and navigational abilities also operate at night, the time when most birds migrate. Directional fluttering, similar to that seen during the day, can be observed under a clear, starry sky but not on a cloudy night. Equivalent behavior was observed when birds were exposed to a normal autumn sky in a planetarium, and, interestingly, compensatory

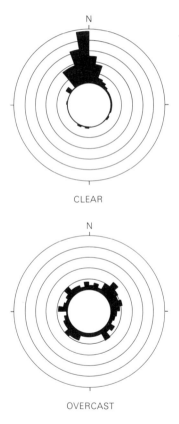

CLEAR

OVERCAST

Figure 4. Flight directions taken by mallard ducks that had been displaced prior to release. Under clear skies, most flew north, but if the skies were cloudy, flight orientation was incorrect. Equivalent data were obtained when mallards were released on clear or overcast nights.
[After Bellrose, F. C.; in R. M. Storm, ed., *Animal Orientation and Navigation: Oregon State University Biology Colloquium, 1966,* Oregon State University Press, 1967.]

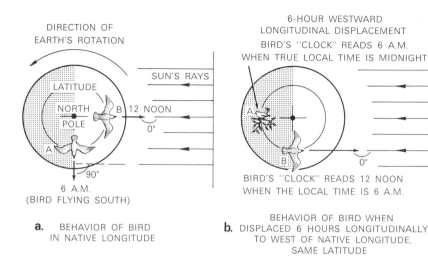

DIRECTION OF
EARTH'S ROTATION

LATITUDE

NORTH
POLE

B 12 NOON

SUN'S RAYS

0°

A

90°

6 A.M.
(BIRD FLYING SOUTH)

a. BEHAVIOR OF BIRD
IN NATIVE LONGITUDE

6-HOUR WESTWARD
LONGITUDINAL DISPLACEMENT
BIRD'S "CLOCK" READS 6 A.M.
WHEN TRUE LOCAL TIME IS MIDNIGHT

A

B

0°

BIRD'S "CLOCK" READS 12 NOON
WHEN THE LOCAL TIME IS 6 A.M.

b. BEHAVIOR OF BIRD WHEN
DISPLACED 6 HOURS LONGITUDINALLY
TO WEST OF NATIVE LONGITUDE,
SAME LATITUDE

Figure 5. Results of longitude displacement on bird orientation. Imagine that you are in space directly above the North Pole and the earth is rotating counterclockwise beneath you. In Figure *a* the bird flies directly away from the North Pole toward the south. If it begins its journey at 6:00 A.M., it displaces its angle of flight 90° with respect to the rays of the sun (position A). Six hours later, the earth has rotated on its axis, and now in order to fly "south," the bird has a 0° angle between flight path and sun's rays. In fact, all day long he compensates for the continually shifting sun and so maintains a constant course on the earth's surface.

In *b* the bird is displaced to the west. As shown in the illustration, when its clock reads 6:00 A.M. the bird is perched on a branch in darkness. During the ensuing 6 hours, the earth rotates and the bird's clock advances. Then the sun appears. It is 6:00 A.M. local time, but the bird's clock reads 12 noon. He orients just as he did at position B of Figure *a* at noon; that is, he flies directly toward the sun. Thus, instead of flying south on the earth's surface, he flies eastward.

shifts in orientation were made when the sky was "advanced" or "retarded" from 1 to 6 hours.

Experiments by Stephen T. Emlen at Cornell University have added significant new ideas to our knowledge of orientation to the stars. First, Emlen has shown that celestial rotation, "movement" of the stars around the pole star Polaris (in the Northern Hemisphere), can be used by hand-reared indigo buntings to define "north." If the celestial sphere is caused to rotate about a new "pole star" (*Betelgeuse* near the constellation Orion) by modifying the projection equipment in a planetarium, then that star becomes the "north" reference point for migrational orientation. This was the case even though Polaris, Ursa Major, and other normal constellations near the real celestial pole were plainly visible to the young birds.

The results imply very strongly that there is no preset, genetically inherited pattern of stars in these animals. Instead, learning of stellar configurations relative to the axis of celestial rotation seems to be a crucial aspect of this phase of behavior. One can predict that, once celestial north is learned from the rotational data, an adult bird should be able to locate north from the "geometry of star configurations alone"; in other words the placement of stars and constellations relative to north would allow ascertainment of that direction even if only patches of sky elsewhere were visible through broken clouds. This agrees with observations on adult birds in a planetarium with a fixed sky.

The concept of a "learned" north is reassuring with respect to an important property of the Earth's motion: the Earth precesses constantly, its axis making a complete circuit every 25,800 years. In about half that time, 13,000 years, Polaris will shift from 90° to 43° north, as Vega becomes the new "pole star." Think of the difficulties for migratory animals if a preset star map was inherited. Time spans of 13,000 years are minute in evolutionary terms; the genetic systems of such organisms would have to be capable of undergoing rapid change in order to compensate for the shifting pattern of stars relative to the pole. Emlen's findings provide a good alternative to such problems. Despite that reassurance, we are still in the dark about how a given bird chooses the correct direction for migration once the pole position is learned.

Another major area of interest with respect to navigation and homing is the fact that migrants are sometimes seen to orient correctly on overcast days or in ground-level fog. William T. Keeton of Cornell University has made the interesting discovery that pigeons use the sun for homing, if it can be seen, and show the typical clock effect on direction, if such an experimental regime is employed. Surprisingly, however, on fully overcast days, birds homed perfectly well! There must be an alternative to using the sun for orientation. When Keeton placed tiny magnets on the pigeons' heads and released them on heavily overcast days, orientation from the release site was random. Apparently, magnetism may play a role in orientation when the sun is not available. Keeton has established that the sun-based orientation is dependent on the organism's biological clock, whereas the magnetism-based type of orientation is not. This might be expected since the Earth's magnetic field does not vary on a 24-hour cycle.

Related observations have been made by several research groups in Germany; the general conclusion has been that migrational direction of robins can be influenced by a magnetic field. After the birds' cage is put in a magnetic field that can be controlled precisely in orientation, compensatory alterations in directional fluttering are observed. Apparently, the axis of the magnetic field (the way lines of magnet force run) can be sensed, but the polarity (which direction in the field is "north" or "south") cannot be detected. It seems possible that the angle of declination, or tilt downward of the lines of magnetic force, relative to the surface of the earth (gravity) can be used to indicate north in the magnetic field.

These recent findings provide a strong lesson to biologists about preconceptions of what is "possible" in the living world. Just a few years ago the idea that any sensory system could detect and measure the Earth's magnetic field was met with derision by many scientists. Now the question is: how is it done? Similarly, we need to know the relationship between genetic factors, learning, and the interplay of alternative forms of navigational control, if we are truly to understand navigation in animals.

Although many statistical problems make strict interpretation of some of these experiments difficult, the results, in general, concur with what is known of normal field behavior. In one of the most important sets of observations to date, A. C. Perdeck displaced a large number of marked adult and young migrating starlings from Holland several hundred kilometers southeast to Switzerland. The adults that were released after such displacement compensated, and they were recovered in their normal wintering area on the northwestern coast of France. They must have "navigated" to get there since they had to go in an abnormal *direction* to complete the trip. The young birds did not go northwest but instead flew to southwestern France and even to Spain and Portugal independently of the old birds, on a route parallel to the normal migratory route from Holland. Furthermore, some of the young birds kept flying farther than they would have on the normal route; they did not stop at a specific longitude. The next spring, both the young and the old birds returned to the original nesting grounds in Holland. Then, the following autumn, the younger birds again went their separate way back to the first year's wintering ground in the southwest of France and in Spain. Related observations indicate that "early" migrants that are captured and displaced to a region in Spain suitable for wintering, continue to migrate. But "late" migrants displaced to the same region soon search for nesting sites. Analogous results have been observed in other contexts, and it seems correct to conclude that the early migrant must migrate for a minimum time before it can enter a behavioral stage of site selection. These experiments demonstrate that untrained young birds have an innate migratory direction that they follow, and that they fly an approximate distance (or time). Obviously learning, too, occurs since the site of hatching and upbringing continued to be the home breeding ground, and the first-year wintering ground was equally permanent.

28

THE HOMING SALMON

ARTHUR D. HASLER AND JAMES A. LARSEN
August 1955

How do salmon find their way back to the waters of their birth? Recent experiments in the laboratory and in the field indicate that they do so by means of a remarkably refined sense of smell

A learned naturalist once remarked that among the many riddles of nature, not the least mysterious is the migration of fishes. The homing of salmon is a particularly dramatic example. The Chinook salmon of the U. S. Northwest is born in a small stream, migrates downriver to the Pacific Ocean as a young smolt and, after living in the sea for as long as five years, swims back unerringly to the stream of its birth to spawn. Its determination to return to its birthplace is legendary. No one who has seen a 100-pound Chinook salmon fling itself into the air again and again until it is exhausted in a vain effort to sur-mount a waterfall can fail to marvel at the strength of the instinct that draws the salmon upriver to the stream where it was born.

How do salmon remember their birthplace, and how do they find their way back, sometimes from 800 or 900 miles away? This enigma, which has fascinated naturalists for many years, is the subject of the research to be reported here. The question has an economic as well as a scientific interest, because new dams which stand in the salmon's way have cut heavily into salmon fishing along the Pacific Coast. Before long nearly every stream of any appreciable size in the West will be blocked by dams. It is true that the dams have fish lifts and ladders designed to help salmon to hurdle them. Unfortunately, and for reasons which are different for nearly every dam so far designed, salmon are lost in tremendous numbers.

There are six common species of salmon. One, called the Atlantic salmon, is of the same genus as the steelhead trout. These two fish go to sea and come back upstream to spawn year after year. The other five salmon species, all on the Pacific Coast, are the Chinook (also called the king salmon), the sockeye, the silver, the humpback and the chum. The

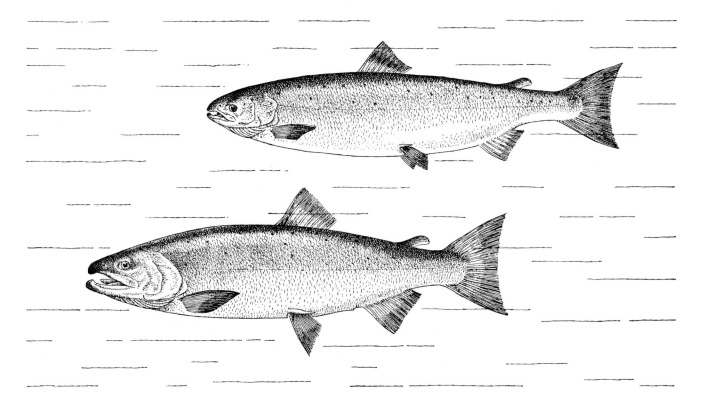

TWO COMMON SPECIES of salmon are (*top*) the Atlantic salmon (*Salmo salar*) and (*bottom*) the silver salmon (*Oncorhynchus kisutch*). The Atlantic salmon goes upstream to spawn year after year; the silver salmon, like other Pacific species, spawns only once.

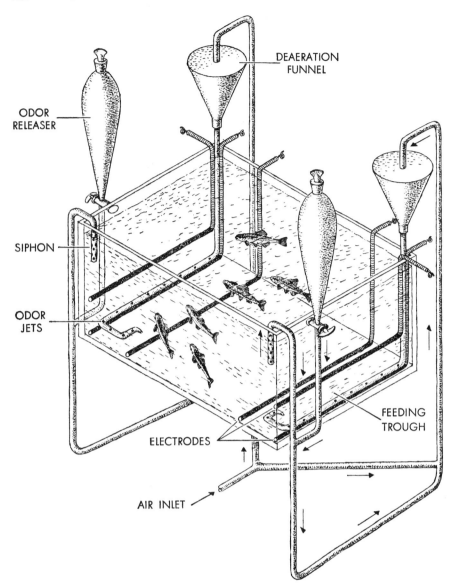

ODOR
RELEASER

DEAERATION
FUNNEL

SIPHON

ODOR
JETS

ELECTRODES

FEEDING
TROUGH

AIR INLET

EXPERIMENTAL TANK was built in the Wisconsin Lake Laboratory to train fish to discriminate between two odors. In this isometric drawing the vessel at the left above the tank contains water of one odor. The vessel at the right contains water of another odor. When the valve below one of the vessels was opened, the water in it was mixed with water siphoned out of the tank. The mixed water was then pumped into the tank by air. When the fish (minnows or salmon) moved toward one of the odors, they were rewarded with food. When they moved toward the other odor, they were punished with a mild electric shock from the electrodes mounted inside the tank. Each of the fish was blinded to make sure that it would not associate reward and punishment with the movements of the experimenters.

Pacific salmon home only once: after spawning they die.

A young salmon first sees the light of day when it hatches and wriggles up through the pebbles of the stream where the egg was laid and fertilized. For a few weeks the fingerling feeds on insects and small aquatic animals. Then it answers its first migratory call and swims downstream to the sea. It must survive many hazards to mature: an estimated 15 per cent of the young salmon are lost at every large dam, such as Bonneville, on the downstream trip; others die in polluted streams; many are swallowed up by bigger fish in the ocean. When, after several years in the sea, the salmon is ready to spawn, it responds to the second great migratory call. It finds the mouth of the river by which it entered the ocean and then swims steadily upstream, unerringly choosing the correct turn at each tributary fork, until it arrives at the stream where it was hatched. Generation after generation, families of salmon return to the same rivulet so consistently that populations in streams not far apart follow distinctly separate lines of evolution.

The homing behavior of the salmon has been convincingly documented by many studies since the turn of the century. One of the most elaborate was made by Andrew L. Pritchard, Wilbert A. Clemens and Russell E. Foerster in Canada. They marked 469,326 young sockeye salmon born in a tributary of the Fraser River, and they recovered nearly 11,000 of these in the same parent stream after the fishes' migration to the ocean and back. What is more, not one of the marked fish was ever found to have strayed to another stream. This remarkable demonstration of the salmon's precision in homing has presented an exciting challenge to investigators.

At the Wisconsin Lake Laboratory during the past decade we have been studying the sense of smell in fish, beginning with minnows and going on to salmon. Our findings suggest that the salmon identifies the stream of its birth by odor and literally smells its way home from the sea.

Fish have an extremely sensitive sense of smell. This has often been observed by students of fish behavior. Karl von Frisch showed that odors from the injured skin of a fish produce a fright reaction among its schoolmates. He once noticed that when a bird dropped an injured fish in the water, the school of fish from which it had been seized quickly dispersed and later avoided the area. It is well known that sharks and tuna are drawn to a vessel by the odor of bait in the water. Indeed, the time-honored custom of spitting on bait may be founded on something more than superstition; laboratory studies have proved that human saliva is quite stimulating to the taste buds of a bullhead. The sense of taste of course is closely allied to the sense of smell. The bullhead has taste buds all over the surface of its body; they are especially numerous on its whiskers. It will quickly grab for a piece of meat that touches any part of its skin. But it becomes insensitive to taste and will not respond in this way if a nerve serving the skin buds is cut.

The smelling organs of fish have evolved in a great variety of forms. In the bony fishes the nose pits have two separate openings. The fish takes water into the front opening as it swims or breathes (sometimes assisting the intake with cilia), and then the water passes out through the second opening, which may be opened and closed rhythmically by the fish's breathing. Any odorous substances in the water stimulate the nasal receptors chemically, perhaps by an effect on enzyme reactions, and the re-

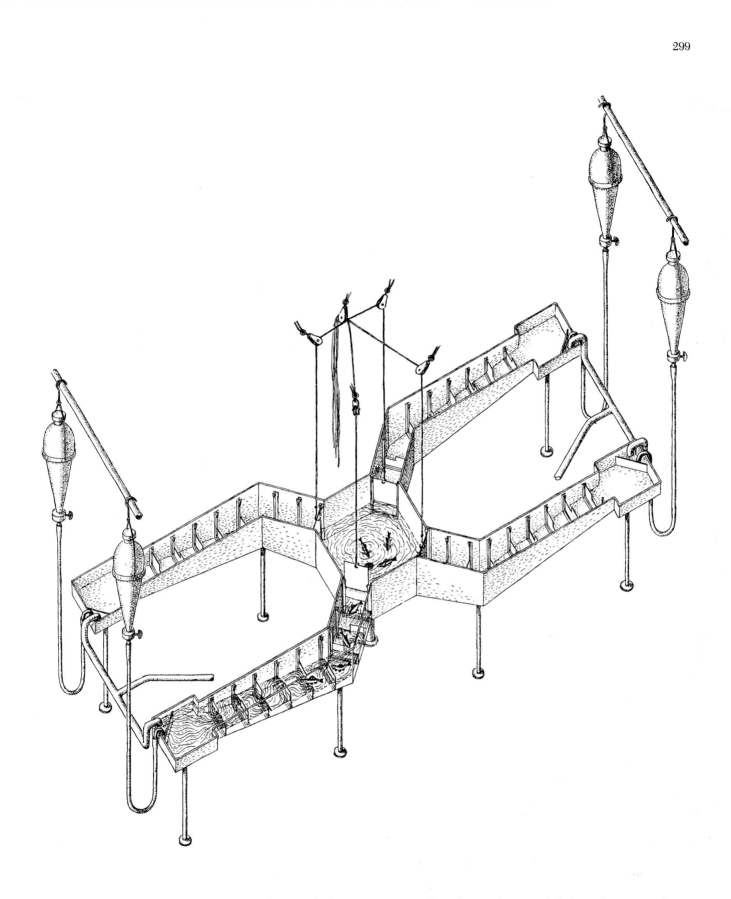

FOUR RUNWAYS are used to test the reaction of untrained salmon fingerlings to various odors. Water is introduced at the outer end of each runway and flows down a series of steps into a central compartment, where it drains. In the runway at the lower left the water cascades down to the central compartment in a series of miniature waterfalls; in the other runways the water is omitted to show the construction of the apparatus. Odors may be introduced into the apparatus from the vessels suspended above the runways. In an experiment salmon fingerlings are placed in the central compartment and an odor is introduced into one of the runways. When the four doors to the central compartment are opened, the fingerlings tend to enter the arms, proceeding upstream by jumping the waterfalls. Whether an odor attracts them, repels or has no effect is judged by the observed distribution of the fish in the runways.

sulting electrical impulses are relayed to the central nervous system by the olfactory nerve.

The human nose, and that of other land vertebrates, can smell a substance only if it is volatile and soluble in fat solvents. But in the final analysis smell is always aquatic, for a substance is not smelled until it passes into solution in the mucous film of the nasal passages. For fishes, of course, the odors are already in solution in their watery environment. Like any other animal, they can follow an odor to its source, as a hunting dog follows the scent of an animal. The quality or effect of a scent changes as the concentration changes; everyone knows that an odor may be pleasant at one concentration and unpleasant at another.

When we began our experiments, we first undertook to find out whether fish could distinguish the odors of different water plants. We used a specially developed aquarium with jets which could inject odors into the water. For responding to one odor (by moving toward the jet), the fish were rewarded with food; for responding to another odor, they were punished with a mild electric shock. After the fish were trained to make choices between odors, they were tested on dilute rinses from 14 different aquatic plants. They proved able to distinguish the odors of all these plants from one another.

Plants must play an important role in the life of many freshwater fish. Their odors may guide fish to feeding grounds when visibility is poor, as in muddy water or at night, and they may hold young fish from straying from protective cover. Odors may also warn fish away from poisons. In fact, we discovered that fish could be put to use to assay industrial pollutants: our trained minnows were able to detect phenol, a common pollutant, at concentrations far below those detectable by man.

All this suggested a clear-cut working hypothesis for investigating the mystery of the homing of salmon. We can suppose that every little stream has its own characteristic odor, which stays the same year after year; that young salmon become conditioned to this odor before they go to sea; that they remember the odor as they grow to maturity, and that they are able to find it and follow it to its source when they come back upstream to spawn.

Plainly there are quite a few ifs in this theory. The first one we tested was the question: Does each stream have its own odor? We took water from two creeks in Wisconsin and investigated whether fish could learn to discriminate between them. Our subjects, first minnows and then salmon, were indeed able to detect a difference. If, however, we destroyed a fish's nose tissue, it was no longer able to distinguish between the two water samples.

Chemical analysis indicated that the only major difference between the two waters lay in the organic material. By testing the fish with various fractions of the water separated by distillation, we confirmed that the identifying material was some volatile organic substance.

The idea that fish are guided by odors in their migrations was further supported by a field test. From each of two different branches of the Issaquah River in the State of Washington we took a number of sexually ripe silver salmon which had come home to spawn. We then plugged with cotton the noses of half the fish in each group and placed all the salmon in the river below the fork to make the upstream run again. Most of the fish with unplugged noses swam back to the stream they had selected the first time. But the "odor-blinded" fish migrated back in random fashion, picking the wrong stream as often as the right one.

In 1949 eggs from salmon of the Horsefly River in British Columbia were hatched and reared in a hatchery in a tributary called the Little Horsefly. Then they were flown a considerable distance and released in the main Horsefly River, from which they migrated to the sea. Three years later 13 of them had returned to their rearing place in the Little Horsefly, according to the report of the Canadian experimenters.

In our own laboratory experiments we tested the memory of fish for odors and found that they retained the ability to differentiate between odors for a long period after their training. Young fish remembered odors better than the old. That animals "remember" conditioning to which they have been exposed in their youth, and act accordingly, has been demonstrated in other fields. For instance, there is a fly which normally lays its eggs on the larvae of the flour moth, where the fly larvae then hatch and develop. But if larvae of this fly are raised on another host, the beeswax moth, when the flies mature they will seek out beeswax moth larvae on which to lay their eggs, in preference to the traditional host.

With respect to the homing of salmon we have shown, then, that different streams have different odors, that salmon respond to these odors and that they remember odors to which they have been conditioned. The next question is: Is a salmon's homeward migration guided solely by its sense of smell? If we could decoy homing salmon to a stream other than their birthplace, by means of an odor to which they were conditioned artificially, we might have not only a solution to the riddle that has puzzled scientists but also a practical means of saving the salmon—guiding them to breeding streams not obstructed by dams.

We set out to find a suitable substance to which salmon could be conditioned. A student, W. J. Wisby, and I [Arthur Hasler] designed an apparatus to test the reactions of salmon to various organic odors. It consists of a compartment from which radiate four runways, each with several steps which the fish must jump to climb the runway. Water cascades down each of the arms. An odorous substance is introduced into one of the arms, and its effect on the fish is judged by whether the odor appears to attract fish into that arm, to repel them or to be indifferent to them.

We needed a substance which initially would not be either attractive or repellent to salmon but to which they could be conditioned so that it would attract them. After testing several score organic odors, we found that dilute solutions of morpholine neither attracted nor repelled salmon but were detectable by them in extremely low concentrations— as low as one part per million. It appears that morpholine fits the requirements for the substance needed: it is soluble in water; it is detectable in extremely low concentrations; it is chemically stable under stream conditions. It is neither an attractant nor a repellent to unconditioned salmon, and would have meaning only to those conditioned to it.

Federal collaborators of ours are now conducting field tests on the Pacific Coast to learn whether salmon fry and fingerlings which have been conditioned to morpholine can be decoyed to a stream other than that of their birth when they return from the sea to spawn. Unfortunately this type of experiment may not be decisive. If the salmon are not decoyed to the new stream, it may simply mean that they cannot be drawn by a single substance but will react only to a combination of subtle odors in their parent stream. Perhaps adding morpholine to the water is like adding the whistle of a freight train to the quiet strains of a violin, cello and flute. The salmon may still seek out the subtle harmonies of an odor combination to which they have been reacting by instinct for centuries. But there is still hope that they may respond to the call of the whistle.

29

ELECTRIC LOCATION BY FISHES

H. W. LISSMANN

March 1963

It is well known that some fishes generate strong electric fields to stun their prey or discourage predators. Gymnarchus niloticus produces a weak field for the purpose of sensing its environment

Study of the ingenious adaptations displayed in the anatomy, physiology and behavior of animals leads to the familiar conclusion that each has evolved to suit life in its particular corner of the world. It is well to bear in mind, however, that each animal also inhabits a private subjective world that is not accessible to direct observation. This world is made up of information communicated to the creature from the outside in the form of messages picked up by its sense organs. No adaptation is more crucial to survival; the environment changes from place to place and from moment to moment, and the animal must respond appropriately in every place and at every moment. The sense organs transform energy of various kinds—heat and light, mechanical energy and chemical energy—into nerve impulses. Because the human organism is sensitive to the same kinds of energy, man can to some extent visualize the world as it appears to other living things. It helps in considering the behavior of a dog, for example, to realize that it can see less well than a man but can hear and smell better. There are limits to this procedure; ultimately the dog's sensory messages are projected onto its brain and are there evaluated differently.

Some animals present more serious obstacles to understanding. As I sit writing at my desk I face a large aquarium that contains an elegant fish about 20 inches long. It has no popular name but is known to science as *Gymnarchus niloticus*. This same fish has been facing me for the past 12 years, ever since I brought it from Africa. By observation and experiment I have tried to understand its behavior in response to stimuli from its environment. I am now convinced that *Gymnarchus* lives in a world totally alien to man: its most important

sense is an electric one, different from any we possess.

From time to time over the past century investigators have examined and dissected this curious animal. The literature describes its locomotive apparatus, central nervous system, skin and electric organs, its habitat and its family relation to the "elephant-trunk fishes," or mormyrids, of Africa. But the parts have not been fitted together into a functional pattern, comprehending the design of the animal as a whole and the history of its development. In this line of biological research one must resist the temptation to be deflected by details, to follow the fashion of putting the pieces too early under the electron microscope. The magnitude of a scientific revelation is not always paralleled by the degree of magnification employed. It is easier to select the points on which attention should be concentrated once the plan is understood. In the case of *Gymnarchus*, I think, this can now be attempted.

A casual observer is at once impressed by the grace with which *Gymnarchus* swims. It does not lash its tail from side to side, as most other fishes do, but keeps its spine straight. A beautiful undulating fin along its back propels its body through the water—forward or backward with equal ease. *Gymnarchus* can maintain its rigid posture even when turning, with complex wave forms running hither and thither over different regions of the dorsal fin at one and the same time.

Closer observation leaves no doubt that the movements are executed with great precision. When *Gymnarchus* darts after the small fish on which it feeds, it never bumps into the walls of its tank, and it clearly takes evasive action at some distance from obstacles placed in

its aquarium. Such maneuvers are not surprising in a fish swimming forward, but *Gymnarchus* performs them equally well swimming backward. As a matter of fact it should be handicapped even when it is moving forward: its rather degenerate eyes seem to react only to excessively bright light.

Still another unusual aspect of this fish and, it turns out, the key to all the puzzles it poses, is its tail, a slender, pointed process bare of any fin ("gymnarchus" means "naked tail"). The tail was first dissected by Michael Pius Erdl of the University of Munich in 1847. He found tissue resembling a small electric organ, consisting of four thin spindles running up each side to somewhere beyond the middle of the body. Electric organs constructed rather differently, once thought to be "pseudoelectric," are also found at the hind end of the related mormyrids.

Such small electric organs have been an enigma for a long time. Like the powerful electric organs of electric eels and some other fishes, they are derived from muscle tissue. Apparently in the course of evolution the tissue lost its power to contract and became specialized in various ways to produce electric discharges [see "Electric Fishes," by Harry Grundfest; SCIENTIFIC AMERICAN, October, 1960]. In the strongly electric fishes this adaptation serves to deter predators and to paralyze prey. But the powerful electric organs must have evolved from weak ones. The original swimming muscles would therefore seem to have possessed or have acquired at some stage a subsidiary electric function that had survival value. Until recently no one had found a function for weak electric organs. This was one of the questions on my mind when I began to study *Gymnarchus*.

I noticed quite early, when I placed a

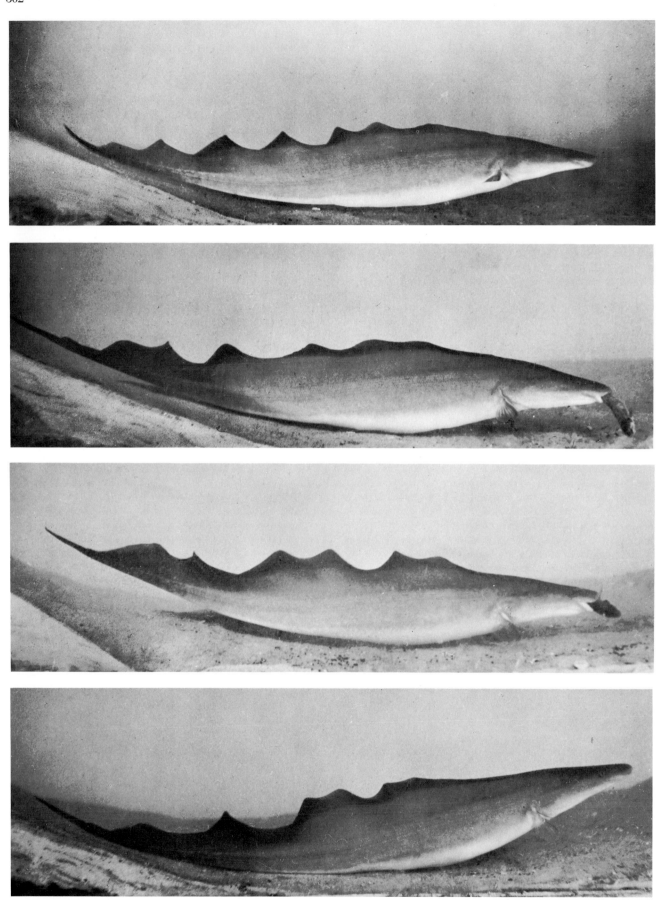

ELECTRIC FISH *Gymnarchus niloticus*, from Africa, generates weak discharges that enable it to detect objects. In this sequence the fish catches a smaller fish. *Gymnarchus* takes its name, which means "naked tail," from the fact that its pointed tail has no fin.

new object in the aquarium of a well-established *Gymnarchus,* that the fish would approach it with some caution, making what appeared to be exploratory movements with the tip of its tail. It occurred to me that the supposed electric organ in the tail might be a detecting mechanism. Accordingly I put into the water a pair of electrodes, connected to an amplifier and an oscilloscope. The result was a surprise. I had expected to find sporadic discharges co-ordinated with the swimming or exploratory motions of the animal. Instead the apparatus recorded a continuous stream of electric discharges at a constant frequency of about 300 per second, waxing and waning in amplitude as the fish changed position in relation to the stationary electrodes. Even when the fish was completely motionless, the electric activity remained unchanged.

This was the first electric fish found to behave in such a manner. After a brief search I discovered two other kinds that emit an uninterrupted stream of weak discharges. One is a mormyrid relative of *Gymnarchus;* the other is a gymnotid, a small, fresh-water South American relative of the electric eel, belonging to a group of fish rather far removed from *Gymnarchus* and the mormyrids.

It had been known for some time that the electric eel generates not only strong discharges but also irregular series of weaker discharges. Various functions had been ascribed to these weak dis-charges of the eel. Christopher W. Coates, director of the New York Aquarium, had suggested that they might serve in navigation, postulating that the eel somehow measured the time delay between the output of a pulse and its reflection from an object. This idea was untenable on physical as well as physiological grounds. The eel does not, in the first place, produce electromagnetic waves; if it did, they would travel too fast to be timed at the close range at which such a mechanism might be useful, and in any case they would hardly penetrate water. Electric current, which the eel does produce, is not reflected from objects in the surrounding environment.

Observation of *Gymnarchus* suggested another mechanism. During each discharge the tip of its tail becomes momentarily negative with respect to the head. The electric current may thus be pictured as spreading out into the surrounding water in the pattern of lines that describes a dipole field [*see illustration on the next page*]. The exact configuration of this electric field depends on the conductivity of the water and on the distortions introduced in the field by objects with electrical conductivity different from that of the water. In a large volume of water containing no objects the field is symmetrical. When objects are present, the lines of current will converge on those that have better conductivity and diverge from the poor conductors [*see top illustration on page 305*]. Such objects alter the distribution of electric potential over the surface of the fish. If the fish could register these changes, it would have a means of detecting the objects.

Calculations showed that *Gymnarchus* would have to be much more sensitive electrically than any fish was known to be if this mechanism were to work. I had observed, however, that *Gymnarchus* was sensitive to extremely small external electrical disturbances. It responded violently when a small magnet or an electrified insulator (such as a comb that had just been drawn through a person's hair) was moved near the aquarium. The electric fields produced in the water by such objects must be very small indeed, in the range of fractions of a millionth of one volt per centimeter. This crude observation was enough to justify a series of experiments under more stringent conditions.

In the most significant of these experiments Kenneth E. Machin and I trained the fish to distinguish between objects that could be recognized only by an electric sense. These were enclosed in porous ceramic pots or tubes with thick walls. When they were soaked in water, the ceramic material alone had little effect on the shape of the electric field. The pots excluded the possibility of discrimination by vision or, because each test lasted only a short time, by a chemical sense such as taste or smell.

The fish quickly learned to choose between two pots when one contained aquarium water or tap water and the other paraffin wax (a nonconductor). After training, the fish came regularly to pick a piece of food from a thread suspended behind a pot filled with aquarium or tap water and ignored the pot filled with wax [*see bottom illustration on page 305*]. Without further conditioning it also avoided pots filled with air, with distilled water, with a close-fitting glass tube or with another nonconductor. On the other hand, when the electrical conductivity of the distilled water was matched to that of tap or aquarium water by the addition of salts or acids, the fish would go to the pot for food.

A more prolonged series of trials showed that *Gymnarchus* could distinguish mixtures in different proportions of tap water and distilled water and perform other remarkable feats of discrimination. The limits of this performance can best be illustrated by the fact that the fish could detect the presence of a glass rod two millimeters in diameter and would fail to respond to a glass rod .8 millimeter in diameter, each hidden in a

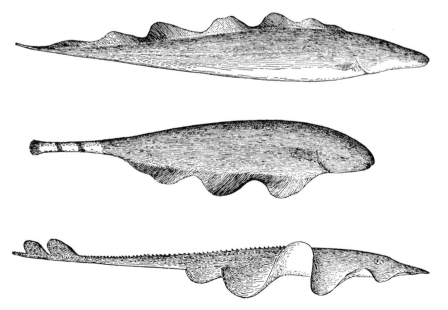

UNUSUAL FINS characterize *Gymnarchus* (*top*), a gymnotid from South America (*middle*) and sea-dwelling skate (*bottom*). All swim with spine rigid, probably in order to keep electric generating and detecting organs aligned. *Gymnarchus* is propelled by undulating dorsal fin, gymnotid by similar fin underneath and skate by lateral fins resembling wings.

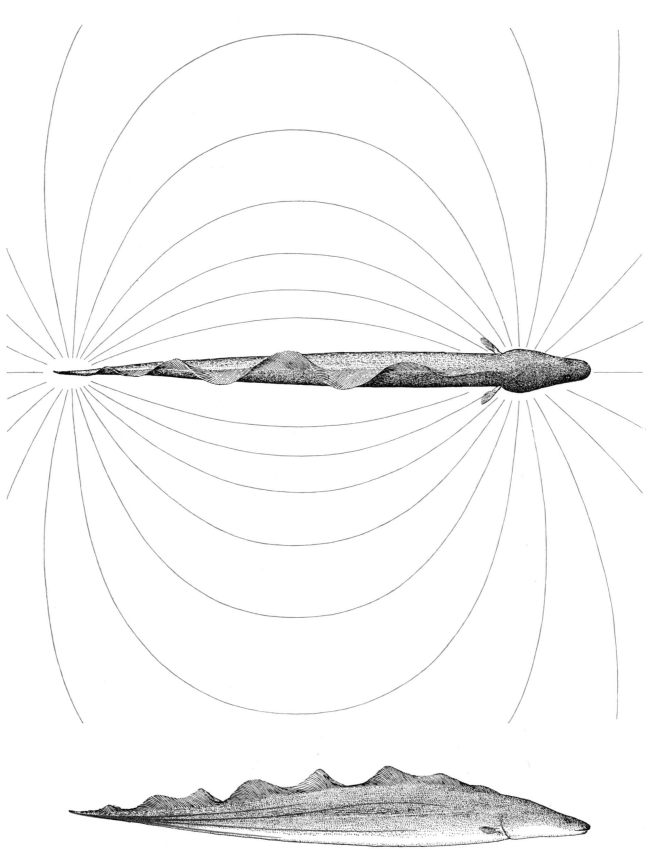

ELECTRIC FIELD of *Gymnarchus* and location of electric generating organs are diagramed. Each electric discharge from organs in rear portion of body (*color in side view*) makes tail negative with respect to head. Most of the electric sensory pores or organs are in head region. Undisturbed electric field resembles a dipole field, as shown, but is more complex. The fish responds to changes in the distribution of electric potential over the surface of its body. The conductivity of objects affects distribution of potential.

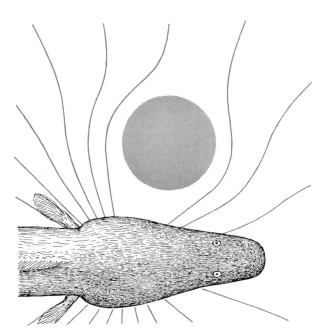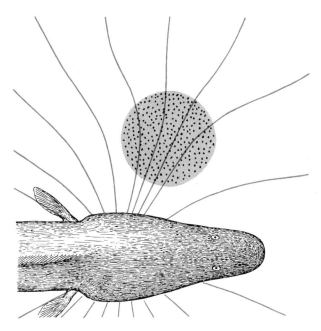

OBJECTS IN ELECTRIC FIELD of *Gymnarchus* distort the lines of current flow. The lines diverge from a poor conductor (*left*) and converge toward a good conductor (*right*). Sensory pores in the head region detect the effect and inform the fish about the object.

pot of the same dimensions. The threshold of its electric sense must lie somewhere between these two values.

These experiments seemed to establish beyond reasonable doubt that *Gymnarchus* detects objects by an electrical mechanism. The next step was to seek the possible channels through which the electrical information may reach the brain. It is generally accepted that the tissues and fluids of a fresh-water fish are relatively good electrical conductors enclosed in a skin that conducts poorly. The skin of *Gymnarchus* and of many mormyrids is exceptionally thick, with layers of platelike cells sometimes arrayed in a remarkable hexagonal pattern [*see top illustration on page 308*]. It can therefore be assumed that natural selection has provided these fishes with better-than-average exterior insulation.

In some places, particularly on and around the head, the skin is closely perforated. The pores lead into tubes often filled with a jelly-like substance or a loose aggregation of cells. If this jelly is a good electrical conductor, the arrangement would suggest that the lines of electric current from the water into the body of the fish are made to converge at these pores, as if focused by a lens. Each jelly-filled tube widens at the base into

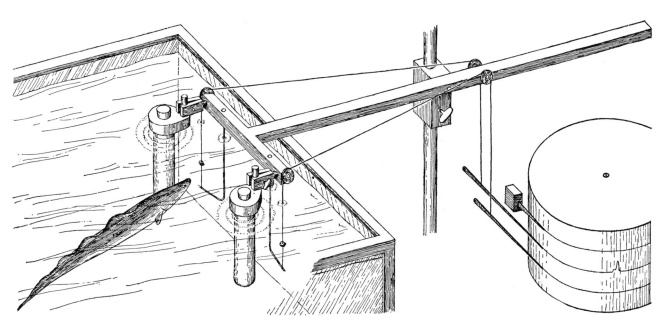

EXPERIMENTAL ARRANGEMENT for conditioned-reflex training of *Gymnarchus* includes two porous pots or tubes and recording mechanism. The fish learns to discriminate between objects of different electrical conductivity placed in the pots and to seek bait tied to string behind the pot holding the object that conducts best. *Gymnarchus* displays a remarkable ability to discriminate.

a small round capsule that contains a group of cells long known to histologists by such names as "multicellular glands," "mormyromasts" and "snout organs." These, I believe, are the electric sense organs.

The supporting evidence appears fairly strong: The structures in the capsule at the base of a tube receive sensory nerve fibers that unite to form the stoutest of all the nerves leading into the brain. Electrical recording of the impulse traffic in such nerves has shown that they lead away from organs highly sensitive to electric stimuli. The brain centers into which these nerves run are remarkably large and complex in *Gymnarchus,* and in some mormyrids they completely cover the remaining portions of the brain [*see illustration on next page*].

If this evidence for the plan as well as the existence of an electric sense does not seem sufficiently persuasive, corroboration is supplied by other weakly electric fishes. Except for the electric eel, all species of gymnotids investigated so far emit continuous electric pulses. They are also highly sensitive to electric fields. Dissection of these fishes reveals the expected histological counterparts of the structures found in the mormyrids: similar sense organs embedded in a similar skin, and the corresponding regions of the brain much enlarged.

Skates also have a weak electric organ in the tail. They are cartilaginous fishes, not bony fishes, or teleosts, as are the mormyrids and gymnotids. This means that they are far removed on the family line. Moreover, they live in the sea, which conducts electricity much better than fresh water does. It is almost too much to expect structural resemblances to the fresh-water bony fishes, or an electrical mechanism operating along similar lines. Yet skates possess sense organs, known as the ampullae of Lorenzini, that consist of long jelly-filled tubes opening to the water at one end and terminating in a sensory vesicle at the other. Recently Richard W. Murray of the University of Birmingham has found that these organs respond to very delicate electrical stimulation. Unfortunately, either skates are rather uncooperative animals or we have not mastered the trick of training them; we have been unable to repeat with them the experiments in discrimination in which *Gymnarchus* performs so well.

Gymnarchus, the gymnotids and skates all share one obvious feature: they swim in an unusual way. *Gymnarchus* swims with the aid of a fin on its back; the gymnotids have a similar fin on their

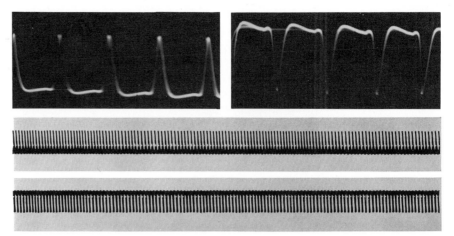

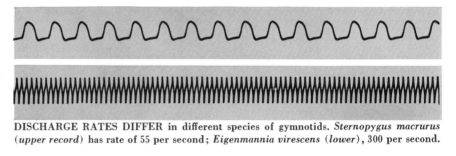

ELECTRIC DISCHARGES of *Gymnarchus* show reversal of polarity when detecting electrodes are rotated 180 degrees (*enlarged records at top*). The discharges, at rate of 300 per second, are remarkably regular even when fish is resting, as seen in lower records.

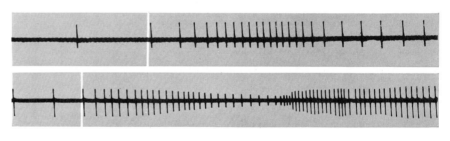

DISCHARGE RATES DIFFER in different species of gymnotids. *Sternopygus macrurus* (upper record) has rate of 55 per second; *Eigenmannia virescens* (lower), 300 per second.

VARIABLE DISCHARGE RATE is seen in some species. Tap on tank (*white line in upper record*) caused mormyrid to increase rate. Tap on fish (*lower record*) had greater effect.

underside; skates swim with pectoral fins stuck out sideways like wings [*see illustration on page 303*]. They all keep the spine rigid as they move. It would be rash to suggest that such deviations from the basic fish plan could be attributed to an accident of nature. In biology it always seems safer to assume that any redesign has arisen for some reason, even if the reason obstinately eludes the investigator. Since few fishes swim in this way or have electric organs, and since the fishes that combine these features are not related, a mere coincidence would appear most unlikely.

A good reason for the rigid swimming posture emerged when we built a model to simulate the discharge mecha-

nism and the sensory-perception system. We placed a pair of electrodes in a large tank of water; to represent the electric organ they were made to emit repetitive electric pulses. A second pair of electrodes, representing the electric sense organ, was placed some distance away to pick up the pulses. We rotated the second pair of electrodes until they were on a line of equipotential, where they ceased to record signals from the sending electrodes. With all the electrodes clamped in this position, we showed that the introduction of either a conductor or a nonconductor into the electric field could cause sufficient distortion of the field for the signals to reappear in the detectors.

In a prolonged series of readings the

slightest displacement of either pair of electrodes would produce great variations in the received signal. These could be smoothed to some extent by recording not the change of potential but the change in the potential gradient over the "surface" of our model fish. It is probable that the real fish uses this principle, but to make it work the electrode system must be kept more or less constantly aligned. Even though a few cubic centimeters of fish brain may in some respects put many electronic computers in the shade, the fish brain might be unable to obtain any sensible information if the fish's electrodes were to be misaligned by the tail-thrashing that propels an ordinary fish. A mode of swimming that keeps the electric field symmetrical with respect to the body most of the time would therefore offer obvious advantages. It seems logical to assume that *Gymnarchus*, or its ancestors, acquired the rigid mode of swimming along with the electric sensory apparatus and subsequently lost the broad, oarlike tail fin.

Our experiments with models also showed that objects could be detected only at a relatively short distance, in spite of high amplification in the receiving system. As an object was moved farther and farther away, a point was soon reached where the signals arriving at the oscilloscope became submerged in the general "noise" inherent in every detector system. Now, it is known that minute amounts of energy can stimulate a sense organ: one quantum of light registers on a visual sense cell; vibrations of subatomic dimensions excite the ear; a single molecule in a chemical sense organ can produce a sensation, and so on. Just how such small external signals can be picked out from the general noise in and around a metabolizing cell represents one of the central questions of sensory physiology. Considered in connection with the electric sense of fishes, this question is complicated further by the high frequency of the discharges from the electric organ that excite the sensory apparatus.

In general, a stimulus from the environment acting on a sense organ produces a sequence of repetitive impulses in the sensory nerve. A decrease in the strength of the stimulus causes a lower frequency of impulses in the nerve. Conversely, as the stimulus grows stronger, the frequency of impulses rises, up to a certain limit. This limit may vary from one sense organ to another, but 500 impulses per second is a common upper limit, although 1,000 per second have been recorded over brief intervals.

In the case of the electric sense organ of a fish the stimulus energy is provided by the discharges of the animal's electric organ. *Gymnarchus* discharges at the rate of 300 pulses per second. A change in the amplitude—not the rate—of these pulses, caused by the presence of an object in the field, constitutes the effective stimulus at the sense organ. Assuming that the reception of a single discharge of small amplitude excites one impulse in a sensory nerve, a discharge of larger amplitude that excited two impulses would probably reach and exceed the upper limit at which the nerve can generate impulses, since the nerve would now be firing 600 times a second (twice the rate of discharge of the electric organ). This would leave no room

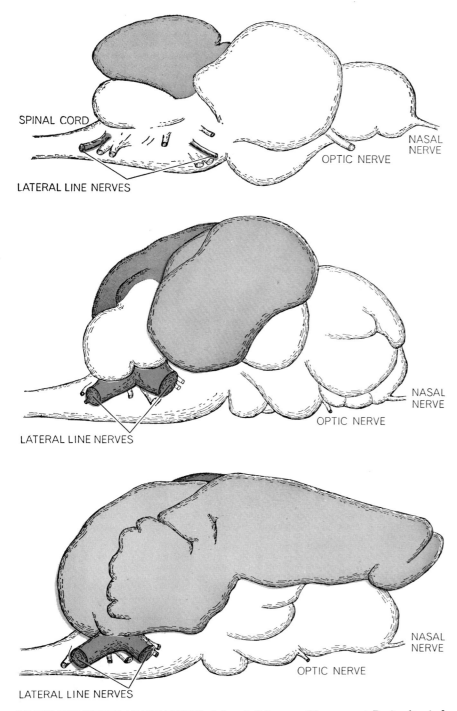

BRAIN AND NERVE ADAPTATIONS of electric fish are readily apparent. Brain of typical nonelectric fish (*top*) has prominent cerebellum (*gray*). Regions associated with electric sense (*color*) are quite large in *Gymnarchus* (*middle*) and even larger in the mormyrid (*bottom*). Lateral-line nerves of electric fishes are larger, nerves of nose and eyes smaller.

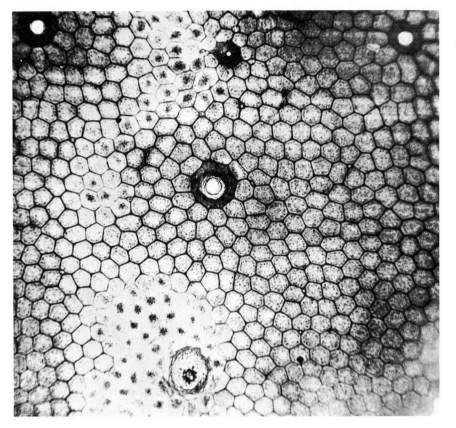

SKIN OF MORMYRID is made up of many layers of platelike cells having remarkable hexagonal structure. The pores contain tubes leading to electric sense organs. This photomicrograph by the author shows a horizontal section through the skin, enlarged 100 diameters.

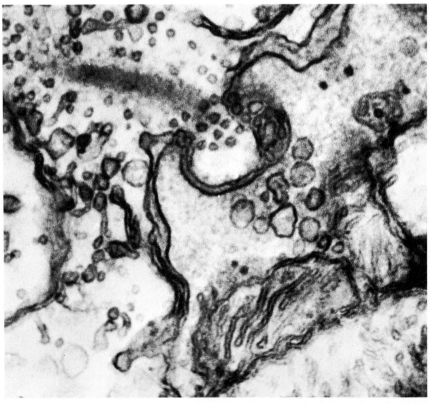

MEETING POINT of electric sensory cell (*left*) and its nerve (*right*) is enlarged 120,000 diameters in this electron micrograph by the author and Ann M. Mullinger. Bulge of sensory cell into nerve ending displays the characteristic dense streak surrounded by vesicles.

to convey information about gradual changes in the amplitude of incoming stimuli. Moreover, the electric organs of some gymnotids discharge at a much higher rate; 1,600 impulses per second have been recorded. It therefore appears unlikely that each individual discharge is communicated to the sense organs as a discrete stimulus.

We also hit on the alternative idea that the frequency of impulses from the sensory nerve might be determined by the mean value of electric current transmitted to the sense organ over a unit of time; in other words, that the significant messages from the environment are averaged out and so discriminated from the background of noise. We tested this idea on *Gymnarchus* by applying trains of rectangular electric pulses of varying voltage, duration and frequency across the aquarium. Again using the conditioned-reflex technique, we determined the threshold of perception for the different pulse trains. We found that the fish is in fact as sensitive to high-frequency pulses of short duration as it is to low-frequency pulses of identical voltage but correspondingly longer duration. For any given pulse train, reduction in voltage could be compensated either by an increase in frequency of stimulus or an increase in the duration of the pulse. Conversely, reduction in the frequency required an increase in the voltage or in the duration of the pulse to reach the threshold. The threshold would therefore appear to be determined by the product of voltage times duration times frequency.

Since the frequency and the duration of discharges are fixed by the output of the electric organ, the critical variable at the sensory organ is voltage. Threshold determinations of the fish's response to single pulses, compared with quantitative data on its response to trains of pulses, made it possible to calculate the time over which the fish averages out the necessarily blurred information carried within a single discharge of its own. This time proved to be 25 milliseconds, sufficient for the electric organ to emit seven or eight discharges.

The averaging out of information in this manner is a familiar technique for improving the signal-to-noise ratio; it has been found useful in various branches of technology for dealing with barely perceptible signals. In view of the very low signal energy that *Gymnarchus* can detect, such refinements in information processing, including the ability to average out information picked up by a large number of separate sense organs, appear to be essential. We have found that *Gymnarchus* can respond to a continuous direct-current electric stimulus of about .15 microvolt per centimeter, a value that agrees reasonably well with the calculated sensitivity required to recognize a glass rod two millimeters in diameter. This means that an individual sense organ should be able to convey information about a current change as small as .003 micromicroampere. Extended over the integration time of 25 milliseconds, this tiny current corresponds to a movement of some 1,000 univalent, or singly charged, ions.

The intimate mechanism of the single sensory cell of these organs is still a complete mystery. In structure the sense organs differ somewhat from species to species and different types are also found

in an individual fish. The fine structure of the sensory cells, their nerves and associated elements, which Ann M. Mullinger and I have studied with both the light microscope and the electron microscope, shows many interesting details. Along specialized areas of the boundary between the sensory cell and the nerve fiber there are sites of intimate contact where the sensory cell bulges into the fiber. A dense streak extends from the cell into this bulge, and the vesicles alongside it seem to penetrate the intercellular space. The integrating system of the sensory cell may be here.

These findings, however, apply only to *Gymnarchus* and to about half of the species of gymnotids investigated to date. The electric organs of these fishes emit pulses of constant frequency. In the other gymnotids and all the mormyrids the discharge frequency changes with the state of excitation of the fish. There is therefore no constant mean value of current transmitted in a unit of time; the integration of information in these species may perhaps be carried out in the brain. Nevertheless, it is interesting that both types of sensory system should have evolved independently in the two different families, one in Africa and one in South America.

The experiments with *Gymnarchus*, which indicate that no information is carried by the pulse nature of the discharges, leave us with a still unsolved problem. If the pulses are "smoothed out," it is difficult to see how any one fish can receive information in its own frequency range without interference from its neighbors. In this connection Akira Watanabe and Kimihisa Takeda at the University of Tokyo have made the potentially significant finding that the gymnotids respond to electric oscillations close in frequency to their own by shifting their frequency away from the applied frequency. Two fish might thus react to each other's presence.

For reasons that are perhaps associated with the evolutionary origin of their electric sense, the electric fishes are elusive subjects for study in the field. I have visited Africa and South America in order to observe them in their natural habitat. Although some respectable specimens were caught, it was only on rare occasions that I actually saw a *Gymnarchus*, a mormyrid or a gymnotid in the turbid waters in which they live. While such waters must have favored the evolution of an electric sense, it could not have been the only factor. The same waters contain a large number of

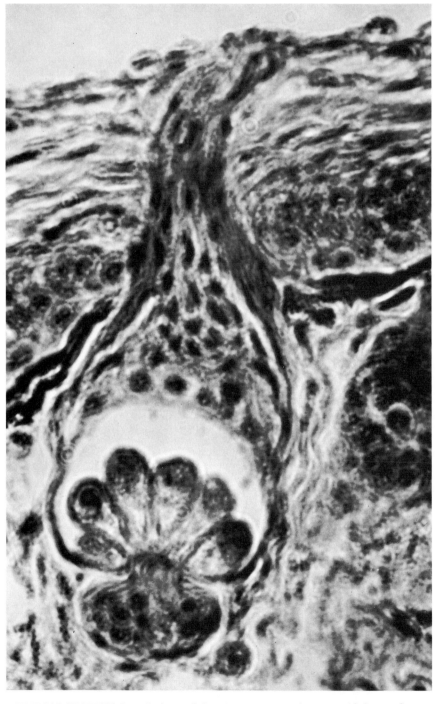

VERTICAL SECTION through skin and electric sense organ of a gymnotid shows tube containing jelly-like substance widening at base into a capsule, known as multicellular gland, that holds a group of special cells. Enlargement of this photomicrograph is 1,000 diameters.

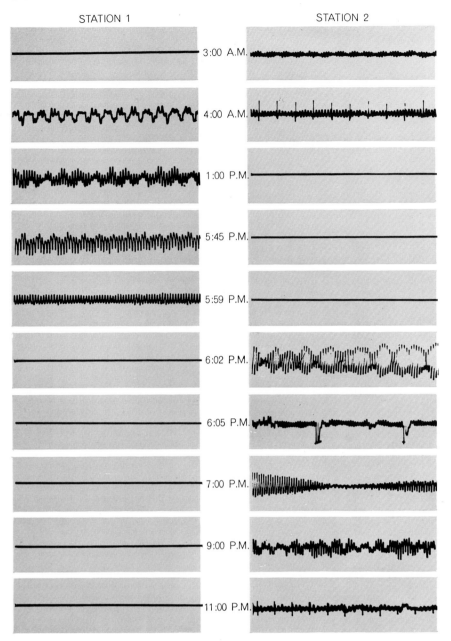

STATION 1   STATION 2

3:00 A.M.

4:00 A.M.

1:00 P.M.

5:45 P.M.

5:59 P.M.

6:02 P.M.

6:05 P.M.

7:00 P.M.

9:00 P.M.

11:00 P.M.

TRACKING ELECTRIC FISH in nature involves placing electrodes in water they inhabit. Records at left were made in South American stream near daytime hiding place of gymnotids, those at right out in main channel of stream, where they seek food at night.

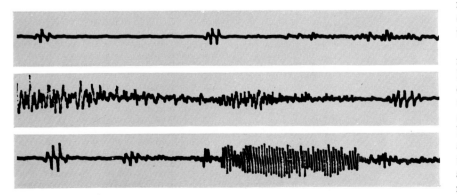

AFRICAN CATFISH, supposedly nonelectric, produced the discharges shown here. Normal action potentials of muscles are seen, along with odd regular blips and still other oscillations of higher frequency. Such fish may be evolving an electric sense or may already have one.

other fishes that apparently have no electric organs.

Although electric fishes cannot be seen in their natural habitat, it is still possible to detect and follow them by picking up their discharges from the water. In South America I have found that the gymnotids are all active during the night. Darkness and the turbidity of the water offer good protection to these fishes, which rely on their eyes only for the knowledge that it is day or night. At night most of the predatory fishes, which have well-developed eyes, sleep on the bottom of rivers, ponds and lakes. Early in the morning, before the predators wake up, the gymnotids return from their nightly excursions and occupy inaccessible hiding places, where they often collect in vast numbers. In the rocks and vegetation along the shore the ticking, rattling, humming and whistling can be heard in bewildering profusion when the electrodes are connected to a loudspeaker. With a little practice one can begin to distinguish the various species by these sounds.

When one observes life in this highly competitive environment, it becomes clear what advantages the electric sense confers on these fishes and why they have evolved their curiously specialized sense organs, skin, brain, electric organs and peculiar mode of swimming. Such well-established specialists must have originated, however, from ordinary fishes in which the characteristics of the specialists are found in their primitive state: the electric organs as locomotive muscles and the sense organs as mechanoreceptors along the lateral line of the body that signal displacement of water. Somewhere there must be intermediate forms in which the contraction of a muscle, with its accompanying change in electric potential, interacts with these sense organs. For survival it may be important to be able to distinguish water movements caused by animate or inanimate objects. This may have started the evolutionary trend toward an electric sense.

Already we know some supposedly nonelectric fishes from which, nevertheless, we can pick up signals having many characteristics of the discharges of electric fishes. We know of sense organs that appear to be structurally intermediate between ordinary lateral-line receptors and electroreceptors. Furthermore, fishes that have both of these characteristics are also electrically very sensitive. We may hope one day to piece the whole evolutionary line together and express, at least in physical terms, what it is like to live in an electric world.

30

THE INFRARED RECEPTORS OF SNAKES

R. IGOR GAMOW AND JOHN. F. HARRIS

May 1973

*The snakes of two large families have sensitive organs that
can detect the heat radiation emitted by their prey. The
performance of these detectors is investigated with the aid of
an infrared laser*

A boa constrictor will respond in 35 milliseconds to diffuse infrared radiation from a carbon dioxide laser. A sensitive man-made instrument requires nearly a minute to make what is essentially the same measurement. Reflecting on this comparison, one wonders what feat of bioengineering nature has performed to make the snake's sensor so efficient. One also wonders if a better understanding of the animal's heat-sensing apparatus would provide a basis for improving the man-made ones. It was the pursuit of these questions that gave rise to the somewhat unusual situation in which a group of workers in an aerospace engineering laboratory (our laboratory at the University of Colorado) was investigating snakes.

Snakes belong to one of the four large orders that comprise the living members of the class Reptilia. The order Testudinata contains such members as the turtles, the tortoises and the terrapins. The order Crocodilia contains the crocodiles and the alligators. The third order, the Rhynchocephalia, has only one member: the tuatara of New Zealand. In the fourth order, the Squamata, are the lizards and the snakes.

On the basis of outward appearance one might suppose that the lizards are more closely related to the crocodiles and the alligators than to the snake. Evolutionary evidence, however, clearly indicates that the snakes arose from the lizard line. Although the lizard is therefore the snake's closest relative, the two animals have developed pronounced differences during the course of evolution. Most lizards have limbs and no snakes have limbs, although vestigial ones are found in certain snakes. Most lizards have two functional lungs, whereas most snakes have only one. Again a few snakes have a small second lung, which

is another indication of the direction of evolution from the lizard to the snake.

Today most herpetologists would agree that the first step in the evolution of the snake occurred when the animal's ancestral form became a blind subterranean burrower. In evolving from their lizard-like form the ancestral snakes lost their limbs, their eyesight and their hearing as well as their ability to change coloration. Later, when the animals reappeared on the surface, they reevolved an entire new visual system but never regained their limbs or their sense of hearing.

Today the snakes constitute one of the most successful of living groups, being found in almost every conceivable habitat except polar regions and certain islands. They live in deep forests and in watery swamps. Some are nocturnal, others diurnal. Some occupy freshwater habitats, others marine habitats. Certain snakes are arboreal and survive by snatching bats from the air, others live in the inhospitable environment of the desert. Their success is indicated by the fact that their species, distributed among 14 families, number more than 2,700.

Two of the 14 families are distinguished by the fact that all their members have heat sensors that respond to minute changes of temperature in the snake's environment. The snake employs these sensors mainly to seek out and capture warm-bodied prey in the dark. It seems probable that the snake also uses the sensors to find places where it can maintain itself comfortably. Although snakes, like all reptiles, are cold-blooded, they are adept at regulating their body temperature by moving from place to place. Indeed, a snake functions well only within a rather narrow range of temperatures and must actively seek environments of the proper temperature.

A case in point is the common sidewinder, which maintains its body temperature in the range between 31 and 32 degrees Celsius (87.8 and 89.6 degrees Fahrenheit). One advantage of a heat sensor is that it enables the snake to scan the temperature of the terrain around it to find the proper environment.

One of the families with heat receptors is the Crotalidae: the pit vipers, including such well-known snakes as the rattlesnake, the water moccasin and the copperhead. The other family is the Boidae, which includes such snakes as the boa constrictor, the python and the anaconda. Although all members of both families have these heat receptors, the anatomy of the receptors differs so much between the families as to make it seem likely that the two types evolved independently.

In the pit vipers the sensor is housed in the pit organ, for which these snakes are named. There are two pits; they are located between the eye and the nostril and are always facing forward. In a grown snake the pit is about five millimeters deep and several millimeters in diameter. The inner cavity of the pit is larger than the external opening.

The inner cavity itself is divided into an inner chamber and an outer one, separated by a thin membrane. A duct between the inner chamber and the skin of the snake may prevent differential changes in pressure from arising between the two chambers. Within the membrane separating the chambers two large branches of the trigeminal nerve (one of the cranial nerves) terminate. In both snake families this nerve is primarily responsible for the input from the heat sensor to the brain. Near the terminus the nerve fibers lose their sheath of myelin and fan out into a broad, flat,

312

GREEN PYTHON of New Guinea (*Chondropython viridis*) is a member of the family Boidae that has visible pits housing its infra-red detectors. The pits extend along the jaws. Photograph was made by Richard G. Zweifel of the American Museum of Natural History.

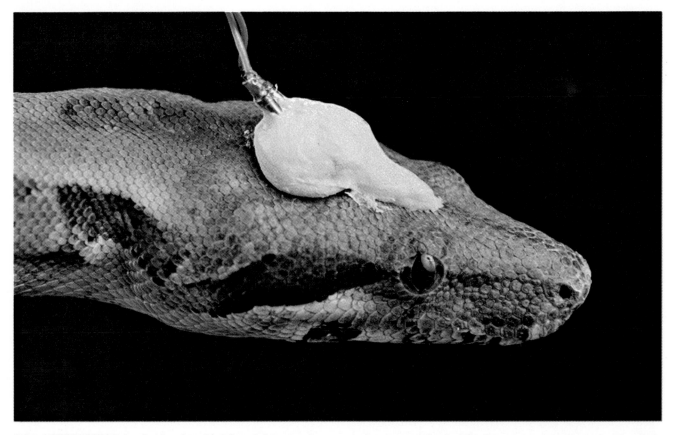

BOA CONSTRICTOR is a boid snake with infrared detectors that are not visible externally, although they are in the same location as the green python's. This boa wears an apparatus with which the authors recorded responses of the brain to infrared stimuli.

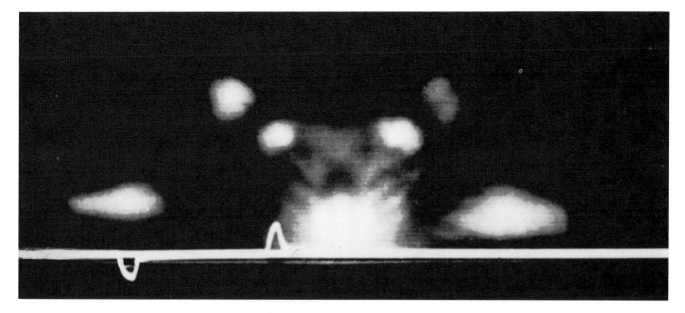

INFRARED VIEW OF RAT suggests what a snake "sees" through its infrared detectors when it is stalking prey. Snakes with such detectors prey on birds and small mammals. This view was obtained with a Barnes thermograph, which detects infrared radiation. In a thermogram the coolest areas have the darkest appearance and the warmest areas, such as the nose of the rat, appear as white spots.

BOA IN LASER BEAM was tested in the authors' laboratory at the University of Colorado for responses to infrared radiation. The carbon dioxide laser, emitting in the infrared at 10.6 microns, appears as a glowing area in the upper right background. Its beam is spread by a lens so that the snake, even when moving about in the cage, is doused in infrared radiation delivered in occasional pulses lasting eight milliseconds each. A brain signal recorded by the electrode assembly on the snake's head goes to a preamplifier and then to an oscilloscope and to a signal-averager. Electroencephalograms recorded in this way appear in the illustration on page 317.

palmate structure. In this structure the nerve endings are packed full of the small intracellular bodies known as mitochondria. Evidence obtained recently by Richard M. Meszler of the University of Maryland with the electron microscope strongly suggests that the mitochondria change morphologically just after receiving a heat stimulus. This finding has led to the suggestion that the mitochondria themselves may constitute the primary heat receptor.

In the family Boidae there are no pit organs of this type, although somewhat different pits are often found along the snakes' upper and lower lips. Indeed, it was once thought that only the boid snakes with labial pits had heat sensors. An extensive study by Theodore H. Bullock and Robert Barrett at the University of California at San Diego has shown, however, that boid snakes without labial pits nonetheless have sensitive heat receptors. One such snake is the boa constrictor.

For experimental purposes the boid snakes are preferable to the pit vipers because the viper is certain to bite sooner or later, and the bite can be deadly. The boids, in contrast, can be described as friendly, and they get along well in a laboratory. When our laboratory became interested several years ago in the possibility of using an infrared laser as a tool to help unravel the secrets of the mode of operation of the snake's heat sensor, we chose to work with boid snakes.

Bullock and his collaborators have done most of the pioneering work on the heat receptors of snakes. In their original experiments, using the rattlesnake, they first anesthetized the animal and then dissected out the bundle of large nerves that constitute the main branches of the trigeminal nerve. It is these branches that receive the sensory information from the receptor.

Bullock and his colleagues found by means of electrical recording that the frequency of nerve impulses increased as the receptor was warmed up and decreased as it was cooled. The changes were independent of the snake's body temperature; they were related only to changes of temperature in the environment. The Bullock group also determined that the operation of the sensor is phasic, meaning that the receptor gives a maximum response when the stimulus is initiated and that the response quickly subsides even if the stimulus is continued. (Many human receptors, such as the ones that sense pressure on the skin, are phasic; if they were not, one would be constantly conscious of such things as a wristwatch or a shirt.)

Our work was built on the foundations laid by Bullock and his associates. In addition we had in mind certain considerations about electromagnetic receptors in general. Biological systems utilize electromagnetic radiation both as a source of information and as a primary source of energy. Vision is an example of electromagnetic radiation as a source of information, and photosynthesis is a process that relies on electromagnetic radiation for energy.

All green plants utilize light as the source of the energy with which they build molecules of carbohydrate from carbon dioxide and water. To collect this energy the plants have a series of pigments (the various species of chlorophyll molecules) that absorb certain frequencies of electromagnetic radiation. Indeed, green plants are green because they absorb the red part of the spectrum and reflect the green part. Because the chlorophyll molecule absorbs only a rather narrow spectral frequency, it can

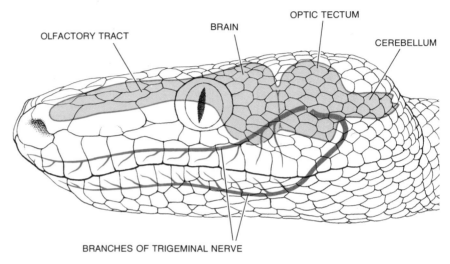

ANATOMY OF RECEPTOR in a boa is indicated. The scales along the upper and lower jaws have behind them an elaborate network of nerves, which lead into the two branches of the trigeminal nerve shown here. When the system detects an infrared stimulus, the trigeminal nerve carries a signal to the brain. A response can be recorded from the brain within 35 milliseconds after a boid snake receives a brief pulse of infrared radiation.

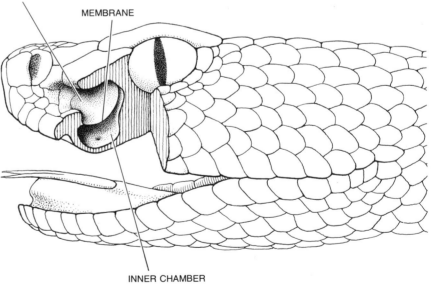

STRUCTURE OF PIT in a pit viper, the rattlesnake *Crotalus viridis*, differs from the anatomy of the infrared receptor in boid snakes. A pit viper has two pits, located between the eye and the nostril and facing forward. Each pit is about five millimeters deep, with the opening narrower than the interior. The elaborate branching of the trigeminal nerve is in the thin membrane that separates the inner and outer chambers of the pit organ.

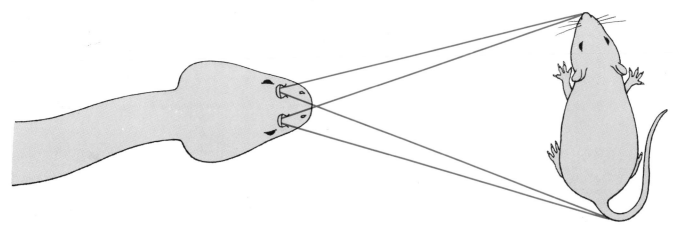

DIRECTIONALITY OF SENSOR in a pit viper is indicated by the location of the two pits on the snake's head and by the geometry of the pits. It appears certain that in stalking its prey, which include birds and small mammals, such a snake can establish the direction in which the prey lies by shifting its head as it does in using eyesight. Rattlesnake and copperhead are among the pit vipers.

be called a frequency (or wavelength) detector.

The eye is also a frequency detector, but it does not use radiation as an energy source. The incoming radiation triggers the release of energy that has been stored in the nerves previously, having been produced by normal metabolism. The eye, like other frequency detectors, operates within a narrow band of the electromagnetic spectrum, namely at wavelengths from about 300 to about 1,000 nanometers (billionths of a meter). One can see how narrow the band is by recalling that man-made instruments can detect electromagnetic wavelengths from 10^{-20} meter to 10^5 meters, a full 25 orders of magnitude, whereas the range of the human eye is from $10^{-6.4}$ to about $10^{-6.1}$ meter. Within this range the eye can resolve thousands of different combinations of wavelengths, which are the number of shades of color one can recognize. Although the eye is a good frequency detector, it is a poor energy detector: a dim bulb appears as bright as a bright one to the dark-adapted eye, which is to say that the eye adjusts its sensitivity according to the conditions to which it has become adapted.

Why has nature chosen this frequency range for its photobiology? From an evolutionary point of view the answers seem clear. One reason is that 83 percent of the sunlight that reaches the surface of the earth is in that frequency range. Moreover, it is difficult to imagine a biological sensor that would detect X rays or hard ultraviolet radiation, because the energy of the photons would be higher than the bonding energy of the receiving molecules. The photons would destroy or at least badly disrupt the structure of the sensor. Low-frequency radiation presents just the opposite difficulty. The energy of long-wavelength infrared radiation and of microwaves is so low that the photons cannot bring about specific changes in a molecule of pigment. Hence the sensor must operate in a frequency range that provides enough energy to reliably change biological pigment molecules from one state to another (from a "ground" state to a transitional state) but not so much energy as to destroy the sensor.

Early workers on the heat detectors of snakes had determined that the receptor responded to energy sources in the near-infrared region of the spectrum. The work left unanswered the question of whether the sensor contained a pigment molecule that trapped this long-wave radiation, thus acting as a kind of eye, or whether the sensor merely trapped energy in proportion to the ability of the tissue to absorb a given frequency and was thus acting as an energy detector. We therefore directed our experiments toward trying to resolve this issue.

To make sure that the response we obtained was maximal, we wanted to work with snakes that were functioning as close to their normal physiological level as possible. First we studied the normal feeding behavior of boa constrictors that were healthy and appeared to be well adjusted. The work entailed seeing how the snake sensed, stalked and captured prey animals such as mice and birds. Since the snake can capture prey in complete darkness as well as in light, it is clear that the heat receptors play a crucial role.

Barrett, while he was a graduate student working with Bullock, went further with this type of behavioral study. He found that the snake would strike at a warm sandbag but not at a cold, dead mouse. On the other hand, the snake would swallow the cold mouse (after a great deal of tongue-flicking and examination) if the mouse was put near the snake's mouth, but it never tried to swallow a sandbag. Barrett concluded that the snake has a strike reflex that is triggered by the firing of the heat receptors, whereas another set of sensory inputs determines whether or not the snake will swallow the object.

In searching for a reliable index that would tell us whether or not the heat receptor was responding to an infrared stimulus, we first tried measuring with an electrocardiograph the change in heartbeat after the snake received a stimulus. This venture ran afoul of the difficulty of finding the heart in such a long animal. (It is about a third of the way along the body from the head.) A more serious difficulty was our discovery that a number of outside influences would change the rate of the heartbeat, so that it was hard to establish a definite stimulus-response relation.

We next turned to a method that proved to be much more successful. It entailed monitoring the electrical activity of the snake's brain with an electroencephalograph. A consistent change in the pattern of an electroencephalogram after a stimulus has been received by the peripheral nervous system is called the evoked potential. When a neural signal from a sensory receptor arrives at the cortex of the brain, there is a small perturbation in the brain's electrical activity. When the signal is small, as is usually the case, it must be extracted from the electrical background noise. The process is best accomplished by averaging a substantial number of evoked potentials. This procedure results in a highly sensi-

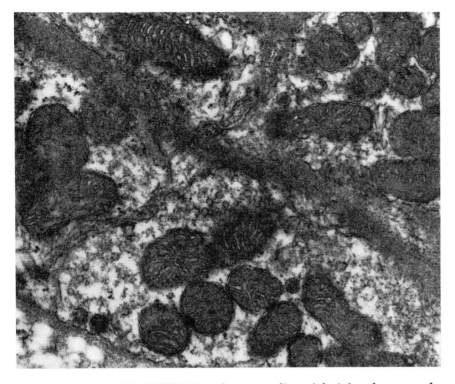

APPEARANCE OF MITOCHONDRIA in the nerve endings of the infrared receptor of a cottonmouth moccasin (*Agkistrodon piscivorus piscivorus*) after exposure of the receptor to an infrared stimulus is shown in this electron micrograph made by Richard M. Meszler of the University of Maryland. The enlargement is 34,000 diameters. In contrast to the mitochondria in the micrograph below, which was made when the receptor was exposed to a cold body, these mitochondria are condensed, as shown by the dense matrix and the organization of the inner membrane. Change in morphology of the mitochondria after a heat stimulus has led to the suggestion that they constitute the primary receptors in the detector.

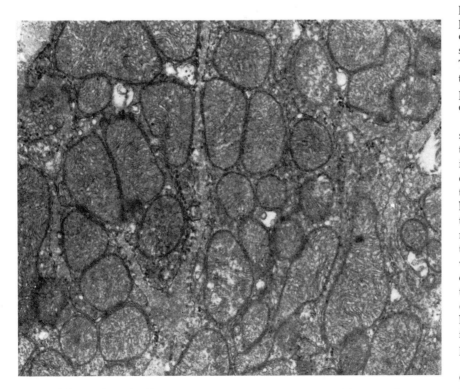

CONTRASTING APPEARANCE of the mitochondria in the infrared detector of a cottonmouth moccasin when the receptor was exposed to a cold body is evident in this electron micrograph made by Meszler. The enlargement is 27,000 diameters. A cold body, in contrast to a warm one, is known to reduce the firing of discharges by the heat-sensitive receptor.

tive measure of a physiological response.

The boa constrictors used in our study ranged from 75 to 145 centimeters in length and from 320 to 1,200 grams in weight. For several weeks before we involved them in experiments they lived under normal conditions in our laboratory. To prepare a snake for the experiments we anesthetized it with pentobarbital and then installed an electrode assembly on its head. After a postoperative recovery period the animal appeared to behave in the same way as snakes that had not been operated on.

A brain signal recorded by this apparatus went to a preamplifier and then to an oscilloscope and to a signal-averager. The signal-averager, which is in essence a small computer, is the workhorse of our system. By averaging the electroencephalogram just before and just after a stimulus it extracts the evoked potential, which would otherwise be buried in the background noise of the brain. In general we average the evoked potentials from about 20 consecutive stimuli.

The birds and mammals that the boa constrictor hunts emit infrared radiation most strongly at wavelengths around 10 microns. A carbon dioxide laser is ideal for our experiments because it produces a monochromatic output at a wavelength of 10.6 microns. We pulse the laser by means of a calibrated camera shutter so that it will deliver a stimulus lasting for eight milliseconds. The opening of the shutter also triggers the signal-averager, thus establishing precisely the time when the stimulus is delivered.

After the beam passes through the shutter it is spread by a special infrared-transmitting lens, so that the entire snake is doused in the radiation. The intensity of the radiation is measured by a sensitive colorimeter placed near the snake's head. This is the instrument we mentioned at the outset that takes nearly a minute to measure the power, whereas the snake gives a maximal response within 35 milliseconds after a single eight-millisecond pulse. Another indication of the sensitivity of the snake's receptor can be obtained by putting one's hand in the diffused laser beam; one feels no heat, even over a considerable period of time.

In order to verify that the responses of the snake resulted directly from stimulation of the heat receptor, we repeated the entire procedure with a common garter snake, which has no heat sensor. Even at laser powers far exceeding the stimulus given to the boas, we found no response in recordings from the garter snake. On the other hand, both species showed clear responses to visible light.

Our data strongly suggested an answer to the question of whether the receptor is a photochemical frequency detector like an eye or is an energy detector. The answer is that the receptor is an energy detector. One argument supporting this conclusion is that the stimulus is so far out in the low-energy infrared region of the spectrum (10.6 microns) that it would not provide enough power to activate an eyelike frequency detector, and yet the snake shows a full response. Another argument has to do with the 35-millisecond interval between the stimulus and the response. Photochemical reactions are quite fast, occur-

ring in periods of less than one millisecond. Although the time a nerve impulse from the eye takes to reach the cortex is about the same (35 milliseconds) as the time the nerve impulse from the heat sensor takes, the neural geometry of the two systems is quite different. The visual pathway incorporates a large number of synapses (connections between neurons), which account for most of the delay. In the trigeminal pathway no synapses are encountered until the signal reaches the brain. We therefore believe the delay found in the heat-receptor response is largely a result of the time required to heat the sensor to its threshold.

We also tested the snake's receptor in the microwave region of the spectrum, where the signals have longer

wavelength, lower frequency and lower energy than in the infrared. The reason was that in view of the many problems that have arisen in contemporary society about exposure to radiation we wanted to see whether an organism experienced physiological or psychological effects after being exposed for various periods of time to low-energy, long-wavelength radiation. There is no question that high-intensity microwave radiation can be detected not only by snakes but also probably by all animals; after all, a microwave oven can cook a hamburger in a matter of seconds. Our concern was with the kind of exposure arising from leaky microwave appliances such as ovens and from the increasing use of radar.

Testing the snakes with microwave radiation as we did with infrared, we obtained a clear-cut response [see illustration on this page]. Our result provides what we believe is the first unambiguous physiological demonstration that a biological system can indeed be influenced by such low-energy microwave radiation. Our conviction that the snake's heat receptor functions entirely as an energy detector is therefore reinforced.

The question of how much energy is required to activate the detector can be answered with certain reservations: it is approximately .00002 (2×10^{-5}) calorie per square centimeter. The reason for the reservations is that it is difficult to obtain an absolute threshold of sensitivity for any biological phenomenon. For one thing, a biological system shows considerable variability at or near its threshold of response. Moreover, there is always a certain amount of variation in the amount of energy put out by our sources of energy. With these reservations we have determined that the snake can easily and reliably detect power densities from the carbon dioxide laser ranging from .0019 to .0034 calorie per square centimeter per second. Since this density is administered in a short time period (eight milliseconds), the total energy that the snake is responding to is about .00002 calorie per square centimeter. The density of microwave power that is needed for a reliable response from the snake is about the same as the amount of laser power.

Our studies have shown that the heat-sensing snakes have evolved an extremely sensitive energy-detecting device giving responses that are proportional to the absorbed energy. It will be interesting to see whether the growing understanding of the snake's heat sensor will point the way toward an improvement in man-made sensors.

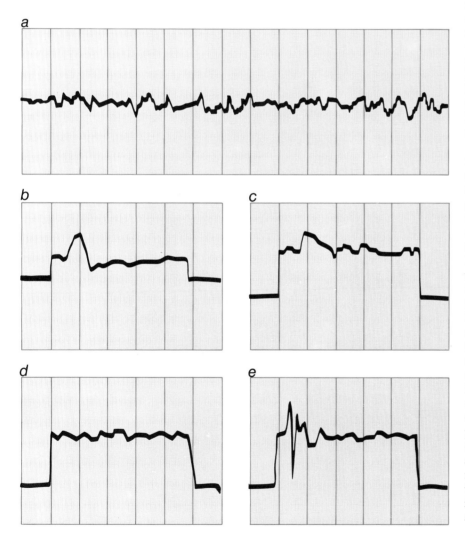

ELECTROENCEPHALOGRAMS of a boa constrictor were recorded under various conditions. The normal activity of a boa's brain (a) was traced directly on a strip recorder; the interval of time between each pair of colored vertical lines was 100 milliseconds. The remaining electroencephalograms were recorded through a signal-averager that reflected both the stimulus and the response. In each case the first rise shows the time of the stimulus and the next rise, if any, shows the response. The time interval is 100 milliseconds in every tracing but b, where it is 50 milliseconds. Traces show the averaged evoked response after an infrared stimulus (b) and a microwave stimulus (c), in the control situation in which the snake was shielded from the stimulus (d) and after a series of visible-light flashes (e).

31

MORE ABOUT BAT "RADAR"

DONALD R. GRIFFIN
July 1958

*A sequel to an earlier article which described the capacity of
bats to locate objects by supersonic echoes. This natural
sonar is now known to incorporate extraordinary refinements*

In these days of technological triumphs
it is well to remind ourselves from
time to time that living mechanisms
are often incomparably more efficient
than their artificial imitations. There is
no better illustration of this rule than the
sonar system of bats. Ounce for ounce
and watt for watt, it is billions of times
more efficient and more sensitive than
the radars and sonars contrived by man
[*see table at bottom of page 320*].

Of course the bats have had some 50
million years of evolution to refine their
sonar. Their physiological mechanisms
for echolocation, based on all this accumulated experience, should therefore
repay our thorough study and analysis.

To appreciate the precision of the
bats' echolocation we must first consider
the degree of their reliance upon it.
Thanks to sonar, an insect-eating bat can
get along perfectly well without eyesight. This was brilliantly demonstrated
by an experiment performed in the late
18th century by the Italian naturalist
Lazaro Spallanzani. He caught some
bats in a bell tower, blinded them and
released them outdoors. Four of these
blind bats were recaptured after they
had found their way back to the bell
tower, and on examining their stomach
contents Spallanzani found that they had
been able to capture and gorge themselves with flying insects in the field. We
know from experiments that bats easily
find insects in the dark of night, even

when the insects emit no sound that can
be heard by human ears. A bat will catch
hundreds of soft-bodied, silent-flying
moths or gnats in a single hour. It will
even detect and chase pebbles or cotton
spitballs tossed into the air.

In our studies of bats engaged in insect-
hunting in the field we use an apparatus which translates the bats' high-
pitched, inaudible sonar signals into audible clicks. When the big brown bat
(*Eptesicus fuscus*) cruises past at 40 or
50 feet above the ground, the clicks
sound like the slow put-put of an old
marine engine. As the bat swoops toward
a moth, the sounds speed up to the
tempo of an idling outboard motor, and

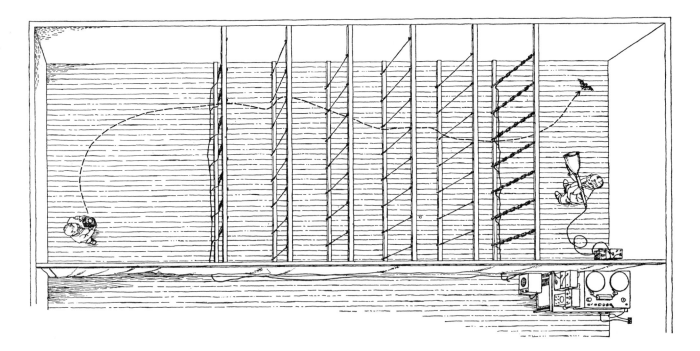

OBSTACLES in the form of thin vertical wires are avoided by a
bat despite the presence of interfering noise. The noise comes
from banks of loudspeakers to left and right of the four sets of
wires. Man at right holds microphone which picks up bat's signals.

when the chase grows really hot they are like the buzz of a model-airplane gasoline engine. It seems almost certain that these adjustments of the pulses are made in order to enable the bat to home on its insect prey.

At the cruising tempo each pulse is about 10 to 15 thousandths of a second long; during the buzz the pulses are shortened to less than a thousandth of a second and are emitted at rates as high as 200 per second. These sound patterns can be visualized by means of a sound spectrogram [see charts on page 321]. Within each individual pulse of sound the frequency drops as much as a whole octave (from about 50,000 to 25,000 cycles per second). As the pitch changes, the wavelength rises from about six to 12 millimeters. This is just the size range of most insects upon which the bat feeds. The bat's sound pulse may sweep the whole octave, because its target varies in size as the insect turns its body and flutters its wings.

The largest bats, such as the flying foxes or Old World fruit bats [see "Bats," by William A. Wimsatt; SCIENTIFIC AMERICAN, November, 1957], have no sonar. As their prominent eyes suggest, they depend on vision; if forced to fly in the dark, they are as helpless as an ordinary bird. One genus of bat uses echolocation in dark caves but flies by vision and emits no sounds in the light. Its orientation sounds are sharp clicks audible to the human ear, like those of the cave-dwelling oil bird of South America [see "Bird Sonar," by Donald R. Griffin; SCIENTIFIC AMERICAN, March, 1954].

On the other hand, all of the small bats (suborder *Microchiroptera*) rely largely on echolocation, to the best of our present knowledge. Certain families of bats in tropical America use only a single wavelength or a mixture of harmonically related frequencies, instead of varying the frequency systematically in each pulse. Those that live on fruit, and the vampire bats that feed on the blood of animals, employ faint pulses of this type.

Another highly specialized group, the horseshoe bats of the Old World, have elaborate nose leaves which act as horns to focus their orientation sounds in a sharp beam; they sweep the beam back and forth to scan their surroundings. The most surprising of all the specialized bats are the species that feed on fish. These bats, like the brown bat and many other species, have a well-developed system of frequency-modulated ("FM") sonar, but since sound loses much of its energy in passing from air into water and *vice*

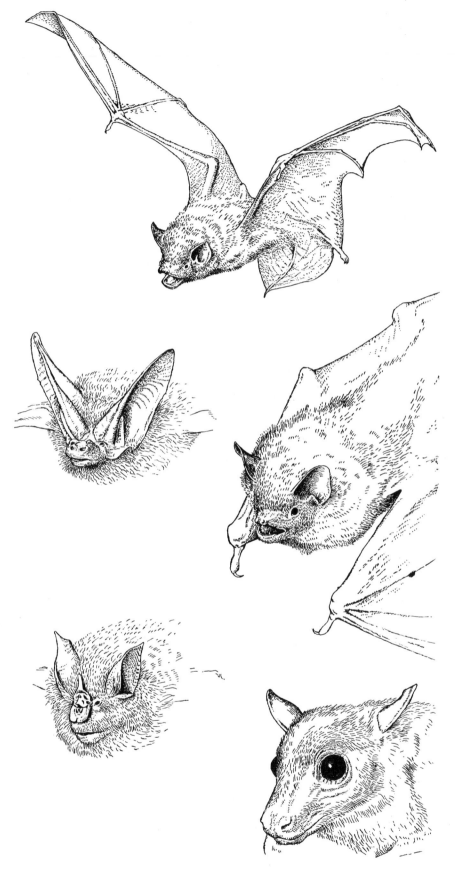

BATS shown in these drawings all use some type of echolocation system except for the fruit bat *Rousettus*, which appears at bottom right. The other species represented are the small brown bat *Myotis lucifugus* (top), the long-eared bat *Plecotus* (left center), the large brown bat *Eptesicus* (right center) and the horseshoe bat *Rhinolophus* (bottom left).

versa, the big puzzle is: How can the bats locate fish under water by means of this system?

Echolocation by bats is still such a new discovery that we have not yet grasped all its refinements. The common impression is that it is merely a crude collision warning device. But the bats' use of their system to hunt insects shows that it must be very sharp and precise, and we have verified this by experiments in the laboratory. Small bats are put through their maneuvers in a room full of standardized arrays of rods or fine wire. Flying in a room with quarter-inch rods spaced about twice their wingspan apart, the bats usually dodge the rods successfully, touching the rods only a small percentage of the time. As the diameter of the rods or wires is reduced, the percentage of success falls off. When the thickness of the wire is considerably less than one tenth the wavelength of the bat's sounds, the animal's sonar becomes ineffective. For example, the little brown bat (*Myotis lucifugus*), whose shortest sound wavelength is about three millimeters, can detect a wire less than two tenths of a millimeter in diameter, but its sonar system fails on wires less than one tenth of a millimeter in diameter.

When obstacles (including insect prey) loom up in the bat's path, it speeds up its emission of sound pulses to help in location. We have made use of this fact to measure the little brown bat's range of detection. Motion pictures, accompanied by a sound track, showed that the bat detects a three-millimeter wire at a distance of about seven feet, on the average, and its range for the finest wires it can avoid at all is about three feet. Considering the size of the bat and of the target, these are truly remarkable distances.

Do the echoes tell the bat anything about the detected object? Some years ago Sven Dijkgraaf at the University of Utrecht in the Netherlands trained some bats to distinguish between two targets which had the form of a circle and a cross respectively. The animals learned to select and land on the target where they had been trained to expect food. Bats can tell whether bars in their path are horizontal or vertical, and they will attempt to get through a much tighter spacing of horizontal bars than of vertical bars. In gliding through a closely spaced horizontal array the bat must decide just how to time its wingbeats so that its wings are level, rather than at the top or bottom of the stroke, at the moment of passage. All in all, we can say that bats obtain a fairly detailed acoustic "picture" of their surroundings by means of echolocation.

Probably the most impressive aspect of the bats' echolocation performance is their ability to detect their targets in spite of loud "noise" or jamming. They have a truly remarkable "discriminator," as a radio engineer would say. Bats are highly gregarious animals, and hundreds fly in and out of the same cave within range of one another's sounds. Yet in spite of all the confusion of signals in the same frequency band, each bat is

INSECT IS LOCATED by means of reflected sound waves (*colored curves*). Variation in the spacing of the curves represents changing wavelength and frequency of the bat's cry.

	BAT	RADARS		SONAR
	EPTESICUS	SCR-268	AN/APS-10	QCS/T
RANGE OF DETECTION (METERS)	2	150,000	80,000	2,500
WEIGHT OF SYSTEM (KILOGRAMS)	.012	12,000	90	450
PEAK POWER OUTPUT (WATTS)	.00001	75,000	10,000	600
DIAMETER OF TARGET (METERS)	.01	5	3	5
ECHOLOCATION EFFICIENCY INDEX	2×10^{9}	6×10^{-5}	3×10^{-2}	2×10^{-3}
RELATIVE FIGURE OF MERIT	1	3×10^{-14}	1.5×10^{-11}	10^{-12}

COMPARISON of the efficiency of the bat's echolocation system with that of man-made devices shows that nature knows tricks which engineers have not yet learned. "Echolocation efficiency index" is range divided by the product of weight times power times target diameter. "Relative figure of merit" compares the echolocation efficiency indexes with the bat as 1.

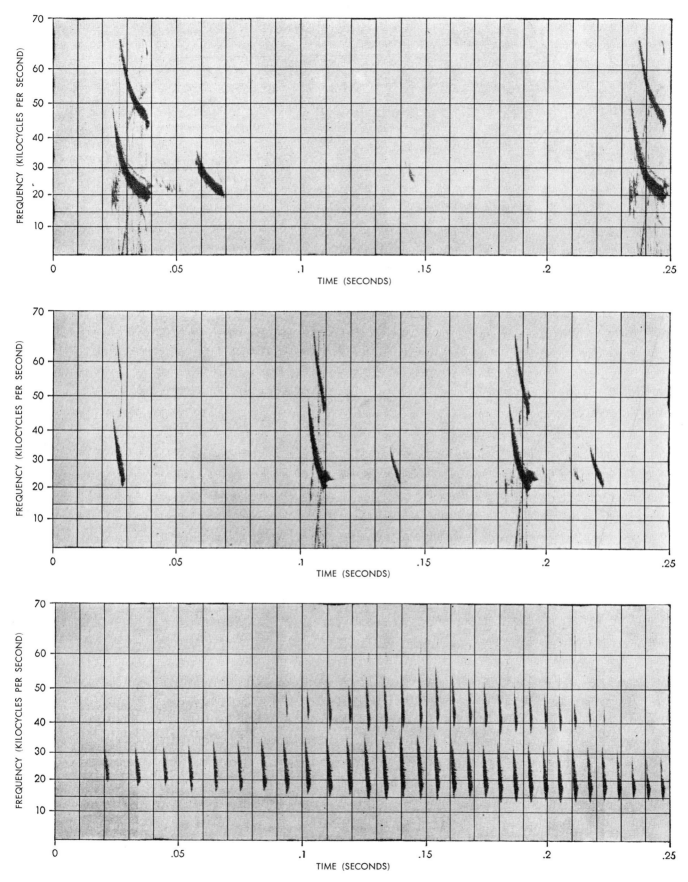

ORIENTATION SOUNDS of the large brown bat were recorded as slanting traces in these spectrograms while the animal was cruising (*top*), beginning pursuit of an insect (*middle*) and closing in on its prey (*bottom*). The traces appearing at .06 seconds in the top spectrogram and at .14 and .22 seconds in the middle spectrogram are echoes, which probably come from nearby buildings.

able to guide itself by the echoes of its own signals. Bats learned long ago how to distinguish the critically important echoes from other distracting sounds having similar properties.

We have recently tested the bats' discriminatory powers by means of special loudspeakers which can generate in-tense sound pulses. We found that a continuous broad-band noise which all but drowned out the bats' cries did not disorient them. They could still evade an insect net with which one tried to catch them; they were able to dodge wires about one millimeter in diameter; they landed wherever they chose.

In some experiments A. D. Grinnell and I did succeed in jamming certain FM bats, but it was not easy, and the effect was only slight. We worked on a species of lump-nosed bat (*Plecotus rafinesquii*) which emits comparatively weak signals. With two banks of loud-speakers we filled the flight room with a noise field of about the same intensity as the bats' echolocation signals. The more skillful individual bats were still able to thread their way through an ar-ray of one-millimeter wires spaced 18 inches apart. Only when we reduced the wires to well below half a millimeter in diameter (less than one tenth the wavelength of the bats' sounds) did the bats fail to detect the wires.

To appreciate the bats' feats of audi-tory discrimination, we must remember that the echoes are very much fainter than the sounds they emit—in fact, fainter by a factor of 2,000. And they must pick out these echoes in a field which is as loud as their emitted sounds. The situation is dramatically illustrated when we play back the recordings at a reduced speed which brings the sounds into the range of human hearing. The bat's outgoing pulses can just barely be heard amid the random noise; the echoes are quite inaudible. Yet the bat is dis-tinguishing and using these signals, some 2,000 times fainter than the back-ground noise!

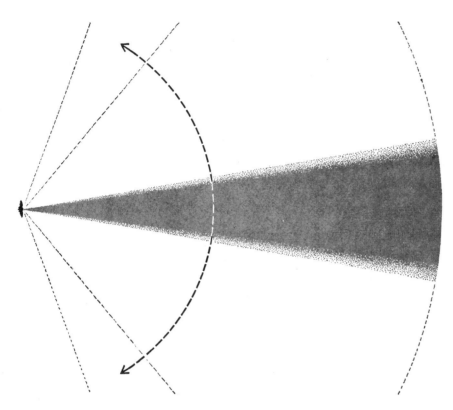

NARROW BEAM which sweeps back and forth is emitted by horseshoe bat in hunting in-sects. Beam is about 20 degrees wide, has a constant frequency and a pulse length of 50 feet.

Much of the modern study of com-munication systems centers on this problem of discriminating information-carrying signals from competing noise. Engineers must find ways to "reach down into the noise" to detect and iden-tify faint signals not discernible by or-dinary methods. Perhaps we can learn something from the bats, which have solved the problem with surprising suc-cess. They have achieved their signal-to-noise discrimination with an auditory system that weighs only a fraction of a gram, while we rely on computing ma-chines which seem grossly cumbersome by comparison.

When I watch bats darting about in pursuit of insects, dodging wires in the midst of the nastiest noise that I can generate, and indeed employing their gift of echolocation in a vast variety of ways, I cannot escape the conviction that new and enlightening surprises still wait upon the appropriate experiments. It would be wise to learn as much as we possibly can from the long and success-ful experience of these little animals with problems so closely analogous to those that rightly command the urgent atten-tion of physicists and engineers.

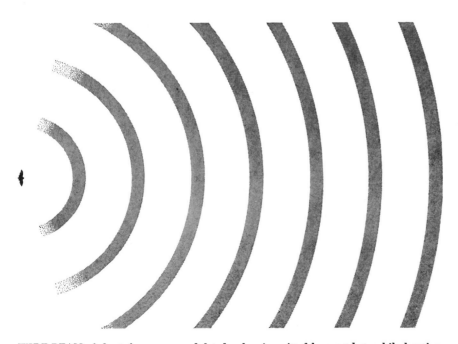

WIDE BEAM of short, frequency-modulated pulses is emitted by most bats while hunting. Each pulse (*gray curves*) is about 1.5 feet long. Beam is fixed with respect to bat's head.

32

THE NAVIGATION OF PENGUINS

JOHN T. EMLEN AND RICHARD L. PENNEY
October 1966

The Adélie penguin travels hundreds of miles from its breeding grounds over virtually featureless land and water and returns to the same nest. How it does so is investigated by experiment

It has long been known that the domesticated pigeon can find its way home after having been transported to a strange location, and experiments conducted in recent years have shown that a number of wild bird species possess the same ability. A series of experiments conducted recently in Antarctica has demonstrated that one wild bird's ability to travel almost straight across hundreds of miles of featureless antarctic landscape depends on a "clock" and a "compass" working in combination. The experiments were conducted with a flightless bird, the Adélie penguin (*Pygoscelis adeliae*). Their results may help to answer the question that inevitably arises in discussions of bird navigation: How do the animals do it? In this article we shall relate why both an unusual field area and an unusual bird were selected for the experiments, relate the outcome of our studies and conclude with a summary of the questions concerning bird navigation to which there now appear to be complete or partial answers.

Early in the nesting season of 1959 five adult male Adélie penguins were captured and banded at an Adélie rookery near Wilkes Station in Antarctica and flown 1,200 miles to McMurdo Sound. The birds were then released; in effect they were being asked if they could find their way home. The majority answer was yes: when the Wilkes rookery penguin population assembled the following spring, three of the five kidnapped males waddled up the beach from the sea to establish their claim at the very nesting sites from which they had been so abruptly taken 10 months earlier. By what route they had traveled home no one knows, but their performance added the Adélie penguin to the lengthening list of bird species able to navigate over long distances. The achievement has an extra dimension; most bird navigators fly, but the penguins had walked, tobogganed on their bellies or swum the entire way.

The observation that migratory birds return to the same nesting sites year after year almost always raises questions that are phrased in terms of how the migrants find their home. The late Gustav Kramer's discovery that captive birds could orient themselves by means of the sun was hailed as a major research breakthrough at the time of his first experiments in Germany 15 years ago. Here was a new insight into a compass-like mechanism that might explain birds' navigational abilities. As Kramer and others pointed out, however, neither bird nor man nor any other animal is able to find the way home from a strange place with the aid of a compass alone. A man who undertakes true bicoordinate navigation needs both a compass and the equivalent of a map on which both his present position and his destination are located; at the very least a bird needs some kind of information about its position with respect to home at some point during its journey. Nonetheless, many investigators who are attempting to discover the cues that may tell navigating birds the direction of home have begun to wonder if the birds have access to such information after all.

One of the more troublesome problems connected with studies of bird navigation gave us a reason for selecting Antarctica as an experimental site. This is the difficulty involved in keeping track of experimental birds and recording their performance. If a bird is placed in a circular cage during its migrating season, it will provide information on some aspects of orientation. But if the bird is released from the cage in the hope of obtaining more information, control of the experiment flies out the window with the bird. Although free-flight experiments making use of birds fitted with telemetering devices that radio their position are currently yielding interesting results, at the time our antarctic experiments were undertaken the advantages offered by a flightless bird that travels overland in an environment free of obstacles and disturbances were compelling.

In addition to being a slow enough traveler on land to be followed easily, the Adélie penguin has a number of other assets as an experimental animal. First, it is a migratory bird with a demonstrated homing ability; each spring

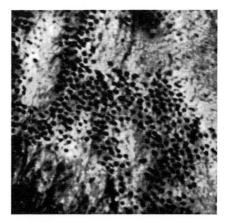

PENGUIN ROOKERY at Cape Crozier in Antarctica has a summer population of some 300,000 Adélie penguins. The aerial photograph on the following page shows a part of the rookery; each of the black dots is the shadow of single or paired birds. The area indicated in color is shown enlarged above. Most of the birds used in the authors' studies of penguin navigation were adult or juvenile inhabitants of the rookery at Cape Crozier.

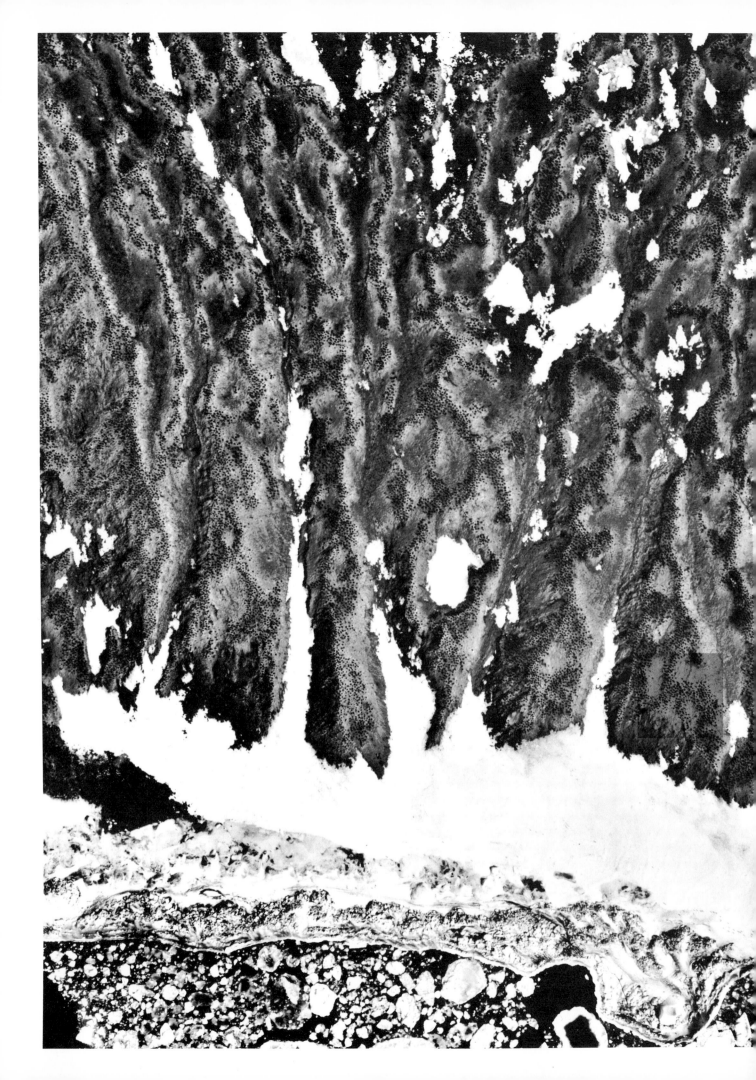

and fall Adélies travel hundreds of miles between the outer fringe of the antarctic pack ice and their rookeries on the coast of the continent. Second, it thrives in subzero weather and shuns all food for weeks on end during the nesting season. Third, it is abundant and easily captured at its rookeries. Fourth, because its normal range is along the coastal fringe of the continent, individual birds are not likely to be familiar with any landmarks in inland areas. Finally, the Adélie's back is broad, black and conspicuous in the polar snowscape,

and on most snow surfaces its tracks can be followed and charted in detail [*see illustration on page 330*].

The experimental advantages of Antarctica itself, in addition to its vast stretches of generally featureless interior, include 24-hour daylight during the summer season. Furthermore, the meridians of longitude come closer and closer together until they meet at the South Pole; therefore a comparatively short journey east or west along the coast may cover dozens of degrees of longitude. With the excellent facilities

provided by the Navy and the scientific-support arm of the Antarctic Research Program, it was possible to move captured penguins readily to locales many miles and many degrees of longitude distant from their home.

The subjects in most of our tests came from a thriving Adélie rookery at Cape Crozier, where some 300,000 birds gather each nesting season. We netted and moved the penguins 20 at a time to our test areas. There they were banded (on the wing) and housed

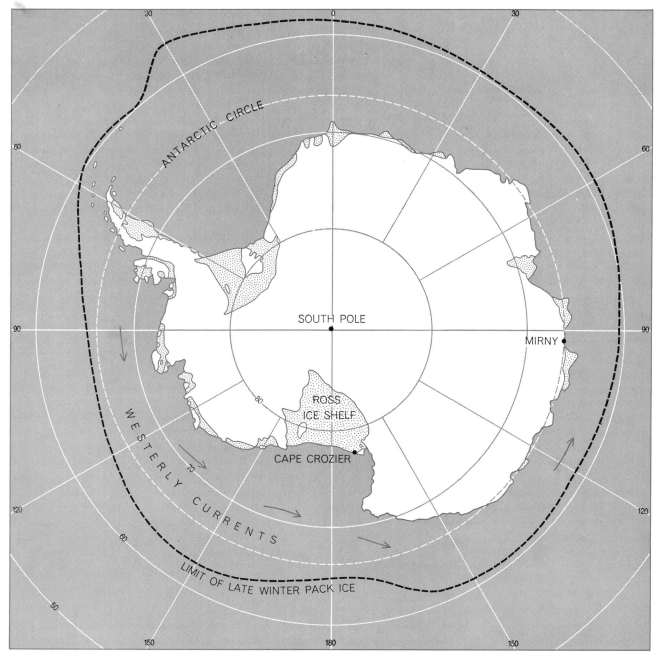

ANTARCTICA was selected as a location for the authors' studies both because Adélie penguins are hardy and accomplished migrants and because moving captive birds a comparatively short distance to the east or west placed them in a completely novel environment that contained a minimum of familiar navigation cues. Each fall the birds travel from any one of a number of coastal rookeries such as Cape Crozier north to the open water along the edge of the pack ice. When spring comes, they return to their home rookeries.

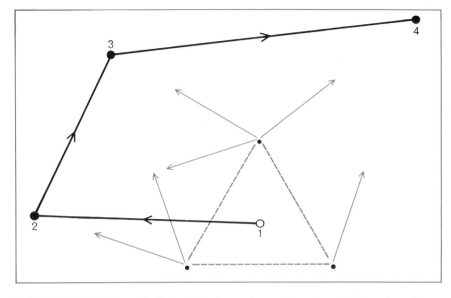

TRACKING SYSTEM required visual checks on the position of each bird at five-minute intervals after it was released from a pit located at the midpoint of an equilateral triangle with sides 200 meters long (1). A transit was located at each point of the triangle; readings from the two best-situated instruments (*arrows*) allowed the bird's position to be recorded until it vanished over the horizon. Some birds were subsequently tracked as far as 16 miles across the snow. A few birds fitted with radio transmitters were monitored up to 50 miles.

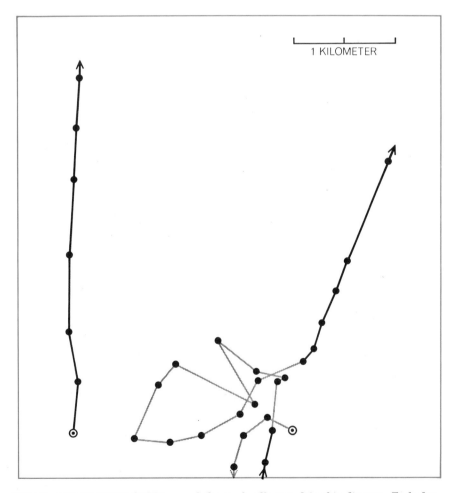

CLEAR AND CLOUDY SKIES caused the tracks illustrated in this diagram. Each dot represents the position of an Adélie penguin at five-minute intervals following its release (*bull's-eye*). In sunlight one bird (*left*) started on and held to a straight course until out of sight. Released under a cloudy sky, a second bird (*right*) headed the opposite way (*gray line*) but reappeared (*black*) when the sky cleared. More clouds left the bird moving at random until sunlight again headed it on much the same course as the first bird's.

in pits excavated in the snow until it was their turn to be tested. Our principal test site was located in the middle of the 150,000-square-mile Ross Ice Shelf, some 180 miles southeast of Cape Crozier, at a point where the 180th meridian and the 80th parallel of south latitude intersect. Here a featureless white landscape stretches northward 105 miles to the Ross Sea and extends unchanged even farther southward and eastward. Only to the west is the flat terrain eventually broken by glacier-crowned mountain ranges, and these were out of sight 180 miles from our camp. Although the snow surface close at hand is wrinkled by irregular wind furrows, the distant line where sky and land meet is monotonously level.

In order to record the direction in which the penguins traveled when they were released we set up three tall tripods so that they formed the points of an equilateral triangle 200 meters on a side. A surveyor's transit was placed at the top of each tripod. At the midpoint of the triangular area we dug a small release pit with a cover that could be pulled clear with a tug on a length of cord that extended to an inconspicuous white shelter some distance away. When all was in readiness, we would remove a bird from its storage pit, put it in the release pit and set the cover in place. We then retired to our shelter and pulled the cord; the bird promptly jumped up to the surface, ready either to travel or, as sometimes happened, to lie down and go to sleep.

Usually the newly freed bird would make a few short dashes in various directions and then stand and peer at the limited scenery: the three transit tripods, our shelter and the antarctic sun. After a few minutes the bird would set off again, this time in a definite direction from which it rarely veered. Every five minutes two observers would take simultaneous azimuth readings, using the two transits that were best situated for the purpose. These readings enabled us to plot the bird's course over a distance of one to three miles and usually for 20 to 40 minutes, by which time our subject would have disappeared over the horizon [*see top illustration at left*]. Sometimes we extended the record by following a bird's tracks for as many as 16 miles from the release site. We also fitted a few penguins with radio transmitters and recorded their direction of travel for 50 miles or so. The bulk of our most useful data, however, was obtained by visually tracking 174 penguins during the first one to three miles of their journey. Most of our subjects were birds from Cape Crozier released at the test site on the Ross Ice Shelf, but we also released birds at this site that had been captured at a rookery near Mirny Station, the Russian antarctic headquarters on the Queen Maud Coast. We also shifted release sites, liberating some Cape Crozier penguins on Marie Byrd Land (at 120 degrees west longitude), on the Victoria Land plateau (at 155 degrees east longitude), at a camp on the pack ice offshore from Cape Crozier and even at the South Pole.

One of the first things we learned was that under clear skies our penguins characteristically selected and followed a straight line of travel for as long as we could see them. They often took small jogs right or left of the line, but these slight deviations were usually corrected in two seconds or so. When we plotted each bird's path as observed at five-minute intervals and compared the length of the plotted path with that of a straight line connecting the starting point and the last observed position, we found that the paths actually traveled by 62 penguins were only 1.025 times longer than the straight-line distances. In fact, nearly 90 percent of the birds achieved a "straightness index" of better than 1.009.

The importance of a clear sky and a good image of the sun to penguin orientation was soon demonstrated. When the sun was veiled by thin clouds (even though we could judge its position in the sky to within a few degrees), the performance of the departing penguins became erratic. When heavy clouds obscured the sun, the birds were completely disoriented and their selection of departure directions became random [see top illustration at right]. A dramatic demonstration of the birds' dependence on the sun was provided by one penguin whose movements we were able to follow for a four-hour interval that included two periods of overcast and two of clear sky. Heavy clouds covered the sun when the bird was released; it soon moved off in a direction exactly opposite to that taken by most birds under a clear sky and was lost from view about a mile away. Half an hour later it reappeared on the horizon, where there was now a patch of sunshine. As the sky cleared above us the bird continued to toboggan in our direction, approaching and passing our observation post. When it had traveled less than half a mile beyond us, clouds covered the sun again; it stopped and after a considerable pause set off at a right angle to its previous heading.

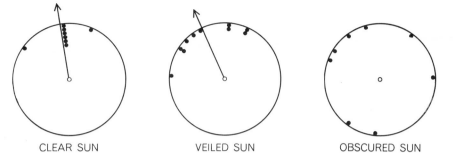

CLEAR SUN   VEILED SUN   OBSCURED SUN

DEPENDENCE ON THE SUN as a compass was demonstrated by recording the penguins' movements in varying weather conditions. Each dot near the edges of the three circles represents the direction in which one bird left the test area; the arrows (*left and center*) show the mean heading. Thin clouds that veiled the sun (*center*) reduced the penguins' ability to orient themselves. When the sun was obscured (*right*), they were totally disoriented.

It wandered irregularly and sometimes hesitantly for the next two hours, which remained cloudy. Then, as the sun broke through again, the bird resumed its interrupted direction of travel and passed out of sight over the horizon.

Navigation that uses the sun as a guide is by no means as simple as it may seem at first. The sun does more than rise in the east and set in the west. Its daily journey across the sky is complicated by an azimuthal motion that continuously alters its position with respect to landmarks on the horizon. In the

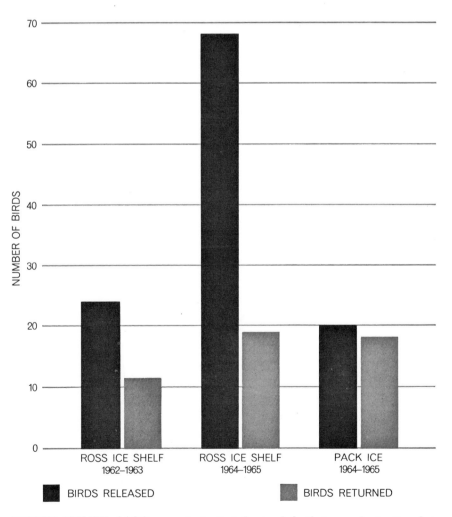

HOMING ABILITY of Adélie penguins is attested not only by their annual migrations but also by the eventual return to Cape Crozier of many of the birds that were transported to areas as far as 200 miles away. Use of the sun as a compass, although it enabled the penguins to orient consistently in unfamiliar territory, is not enough to explain their homing skill.

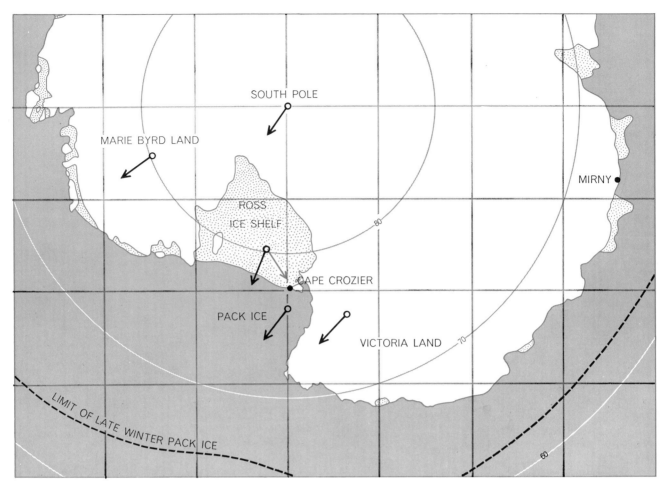

FIVE RELEASE POINTS, situated to the east, west, north and south of Cape Crozier, were used by the authors for their tests. Regardless of where they were released, birds from Cape Crozier all chose escape routes that were roughly parallel and approximately north-northeast with respect to a north-south line passing through Cape Crozier (*black arrows and gray grid*). When birds from the Russian antarctic station of Mirny were transported to the Ross Ice Shelf and released, they first chose escape routes almost perpendicular to those of Cape Crozier birds (*colored arrow*). When plotted with respect to a north-south line passing through Mirny, this heading also proved to be north-northeast. Other penguins from Mirny, kept on the Ross Ice Shelf for a few weeks, readjusted their "biological clock." When released, they chose the same general escape direction taken by the birds from the Cape Crozier rookery.

polar summer, when the sun neither rises nor sets, almost all its motion is azimuthal. To hold a constant course while one's "sun compass" swings around the horizon calls for continuous access to information on the passage of time, yet we found that the penguins were not affected by the sun's shifting azimuth. Birds that were released early in the morning set off northward, even though at that time a northerly heading was about 90 degrees to the left of the sun; at noon their northward departure was straight toward the sun; in the afternoon they still headed north, although by then such a heading was about 90 degrees to the right of the sun. Moreover, once the birds selected a course, they showed no tendency to fix on and follow the sun in its inexorable counterclockwise march along the horizon. This ability to be guided by the sun and yet to compensate for changes in its azimuthal position with the passage of

time is the strongest kind of evidence that the Adélie penguins possess the timing mechanism commonly called a "biological clock."

Although the penguins' general heading after release was northward, it could more accurately be described as north-northeasterly rather than due north. Some considerations of polar geography are a necessary preliminary to discussion of this point. When one is standing on the South Pole, all true compass directions are north. In polar areas, therefore, it is useful to forget about true compass headings and instead to calculate directions on the basis of a rectangular grid. Such grids take the meridian on which the user is located as their central vertical line. In the illustration above, for example, the grid is oriented to the meridian that runs from the South Pole through the Cape Crozier rookery. Analyzing the penguins' travel directions in terms of this grid,

we found that no matter where birds from Cape Crozier were set free they all set off in much the same north-northeast direction. This meant, of course, that they seldom headed for "home" at all. When Cape Crozier penguins were released on the Ross Ice Shelf or at Byrd Station, which are to the east of Cape Crozier, their heading pointed them far to the right of a homing course; when released in Victoria Land, to the west, their heading was about 90 degrees to the left of home. As for the penguins we released on the pack ice some 125 miles offshore, they set off on a north-northeast heading that took them directly away from the Cape Crozier rookery.

The fact that all the Cape Crozier birds, wherever they were released, pursued essentially parallel courses regardless of the actual direction of their home rookery is further evidence that the cue that guided all the birds was far more remote than anything connected with their home or, for that matter, with any localized terrestrial cue. The sun, some 93 million miles away, was their compass, and they could use it unerringly because their biological clocks served to correct for its azimuthal changes. At the same time, it was clear that the Cape Crozier penguins did not head for home. Their movements were oriented in terms of a common direction of escape rather than an approach toward a common goal.

The cooperation of our Russian colleagues at Mirny, 2,000 miles and 90 degrees of longitude away, enabled us to make a further test of our conclusion. Forty Adélie penguins were captured at Mirny and flown to the Ross Ice Shelf. When we released 20 of them, the birds set off north-northeast, but north-northeast only with respect to a grid centered on their home meridian at Mirny. Their actual course with respect to the local grid was northwest, so that their route lay almost at right angles to that of the Cape Crozier penguins.

What functional explanation can be found, in terms of survival value, for the generally northward escape route taken by the Adélie penguins we captured at various rookeries? Northward, to be sure, is always seaward in Antarctica; an attractively simple explanation is that movement northward would carry a lost penguin along the shortest route to the sea—the species' only source of food. There is also another consideration. For migratory birds such as the Adélie penguin a sun compass "set" to a home meridian would tend to reduce lateral dispersal of any one rookery's population and ensure that the same group of penguins returned from the fringes of the pack ice to the same stretch of coast each year. Indeed, the home meridian—regarded as a reasonably broad corridor with one end in the coastal rookery and the other in the food-rich waters at the edge of the pack ice—may be what "home" is to an Adélie penguin. The bird's annual movements inward or outward along the corridor could be the result of physiologically controlled negative or positive responses to the same basic orientation cue or cues.

What accounts for the slight but unmistakable easterly vector that turns the outward journey from true north to north-northeast? At the end of our first season's work in Antarctica we were prepared to dismiss it as a chance variation. Further and stronger evidence of the easterly vector gathered in 1964

POSITION READING is taken by a member of the authors' party, who stands atop a stepladder in the middle of the Ross Ice Shelf. The escape paths pursued by 54 penguins released during clear weather varied from a straight line by a factor of less than 1 percent.

ESCAPING PENGUIN ignores another bird's path as it heads across the Ross Ice Shelf. Adélie penguins travel two to three miles per hour; the fastest did eight miles per hour.

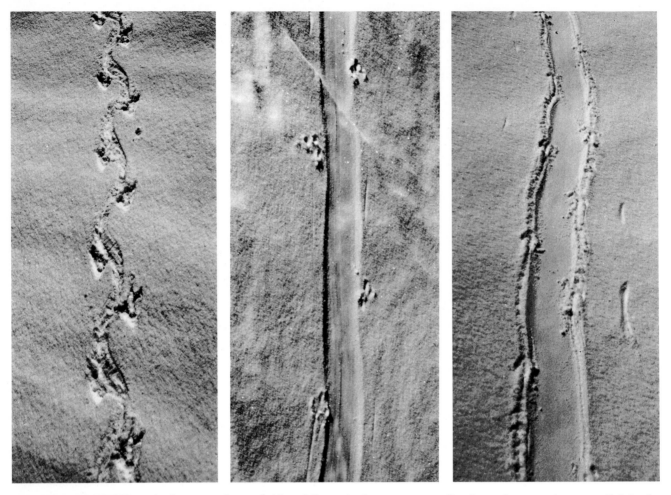

PENGUIN'S PROGRESS on land can vary from a brisk waddle (*left*), in which the bird's oscillating tail leaves a serpentine track in the snow, to a swift tobogganing, sometimes propelled by the feet alone (*center*) and sometimes assisted by wing strokes (*right*).

made it impossible to ignore. One possible answer suggests itself at present: The coastal currents in both the Cape Crozier and the Mirny areas are westerly. A penguin adrift on the offshore pack ice would be slowly carried westward out of its home zone unless it compensated by moving to the east from time to time. Similarly, a bird moving from the coast to the outer fringe of the pack ice would be borne toward the west if its line of march were true north instead of north-northeast. One observation that lends weight to our suggestion that the easterly vector is a compensation for westerly coastal currents is that the Mirny birds released on the Ross Ice Shelf showed a stronger easterly bias than the Cape Crozier birds released there. The westerly currents off the Mirny coast are known to run strong.

The contrast between the escape headings chosen by Cape Crozier and Mirny birds immediately raises two questions. It seems safe to assume that a penguin's ability to select an escape direction when transferred to a strange

environment is something innate. But what about the specific direction selected? Is this innately determined or is it a function of the environment in which the bird was raised? And can external influences alter the setting of the bird's compass?

We decided to see what information the escape behavior of two-month-old, inexperienced penguins might provide concerning the first of these questions. We captured a number of Cape Crozier chicks, some near the nests where they had been hatched and others on the beach that borders the rookery. When they were released on the Ross Ice Shelf, all of them headed off generally north-northeast on a straight course, although they showed somewhat greater variations in their selection of departure directions than did the adult Adélies.

The chicks' possession of a directional preference cannot, however, be taken as proof that the response is innate in the species. From the day they emerge from their eggs the Adélie penguins at

Cape Crozier look down a steep slope that faces the Ross Sea to the northeast. Below, adult penguins are constantly entering and leaving the water, returning with food for the ravenous chicks (who, when their hunger is unsatisfied, will often pursue a parent part of the way back to the water). The same north-easterly direction is the one in which the chicks themselves will move once they become independent of the nesting area. North-northeast thus becomes a significant direction to the birds as they mature. It seems plausible that such an upbringing, in which the topographic cues that come to mean "seaward" are constantly associated with the celestial cues that represent north-northeast, might allow the celestial cues to persist and function in situations where the topographic ones are no longer available.

As for the second question, we would have found it surprising if our birds had proved impervious to the environmental influences that are known to reset the biological clocks of many Temperate Zone animals within a period of a week or so. Yet, compared with the day-and-night cycle of the Temperate Zone, the cues available in the continuous daylight of the antarctic summer are notably weak ones. To put the matter to a test we kept the remaining 20 of the 40 birds from Mirny penned under the open sky on the Ross Ice Shelf for periods that ranged from 18 to 27 days. These birds had arrived with their clock synchronized to a time zone some 88 degrees to the west (equivalent to the six-hour time difference between London and New Orleans). At the end of their three- to four-week detention the 20 Mirny birds selected local north-northeast as their departure direction. Their clock—and thus their compass—had been reset; the resetting turned them 55 degrees to the right and pointed them in the same general escape direction as the one selected by the "local" Cape Crozier birds.

This finding leaves us in something of a dilemma. If all the Adélie penguins in the Antarctic can reset their clock as easily as the 20 birds from Mirny did, our earlier speculation that a constant local escape orientation would be useful in maintaining population-specific coastal zones may not be warranted. Any birds that wandered toward the lateral boundary of their population's corridor would be in danger of adapting to the new conditions there. The unifying interrelation of breeding grounds and feeding area that is provided by a common compass setting would therefore be lost to them. For the moment we can only suggest that unconfined birds may not reset their clock as readily as birds that are artificially detained do.

We have seen that the Adélie penguin can navigate with the sun as a directional cue and with an inherent timing mechanism as compensation for the sun's position changes from hour to hour. We have also seen that the consistent escape direction chosen by released birds is a northerly one that under normal circumstances would possess considerable survival value by leading a lost bird to the nearest feeding ground at sea. Unfortunately these two observations tell us little about the question that faced us at the outset: How do the birds find their way home? The fact that so many of our experimental subjects did reach their home rookeries indicates that Adélie penguins possess something by way of navigational equipment in addition to the sun-compass orientation they demonstrated during the release experiments. Cape Crozier birds released on the Ross Ice Shelf consistently headed far to the right of their true homeward direction, but at least half of them were back at Cape Crozier within 25 days. Of the 20 Cape Crozier birds we turned loose on the pack ice many miles offshore, 15 were back home within two weeks even though all had started off in exactly the opposite direction. How did the three birds from Wilkes Station and at least two of the birds from Mirny find their way home over thousands of intervening miles in time for the next nesting season? For that matter, how do millions of Adélie penguins do much the same thing in their annual migrations?

We can only speculate that a homing bird must receive new information at some point on its long trek. We suspect that the birds receive this new information when they reach the point at which open water enables them to swim and feed after a long period of deprivation. What a homing bird does and what cues it uses at this or any other critical point in its travels, however, are unknown. A simple 180-degree reversal in direction of travel, still depending on the sun compass, would be a maneuver useful to a penguin that had finally penetrated to the seaward fringe of the pack ice. Birds within 75 miles or so of the coast might also make use of visual cues such as the 13,000-foot peak of Mount Erebus on Ross Island. But at this stage in our investigations it is fruitless to speculate further.

To sum up, we have shown that the Adélie penguin uses a sun compass for orientation when released in a strange environment. We surmise that the same sun compass may guide the birds back from the open water to their breeding grounds after some other as yet undetermined cue or cues have reversed their direction of travel. A number of questions remain to be answered. We may, however, already have the answer to one of them. The antarctic winter is the season during which Adélie penguins spend their time away from land; when they leave the polar ocean at all, they perch on the ceaselessly shifting and drifting floes of the offshore pack ice. The antarctic winter is also a time of nearly total darkness, just as summer is a time of 24-hour-long daylight. How do these competent little navigators keep track of their position during the long dark period? We suspect that in the Adélie penguins' winter habitat the sun's brief appearance low on the horizon each noon is a strong enough cue to keep the birds' sun compass correctly set throughout the long polar night.

VII

HORMONES AND INTERNAL REGULATION

There is another remarkable fact about songbirds; they usually change their colour and note with the season, and suddenly become different—which among the larger class of birds only cranes do, for these grow black in old age. The blackbird changes from black to red; and it sings in the summer, and chirps in winter, but at midsummer is silent; also the beak of yearling blackbirds, at all events the cocks, is turned to ivory colour. Thrushes are of a speckled colour round the neck in summer but self-coloured in winter.

Pliny
NATURAL HISTORY, X, xli.

VII

HORMONES AND INTERNAL REGULATION

INTRODUCTION

Research during the past decade has truly revolutionized our understanding of the hormonal system of vertebrates. Two major areas of information involve: first, the central role of the hypothalamus in controlling endocrine activities; and, second, the mode of hormone action at the cell level.

There are two groups of neurosecretory cells in the hypothalamus. Those of the first group have cell bodies at various points in that portion of the brain and neurosecretory endings in the median eminence or stalk region near the anterior pituitary gland (the adenohypophysis). The other neurosecretory cells have axonal terminations in the posteror pituitary gland (the neurohypophysis). "The Hormones of the Hypothalamus" by Roger Guillemin and Roger Burgus summarizes the ways that the first group of cells produces, stores, and releases specific molecules (all of which may be polypetides) that act as releasing agents for anterior pituitary hormones: growth hormone, leuteinizing hormone, corticotropin (ACTH), follicle-stimulating hormone, and thyrotropin are all released by these agents. Neuronal discharge in the hypothalamus apparently stimulates appropriate neurosecretory discharge of the agents, which then travel by special portal blood vessels to the adenohypophysis where they cause secretion of the hormones. Interestingly, the secretion of prolactin and probably of melanocyte-stimulating hormone (MSH) is apparently inhibited by certain hypothalamic substances, though for both hormones actual releasing factors may also operate.

The second group of neurosecretory cells, extending into the posterior pituitary, produces, stores, and releases vasopressin (ADH or antidiuretic hormone) and oxytocin. It is worth noting that there is little strong evidence favoring the two-step secretory mechanism illustrated as part *C* of the diagram on page 346. Instead, the direct release from nerve terminals, as in part *D*, may be the more common mode. Within neurosecretory cells, both vasopressin and oxytocin are apparently bound to a protein called neurophysin, so that discharge of the hormones requires dissociation from this complex. When nerve impulses reach the secretory endings, calcium ions (Ca^{++}) enter the cells and somehow participate in the release of the hormones.

The steps leading to release of the neurohypophyseal hormones are simpler than those causing discharge of adenohypophyseal hormones. Thus, nerve discharge in the hypothalamus causes the release of the neuro-hypophyseal hormone ADH, which acts directly on the target organs, the kidney tubules, to produce physiological effects. In contrast, when hypothalamic activity stimulates the production of the ACTH-releasing agent, that chemical must go to the adenohypophysis to cause discharge of ACTH, which in turn passes to the adrenal gland to evoke increased secretion of such compounds as cortisol (one of the adrenal steriods). The cortisol then acts on the target organ to produce a physiological response; the response may then affect in some way the hypothalamus, and so complete the loop. Thus, three agents and three cell types are involved in the complete process from nerve discharge to target-organ response. Other adenohypophyseal hormones involve only two agents and cell types in their loops.

It is not certain which of the two types of secretory control is more primitive. Although the neurohypophyseal system appears to be simpler, it is difficult to assign specific and consistent functions to ADH or to oxytocin in the fishes. ADH, for instance, seems to affect ion (particularly sodium) balance in fishes, but only influences water balance in terrestrial vertebrates. The endocrinologists Aubrey Gorbman and Howard A. Bern have suggested that, in the earliest vertebrate endocrine system, hypothalamic neurosecretion may have acted on the adenohypophysis. Then, hypothalamic cell secretions began acting directly elsewhere in the body. The cells producing these neurosecretions (ADH, oxytocin) could next have acquired a more efficient means of secreting directly into the systemic circulation; thus, the discrete anatomical area, the neurohypophysis (pars nervosa) and its blood supply appeared.

Further studies have shown that there are close chemical relationships between endocrine agents performing quite different functions. Thus, ADH is very similar to the ACTH-releasing polypeptide that originates in the median eminence near the adenohypophysis. In fact, injected ADH can cause release of ACTH from the adenohypophysis (it is not absolutely certain here whether the ADH acts directly upon the ACTH, or indirectly by causing release of ACTH-releasing factor). Mammalian oxytocin differs from vasopressin in only two of its eight amino acids; and, mammalian vasopressin differs from the vasopressin of other terrestrial vertebrates (arginine "vasotocin") by a single amino acid substitution (see the following paragraph for details). Still another similarity in structure is that between MSH and ACTH: the amino acid sequence of the first compound is found in the longer peptide chain of the second. Finally, the whole family of steroids produced by the adrenal glands (aldosterone, cortisol, and so on) are built around a single basic chemical backbone. We could list other examples of this phenomenon, but these should make it clear that at all levels in the endocrine system the same basic molecules are adapted for different purposes. This, of course, is one reason why different hormones may cause similar effects: for example, an ACTH injection will cause pigment cells to darken; recall its structural similarity to MSH, the hormone that normally controls this process. If in fact the same molecules are used in many ways, then it is obvious that the critical variable in the endocrine system is the sensitivity and specificity of target organs. Thus, some anterior pituitary cells respond to a particular hypothalamic peptide by secreting a hormone, and certain kidney tubular cells respond to a very closely related hypothalamic peptide by becoming permeable to water; the difference is due to the target organ, and appropriately so, for that is where the physiologically useful response occurs. In summary, two properties are critical to the development of the endocrine system: (1) the production of specific chemicals by hypothalamic cells; (2) the ability of

Figure 1. Representations of the pituitary gland of a bird (left) and of that of a mammal (right). The median eminence is the source of the various "releasing factors." In a bird, none of the neurosecretory cells originating in the hypothalamus end in the median eminence; instead, blood vessels course near the axons of the neurosecretory cells and then carry the blood and releasing factors downward to the adenohypophysis. In contrast, in a mammal, some neurosecretory cells actually terminate near the capillary loops that pick up the releasing factor peptides. Note that the blood supply of the neural lobe of the neurohypophysis is separate from that of the adenohypophysis in both kinds of organisms. This separate circulatory system, and the region it drains in the neurohypophysis, are not found in fishes; hence the presence of these structures is correlated with life on land and with the ready release of vasopressin and oxytocin.

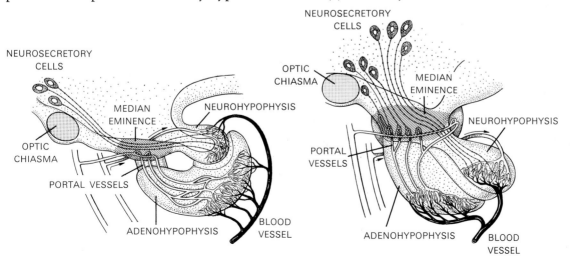

different cell types to respond physiologically, in different ways, to such chemicals. Evolutionary changes in the endocrine system have centered on variations in molecular hormone structures, and on alterations in sensitivity and selectivity among different potential target cell types.

A good example of the mechanisms of evolutionary change has been provided by analyses of the amino acid sequences of the posterior pituitary hormones in various vertebrates. Professor R. Archer of the University of Paris has recently summarized the data. It seems very likely that the gene coding for the ancestral molecule was duplicated and that each copy subsequently evolved separately by modifications in the amino acids located at positions 3, 4, and 8 of the peptide. Thus, an oxytocin "line" and a vasopressin "line" can be traced through the vertebrates (neither hormone has been detected yet in cyclostomes). The table shows the amino acid variations in the posterior pituitary hormones of several classes of vertebrate animals (the numbers in superscript refer to the positions in the peptide chains). In order to use the table, assume that, for the "oxytocin" line, position number 3 is isoleucine in all forms, with the indicated variations occurring in positions 4 and 8; for the "vasopressin" line, assume glutamine occupies position 4 throughout, and variations occur at positions 3 and 8. The invariability of the other amino acids suggests their critical nature in the shape of the molecules, whereas the substitutions at positions 3, 4 and 8 are obviously sources of difference in physiological action.

position	1	2	3	4	5	6	7	8	9
amino acid	cystine	tyrosine	☐	☐	asparagine	cystine	proline	☐	glycine

	OXYTOCIN LINE		VASOPRESSIN LINE	
bony fish	*isotocin*	serine[4], isoleucine[8]	*vasotocin*	isoleucine[3], arginine[8]
amphibians	*mesotocin*	glutamine[4], isoleucine[8]	*vasotocin*	isoleucine[3], arginine[8]
mammals	*oxytocin*	glutamine[4], leucine[8]	*vasopressin*	phenylalanine[3], arginine[8]
pig	*oxytocin*	glutamine[4], leucine[8]	*vasopressin*	phenylalanine[3], lysine[8]

The second major area of new information about hormones and the endocrine system centers on the mode of hormone action on cells. On the one hand, peptide and protein hormones appear to elicit formation of cyclic AMP in responsive cells; this "second messenger" then evokes an appropriate physiological response to the original hormone. A typical instance of this process is discussed in the article "Calcitonin" by Rasmussen and Pechet. In contrast to data concerning peptide hormones, increasingly conclusive data implies that steroid hormones may actually be transported into cells or even into nuclei where direct effects on the genetic apparatus or other cellular organelles are possible. These observations on the interactions of peptide and steroid hormones with cells emphasize the crucial role of receptor molecules. The presence or absence of receptor molecules may well be the key to "specificity" in the endocrine system, the reason why only certain cells respond even though hormones travel everywhere in the blood system and body fluids.

Two points must be made to modify "The Pineal Gland" by Richard J. Wurtman and Julius Axelrod. First, it is now known that the control of MSH release is complicated by the role of melatonin. When the pineal discharges melatonin, stores of MSH in the adenohypophysis are depleted. Therefore, both the hypothalamic releasing agent and melatonin can control MSH levels in the blood. Second, there are apparently significant differences among vertebrates with respect to the relationship between the "biological clock" and the pineal. In birds, the pineal is clearly associated with clock-driven phenomena; in mammals, however, an area of the brain above the optic chiasma may perform equivalent functions.

The pineal organ, which is usually assumed to participate in control of pigmentation or cyclical biological activities, may also affect responses to altered ion balance and osmotic pressure. When abnormally low body fluid volumes, or high extracellular potassium ion (K^+) concentration is present, a substance called glomerulotropin is apparently released from the pineal gland. This hormone acts on the adrenal gland to adjust aldosterone levels and so bring about sodium retention and potassium excretion. Evidence that such a hormone exists comes from experiments in which pineal extracts were injected into operated animals.

Analogous experiments have recently been peformed on rats, and they demonstrate that the pituitary gland has a role in aldosterone secretion. If its pituitary gland is missing, a rat fails to respond to sodium depletion by releasing aldosterone. Injection of whole frozen pituitaries restores this ability, but injection of various known pituitary hormones (such as ACTH and the growth hormone, STH) does not. Perhaps a heretofore unidentified hormone is involved in the reaction.

If they are corroborated, these results will imply that the renin-aldosterone regulatory system of the kidney can be influenced by central nervous system tissues (the pineal and the hypothalamus-pituitary tissues), just as other peripheral endocrine systems can.

Our knowledge of the intricate workings of calcium and phosphate regulation has been modified drastically by work on the recently discovered hormone "Calcitonin" (see article by Rasmussen and Pechet). For years it has been clear that abnormally low calcium levels in the blood lead to release of parathyroid hormone from the parathyroid gland (see "The Parathyroid Hormone" by H. Rasmussen, *Scientific American*; Offprint 86). Parathyroid hormone acts on bone cells called osteoclasts to stimulate their bone-destroying activities, one result being an increase in calcium ions in the blood. Calcitonin acts in just the reverse sense: when calcium levels are too high, calcitonin is released so that it can inhibit osteoclast function. As a result calcium levels fall.

The system to regulate calcium ions is more complex than this brief treatment implies. For instance, some evidence suggests that a thyrocalcitonin-releasing factor comes from the parathyroid gland to act upon the thyroid and cause discharge of active calcitonin; if so, this relationship is like that between the hypothalamus and the anterior pituitary cells. In summary, calcium regulation results from a balance of positive (parathormone) and negative (thyrocalcitonin) hormones acting concordantly with vitamin D.

One of the main obstacles to our understanding of this regulatory complex is its high variability among different vertebrates. A dog or a pigeon is in immediate distress after the parathyroids are removed, but a rat or a rooster is hardly affected. Different amphibia are equally variable, and no parathyroids are known in fishes. In fishes, however, another derivative of the anterior embryonic gut is present and may be a source of a calcium-regulating hormone. This tissue is called the ultimobranchial body, and it arises as a pouch, as do the thyroid and parathyroid glands. In terrestrial vertebrates the ultimobranchial body may regress (in birds), or it may persist (in mammals) and become a part of the region in which thyrocalcitonin-secreting cells are found. Thyrocalcitonin has been isolated from several birds, a reptile, and a dogfish shark. The presence of a discrete ultimobranchial body in lower vertebrates, and the probable positive correlation between its presence and the production of thyrocalcitonin, suggest that the negative regulation by thyrocalcitonin may be the primitive means of controlling calcium levels in vertebrates, and that the parathyroid gland and parathormone came later in terrestrial forms as a refinement in control. Verification of this hypothesis must wait until thyrocalcitonin has been identified in various types of bony fishes.

An interesting sidelight on vitamin D action in man has been suggested by W. Farnsworth Loomis of Brandeis University (see "Rickets," *Scientific American*; Offprint 1207). Vitamin D can, in fact, be manufactured in vertebrate skin when ultraviolet light of the correct wavelength and intensity strikes the body; thus, the compound might be called a "hormone," since it is produced by one cell type and acts on another. The amounts of vitamin D in the human diet vary, being generally very low in winter time. Consequently, the sunlight-induced conversion of 7-dehydrocholesterol into vitamin D is the sole source of the agent if dietary supplementation is not employed. Normal amounts of Vitamin D within the body are essential: rickets and "softening" of the adult bones result from a deficiency in the vitamin; abnormal calcifications and kidney stones may be caused by an overabundance.

Loomis proposes that the quantities of vitamin D synthesized in skin are controlled by the amounts of melanin pigmentation (browns, blacks) or of keratinization (yellows) in the outer cell layers of the epidermis. Among human populations, such pigmentation varies with latitudes, the lighter, unpigmented skins being common in northern latitudes, where the intensity of ultraviolet light is very low for portions of the year. But in the latitudes near the equator, where the intensity of sunlight is high, darker skins that produce relatively less vitamin D are found. As might be expected, in the populations of the northern latitudes reversible pigmentation —suntanning—can occur to prevent excess production of vitamin D during the summer when the sun is high in the sky. Perhaps it is no coincidence that the melanization and the keratinization that constitute suntanning are elicited by the same wavelengths of light that produce vitamin D. Data show that the amount of ultraviolet radiation transmitted by the skin of an albino African Negro is close to that transmitted by the skin of caucasian northern Europeans, and it is about three times the amount transmitted by normal Negro skin (i.e., 53 percent of incident ultraviolet is passed through the skin of an albino African Negro; 18 percent, through the skin of a normal Negro).

Some of the implications of these relations are borne out by medical data: negroes tend to be particularly susceptible to rickets, particularly in high latitudes. Although less information is available, it also seems likely that white northerners may be disposed to have kidney stones when living in tropic regions. Evolutionary implications are equally obvious: presumably the early hominids of Africa were hairy and pigmented and had no problem of too much Vitamin D. But some of their descendants migrated north, and those early men became subject to Vitamin D deficiency and, to counteract this, depigmentation may have resulted. The only northern people having reasonably heavy pigmentation are the Eskimos, and their diet is heavily supplemented with fish oils containing Vitamin D.

Although Loomis' arguments are controversial, they offer a logical and cohesive explanation of pigmentation differences among men. They also suggest an answer to the long-standing enigma of pigmentation and heat absorption. It has never been clear why darkly pigmented skin, which absorbs heat so well, should be found in the tropics, while white skin, which tends to reflect heat rather than absorbing it, should be found in cooler climates. Just the opposite might be expected. However, if ultraviolet absorption and regulation of vitamin D synthesis is more critical than problems of heat balance, then an explanation is available.

The last article in this section provides a comprehensive illustration of the way that a variety of hormones interact during an important biological process. In "How an Eggshell Is Made," T. G. Taylor emphasizes the complexity of interactions between bones, oviduct, pituitary, and gonads. The inescapable conclusion is that the evolutionary appearance of the cleidoic egg (see "An Essay on Vertebrates") must have involved numerous mutational events taking place over eons—not a precipitous acquisition of the new mode of reproduction.

33

THE HORMONES OF THE HYPOTHALAMUS

ROGER GUILLEMIN AND ROGER BURGUS
November 1972

The anterior pituitary gland, which controls the peripheral endocrine glands, is itself regulated by "releasing factors" originating in the brain. Two of these hormones have now been isolated and synthesized

The pituitary gland is attached by a stalk to the region in the base of the brain known as the hypothalamus. Within the past year or so, after nearly 20 years of effort in many laboratories throughout the world, two substances have been isolated from animal brain tissue that represent the first of the long sought hypothalamic hormones. Because the molecular structure of the new hormones is fairly simple the substances can readily be synthesized in large quantities. Their availability and their high activity in humans has led physiologists and clinicians to consider that the hypothalamic hormones will open a new chapter in medicine.

It has long been known that the pituitary secretes several complex hormones that travel through the bloodstream to target organs, notably the thyroid gland, the gonads and the cortex of the adrenal glands. There the pituitary hormones stimulate the secretion into the bloodstream of the thyroid hormones, of the sex hormones by the gonads and of several steroid hormones such as hydrocortisone by the adrenal cortex. The secretion of the thyroid, sex and adrenocortical hormones thus has two stages beginning with the release of pituitary hormones. Studies going back some 50 years culminated in the demonstration that the process actually has three stages: the release of the pituitary hormones requires the prior release of another class of hormones manufactured in the hypothalamus. It is two of these hypothalamic hormones that have now been isolated, chemically identified and synthesized.

One of the hypothalamic hormones acts as the factor that triggers the release of the pituitary hormone thyrotropin, sometimes called the thyroid-stimulating hormone, or TSH. Thus the hypothalamic hormone associated with TSH is called the TSH-releasing factor, or TRF. The other hormone is LRF. Here again "RF" stands for "releasing factor"; the "L" signifies that the substance releases the gonadotropic pituitary hormone LH, the luteinizing hormone. A third gonadotropic hormone, FSH (follicle-stimulating hormone), may have its own hypothalamic releasing factor, FRF, but that has not been demonstrated. It is known, however, that the hypothalamic hormone LRF stimulates the release of FSH as well as LH.

Studies are continuing aimed at characterizing several other hypothalamic hormones that are known to exist on the basis of physiological evidence but that have not yet been isolated. One of them regulates the secretion of adrenocorticotropin (ACTH), the pituitary hormone whose target is the adrenal cortex. Another hormone (possibly two hormones with opposing actions) regulates the release of prolactin, the pituitary hormone involved in pregnancy and lactation. Still another hormone (again possibly two hormones with opposing actions) regulates the release of the pituitary hormone involved in growth and structural development (growth hormone).

That the hypothalamus and the pituitary act in concert can be suspected not only from their physical proximity at the base of the brain but also from their development in the embryo. During the early embryological development of all mammals a small pouch forms in the upper part of the developing pharynx and migrates upward toward the developing brain. There it meets a similar formation, resembling the finger of a glove, that springs from the base of the primordial brain. Several months later the first pouch, now detached from the upper oral cavity, has filled into a solid mass of cells differentiated into glandular types. At this point the second pouch, still connected to the base of the brain, is rich with hundreds of thousands of nerve fibers associated with a modified type of glial cell, not too unlike the glial cells found throughout the brain. The two organs are now enclosed in a single receptacle that has formed as an open spherical cavity within the sphenoid bone, on which the brain rests.

This double organ, now ensconced in the sphenoidal bone, is the pituitary gland, or hypophysis. The part that migrated from the brain is the posterior lobe, or neurohypophysis; the part that migrated from the pharynx is the anterior lobe, or adenohypophysis. Both parts of the gland remain connected to the brain by a common stalk that goes through the covering flap of the sphenoidal cavity. For many years after the double embryological origin of the pituitary gland was recognized the role of the gland was no more clearly understood than it had been in the old days. Indeed, the name "pituitary" had been given to it in the 16th century by Vesalius, who thought that the little organ had to do with secretion of *pituita:* the nasal fluid.

We know now that the anterior lobe of the pituitary gland controls the secretion and function of all the "periph-

HYPOTHALAMIC FRAGMENTS of sheep brains were the source from which the authors' laboratory extracted one milligram of TRF, the first hypothalamic hormone to be characterized and synthesized. The photograph is of about 30 frozen hypothalamic fragments; some five million such fragments, dissected from 500 tons of sheep brain tissue, were processed over a period of four years.

342

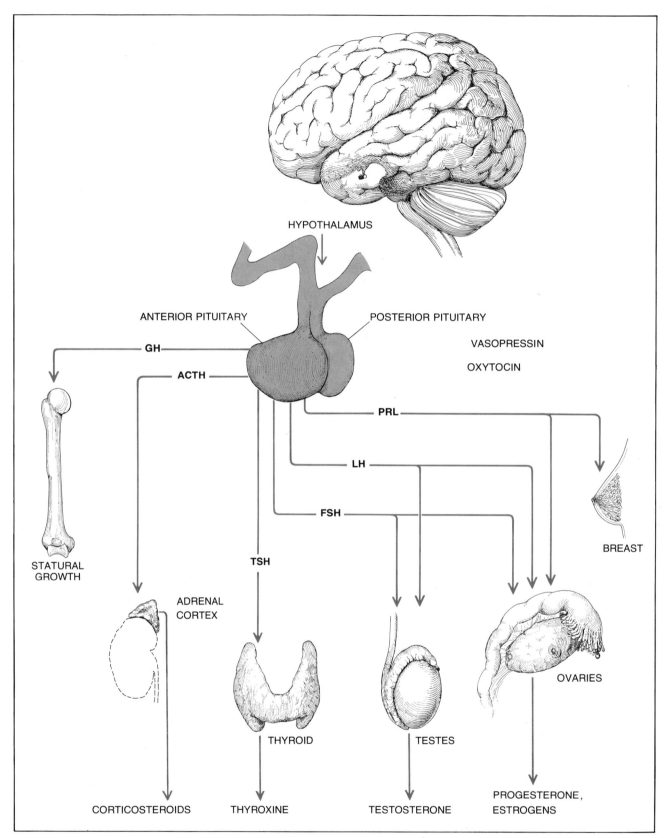

HYPOTHALAMUS

ANTERIOR PITUITARY

POSTERIOR PITUITARY

GH

ACTH

VASOPRESSIN

OXYTOCIN

PRL

LH

FSH

TSH

STATURAL
GROWTH

ADRENAL
CORTEX

THYROID

TESTES

BREAST

OVARIES

CORTICOSTEROIDS

THYROXINE

TESTOSTERONE

PROGESTERONE,
ESTROGENS

PITUITARY GLAND, connected to the hypothalamus at the base of the brain, has two lobes and two functions. The posterior lobe of the pituitary stores and passes on to the general circulation two hormones manufactured in the hypothalamus: vasopressin and oxytocin. The anterior lobe secretes a number of other hormones: growth hormone (GH), which promotes statural growth; adreno-corticotropic hormone (ACTH), which stimulates the cortex of the adrenal gland to secrete corticosteroids; thyroid-stimulating hormone (TSH), which stimulates secretions by the thyroid gland, and follicle-stimulating hormone (FSH), luteinizing hormone (LH) and prolactin (PRL), which in various combinations regulate lactation and the functioning of the gonads. Several of these anterior pituitary hormones are known to be controlled by releasing factors from the hypothalamus, two of which have now been synthesized.

eral" endocrine glands (the thyroid, the gonads and the adrenal cortex). It also controls the mammary glands and regulates the harmonious growth of the individual. It accomplishes all this by the secretion of a series of complex protein and glycoprotein hormones. All the pituitary hormones are manufactured and secreted by the anterior lobe. Why should this master endocrine gland have migrated so far in the course of evolution (a journey recapitulated in the embryo) to make contact with the brain? As we shall see, recent observations have answered the question.

The posterior lobe of the pituitary has been known for the past 50 years to secrete substances that affect the reabsorption of water from the kidney into the bloodstream. These secretions also stimulate the contraction of the uterus during childbirth and the release of milk during lactation. In the early 1950's Vincent Du Vigneaud and his co-workers at the Cornell University Medical College resolved a controversy of many years' standing by showing that the biological activities of the posterior lobe are attributable to two different molecules: vasopressin (or antidiuretic hormone) and oxytocin. The two molecules are octapeptides: structures made up of eight amino acids. Du Vigneaud's group showed that six of the eight amino acids in the two molecules are identical, which explains their closely related physicochemical properties and similar biological activity. Both hormones exhibit (in different ratios) all the major biological effects mentioned above: the reabsorption of water, the stimulation of uterine contractions and the release of milk.

As early as 1924 it was realized that the hormones secreted by the posterior lobe of the pituitary are also found in the hypothalamus: that part of the brain with which the lobe is connected by nerve fibers through the pituitary stalk. Later it was shown that the two hormones of the posterior pituitary are actually manufactured in some specialized nerve cells in the hypothalamus. They flow slowly down the pituitary stalk to the posterior pituitary through the axons, or long fibers, of the hypothalamic nerve cells [*see top illustration on page 345*]. They are stored in the posterior pituitary, which is now reduced to a storage organ rather than a manufacturing one. From it they are secreted into the bloodstream on the proper physiological stimulus.

These observations had led several

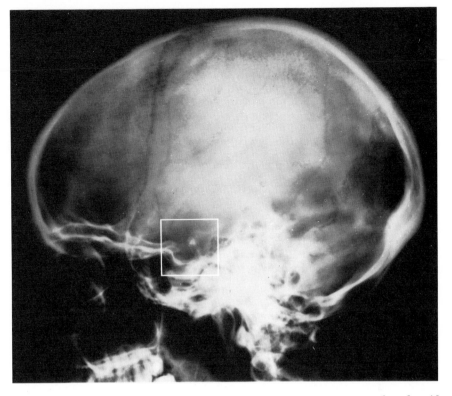

BONY RECEPTACLE in which the pituitary gland is enclosed is a cavity in the sphenoid bone, on which the base of the brain rests. White rectangle shows area diagrammed below.

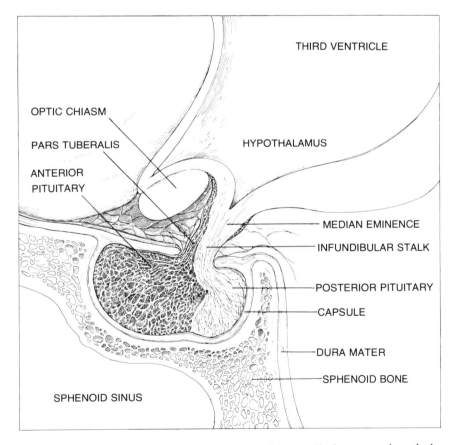

HYPOTHALAMUS AND PITUITARY are connected by a stalk that passes through the membranous lid of the receptacle in the sphenoid bone in which the pituitary rests. The double embryological origin of the two lobes of the pituitary is reflected in their differing tissues and functions and in the different ways that each is connected to the hypothalamus.

biologists, notably Ernst and Berta Scharrer, to the striking new concept of neurosecretion (the secretion of hormones by nerve cells). They suggested that specialized nerve cells might be able to manufacture and secrete true hormones, which would then be carried by the blood and would exert their effects in some target organ or tissue remote from their point of origin. The ability to manufacture hormones had traditionally been assigned to the endocrine glands: the thyroid, the gonads, the adrenals and so on. The suggestion that nerve cells could secrete hormones would endow them with a capacity far beyond their ability to liberate neurotransmitters such as epinephrine and acetylcholine at the submicroscopic regions (synapses) where they make contact with other nerve cells.

Even as these studies were in progress and these new concepts were being formulated other laboratories were reporting evidence that functions of the anterior lobe of the pituitary were somehow dependent on the structural integrity of the hypothalamic area and on a normal relation between the hypothalamus and the pituitary gland. For example, minute lesions of the hypothalamus, such as can be created by introducing small electrodes into the base of the brain in an experimental animal and producing localized electrocoagulation, were found to abolish the secretion of anterior pituitary hormones. On the other hand, the electrical stimulation of nerve cells in the same regions dramatically increased the secretion of the hormones [see illustration below].

Thus the question was presented: Precisely how does the hypothalamus regulate the secretory activity of the anterior pituitary? The results produced by electrocoagulation and electrical stimulation of the hypothalamus suggested some kind of neural mechanism. One objection to this theory was rather hard to overcome. Careful anatomical studies over many years had clearly established that there were no nerve fibers extending from the hypothalamus to the anterior pituitary. The only nerve fibers found in the pituitary stalk were those that terminate in the posterior lobe.

A way out of the dilemma was provided by an entirely different working hypothesis, suggested by the discovery in 1936 of blood vessels of a peculiar type that were shown to extend from the floor of the hypothalamus through the pituitary stalk to the anterior pituitary [see bottom illustration on opposite page]. If these tiny blood vessels were cut, the secretions of the anterior pituitary would instantly decrease. If the capillary vessels regenerated across the surgical cut, the secretions resumed.

Accordingly a new hypothesis was put forward about 1945 with which the name of the late G. W. Harris of the University of Oxford will remain associated. The hypothesis proposed that hypothalamic control of the secretory activity of the anterior pituitary could be neurochemical: some substance manufactured by nerve cells in the hypothalamus could be released into the capillary vessels that run from the hypothalamus to the anterior pituitary, where it could be delivered to the endocrine cells of the gland. On reaching these endocrine cells the substance of hypothalamic origin would somehow stimulate the secretion of the various anterior pituitary hormones.

The hypothesis that pituitary function is controlled by neurohormones originating in the hypothalamus was soon well established on the basis of intensive physiological studies in several laboratories. The next problem was therefore to isolate and characterize the postulated hypothalamic hormones. It was logical to guess that the hormones might be polypeptides of small molecular weight, since it had been well established that the two known neurosecretory products of hypothalamic origin, oxytocin and vasopressin, are each composed of eight amino acids. Indeed, in 1955 it was reported that crude aqueous hypothalamic extracts designed to contain polypeptides were able specifically to stimulate the secretion of ACTH, the pituitary hormone that controls the secretion of the steroid hormones of the adrenal cortex.

It was quickly demonstrated that none of the substances known to originate in the central nervous system (such as epinephrine, acetylcholine, vasopressin and oxytocin) could account for the ACTH-releasing activity observed in the extract of hypothalamic tissue. It therefore seemed reasonable to postulate the existence and involvement in this phenomenon of a new substance designated (adreno)corticotropin-releasing factor, or CRF. Several laboratories then undertook the apparently simple task of purifying CRF from hypothalamic extracts, with the final goal of isolating it and establishing its chemical structure. Seventeen years later the task still remains to be accomplished. Technical difficulties involving the methods

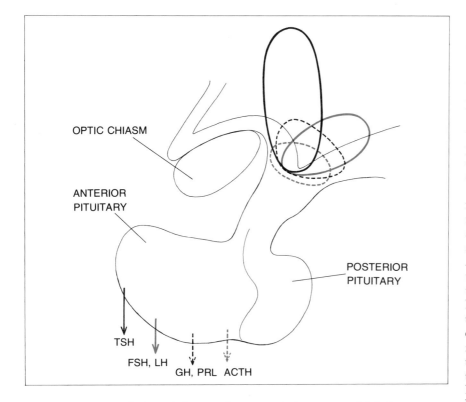

RELATION between the hypothalamus and anterior pituitary was established experimentally. Lesions in specific regions of the hypothalamus interfere with secretion by the anterior lobe of specific hormones; electrical stimulation of those regions stimulates secretion of the hormones. The regions associated with each hormone are mapped schematically.

of assaying for CRF, together with certain peculiar characteristics of the molecule, have defied the enthusiasm, ingenuity and hard work of several groups of investigators.

More rewarding results were obtained in a closely related effort. About 1960 it was clearly established that the same crude extracts of hypothalamic tissue were able to stimulate the secretion of not only ACTH but also the three other pituitary hormones mentioned above: thyrotropin (TSH) and the two gonadotropins (LH and FSH). TSH is the pituitary hormone that controls the function of the thyroid gland, which in turn secretes the two hormones thyroxine and triiodothyronine. LH controls the secretion of the steroid hormones responsible for the male or female sexual characteristics; it also triggers ovulation. FSH controls the development and maturation of the germ cells: the spermatozoa and the ova. In reality the way in which LH and FSH work together is considerably more complicated than this somewhat simplistic description suggests.

Results obtained between 1960 and 1962 were best explained by proposing the existence of three separate hypothalamic releasing factors: TRF (the TSH-releasing factor), LRF (the LH-releasing factor) and FRF (the FSH-releasing factor). The effort began at once to isolate and characterize TRF, LRF and FRF. Whereas it was difficult to find a good assay for CRF, a simple and highly reliable biological assay was devised for TRF. At first, however, the assays for LRF and FRF still left much to be desired.

With a good method available for assaying TRF, progress was initially rapid. Within a few months after its discovery TRF had been prepared in a form many thousands of times purer. Preparations of TRF obtained from the brains of sheep showed biological activity in doses as small as one microgram. A great deal of physiological information was obtained with those early preparations. For example, the thyroid hormones somehow inhibit their own secretion when they reach a certain level in the blood. This fact had been known for 40 years and was the first evidence of a negative feedback in endocrine regulation. Studies with TRF showed that the feedback control takes place at the level of the pituitary gland as the result of some kind of competition between the number of available molecules of thyroid hormones and of TRF. Other significant observations were made on the gonadotropin-releasing factors when

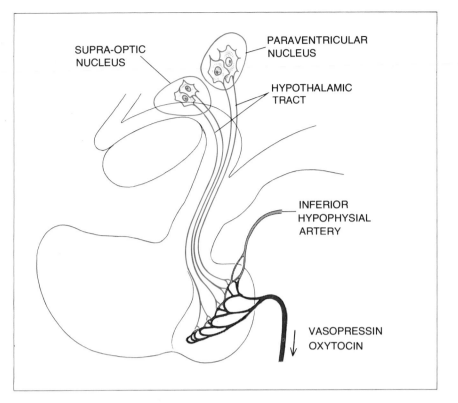

NEURAL CONNECTIONS could not explain the relation of the hypothalamus and the anterior lobe. The only significant nerve fibers connecting hypothalamus and pituitary run from two hypothalamic centers to the posterior lobe. They transmit oxytocin and vasopressin, two hormones manufactured in the hypothalamus and stored in the posterior lobe.

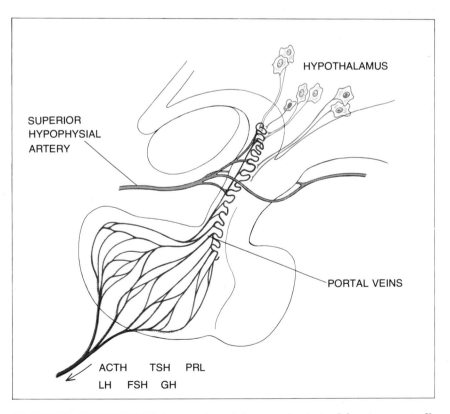

VASCULAR CONNECTIONS between hypothalamus and anterior lobe were eventually discovered: a network of capillaries reaching the base of the hypothalamus supplies portal veins that enter the anterior pituitary. Small hypothalamic nerve fibers apparently deliver to the capillaries releasing factors that stimulate secretion of the anterior-lobe hormones.

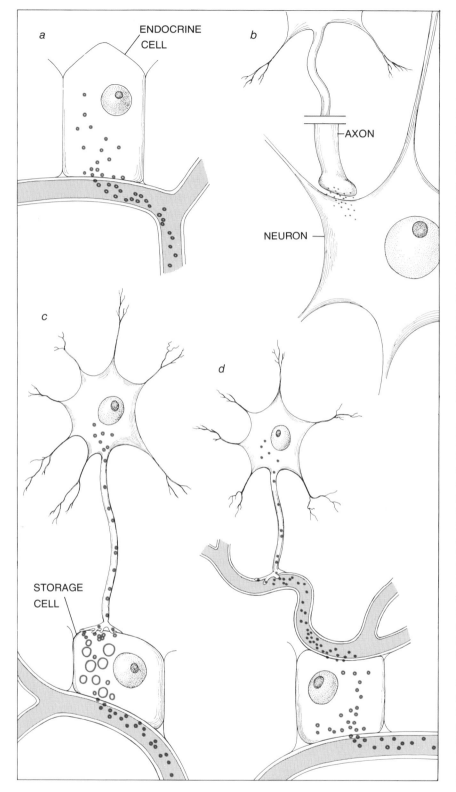

NEUROHUMORAL SECRETIONS involved in hypothalamic-pituitary interactions differ from classical hormone secretion and classical nerve-cell communication. A classical endocrine cell (such as those in the anterior pituitary or the adrenal cortex, for example) secretes its hormonal product directly into the bloodstream (*a*). At a classical synapse, the axon, or fiber, from one nerve cell releases locally a transmitter substance that activates the next cell (*b*). In neurosecretion of oxytocin or vasopressin the hormones are secreted by nerve cells and pass through their axons to storage cells in the posterior pituitary, eventually to be secreted into the bloodstream (*c*). Hypothalamic (releasing factor) hormones go from the neurons that secrete them into local capillaries, which carry them through portal veins to endocrine cells in the anterior lobe, whose secretions they in turn stimulate (*d*).

purified preparations, also active at microgram levels, were injected in experimental animals, for instance to produce ovulation.

It soon became apparent, however, that the isolation and chemical characterization of TRF, LRF and FRF would not be simple. The preparations active in microgram doses were chemically heterogeneous; they showed no clear-cut indication of a major component. It was also realized that each fragment of hypothalamus obtained from the brain of a sheep or another animal contained nearly infinitesimal quantities of the releasing factors. The isolation of enough of each factor to make its chemical characterization possible would therefore require the processing of an enormous number of hypothalamic fragments. Two groups of workers in the U.S. undertook this challenge: a group headed by A. V. Schally at the Tulane University School of Medicine and our own group, first at the Baylor University College of Medicine in Houston and then at the Salk Institute in La Jolla, Calif.

Over a period of four years the Tulane group worked with extracts from perhaps two million pig brains. Our laboratory collected, dissected and processed close to five million hypothalamic fragments from the brains of sheep. Since one sheep brain has a wet weight of about 100 grams, this meant handling 500 tons of brain tissue. From this amount we removed seven tons of hypothalamic tissue (about 1.5 grams per brain). Semi-industrial methods had to be developed in order to handle, extract and purify such large quantities of material. Finally in 1968 one milligram of a preparation of TRF was obtained that appeared to be homogeneous by all available criteria.

On careful measurement the entire milligram could be accounted for by the sole presence of three amino acids: histidine, glutamic acid and proline. Moreover, the three amino acids were present in equal amounts, which suggested that we were dealing with a relatively simple polypeptide perhaps as small as a tripeptide. In the determination of peptide sequences it is customary to subject the sample to attack by proteolytic enzymes, which cleave the peptide bonds holding the polypeptide chain together in well-established ways. Pure TRF, however, was shown to be resistant to all the proteolytic enzymes used. Since we could spare only a tiny amount of our precious one-milligram sample for studies of molecular weight, we could not obtain a

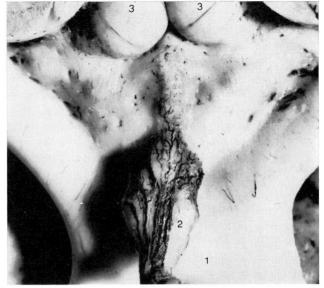

BLOOD VESSELS linking the hypothalamus and the anterior pituitary are seen in photographs made by Henri Duvernoy of the University of Besancon. The photomicrograph (*left*) shows some of the individual loops that characterize the capillary network at the base of the hypothalamus of a dog. The ascending branch of one loop is clearly seen (*1*); the loop comes close to the floor of the third ventricle (*2*) and then descends (*3*), carrying with it the releasing factors that are secreted by this region of the hypothalamus and entering the pars tuberalis of the anterior lobe (*4*). The photograph of the floor of the human hypothalamus (*right*) shows the optic chiasm (*1*), the posterior side of the pituitary stalk with its portal veins (*2*) and the mammillary bodies of the brain (*3*).

precise value for that important measurement. On the basis of inferential evidence, however, it seemed to be reasonable to assume that the molecular weight of TRF could not be more than 1,500.

With small molecules it is often possible to use methods based on the technique of mass spectrometry to obtain in a matter of hours the complete molecular structure of the compound under investigation. Because of the minute quantities of TRF available such efforts on our part were frustrated; the mass-spectrometric methods available to us in 1969 were not sensitive enough to indicate the structure of our unknown substance. Other approaches involve the use of infrared or nuclear magnetic-resonance spectrometry, which can provide direct insight into molecular structure. Here too the techniques then available were inadequate for providing clear-cut information about polypeptide samples that weighed only a few micrograms.

Confronted with nothing but dead ends, we decided on an entirely different approach to finding the structure of TRF. That approach was first to synthesize each of six possible tripeptides composed of the three amino acids known to be present in TRF: histidine (abbreviated His), glutamic acid (Glu) and proline (Pro). The six tripeptides were then assayed for their biological activity. None showed any activity when they were injected at doses of up to a million times the level of the active natural TRF.

Was this another dead end? Not quite. Our synthetic polypeptides all had a

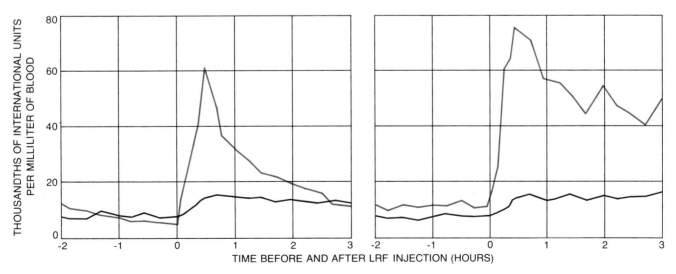

HYPOTHALAMIC HORMONES have clinical implications and applications. For example, women with no pituitary or ovarian defect respond to the administration of synthetic LRF by secreting normal amounts of the hormones LH and FSH. Curves show effect of LRF on secretion of LH (*color*) and FSH (*black*) in a normal woman on the third (*left*) and 11th (*right*) day of menstrual cycle.

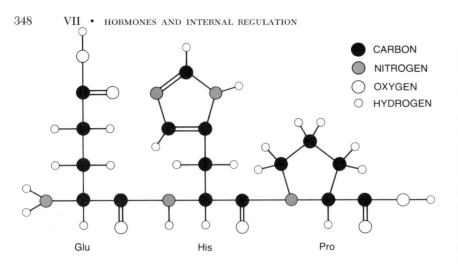

CARBON
NITROGEN
OXYGEN
HYDROGEN

Glu   His   Pro

AMINO ACID CONTENT of TRF, the releasing factor for thyroid-stimulating hormone (TSH), was established: glutamic acid, histidine and proline in equal proportions. Each of six possible tripeptides was synthesized; one is diagrammed. None was active biologically.

free amino group (NH$_2$) at the end of the molecule designated the N terminus. We knew that in several well-characterized hormones the N-terminus end was not free; it was blocked by a small substitute group of some kind. Indeed, we had evidence from the small quantity of natural TRF that its N terminus was also blocked. To block the N terminus of our six candidate polypeptides was not difficult: we heated them in the presence of acetic anhydride, which typically couples an acetyl group (CH$_3$CO) to the N terminus. When these "protected" tripeptides were tested, the results were unequivocal. The biological activity of the sequence Glu-His-Pro, and that sequence alone, was qualitatively indistinguishable from the activity of natural TRF. Quantitatively, how-

ever, there was still a considerable difference between the synthetic product and natural TRF. Next it was shown that the protective effect of heating Glu-His-Pro with the acetic anhydride had been to convert the glutamic acid at the N terminus into a ring-shaped form known as pyroglutamic acid (pGlu).

We now had available gram quantities of the synthetic tripeptide pGlu-His-Pro-OH. (The OH is a hydroxyl group at the end of the molecule opposite the N terminus.) Accordingly we could bring into play all the methods that had yielded no information with the microgram quantities of natural TRF. Several of the techniques were modified, particularly with the aim of obtaining mass spectra of the synthetic

peptide at levels of only a few micrograms.

Meanwhile, armed with knowledge about the structure of other hormones, we modified the synthetic pGlu-His-Pro-OH to pGlu-His-Pro-NH$_2$ by replacing the hydroxyl group with an amino group (NH$_2$) to produce the primary amide [*see top illustration on opposite page*]. This substance proved to have the same biological activity as the natural TRF. At length the complete structure of the natural TRF was obtained by high-resolution mass spectrometry. It turned out to be the structure pGlu-His-Pro-NH$_2$. The time was late 1969. Thus TRF not only was the first of the hypothalamic hormones to be fully characterized but also was immediately available by synthesis in amounts many millions of times greater than the hormone present in one sheep hypothalamus. TRF from pig brains was subsequently shown to have the same molecular structure as TRF from sheep brains.

Characterization of the hypothalamic releasing factor LRF, which controls the secretion of the gonadotropin LH, followed rapidly. Isolated from the side fractions of the programs for the isolation of TRF, LRF was shown in 1971 to be a polypeptide composed of 10 amino acids. Six of the amino acids are not found in TRF: tryptophan (Trp), serine (Ser), tyrosine (Tyr), glycine (Gly), leucine (Leu) and arginine (Arg). The full sequence of LRF is pGlu-His-Trp-Ser-Tyr-Gly-Leu-Arg-Pro-Gly-NH$_2$ [*see bottom illustration on these two pages*]. Although this structure is more compli-

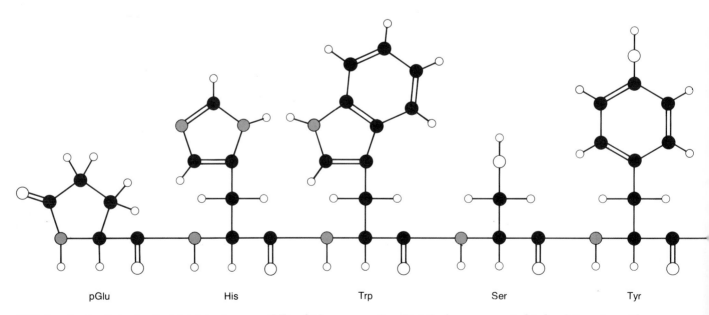

pGlu   His   Trp   Ser   Tyr

LRF, the releasing factor for the luteinizing hormone (LH), which affects the activity of the gonads, was characterized and synthesized

soon after. First the hormone was isolated and its amino acid content was determined. Then their intramolecular sequence was es-

cated than the structure of TRF, it begins with the same two amino acids (pGlu-His) and has the same group at the other terminus (NH₂).

It turns out that LRF also stimulates the secretion of the other gonadotropin, FSH, although not as powerfully as it stimulates the secretion of LH. It has been proposed that LRF may be the sole hypothalamic controller of the secretion of the two gonadotropins: LH and FSH.

There is good physiological evidence that the hypothalamus is also involved in the control of the secretion of the other two important pituitary hormones: prolactin and growth hormone. Curiously, prolactin is as plentiful in males as in females, but its role in male physiology is still a mystery. The hypothalamic mechanism involved in the control of the secretion of prolactin or growth hormone is not fully understood. It is quite possible that the secretion of these two pituitary hormones is controlled not by releasing factors alone but perhaps jointly by releasing factors and specific hypothalamic hormones that somehow act as inhibitors of the secretion of prolactin or growth hormone. If it should turn out that inhibitory hormones rather than stimulative ones are involved in the regulation of prolactin and growth hormone, one should not be too surprised. The brain provides many examples of inhibitory and stimulative systems working in parallel.

The hypothalamic hormones TRF and LRF are both now available by synthesis in unlimited quantities. Both are highly active in stimulating pituitary functions in humans. TRF is already a

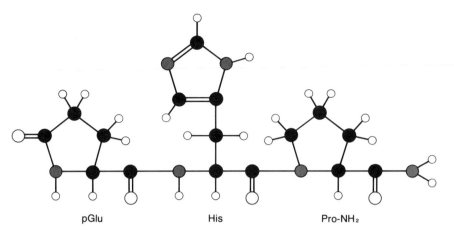

TRIPEPTIDES were modified in an effort to characterize the releasing factor. When the sequence glutamic acid–histidine-proline was modified by forming the glutamic acid into a ring and converting the proline end (*right*) to an amide, it was found to be TRF.

powerful tool for exploring pituitary functions in several diseases characterized by the abnormality of one or several of the pituitary secretions. There is increasing evidence that most patients with such abnormalities (primarily children) actually have normally functioning glands, since they respond promptly to the administration of synthetic hypothalamic hormones. Evidently their abnormalities are due to hypothalamic rather than pituitary deficiencies. These deficiencies can now be successfully treated by the administration of the hypothalamic polypeptide TRF.

Similarly, an increasing number of women who have no ovulatory menstrual cycle and who show no pituitary or ovarian defect begin to secrete normal amounts of the gonadotropins

LH and FSH after the administration of LRF. The administration of synthetic LRF should therefore be the method of choice for the treatment of those cases of infertility where the functional defect resides in the hypothalamus-pituitary system. Indeed, ovulation can be induced in women by the administration of synthetic LRF. On the other hand, knowledge of the structure of the LRF molecule may open up an entirely novel approach to fertility control. Synthetic compounds closely related to LRF in structure may act as inhibitors of the native LRF. Two such analogues of LRF, made by modifying the histidine in the hormone, have been reported as antagonists of LRF. It is therefore possible that LRF antagonists will be used as contraceptives.

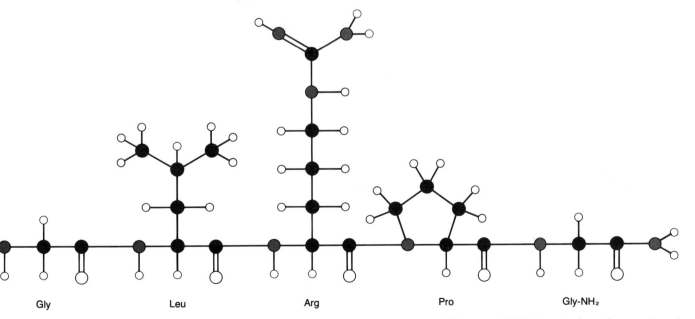

tablished and reproduced by synthesis; the synthetic replicate shown here was found to have full biological activity. In addition to stimulating LH activity, LRF also stimulates the secretion of another gonadotropic hormone, FSH, although not so powerfully.

34

THE PINEAL GLAND

RICHARD J. WURTMAN AND JULIUS AXELROD
July 1965

*The function of this small organ near the center of the
mammalian brain has long been a mystery. Recent studies
indicate that it is a "biological clock" that regulates the
activity of the sex glands*

Buried nearly in the center of the brain in any mammal is a small white structure, shaped somewhat like a pinecone, called the pineal body. In man this organ is roughly a quarter of an inch long and weighs about a tenth of a gram. The function of the pineal body has never been clearly understood. Now that the role of the thymus gland in establishing the body's immunological defenses has been demonstrated, the pineal has become perhaps the last great mystery in the physiology of mammalian organs. This mystery may be nearing a solution: studies conducted within the past few years indicate that the pineal is an intricate and sensitive "biological clock," converting cyclic nervous activity generated by light in the environment into endocrine—that is, hormonal—information. It is not yet certain what physiological processes depend on the pineal clock for cues, but the evidence at hand suggests that the pineal participates in some way in the regulation of the gonads, or sex glands.

A Fourth Neuroendocrine Transducer

Until quite recently most investigators thought that the mammalian pineal was simply a vestige of a primitive light-sensing organ: the "third eye" found in certain cold-blooded vertebrates such as the frog. Other workers, noting the precocious sexual development of some young boys with pineal tumors, had proposed that in mammals the pineal was a gland. When the standard endocrine tests were applied to determine the possible glandular function of the pineal, however, the results varied so much from experiment to experiment that few positive conclusions seemed justified. Removal of the pin-

eal in young female rats was frequently followed by an enlargement of the ovaries, but the microscopic appearance of the ovaries did not change consistently, and replacement of the extirpated pineal by transplantation seemed to have little or no physiological effect. Most experimental animals could survive the loss of the pineal body with no major change in appearance or function.

In retrospect much of the difficulty early workers had in exploring and defining the glandular function of the pineal arose from limitations in the traditional concept of an endocrine organ. Glands were once thought to be entirely dependent on substances in the bloodstream both for their own control and for their effects on the rest of the body: glands secreted hormones into the blood and were themselves regulated by other hormones, which were delivered to them by the circulation. The secretory activity of a gland was thought to be maintained at a fairly constant level by homeostatic mechanisms: as the level of a particular hormone in the bloodstream rose, the gland invariably responded by decreasing its secretion of that hormone; when the level of the hormone fell, the gland increased its secretion.

In the past two decades this concept of how the endocrine system works has proved inadequate to explain several kinds of glandular response, including changes in hormone secretion brought about by changes in the external environment and also regular cyclic changes in the secretion of certain hormones (for example, the hormones responsible for the menstrual cycle and the steroid hormones that are produced on a daily cycle by the adrenal gland). Out of the realization that these and other endocrine responses must depend

in some way on interactions between the glands and the nervous system the new discipline of neuroendocrinology has developed.

In recent years much attention has centered on the problem of locating the nervous structures that participate in the control of glandular function. It has been known for some time that special types of organs would be needed to "transduce" neural information into endocrine information. Nervous tissue is specialized to receive and transmit information directly from cell to cell; according to the traditional view, glands are controlled by substances in the bloodstream and dispatch their messages to target organs by the secretion of hormones into the bloodstream. In order to transmit information from the nervous system to an endocrine organ a hypothetical "neuroendocrine transducer" would require some of the special characteristics of both neural and endocrine tissue. It should respond to substances (called neurohumors) released locally from nerve endings, and it should contain the biochemical machinery necessary for synthesizing a hormone and releasing it into the bloodstream. Three such neurosecretory systems have so far been identified. They are (1) the hypothalamus–posterior-pituitary system, which secretes the antidiuretic hormone and oxytocin, a hormone that causes the uterus to contract during labor; (2) the pituitary-releasing-factor system, also located in the hypothalamus, which secretes polypeptides that control the function of the pituitary gland, and (3) the adrenal medulla, whose cells respond to a nervous input by releasing adrenaline into the bloodstream.

The advent of neuroendocrinology has provided a conceptual framework

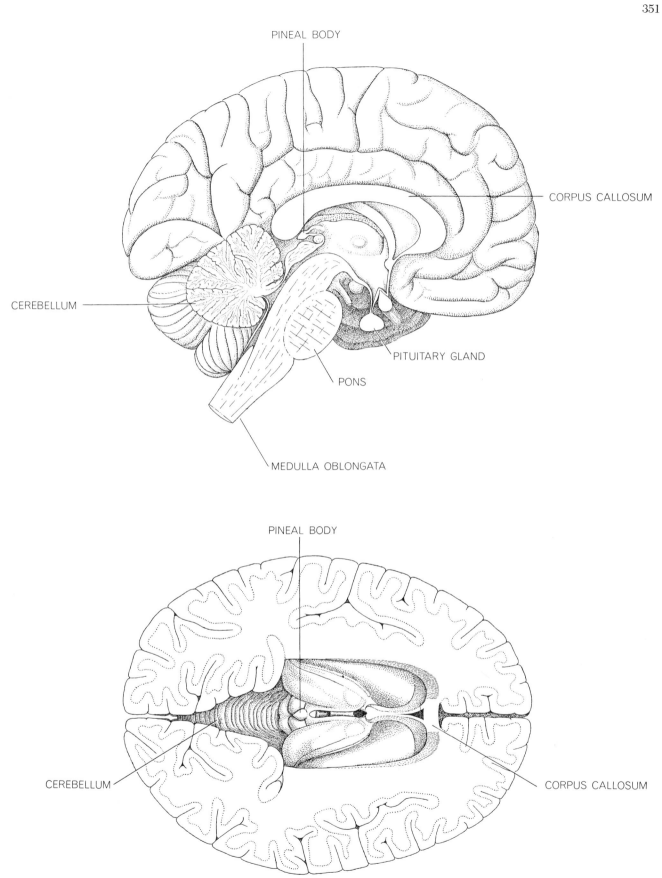

PINEAL BODY

CORPUS CALLOSUM

CEREBELLUM

PITUITARY GLAND

PONS

MEDULLA OBLONGATA

PINEAL BODY

CEREBELLUM

CORPUS CALLOSUM

TWO VIEWS of the human brain reveal the central position of the pineal body. Section at top is cut in the median sagittal plane and is viewed from the side. Section at bottom is cut in a horizontal plane and is viewed from above; an additional excision has been made in this view to reveal the region immediately surrounding the pineal. In mammals the pineal is the only unpaired midline organ in the brain. The name "pineal" comes from the organ's resemblance to a pinecone, the Latin equivalent of which is *pinea*.

that has been most helpful in characterizing the role of the pineal gland. On the basis of recent studies conducted by the authors and their colleagues at the National Institute of Mental Health, as well as by investigators at other institutions, it now appears that the pineal is not a gland in the traditional sense but is a fourth neuroendocrine transducer; it is a gland that converts a nervous input into a hormonal output.

A Prophetic Formulation

The existence of the pineal body has been known for at least 2,000 years. Galen, writing in the second century A.D., quoted studies of earlier Greek anatomists who were impressed with the fact that the pineal was perched atop the aqueduct of the cerebrum and was a single structure rather than a paired one; he concluded that it served as a valve to regulate the flow of thought out of its "storage bin" in the lateral ventricles of the brain. In the 17th century René Descartes embellished this notion; he believed that the pineal housed the seat of the rational soul. In his formulation the eyes perceived the events of the real world and transmitted what they saw to the pineal by way of "strings" in the brain [*see illustration below*]. The pineal responded by allowing humors to pass down hollow tubes to the muscles, where they produced the appropriate responses. With the hindsight of 300 years of scientific development, we can admire this prophetic formulation of the pineal as a neuroendocrine transducer!

In the late 19th and early 20th centuries the pineal fell from its exalted metaphysical state. In 1898 Otto Heubner, a German physician, published a case report of a young boy who had shown precocious puberty and was also found to have a pineal tumor. In the course of the next 50 years many other children with pineal tumors and precocious sexual development were described, as well as a smaller number of patients whose pineal tumors were associated with delayed sexual development. Inexplicably almost all the cases of precocious puberty were observed in boys.

In a review of the literature on pineal tumors published in 1954 Julian I. Kitay, then a fellow in endocrinology at the Harvard Medical School, found that most of the tumors associated with precocious puberty were not really pineal in origin but either were tumors of supporting tissues or were teratomas (primitive tumors containing many types of cells). The tumors associated with delayed puberty, however, were in most cases true pineal tumors. He concluded that the cases of precocious puberty resulted from reduced pineal function due to disease of the surrounding tissue, whereas delayed sexual development in children with true pineal tumors was a consequence of increased pineal activity.

The association of pineal tumors and sexual malfunction gave rise to hundreds of research projects designed to test the hypothesis that the pineal was a gland whose function was to inhibit the gonads. Little appears to have resulted from these early efforts. Later in 1954 Kitay and Mark D. Altschule, director of internal medicine at McLean Hospital in Waverly, Mass., reviewed the entire world literature on the pineal: some 1,800 references, about half of which dealt with the pineal-gonad question. They concluded that of all the studies published only two or three had used enough experimental animals and adequate controls for their data to be analyzed statistically. These few papers suggested a relation between the pineal and the gonads but did little to characterize it. After puberty the human pineal is hardened by calcification; this change in the appearance of the pineal led many investigators to assume that the organ was without function and further served to discourage research in the field. (Actually calcification appears to be unrelated to the pineal functions we have measured.)

As long ago as 1918 Nils Holmgren, a Swedish anatomist, had examined the pineal region of the frog and the dogfish with a light microscope. He was surprised to find that the pineal contained distinct sensory cells; they bore a marked resemblance to the cone cells of the retina and were in contact with nerve cells. On the basis of these obser-

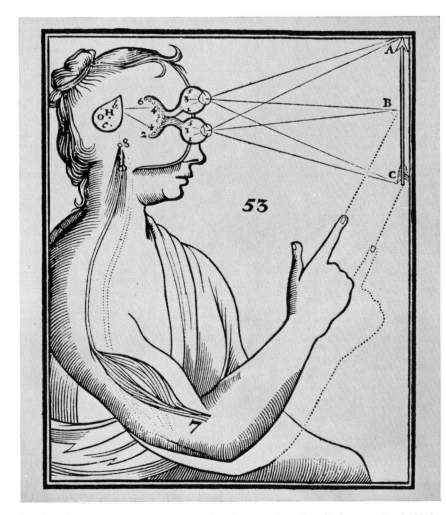

SEAT OF THE RATIONAL SOUL was the function assigned to the human pineal (*H*) by René Descartes in his mechanistic theory of perception. According to Descartes, the eyes perceived the events of the real world and transmitted what they saw to the pineal by way of "strings" in the brain. The pineal responded by allowing animal humors to pass down hollow tubes to the muscles, where they produced the appropriate responses. The size of the pineal has been exaggerated in this wood engraving, which first appeared in 1677.

vations he suggested that the pineal might function as a photoreceptor, or "third eye," in cold-blooded vertebrates. In the past five years this hypothesis has finally been confirmed by electrophysiological studies: Eberhardt Dodt and his colleagues in Germany have shown that the frog pineal is a wavelength discriminator: it converts light energy of certain wavelengths into nervous impulses. In 1927 Carey P. McCord and Floyd P. Allen, working at Johns Hopkins University, observed that if they made extracts of cattle pineals and added them to the media in which tadpoles were swimming, the tadpoles' skin blanched, that is, became lighter in color.

Such was the state of knowledge about the pineal as late as five or six years ago. It appeared to be a photoreceptor in the frog, had something to do with sexual function in rats and in humans (at least those with pineal tumors) and contained a factor (at least in cattle) that blanched pigment cells in tadpoles.

The Discovery of Melatonin

Then in 1958 Aaron B. Lerner and his co-workers at the Yale University School of Medicine identified a unique compound, melatonin, in the pineal gland of cattle [see "Hormones and Skin Color," by Aaron B. Lerner; SCIENTIFIC AMERICAN, July, 1961]. During the next four years at least half a dozen other major discoveries were made about the pineal by investigators representing many different disciplines and institutions. Lerner, a dermatologist and biochemist, was interested in identifying the substance in cattle pineal extracts that blanched frog skin. He and his colleagues prepared and purified extracts from more than 200,000 cattle pineals and tested the ability of the extracts to alter the reflectivity of light by pieces of excised frog skin. After four years of effort they succeeded in isolating and identifying the blanching agent and found that it was a new kind of biological compound: a methoxylated indole, whose biological activity requires a methyl group (CH_3) attached to an oxygen atom [see illustration on next two pages].

Methoxylation had been noted previously in mammalian tissue, but the products of this reaction had always appeared to lose their biological activity as a result. The new compound, named melatonin for its effect on cells containing the pigment melanin, appeared to lighten the amphibian skin by causing

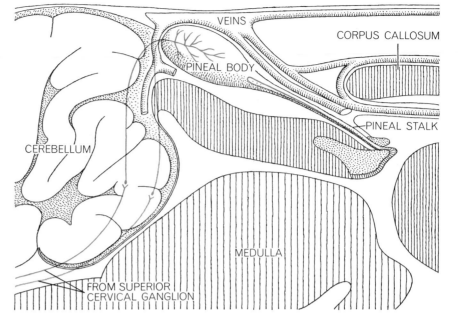

INNERVATION OF RAT PINEAL was the subject of a meticulous study by the Dutch neuroanatomist Johannes Ariëns Kappers in 1961. He demonstrated that the pineal of the adult rat is extensively innervated by nerves from the sympathetic nervous system. The sympathetic nerves to the pineal originate in the neck in the superior cervical ganglion, enter the skull along the blood vessels and eventually penetrate the pineal at its blunt end (top). Aberrant neurons from the central nervous system sometimes run up the pineal stalk from its base, but these generally turn and run back down the stalk again without synapsing. The pineal is surrounded by a network of great veins, into which its secretions probably pass. According to Ariëns Kappers, the innervation of the human pineal is quite similar.

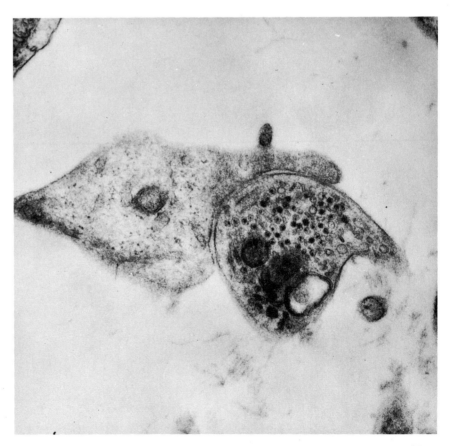

SYMPATHETIC NERVE terminates directly on a pineal cell, instead of on a blood vessel or smooth muscle cell, in this electron micrograph of a portion of a rat pineal made by David Wolfe of the Harvard Medical School. The nerve ending is characterized by dark vesicles, or sacs, that contain neurohumors. Magnification is about 12,500 diameters.

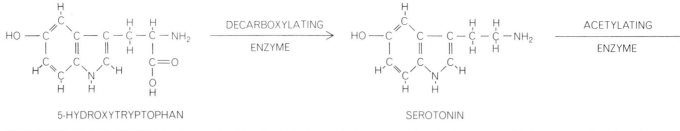

5-HYDROXYTRYPTOPHAN

SEROTONIN

SYNTHESIS OF MELATONIN in the rat pineal begins with the removal of a carboxyl (COOH) group from the amino acid 5-hydroxytryptophan by the enzyme 5-hydroxytryptophan decarboxylase. Serotonin, the product of this reaction, is then enzymatically

the aggregation of melanin granules within the cells. It was effective in a concentration of only a trillionth of a gram per cubic centimeter of medium. No influence of melatonin could be demonstrated on mammalian pigmentation, nor could the substance actually be identified in amphibians, in which it exerted such a striking effect. It remained a biological enigma that the mammalian pineal should produce a substance that appeared to have no biological activity in mammals but was a potent skin-lightening agent in amphibians, which were unable to produce it!

Both aspects of the foregoing enigma have now been resolved. Subsequent research has shown that melatonin does in fact have a biological effect in mammals and can be produced by amphibians. Spurred by Lerner's discovery of this new indole in the cattle pineal, Nicholas J. Giarman, a pharmacologist at the Yale School of Medicine, analyzed pineal extracts for their content of other biologically active compounds. He found that both cattle and human pineals contained comparatively high levels of sero-

tonin, an amine whose molecular structure is similar to melatonin and whose function in nervous tissue is largely unknown. Studies by other investigators subsequently showed that the rat pineal contains the highest concentration of serotonin yet recorded in any tissue of any species.

A year before the discovery of melatonin one of the authors (Axelrod) and his co-workers had identified a methoxylating enzyme (catechol-O-methyl transferase) in a number of tissues. This enzyme acted on a variety of catechols (compounds with two adjacent hydroxyl, or OH, groups on a benzene ring) but showed essentially no activity with respect to single-hydroxyl compounds such as serotonin, the most likely precursor of melatonin. In 1959 Axelrod and Herbert Weissbach studied cattle pineal tissue to see if it might have the special enzymatic capacity to methoxylate hydroxyindoles. They incubated N-acetylserotonin (melatonin without the methoxyl group) with pineal tissue and a suitable methyl donor and observed that melatonin was indeed formed. Sub-

sequently they found that all mammalian pineals shared this biochemical property but that no tissue other than pineal could make melatonin. Extensive studies of a variety of mammalian species have confirmed this original observation that only the pineal appears to have the ability to synthesize melatonin. (In amphibians and some birds small amounts of melatonin are also manufactured by the brain and the eye.) Other investigators have found that the pineal contains all the biochemical machinery needed to make melatonin from an amino acid precursor, 5-hydroxytryptophan, which it obtains from the bloodstream. It was also found that circulating melatonin is rapidly metabolized in the liver to form 6-hydroxymelatonin.

Anatomy of the Pineal

While these investigations of the biochemical properties of the pineal were in progress, important advances were being made in the anatomy of the pineal by the Dutch neuroanatomist Johannes Ariëns Kappers and by several

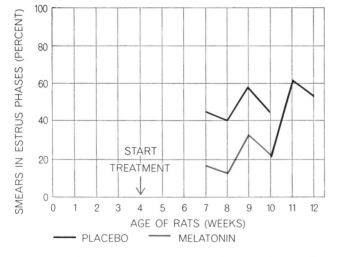

PLACEBO   MELATONIN

EFFECT OF MELATONIN on the estrus cycles of female rats is depicted here. Rats that had been given daily injections of melatonin starting in their fourth week of life developed a longer estrus cycle than rats that had been similarly treated with a placebo. When the melatonin-treated animals were 10 weeks old, a placebo was substituted for the melatonin and the estrus cycle returned to normal.

DARK   LIGHT

EFFECTS OF LIGHTING on the estrus cycles of three groups of female rats are shown in the graphs on these two pages. The groups, each consisting of about 20 rats, were subjected respectively to a sham operation (*left*), removal of their superior cervical ganglion (*middle*) and removal of their eyes (*right*). Each group was then further subdivided, with about half being placed in constant light

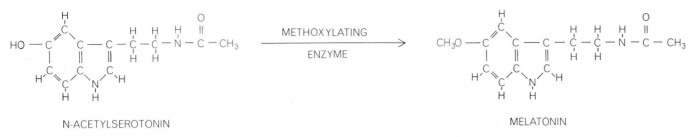

N-ACETYLSEROTONIN   MELATONIN

acetylated to form N-acetylserotonin. This compound in turn is methoxylated by the enzyme hydroxyindole-O-methyl transferase (HIOMT) to yield melatonin. In mammals HIOMT is found only in the pineal. Changes in basic molecule are indicated by color.

American electron microscopists, including Douglas E. Kelly of the University of Washington, Aaron Milofsky of the Yale School of Medicine and David Wolfe, then at the National Institute of Neurologic Diseases and Blindness. In 1961 Ariëns Kappers published a meticulous study of the nerve connections in the rat pineal. He demonstrated clearly that although this organ originates in the brain in the development of the embryo, it loses all nerve connections with the brain soon after birth. There is thus no anatomical basis for invoking "tracts from the brain" as the pathway by which neural information is delivered to the pineal.

Ariëns Kappers showed that instead the pineal of the adult rat is extensively penetrated by nerves from the sympathetic portion of the autonomic nervous system. The sympathetic nervous system is involuntary and is concerned with adapting to rapid changes in the internal and external environments; the sympathetic nerves to the pineal originate in the superior cervical ganglion in the neck, enter the skull along the blood vessels and eventually penetrate the pineal [see top illustration on page 353]. Electron microscope studies later showed that within the pineal many sympathetic nerve endings actually terminate directly on the pineal cells, instead of on blood vessels or smooth-muscle cells, as in most other organs [see bottom illustration on page 353]. Among endocrine structures the organization of nerves in the mammalian pineal appeared to be most analogous to that of the adrenal medulla, one of the three demonstrated neuroendocrine transducers.

Meanwhile electron microscope studies by other workers on the pineal regions of frogs had confirmed many of Holmgren's speculations. It was found that the pineal cells of amphibians contained light-sensitive elements that were practically indistinguishable from those found in the cone cells of the retina, but that the pineal cells of mammals did not contain such elements. By 1962 it could be stated with some assurance that the mammalian pineal was not simply a vestige of the frog "third eye," since the "vestige" had undergone profound anatomical changes with evolution.

The Melatonin Hypothesis

Even though the mammalian pineal no longer seemed to respond directly to light, there now appeared good evidence that its function continued to be related somehow to environmental light. In 1961 Virginia Fiske, working at Wellesley College, reported that the exposure of rats to continuous environmental illumination for several weeks brought about a decrease in the weight of their pineals. She had been interested in studying the mechanisms by which the exposure of rats to light for long periods induces changes in the function of their gonads. (For example, continous light increased the weight of the ovaries and accelerated the estrus cycle). At the same time one of the authors (Wurtman, then at the Harvard Medical School), in collaboration with Altschule and Willard Roth, was studying the conditions under which the administration

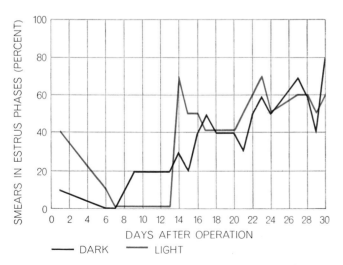

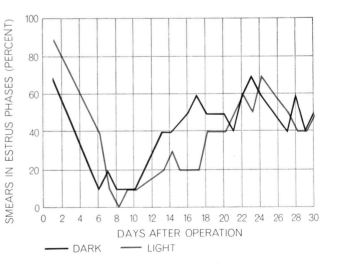

and the other half in constant darkness beginning one day after their respective operations. Daily vaginal smears were taken on the first day and on the sixth through the 30th days after the operations. Results were plotted as the percentage of all the smears in a treatment group showing estrus phases each day. In general it was found that interference with the transmission of light information to the pineal gland (either by blinding or by cutting the sympathetic nerves) also abolished most of the gonadal response to light. These findings supported the authors' melatonin hypothesis, which holds that one mechanism whereby light is able to accelerate the estrus cycle in normal animals is by inhibiting the synthesis in the pineal of melatonin, a compound that in turn inhibits estrus.

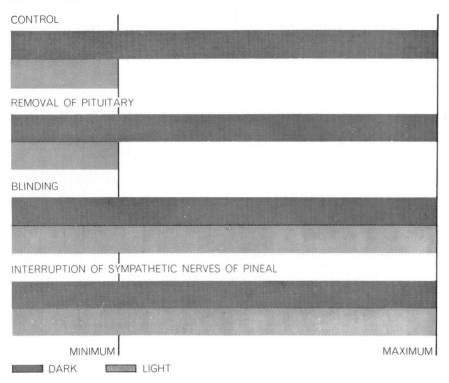

CONTROL

REMOVAL OF PITUITARY

BLINDING

INTERRUPTION OF SYMPATHETIC NERVES OF PINEAL

MINIMUM   MAXIMUM

▰▰▰ DARK   ▰▰▰ LIGHT

RESPONSE OF MELATONIN-FORMING ENZYME hydroxyindole-O-methyl transferase (HIOMT) to continuous light or darkness is shown under four different circumstances. In the control, or normal, animal continuous darkness induces an increase in HIOMT activity, whereas exposing the animal to continuous light has the opposite effect. The ability of the pineal gland to respond to environmental lighting is unaffected by the removal of the pituitary gland but is abolished following blinding or sympathetic denervation of the pineal.

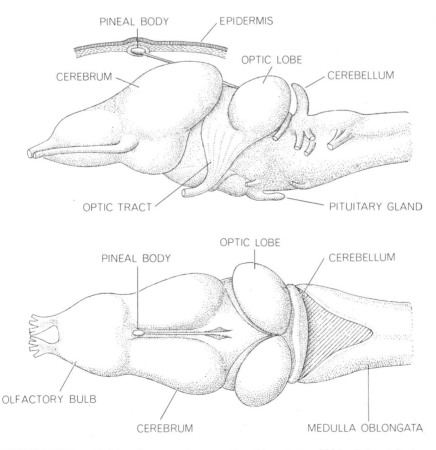

PINEAL BODY   EPIDERMIS

CEREBRUM   OPTIC LOBE   CEREBELLUM

OPTIC TRACT   PITUITARY GLAND

OPTIC LOBE

PINEAL BODY   CEREBELLUM

OLFACTORY BULB

CEREBRUM   MEDULLA OBLONGATA

PINEAL EYE is a primitive photoreceptive organ found in certain cold-blooded vertebrates such as the frog. Frog's brain is shown from the side (*top*) and from above (*bottom*).

of cattle pineal extracts decreased ovary weight and slowed the estrus cycle.

We soon confirmed Mrs. Fiske's findings, and we were also able to show that the exposure of female rats to continuous light or the removal of their pineals had similar, but not additive, effects on the weight of their ovaries. These experiments suggested that perhaps one way in which light stimulates ovary function in rats is by inhibiting the action of an inhibitor found in pineal extracts. It now became crucial to identify the gonad-inhibiting substance in pineal extracts and to see if its synthesis or its actions were modified by environmental lighting.

In 1962 we began to work together on isolating the anti-gonadal substance present in pineal extracts. Our plan was to subject extracts of cattle pineal glands to successive purification steps and test the purified material for its ability to block the induction by light of an accelerated estrus cycle in the rat. Before undertaking the complicated and time-consuming procedure of isolating the active substance in the pineal glands of cattle, we first tested a mixture of all the constituents that had already been identified in this tissue. The mixture was found to block the effects of light on the estrus cycle.

Next we tested melatonin alone, since it was apparently the only compound produced uniquely by the pineal. To our good fortune we found that when rats were given tiny doses (one to 10 micrograms per day) of melatonin by injection, starting before puberty and continuing for a month thereafter, the estrus cycle was slowed and the ovaries lost weight—just as though the animals had been treated with pineal extracts. In later studies we found that this effect of melatonin was chemically specific: it was simulated by neither N-acetylserotonin, the immediate precursor of melatonin, nor 6-hydroxymelatonin, the major product of its metabolism. Moreover, it was possible to accelerate the estrus cycle by removal of the pineal and to block this response by the injection of melatonin.

On the basis of these studies, performed in collaboration with Elizabeth Chu of the National Cancer Institute, we postulated that melatonin was a mammalian hormone, since it is produced uniquely by a single gland (the pineal), is secreted into the bloodstream and has an effect on a distant target organ (the vagina and possibly also the ovaries). We were not able to identify the precise site of action of melatonin in affecting the gonads. The slowing of the estrus cycle could be produced by actions at any of several sites in the neuroendocrine apparatus, including the brain, the pituitary, the ovaries or the vagina itself. When melatonin was labeled with radioactive atoms and injected into cats, it was taken up by all these organs and was selectively concentrated by the ovaries.

William M. McIsaac and his colleagues at the Cleveland Clinic have confirmed the effects of melatonin on the estrus cycle and have identified another pineal methoxyindole—methoxytryptophol—that has similar effects. It appears likely that pineal extracts contain a family of hormones: the methoxyindoles, all of which have in common the fact that they can be synthesized by the methoxylating enzyme found only in the mammalian pineal.

We next set out to determine whether or not these effects of injected melatonin were physiological. Could the rat pineal synthesize melatonin and, if so, in what quantities? When rat pineal glands were examined for their ability to make melatonin, we were disappointed to find that the activity of the melatonin-forming enzyme (hydroxyindole-O-methyl transferase, or HIOMT) in the rat was much lower than in most other species; the maximum amount of melatonin that the rat could make was probably on the order of one microgram per day. Our disappointment was soon relieved, however, when we realized that the low activity of this enzyme made it likely that it was controlling the rate-limiting step in melatonin synthesis in the intact animal. Knowing that continuous exposure to light decreased pineal weight, as well as the amount of ribonucleic acid (RNA) and protein in the pineal, we next explored what effect illumination might have on HIOMT activity and thus on melatonin synthesis.

Since the rat pineal gland was so small (about a milligram in weight) and had so little enzymatic activity, it was necessary to devise extremely sensitive techniques to measure this activity. When rats were subjected to constant light for as short a period as a day or two, the rate of melatonin synthesis in their pineals fell to as little as a fifth that of animals kept in continuous darkness. Since this effect of illumination or its absence could be blocked by agents that interfered with protein synthesis, it appeared that light was actually influencing the rate of formation of the enzyme protein itself.

How was information about the state of lighting being transmitted to the rat pineal? Three possible routes suggested

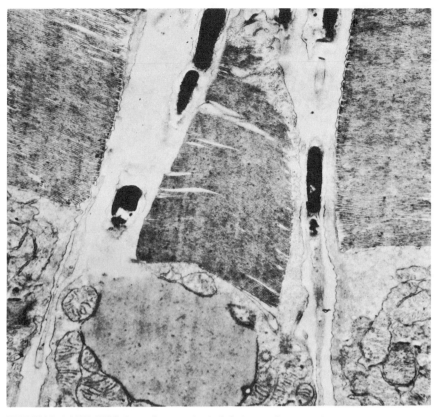

RETINAL CONE CELL from the eye of an adult frog is shown in this electron micrograph made by Douglas E. Kelly of the University of Washington. The photoreceptive outer segment of the cell (*top center*) consists of a densely lamellated membrane. Parts of two larger rod photoreceptors can be seen on each side of cone. Magnification is about 13,000 diameters.

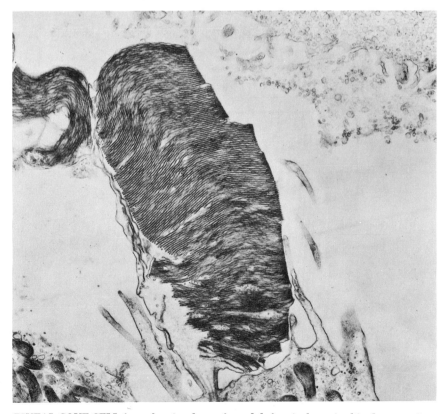

PINEAL CONE CELL from the pineal eye of an adult frog is shown in this electron micrograph made by Kelly at approximately the same magnification as the micrograph at top. The lamellated outer segment of the pineal cell is practically indistinguishable from that of the retinal photoreceptor. Part of the membrane has torn away from the cell (*top left*).

themselves. The first was that light penetrated the skull and acted directly on the pineal; W. F. Ganong and his colleagues at the University of California at Berkeley had already shown that significant quantities of light do penetrate the skulls of mammals. This hypothesis was ruled out, however, by demonstrating that blinded rats completely lost the capacity to respond to light with changed HIOMT activity; hence light had to be perceived first by the retina and was not acting directly on the pineal.

The second possibility was that light altered the level of a circulating hormone, perhaps by affecting the pituitary gland, and that this hormone secondarily influenced enzyme activity in the pineal gland. This hypothesis was also ruled out by demonstrating that the removal of various endocrine organs, including the pituitary and the ovaries, did not interfere with the response of pineal HIOMT to light.

The third possibility was that information about lighting was transmitted to the pineal by nerves. Fortunately Ariëns Kappers had just identified the nerve connections of the rat pineal as coming from the sympathetic nervous system. We found that if the sympathetic pathway to the pineal was interrupted by the removal of the superior cervical ganglion, the ability of melatonin-forming activity to be altered by light was completely lost. Thus it appeared that light was stimulating the retina and then information about this light was being transmitted to the pineal via sympathetic nerves. Within the pineal the sympathetic nerves probably released neurohumors (noradrenaline or serotonin), which acted on pineal cells to induce (or block the induction of) HIOMT; this enzyme in turn regulated the synthesis of melatonin.

Since one way light influences the gonads is by changing the amount of melatonin secreted from the pineal, we reasoned that the effects of light on the gonads might be blocked if the transmission of information about light to the pineal were interrupted. This could be accomplished by cutting the sympathetic nerves to the pineal—a procedure much less traumatic than the removal of the pineal itself. To test this hypothesis we placed groups of rats whose pineals had been denervated along with blinded and untreated animals in continuous light or darkness for a month. Vaginal smears were checked daily for evidence of changes in the estrus cycle, and pineals were tested for melatonin-synthesizing ability at the end of the experiment. It was found that interrupting the transmission of light information to the pineal (by cutting its sympathetic nerves—a procedure that does not interfere with the visual response to light) also abolished most of the gonadal response to light.

Incidentally, the observation that sympathetic nerves control enzyme synthesis in the pineal has provided, and should continue to provide, a useful tool for studies in a number of other biological disciplines. For example, studying the changes in brain enzymes produced by environmental factors offers a useful method for tracing the anatomy

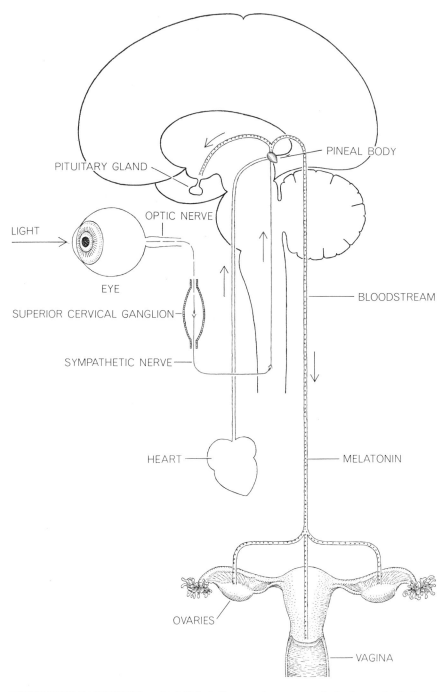

LIGHT

PITUITARY GLAND

OPTIC NERVE

EYE

SUPERIOR CERVICAL GANGLION

SYMPATHETIC NERVE

HEART

PINEAL BODY

BLOODSTREAM

MELATONIN

OVARIES

VAGINA

SUGGESTED PATHWAY by which light influences the estrus cycle in the rat is depicted in this schematic diagram. Light stimuli impinge on the retinas and cause a change in the neural output of the superior cervical ganglion by way of an unknown route. This information is then carried by sympathetic nerves to the pineal gland, where it causes a decrease in the activity of HIOMT and in the synthesis and release of melatonin. This decrease in turn lessens the inhibiting effect of the circulating melatonin on the rate of the estrus cycle. The precise site of action of melatonin in influencing the gonads is unknown; the slowing of the estrus cycle could be produced by actions at any one of several sites in the neuroendocrine apparatus, including the brain, the pituitary, the ovaries and the vagina.

of the nerve tracts involved. The observation that the activity of at least one part of the sympathetic nervous system (the superior cervical ganglion) is affected by environmental lighting raises the possibility that other regions of this neural apparatus are affected similarly. If so, physiological studies of the effects of light on other sympathetically innervated structures (for example the kidneys and fat tissue) may be profitable.

We have also found that light influences the serotonin-forming enzyme in the pineal gland but not in other organs. In contrast to HIOMT, the activity of this enzyme increases when rats are kept in constant light and decreases in darkness. When rats are blinded or when the sympathetic nerves to the pineal are cut, the effect of light and darkness on the serotonin-forming enzyme is also extinguished. Furthermore, certain drugs that block the transmission of sympathetic nervous impulses also abolish the effect of illumination on this enzyme. The fact that lighting influences pineal weight and at least two enzyme systems in this organ suggests that it may regulate many additional, undiscovered biochemical events in the pineal, via the sympathetic nervous system.

Diurnal and Circadian Rhythms

The pineal had been shown to respond and function under quite unusual conditions; for example, when an experimental animal was exposed to continuous light or darkness for several days. In nature, of course, animals that live in the temperate and tropical zones are rarely subjected to such conditions. It became important to determine if the pineal could also respond to naturally occurring changes in the environment.

In nature the level of light exposure changes with both diurnal and annual cycles. Except in polar regions every 24-hour day includes a period of sunlight and a period of darkness; the ratio of day to night varies with an annual rhythm that reaches its nadir at the winter solstice and its zenith on the first day of summer. Lighting cycles have been shown to be important in regulating several types of endocrine function: the increase in sunlight during the winter and spring triggers the annual gonadal growth and breeding cycles in many birds and some mammals that breed yearly, and the daily rhythm of day and night synchronizes a variety of roughly daily rhythms in mammals, such as the cycle of adrenal-steroid secretion. Such rhythms are called circadian, from the Latin phrase meaning "about one day." Could the pineal respond to natural diurnal lighting shifts? If so, it might function to synchronize the endocrine apparatus with these shifts.

In order to determine if normal lighting rhythms influenced the pineal, we kept a large population of rats under controlled lighting conditions (lights on from 7:00 A.M. to 7:00 P.M.) for several weeks and then tested their pineals for melatonin-forming ability at 6:00 A.M., noon, 6:00 P.M. and midnight. In the five hours after the onset of darkness (that is, by midnight) this enzymatic capacity increased between two and three times. Moreover, pineal weight also changed significantly during this period, again indicating that light was affecting many more compounds in the pineal than the single enzyme we were measuring.

All circadian rhythms studied up to this stage had in common the ability to persist for some weeks after animals were deprived of environmental lighting cues (by blinding or being placed in darkness). These rhythms no longer showed a period of precisely 24 hours, but they did fall in a range between 22 and 26 hours and hence were thought to be regulated by some internal mechanism not dependent on, but usually synchronized with, environmental lighting. Such endogenous, or internally regulated, circadian rhythms in rodents include motor activity and rectal temperature, as well as the rhythm in adrenal-steroid secretion. When we blinded rats or placed them in continuous light or darkness, the pineal rhythm in melatonin-forming activity was rapidly extinguished. If instead of turning off the lights at 7:00 P.M. illumination was

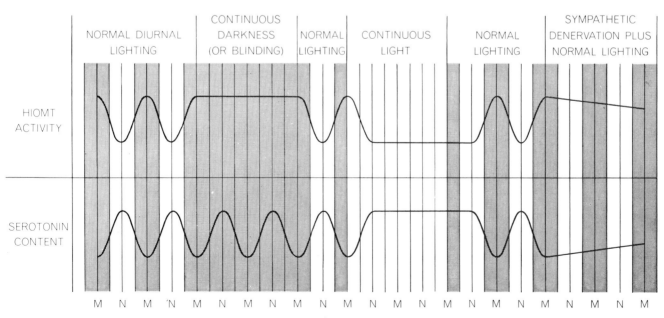

BIOCHEMICAL RHYTHMS in the pineal gland of the rat were recorded under various lighting and other conditions. Normally both the content of serotonin and the activity of the melatonin-forming enzyme (HIOMT) vary with a 24-hour cycle. The serotonin content is greatest at noon (N), whereas the HIOMT activity is greatest at midnight (M). The HIOMT cycle is completely dependent on environmental lighting conditions: it disappears when animals are kept in continuous light or darkness, or when they are blinded. The serotonin cycle persists in continuous darkness or after blinding but can be abolished by keeping the rats in continuous light. Both cycles are depressed when the sympathetic nerves to the pineal gland are cut (*extreme right*). Gray areas signify darkness.

continued for an additional five hours and pineals were examined as usual at midnight, the expected rise in melatonin-forming activity was completely blocked. This pineal rhythm in HIOMT activity thus appears to be truly exogenous, or externally regulated, and is entirely dependent on shifts in environmental lighting. Hence this enzyme rhythm may be more important in carrying information about light to the glands than other circadian rhythms that do not depend on light for their existence.

Recently Wilbur Quay of the University of California at Berkeley has found that the content of serotonin in the rat pineal also undergoes marked circadian rhythms. The highest levels of this amine are found in pineals at noon and the lowest levels at midnight. Serotonin content falls rapidly just at the time that melatonin-forming activity is rising.

In collaboration with Solomon Snyder we studied the mechanism of the serotonin cycle. When rats are kept in continuous light, the serotonin cycle is extinguished. To our surprise, however, when rats are kept continuously in darkness or blinded, this rhythm persists, unlike the rhythm in the melatonin-forming ezyme. When the sympathetic nerves to the pineal are cut, the serotonin and HIOMT cycles are both suppressed. When the nerves from the central nervous system to the superior cervical ganglion are interrupted, the serotonin rhythm is also abolished [*see illustration on previous page*]. Hence the serotonin rhythm in the pineal gland is similar to most other circadian rhythms (and differs from the HIOMT cycle) in that it is endogenous and depends on environmental light only as an external synchronizer. The mechanism that controls the serotonin rhythm appears to

reside within the central nervous system. The pineal gland thus contains at least two distinct biological clocks, one totally dependent on environmental lighting and the other originating within the brain but cued by changes in lighting.

At present little is known about what organs are dependent on the pineal clock for cues. The ability of melatonin to modify gonadal function suggests, but does not prove, that its secretion may have something to do with the timing of the estrus and menstrual cycles—two phenomena about whose mechanisms of control very little is known. One is tempted to argue teleologically that any control mechanism as complicated and sensitive as that found in the mammalian pineal gland must have some place in the economy of the body.

35

CALCITONIN

HOWARD RASMUSSEN AND MAURICE M. PECHET

October 1970

*This recently discovered thyroid hormone plays an important
role in metabolism: it inhibits the breakdown of bone and
thus keeps the calcium in the blood from reaching an
excessively high level*

The endocrinology of mammals has been investigated so intensively and with such a wealth of discoveries over three-quarters of a century that one might suppose the subject by this time would have lost some of its novelty and excitement. In actuality, however, the story of the endocrine glands and their hormones retains an unending fascination. New mammalian hormones are currently being discovered at a high rate, as in the early rush of exploration during the first quarter of this century. The new hormone we shall discuss in this article turns out to be of great importance in the regulation of a crucial phase of human metabolism. Ironically, this hormone escaped attention for nearly 80 years although it was more or less constantly on stage throughout that time. It is a product of the thyroid gland, which over the years has been studied more intensively than any other endocrine organ. Discovered less than a decade ago, the long overlooked hormone, calcitonin, has evoked such interest among physiologists, biochemists and physicians that it has already been isolated in pure form, synthesized completely in the laboratory, investigated in detail as to its functions and used therapeutically in human disease.

The discovery of calcitonin grew out of a puzzling question regarding the regulation of the calcium level in the blood. A constant supply of calcium (and phosphate as well) is required in the circulating blood for the building of bone and the control of certain functions of cells. If the concentration of calcium ions in the blood plasma falls below normal, the nerve and muscle cells, for instance, begin to discharge spontaneously and the voluntary muscles go into continuous contraction—the condition known as tetany. It was known that two agents, vitamin D and a hormone of the parathyroid gland, were active in maintaining the plasma's calcium supply: vitamin D by assisting the uptake of calcium from food by the intestinal cells, the parathyroid hormone by causing the release of calcium to the blood from bone and by inducing the kidney tubules to capture, for return to the circulating blood, calcium that would otherwise be lost in the urine.

Experiments had established that the activity of the parathyroid was governed by a feedback system. When the calcium in the plasma dropped below the normal level, the gland increased its secretion of the hormone; when the calcium rose above normal, secretion of the hormone stopped. The question arose: Was this strictly negative control—the shutoff of the parathyroid hormone—sufficient to protect the animal against a dangerous rise in the blood's calcium concentration? Peter Sanderson and his associates at the Peter Bent Brigham Hospital in Boston and D. Harold Copp at the University of British Columbia began to look into the possibility that there was also some mechanism that exerted a positive control on the accumulation of calcium in the blood. They soon found evidence that such a mechanism was indeed at work.

Sanderson's group performed an elegant series of experiments involving both the thyroid and the parathyroid glands. They first tested the reactions of dogs, with the glands intact, to abrupt experimental changes in the plasma calcium level; they raised the level by injecting calcium salts or lowered it by injecting a chelating agent that bound and inactivated the calcium ions in the plasma. They found that in either case the animals' plasma calcium quickly returned to the normal level after the infusions were stopped. They then removed the thyroid and parathyroid glands surgically from the same animals and retested them with the injection treatments. This time there was no rapid return to normal; the calcium content of the plasma remained elevated or subpar for as long as 36 hours after the injection.

The failure of calcium to rise rapidly to normal in the cases where the level had been depressed could be explained by the absence of the stimulating parathyroid hormone (the source of the hormone having been removed). Obviously, however, lack of the hormone could not account for the fact that the calcium level remained high in the cases where the level had been elevated. It was clear that some other corrective agent must be missing in the animals with the thyroid and parathyroid glands removed—some positive agent that could cut back the delivery of calcium to the blood or speed up its removal. Copp, on the basis of experiments similar to Sanderson's, concluded that the agent must be a hormone that he named calcitonin, signifying that it participated in regulating the tone, or concentration, of calcium in the blood. Two groups of investigators, Philip F. Hirsch and Paul L. Munson at the Harvard Dental School and Iain MacIntyre and his colleagues at Hammersmith Hospital in London, soon located the source of the hormone. They found it not in the parathyroid, as Copp had suggested, but in the thyroid gland.

A. G. E. Pearse of the Postgraduate Medical School of London went on to discover the specific site where calcitonin is synthesized. It had been known for many years that the thyroid gland contains two types of cell: the type that produces the gland's classic hormone, thyroxine, and another type called *C* cells, which stain differently. Although

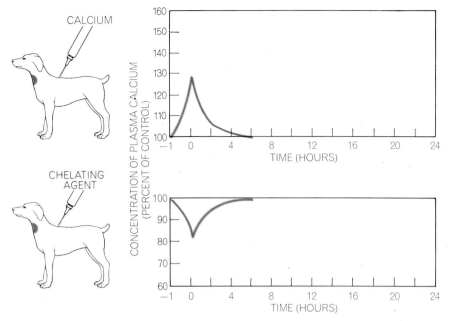

FIRST PHASE OF DOG EXPERIMENT conducted by Peter Sanderson and his colleagues at the Peter Bent Brigham Hospital indicates that there is a positive mechanism preventing excessively high levels of calcium in the blood. At top dog is injected with calcium. Graph at right shows that one hour before injection calcium concentration in the blood is normal. At time of injection concentration rises. About six hours later, however, concentration is normal. Therefore some agent is probably reducing the flow of calcium from the bone into the blood. At bottom same dog is injected with a chelating agent that captures calcium in blood and renders it inactive. The graph shows that one hour before the injection the calcium concentration is normal. After the injection it is depressed. About six hours after the injection the concentration is normal again, as is expected because a hormone secreted by the parathyroid gland is known to elevate the calcium concentration in the blood.

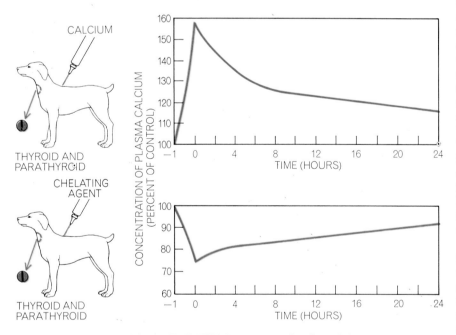

SECOND PHASE OF DOG EXPERIMENT demonstrates that there definitely is a controlling agent. At top a dog with thyroid and parathyroid glands removed is injected with calcium. The graph indicates that the calcium concentration, which was normal one hour before the injection, quickly rises. More than 24 hours later the concentration is still high, indicating that some agent that normally reduces the concentration is no longer present because the glands are missing. At bottom a chelating agent is administered. The graph at right indicates that the calcium concentration, normal one hour earlier, is depressed. More than 24 hours later the concentration is still low, because the parathyroid hormone, which keeps the concentration high, has been eliminated by removal of the gland. D. Harold Copp of the University of British Columbia demonstrated the significance of these experiments.

the C cells were described by Jose F. Nonidez of Cornell University in 1931, their function was unknown until Pearse showed that they produced calcitonin. These cells have had an interesting evolutionary history. In mammals the C cells arise in ultimobranchial glands that lose their distinct identity during development and merge with the thyroid. In fishes, amphibians, reptiles and birds, however, the ultimobranchial glands remain separate and persist as distinct organs in the adult. Stuart Tauber of the University of Texas Southwestern Medical School and Copp established that in these animals the ultimobranchial gland contains a high concentration of calcitonin but that the hormone does not show up at all in the thyroid gland.

Thus by 1967 the source of calcitonin in mammals was well established, and investigations of various aspects of its physiological activity were under way. One of the first aspects to be studied was the system that calls forth the secretion of calcitonin and its counterpart, the parathyroid hormone. Sanderson's and Copp's earlier studies had indicated, as we have seen, that the basic controlling factor was the calcium concentration in the blood. Direct measurements of the levels of calcium, calcitonin and the parathyroid hormone in the circulating blood of experimental animals now gave a quantitative picture of the control system. These measurements were made possible by newly developed assay techniques derived from the fundamental work of Solomon A. Berson and Rosalyn S. Yalow of the Veterans Administration Hospital in the Bronx. These investigators, working with John T. Potts, Jr., and Gerald D. Aurbach of the National Institutes of Health, had devised a radioimmunoassay for measuring the concentration of the parathyroid hormone; later Potts, and independently Claude Arnaud of the Mayo Clinic, had worked out a similar assay for calcitonin.

Using these assays, Potts and Arnaud found that in pigs infused with measured amounts of calcium the secretion of calcitonin increased in direct proportion to the rise in the blood's calcium content, and the secretion of the parathyroid hormone decreased in proportion to the calcium rise. Thus the results showed that the physiological control system responsible for keeping the blood's calcium supply at a stable level consists of two feedback loops: the parathyroid hormone operating to sustain the supply, calcitonin operating to prevent calcium from rising above the required level.

After the discoveries of calcitonin's

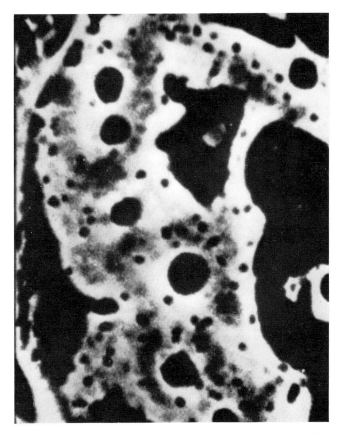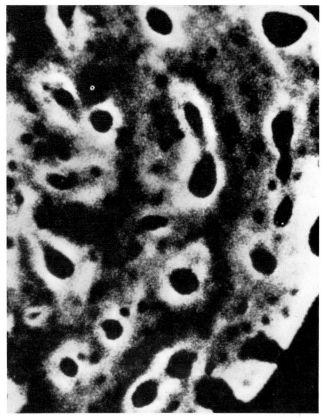

SECTIONS OF SHEEP BONE shown in these microradiographs demonstrate effects of the parathyroid hormone. The microradiographs were made by interposing a thin section of bone between a source of alpha radiation and a photographic plate. In microradiograph at left bright areas are calcified bone. Large dark holes are channels for blood vessels; small dark holes are occupied by osteocytes. The microradiograph at right was made after parathyroid hormone had caused osteocytes to resorb bone tissue. Resorption of bone tissue produces large blurred areas. The osteocyte spaces are enlarged and some of them have merged with one another.

role and the source that produced the hormone, the next challenge was to decipher its chemical nature, as a preliminary to learning how the hormone brings about its striking effect. A number of investigators set out to isolate the hormone, using various assay methods based on measurements of the potency of their preparations. Six laboratories reported late in 1967 and early in 1968 that they had extracted the hormone in pure form from pig thyroid glands. It turned out to be a polypeptide consisting of a single chain of 32 amino acids. Three laboratories soon determined the sequence of the amino acids in the molecule's structure, and shortly thereafter the Lederle Laboratories in Pearl River, N.Y., and the Ciba and Sandoz laboratories in Switzerland announced the synthesis of the hormone from its amino acids. The Ciba group, headed by R. Neher, also succeeded in isolating the human version of calcitonin (from tissue obtained from a patient with a carcinoma of the thyroid gland) and analyzed its amino acid sequence. This proved to be somewhat different from the structure of the pig hormone, indicating that the activity of calcitonin in curtailing the calcium content of blood depends on certain critical features of the amino acid sequence (common to both the pig and the human forms of the molecule) rather than on the structure as a whole.

The way was now open to investigate calcitonin's mode of action. How does the hormone function to lower the concentration of calcium in the blood plasma? Experiments with preparations of purified calcitonin in many laboratories in the U.S. and Europe soon showed that the hormone produces its effect by decreasing the release of calcium to the blood from bone. Calcitonin was found to act on the metabolism of bone in a way that inhibits bone resorption.

Bone is by far the most refractory and durable of all biological tissues, as is evident in the survival of fossil bones that have been buried for hundreds of thousands of years. Because of the apparent obdurability of bone it was supposed until very recently that once the skeleton of a vertebrate has been formed it ceases to partake of metabolism or to have any appreciable breakdown. Definitive evidence that this is by no means the case did not come until the availability of radioactive isotopes made it possible to examine the events actually occurring in the bone of living animals.

Tracer experiments with labeled isotopes of the main elements that go into the makeup of this tissue (calcium, phosphorus and the carbon of amino acids) showed that even in fully mature bone there is a constant turnover of these materials. It then became evident that metabolic activity in the skeleton is a vital necessity for at least two reasons. For one thing, such activity allows remodeling of the skeleton to enable it to deal with the developing mechanical stresses on the body. It is as if each bone were an elaborate Gothic structure in which a resident engineer, in response to changes in stresses, continually directs the replacement of supporting arches with new ones providing a slightly different center of thrust. The other vital function of metabolic activity in bone is that, by allowing an exchange of minerals (calcium and phosphate ions) with the blood, it provides a storehouse of these materials to help meet the body's general requirements. And in fact normally there is an overall balance of supply and de-

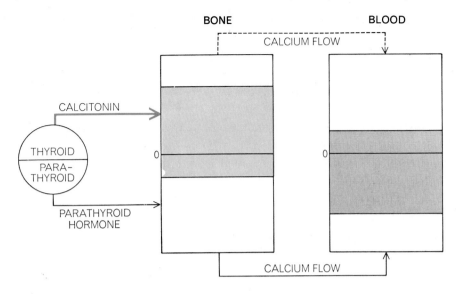

CONTROL LOOP regulating balance of calcium in bone and blood is established by calcitonin and the parathyroid hormone. Thyroid (*left*) secretes calcitonin (*colored arrow*) when blood calcium is high. Calcitonin prevents resorption, so that bone calcium level (*color in upper part of box in middle*) remains high and flow of calcium into blood is reduced (*broken arrow at top*). Therefore calcium level in the blood (*shading in upper part of box at right*) is low. When calcium level gets too low, parathyroid gland (*left*) increases it by secreting hormone (*black arrow*) so that more calcium enters blood (*solid arrow at bottom*).

mand for the bone minerals, new bone being laid down in some parts of the skeleton and old bone being resorbed elsewhere at the same rate.

To understand the operations of calcitonin and the parathyroid hormone in this system we need to look into the events in the formation and the resorption, or destruction, of bone. We are concerned here with what goes on at the surface of existing bone as new tissue is formed or old bone is dissolved away. The building on of new bone is initiated by a surface layer of osteoblasts, the cells that synthesize and extrude molecules of the fibrous protein collagen. The collagen molecules form thin, insoluble fibrils that pack themselves closely in a definite geometric array constituting a matrix, firmly anchored in the extracellular spaces of the bone. The matrix, which takes several days to form, then acts as a template for the deposit of crystals of mineral in an ordered arrangement. Studies by Melvin J. Glimcher and Stephen M. Krane of the Massachusetts General Hospital indicate that the buildup of the minerals begins with the attachment of phosphate groups on specific sites in the matrix. Once the first mineral crystals form, the depositing of phosphate and calcium proceeds rapidly, and within a few days the new bone contains 70 percent or more of all the mineral it will ever possess. As mineral is deposited, it displaces water and even-

tually occupies most of the available space in the matrix. It is then locked in so that (in an adult animal) most of the mineral is not free to take part in exchanges of mineral between the bone and the blood. This, of course, is an essential condition, because unrestricted removal of mineral from the bone could deprive it of its rigidity and mechanical strength.

Exactly where in the matrix is the mineral deposited? Is it encrusted around the collagen fibrils? If this were so, one would expect the mineral to play a large role in maintaining the structural form (shape and size) of the bone. Actually experiments have shown that when the mineral is extracted (by artificial means), the bone is not changed in shape or size but merely softens. A more plausible picture of the location of the mineral crystals, suggested by John A. Petruska and Alan J. Hodge of the California Institute of Technology and fitting the known facts, is that they are lodged in spaces between the ends of the fibrils, which are supposedly arranged in an overlapping array. This model implies that the mechanical properties of a bone are determined by the location of mineral in the matrix, and the bone's shape and size are determined by the three-dimensional configuration of the matrix, which is fixed by strong chemical bonds tying the fibrils together.

We come now to the opposite of the bone-building process: the resorp-

tion of bone that makes possible the continuous exchange of phosphate and calcium ions between bone and the blood. It turns out that the bone-forming cells play a role in resorption as well. As the osteoblasts complete their function of producing collagen, they become entrapped in these fibrils and are continually overlain by new osteoblasts growing on the surface. The buried osteoblasts are now converted into osteocytes and put out thin extensions, forming an extensive cellular network throughout the bone. When the bone has been fully formed, it is covered by a film of fluid topped by a thin layer of "resting" osteoblasts, which are no longer making collagen. These cells, together with the osteocytes, perform the important function of acting as sentinels that patrol the flow of mineral ions between the blood plasma and the special fluid bathing the bone. Were it not for this guarding network, a small change in the bone's chemistry might lead to a flood of calcium into or out of the blood, either of which could be lethal.

The involvement of osteocytes in bone resorption was clearly disclosed recently in radiographic studies by Leonard F. Bélanger of the University of Ottawa. It had long been supposed that bone was broken down only by the giant digesting cells called osteoclasts. Examining thin sections of bone by means of radiography (using alpha radiation and X rays), Bélanger produced pictures showing the matrix and indicating where it was calci-

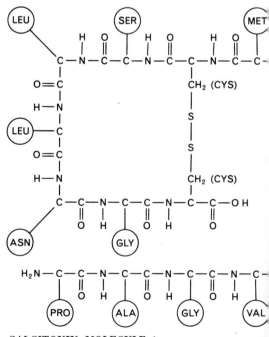

CALCITONIN MOLECULE in man consists of a specific sequence of 32 amino acid

fied and where it was not. He found that in bone that had been treated with the parathyroid hormone both the matrix and the calcium were dissolved away around each osteocyte [*see illustration on page 363*]. Apparently calcium is so tightly trapped in the matrix that it is not readily released unless the matrix structure is broken down.

The experiment thus demonstrated that the parathyroid hormone promotes bone resorption by acting on the osteocytes; in some way it incites these cells to engulf and destroy the bone tissue around them, with a consequent release of dissolved calcium into the bloodstream. We undertook a different set of experiments to probe the counteracting behavior of calcitonin. Conceivably this hormone might lower the blood's calcium level by speeding up the laying down of calcium in new bone, but it seemed much more likely that it acted by inhibiting bone resorption.

In looking into the effects of calcitonin we had the benefit not only of the radiographic techniques for examining what went on in bone tissue but also of a newly developed chemical measure of the rate of bone resorption in living animals. The collagen of bone is a protein rich in a peculiar amino acid: hydroxyproline. Darwin J. Prockop of the University of Pennsylvania School of Medicine had learned that hydroxyproline is not incorporated in a protein in the usual way—with the help of transfer RNA. There is no specific transfer RNA for this amino acid. A precursor molecule, proto-

collagen, is first synthesized and then specific proline units within it are acted on by an enzyme that adds the hydroxyl group to some of the prolines. In this way hydroxyproline is produced at the necessary positions on the collagen molecule. When the collagen molecule is broken down, hydroxyproline shows up in urine. It serves as a distinctive signal; the amount of hydroxyproline in the urine has in fact been found to reflect the rate of bone resorption in an animal. We found it useful in determining whether or not the resorption rate was influenced by calcitonin.

We removed the thyroid and parathyroid glands from rats and then injected the parathyroid hormone, so that the animals had an ample supply of this hormone but no calcitonin. The rats excreted large amounts of calcium and hydroxyproline in their urine, showing a high rate of bone resorption, and radiographic examination of sections from their bones confirmed that the osteocytes had destroyed bone tissue around them and had greatly swelled in size. When we carried out a parallel experiment with another set of rats, this time supplying them with injections of calcitonin as well as the parathyroid hormone after their thyroid and parathyroid glands

were removed, the results were strikingly different. The animals' excretion of calcium and hydroxyproline in the urine *decreased,* and radiographic examination of bone sections showed no sign of bone breakdown around the osteocytes. It is also noteworthy that the osteoclasts, the large cells on the surface of bone that are involved in its remodeling, also increase their resorptive activity in response to parathyroid hormone and decrease it in response to calcitonin.

Here, then, was clear evidence that calcitonin performs its role of controlling the calcium content of the blood by inhibiting bone resorption. A simple further experiment demonstrated how vital this control is for the animal. In animals that had both glands removed and were then injected with a continuous infusion of calcium and the parathyroid hormone, the calcium concentration in the blood rose rapidly to a high level, urine excretion began to fail after 16 hours, and shortly afterward the animals died. The cause of death, as postmortem examination showed, was a massive accumulation of insoluble calcium phosphate in the kidneys, destroying their function. When, on the other hand, we removed only the parathyroid

ALA ALANINE
ASN ASPARAGINE
ASP ASPARTIC ACID
CYS CYSTINE
GLN GLUTAMINE
GLY GLYCINE

HIS HISTIDINE
ILE ISOLEUCINE
LEU LEUCINE
LYS LYSINE
MET METHIONINE
PHE PHENYLALANINE

PRO PROLINE
SER SERINE
THR THREONINE
TYR TYROSINE
VAL VALINE

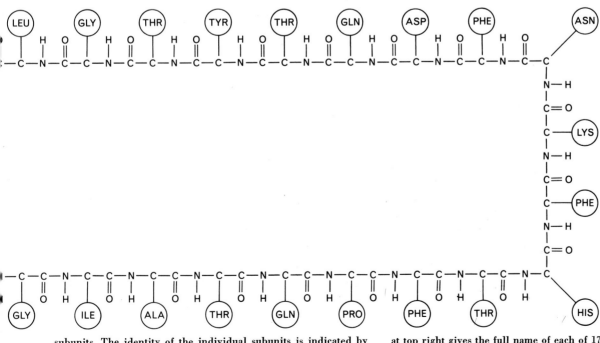

subunits. The identity of the individual subunits is indicated by the abbreviations that mark the amino acid side chains. The key **at top right gives the full name of each of 17 different amino acid constituents of hormone. The sequence differs in other species.**

gland, leaving the thyroid intact, the animals responded to the same infusion of calcium and parathyroid hormone without apparent ill effects. Clearly in this case the thyroid gland saved the day with an output of calcitonin that prevented a marked rise in the blood's calcium level, and we found no damaging deposits of calcium in the kidneys.

In view of calcitonin's now clearly established function as an inhibitor of bone resorption, it is curious to find that some primitive cartilaginous animals, such as the dogfish shark, produce calcitonin (in an ultimobranchial gland) although they have no bony skeleton. We can only speculate about what function the hormone may serve for them. Apparently calcitonin, like some other hormones, is an ancient biological agent that evolution converted from its original function to a new use in the vertebrates.

From the physiological study of calcitonin's action we have now gone on to try to learn how the hormone produces its effect in biochemical terms. This is difficult to get at, because bone tissue is structurally so complex and composed of such a diversity of cells that direct exploration by specific chemical experiments is almost out of the question. On the basis, however, of various items of evidence, mostly indirect, we have arrived at a tentative hypothesis about the biochemistry of calcitonin's control of metabolism in bone cells.

A starting point was provided by recent discoveries concerning the biochemical action of the parathyroid hormone. These discoveries have been made in the years since one of us (Rasmussen) wrote an article in *Scientific American* on that hormone ["The Parathyroid Hormone," by Howard Rasmussen; SCIENTIFIC AMERICAN Offprint 86]. In the urine of rats given parathyroid hormone L. R. Chase and Aurbach of the National Institutes of Health noted an increased content of the remarkable compound known as cyclic 3'5' adenosine monophosphate (cyclic AMP). This substance, first discovered by Earl W. Sutherland, Jr., then at Washington University, had been found to serve as an intermediate in the action of many hormones—the hormone being called the "first messenger" and cyclic AMP the "second messenger."

Sutherland and his co-workers had established that cyclic AMP is produced from adenosine triphosphate (ATP) by an enzyme, adenyl cyclase, located at the surface of cells. From experiments they developed the hypothesis that many hormones interact with specific adenyl cyclases on the surface of specific cells and bring about an increase in the production of cyclic AMP within the cell. For instance, the hormone epinephrine triggers the synthesis of cyclic AMP in liver cells, the parathyroid hormone does so in kidney cells but not in liver cells, and so on. The effect of the second messenger varies according to the special character of the target cell; for example, in the liver it causes an increase in the synthesis and release of glucose; in the adrenal cortex (where cyclic AMP acts as the second messenger for ACTH) it stimulates the synthesis and release of steroid hormones. It is as if cyclic AMP conveyed the messages of the hormones in a generalized form, simply saying to each cell: "Do your thing."

The discovery by Chase and Aurbach that the amount of cyclic AMP in the urine of rats increases following the administration of parathyroid hormone offered an opening for analysis of calcitonin's action. Assume that the parathyroid hormone fosters production of cyclic AMP in bone cells and that the cyclic AMP brings about the resorption of bone. It could be supposed, then, that calcitonin might block the effect of the parathyroid hormone either by preventing the synthesis of cyclic AMP or by inactivating that compound in the cell. (It was known that cyclic AMP could be inactivated by phosphodiesterase, an enzyme in cells that converts it by hydrolysis to 5' AMP.) Looking into the question experimentally, we first infused rats that had been deprived of their thyroid and parathyroid glands with cyclic AMP, using a dibutyryl form of the compound that had been synthesized by Theo Posternak of Case Western Reserve University and had been found to be invulnerable to hydrolysis by phosphodiesterase.

The urine of these animals showed that the injection of the cyclic AMP increased bone resorption just as injections of the parathyroid hormone did. We then found that injections of calcitonin blocked the bone-resorption effect of cyclic AMP, just as it blocked bone resorption induced by parathyroid hormone. The results of these experiments clearly indicated that neither of the hypotheses about calcitonin's action was correct: obviously it did not block bone resorption simply by preventing the parathyroid hormone from producing cyclic AMP, because it was effective against the action of that compound itself in the absence of the parathyroid hormone, and our use of the unhydrolyzable form of the cyclic AMP showed on the other hand that calcitonin did not inactivate cyclic AMP by causing its hydrolysis. Chase and Aurbach produced supporting evidence for these conclusions: they demonstrated that the parathyroid hormone brings about an increase in the concentration of cyclic AMP in bone cells and that this increase is not affected by a simultaneous presence of calcitonin.

Evidently the answer to the problem of how bone resorption was inhibited had to be sought not in some action on the cyclic AMP system itself but in metabolic events in the cell that presumably were influenced by cyclic AMP. Andre B. Borle of the University of Pitts-

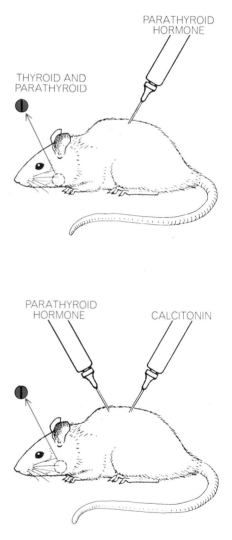

RAT EXPERIMENTS demonstrate clearly that calcitonin controls calcium concentration by inhibiting the resorption of bone induced by the parathyroid hormone. In first experiment (*top*) both the parathyroid and the thyroid glands are removed from the rat (*left*). First graph shows the level of hy-

burgh School of Medicine produced a clue that carried the search forward. Growing cells in a tissue culture, he found that when he added the parathyroid hormone to the culture medium, the cells increased their uptake of calcium. Did this effect also take place in cells growing normally in a living organism? Following up Borle's clue, one of us (Rasmussen) and a co-worker, Naokazu Nagata, established that the hormone did indeed affect calcium uptake in natural conditions, and we learned some details of the associated events.

Our first experiments were on whole animals. After removing the thyroid and parathyroid glands from rats, we injected some of the animals with the para-

thyroid hormone and others with calcium chloride. We then rapidly removed the kidneys, froze them to stop the many metabolic reactions going on in the cells, inactivated all the enzymes with cold perchloric acid and extracted the collection of metabolites, and then by a complex of techniques determined the concentrations of the various important metabolic intermediates.

We found that the infusion of calcium salt produced much the same metabolic picture as that resulting from the parathyroid hormone infusion. In order to obtain a more detailed picture under conditions enabling us to control some of the variables, we followed up with test-tube studies of kidney tissue, broken

down with two enzymes, collagenase and hyaluronidase, that split apart the many thousands of small kidney tubules. The experiments on small segments of isolated tubules confirmed that the pattern of metabolism brought about by adding to the cells' calcium supply was almost identical with that produced by supplying the parathyroid hormone. This indicated that both treatments somehow operated on the same enzymes in the cell. Both, for example, speeded up the rate at which the cells convert lactic or malic acid to glucose. The rate depends on the cells' uptake of calcium, acting as a signal, and it appears that the amount of calcium entering the cells can be enhanced either by increasing

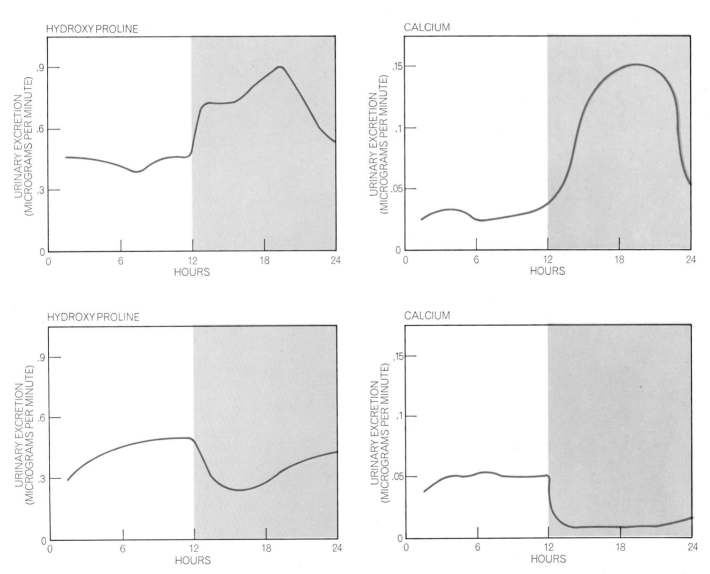

droxyproline, an amino acid produced by the breakdown of bone collagen, in the rat's urine. The second graph shows the calcium levels in the urine. The white area in each graph represents control period. The colored area represents the period during which the rat is constantly infused with the parathyroid hormone. During the control period in the first experiment, hydroxyproline levels and calcium levels are low. During the infusion period, however, both

calcium and hydroxyproline levels are high, indicating considerable release of bone material. In the second experiment (*bottom*) glandless rat receives both parathyroid hormone and calcitonin. The first graph shows that the hydroxyproline level falls during infusions, indicating that calcitonin has blocked resorption of bone tissue. The second graph shows that calcium excretion also falls, confirming that the resorption of bone has been inhibited.

the supply of calcium outside the cell or by providing the parathyroid hormone. Thus in these cells calcium is an important second messenger relaying the hormonal, or first, message into changes in cellular activity.

How does the parathyroid hormone produce that effect on cell calcium? Here we may find an answer to the still mysterious question of cyclic AMP's func-

tion. Alan M. Tenenhouse of McGill University and one of us (Rasmussen) have suggested as a working hypothesis that the parathyroid hormone has two simultaneous effects on responsive cells: it increases the uptake of calcium and it stimulates the production of cyclic AMP. The increased intracellular concentration of cyclic AMP has two important effects related to calcium. It activates certain

enzymes that now become calcium-sensitive, and it alters the intracellular distribution of calcium among various cell organelles.

In the light of this hypothesis and all the experimental findings, we can now see a possible explanation of the counteracting effect of calcitonin. That hormone may block the uptake of calcium by bone cells, a passive process and the one increased by parathyroid hormone, or alternatively it may somehow bring into play energy for active pumping of calcium out of the cells, thereby altering the changes induced by parathyroid hormone in calcium transport and bone resorption. The latter alternative seems more likely, and recently I. Radde of the University of Toronto reported early results in experiments on red blood cells that indicate calcitonin does stimulate the calcium pump in those cells.

The Ciba group of investigators announced several months ago that they had achieved total synthesis of the human form of calcitonin. This has opened the way for large-scale trials of the hormone for the treatment of various bone diseases. Some important uses in diseases characterized by derangement of calcium and bone metabolism had already been established. Unfortunately calcitonin does not appear promising as a cure for osteoporosis, by far the most common of the metabolic bone diseases. (It is particularly common in women over the age of 55 and in severe cases frequently leads to the fracture of bones because they cannot withstand ordinary mechanical stresses.) The discovery of calcitonin has, however, given a great impetus to the study of the fundamental processes of bone metabolism and turnover of the bone substances, and the rapid growth of knowledge in this field may soon bring forth a rational therapy for the disease. If so, the bringing to light of this remarkable new hormone will have served a most important catalytic function in medicine.

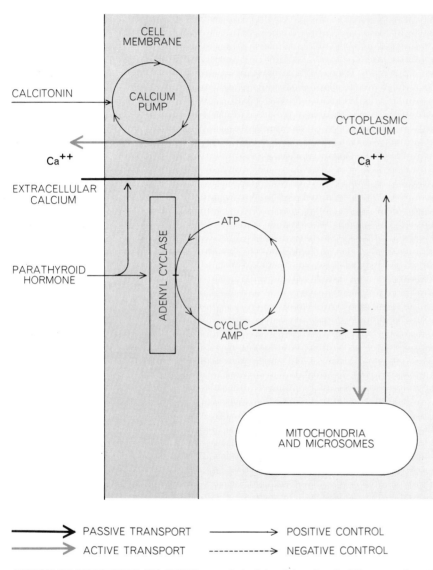

ACTION OF HORMONES ON CELLS controls body's calcium levels. When parathyroid hormone (*left*) reaches a cell wall (*dark colored area at left*), it stimulates the uptake of calcium by the cell from the extracellular space. At the same time the hormone activates adenyl cyclase, an enzyme in the membrane. Adenyl cyclase acts on ATP (adenosine triphosphate) to produce cyclic AMP (3'5' adenosine monophosphate). As calcium enters the cytoplasm (*light colored area at right*), cyclic AMP (*center*) interferes with active movement of calcium out of the cytoplasm and into the mitochondria and other organelles (*right*). Since this active transport is interrupted, the calcium remains in the cytoplasm. Meanwhile calcium continues to flow passively from the organelles back into the cytoplasm. As a result calcium accumulates there. This accumulation is the key signal to the cell to begin resorbing the surrounding matrix of bone tissue. Calcitonin (*left*), however, counteracts this chain of events by activating a "calcium pump" represented by circular arrows at top left. This process pumps calcium out of the cell's cytoplasm, across the membrane and into the extracellular space, in effect "canceling" the message. Key at bottom shows that colored arrows indicate active transport of calcium and heavy black arrows passive transport. Small black arrows represent positive control; broken arrow shows negative control.

"C" CELLS that produce calcitonin are visible as luminous bodies in the micrograph on the opposite page. The tissue is from the thyroid gland of a dog. The *C* cells were made luminous by the injection of a fluorescent antibody that combines with calcitonin. The large dark areas are follicles that contain the principal thyroid hormone, thyroxin. Thyroxin is produced by cells forming grayish perimeter of each follicle. Small dark areas in some of the *C* cells are nuclei. Micrograph was made by A. G. E. Pearse of the Postgraduate Medical School of London.

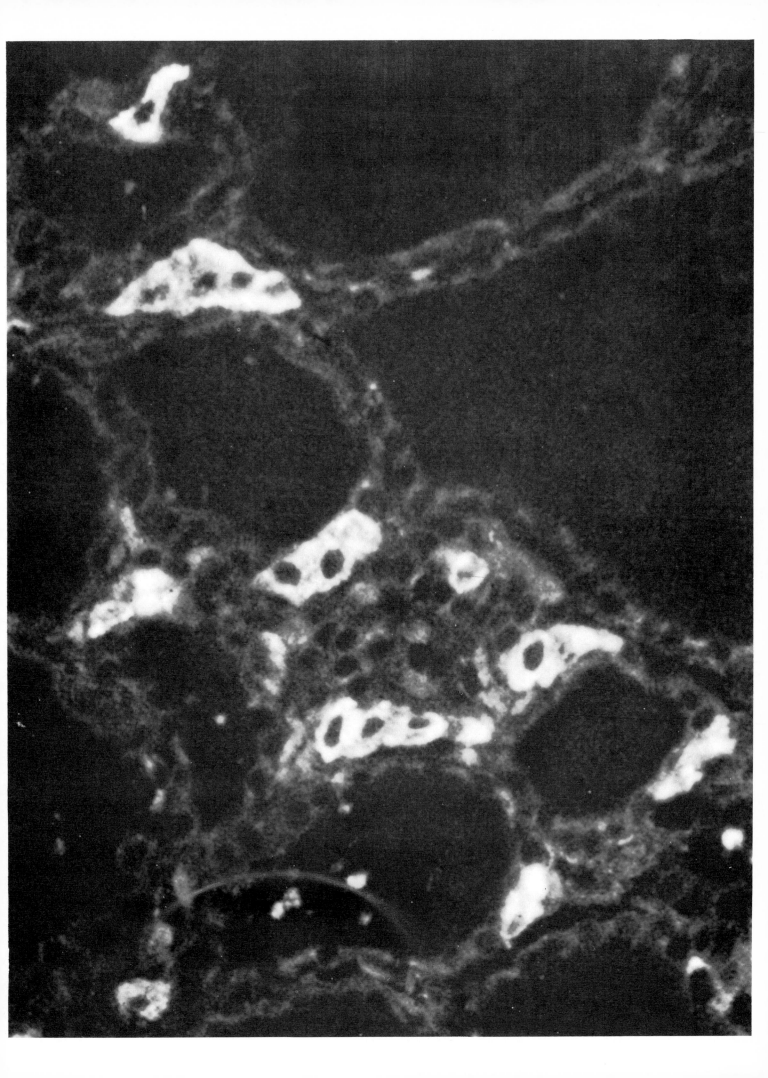

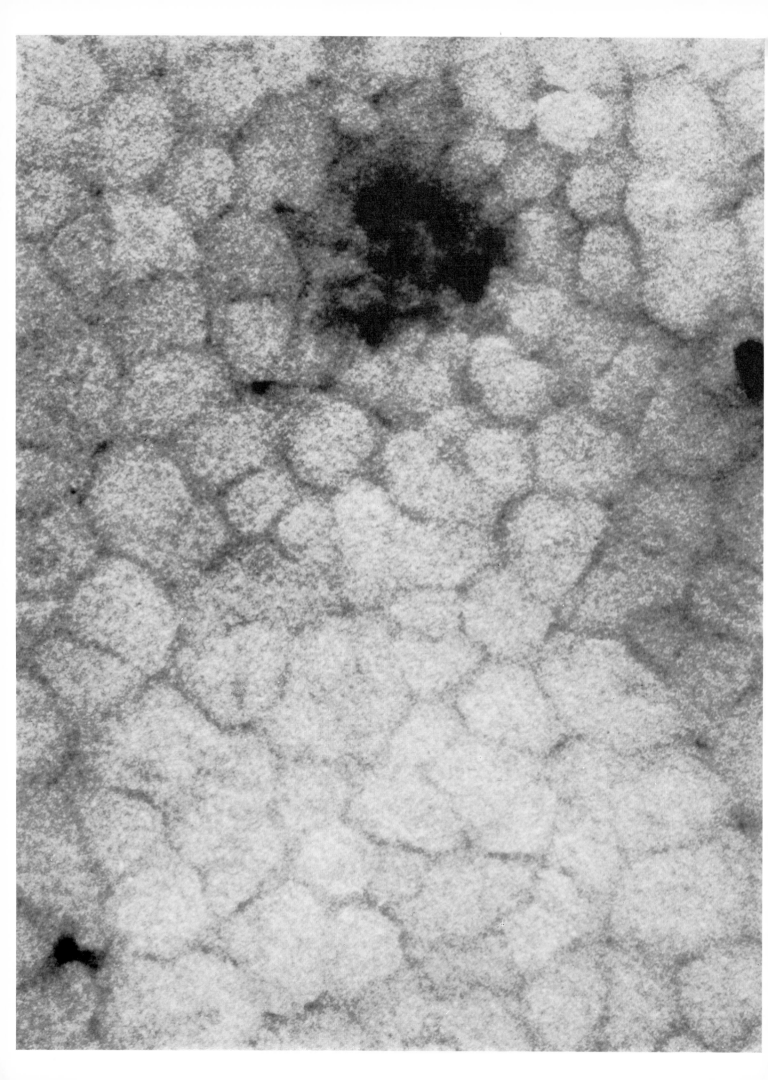

36

HOW AN EGGSHELL IS MADE

T. G. TAYLOR

March 1970

Eggshell is largely crystalline calcium carbonate. The calcium comes partly from the hen's bones, and when necessary the hen can mobilize 10 percent of her bone for this purpose in a day

To a housewife an egg is an article of food, and its shell serves to protect it from physical damage and to prevent the entry of dirt and microorganisms. To the hen an egg is a potential chick, and the shell serves not only as a protective covering but also as a source of calcium for the embryo and as a membrane through which the embryo respires. The eggshell performs its various functions with high efficiency, which is remarkable considering the number of eggs (five to seven a week) that the hen turns out. What is even more remarkable is the process whereby the hen obtains the substantial supply of calcium needed for the formation of the eggshells. The element comes in large part from her bones. Indeed, in extreme cases the hen can mobilize for this purpose as much as 10 percent of her total bone substance in less than a day! The physiology of this unusual process rewards close examination.

In its immature state the egg is one of many oöcytes, or unripened ova, in the ovary of the hen. Each oöcyte is encased in a membrane one cell thick; the entire structure is termed a follicle. At any one time follicles of various sizes, containing yolks at different stages of development, can be found in the ovary. Normally follicles ripen singly at a rate of one a day in hens that are laying regularly. There are occasional pauses. On

CHICKEN'S EGGSHELL consists mainly of columns of calcite, a crystalline form of calcium carbonate. They appear on the opposite page in an X-ray micrograph made through the thickness of a shell by A. R. Terepka of the University of Rochester; enlargement is 370 diameters. Large dark spot at top center is a "glassy" region of less opaque mineral; to its right is a pore.

the other hand, two follicles sometimes ovulate at the same time, giving rise to a double-yolk egg.

Ovulation takes place within six or eight hours after the release of a high level of a hormone produced by the pituitary gland. The release of the hormone is related to the time of onset of darkness, and it normally occurs between midnight and about 8:00 A.M. It follows that the hen always ovulates in daylight. Moreover, since it takes about 24 hours after ovulation to complete the formation of the egg, the egg is also laid during the daylight hours.

Once the yolk is released from the ovary all the remaining stages of egg formation take place in the oviduct, which consists of several distinct regions: the infundibulum, the magnum, the isthmus, the shell gland (uterus) and the vagina [*see illustration on page 373*]. The oviduct, like the ovary, is on the left side of the hen's body; a vestigial ovary and a vestigial oviduct are sometimes found on the right side in a mature bird, but they normally degenerate completely during the development of the embryo. One can only speculate on the evolutionary reason for the disappearance of the right ovary and oviduct. A reasonable guess is that two ovaries were disadvantageous because of the problem of providing enough calcium for the shells of two eggs at once. Birds have enough of a job supplying calcium for one egg a day. Certain species of wild birds have retained two functional ovaries and oviducts. It is not known how ovulation is controlled in these species, but apparently wild birds do not lay two eggs in one day.

After the ovum is released from the follicle it is engulfed by the funnel-like infundibulum of the oviduct. It is

here that the egg is fertilized in hens that have been mated. As the yolk passes along the oviduct, layers of albumen are laid down in the magnum. The proteins of the albumen, which constitute the egg white, are synthesized in the magnum from amino acids removed from the blood. The synthesis is continuous, and in the periods between the passage of yolks down the oviduct albumen is stored in the tissue of the magnum. The addition of the layers of albumen to the yolk takes about four hours.

The next stage in the formation of the egg is the laying down of two shell membranes, an inner one and an outer one, around the albumen. The membranes are formed in the thin, tubular isthmus. When the membranes are first laid down, they cover the albumen tightly, but they soon stretch. By the time the egg enters the shell gland they fit quite loosely.

The egg passes the next five hours in the process known as "plumping." This entails the entry of water and salts through the membranes until the egg is swollen. The plumping period appears to be an essential preliminary to the main process of shell calcification, which occupies the next 15 to 16 hours.

The shell is composed of calcite, which is one of the crystalline forms of calcium carbonate. A sparse matrix of protein runs through the crystals of the shell. The final stage in the formation of the egg is the deposition of a cuticle on the fully calcified shell; this is accomplished just before the egg is laid.

Let us now look at the structure of the eggshell in rather more detail. From the accompanying illustration [*bottom of page 375*] it will be seen that the shell is attached to the outer membrane by hemispherical structures known as mammillary knobs. Histochemical studies have shown that the cores of the knobs consist of a protein-mucopolysaccharide complex rich in acid groups, and that anchoring fibers run from the outer membrane into the knobs.

The cores of the mammillary knobs are laid down as the membrane-covered egg passes through the part of the oviduct called the isthmo-uterine junction; it is between the isthmus and the shell gland. It seems probable that the knobs are calcified soon after they are formed, before the egg enters the shell gland, and that they subsequently act as nuclei for the growth of the calcite crystals comprising the shell. Modern ideas on the mechanism of biological calcification—whether in bones, teeth, eggshells or any of the other places where calcium is deposited in animal bodies—emphasize the importance of crystal growth. Earlier theories seeking to explain the mechanism laid much stress on the role of precipitation of calcium salts from supersaturated solutions, but in the light of more recent evidence this concept no longer seems valid.

The mechanism whereby the mammillary knobs are calcified is not well understood. It is thought to involve the binding of calcium ions to the organic cores of the knobs by means of the sulfonic acid groups on the acid-mucopolysaccharide-protein material of which the cores are composed. It is suggested that the spatial arrangement of the bound calcium ions is the same as it is in the

lattice of the calcite crystal, so that these oriented calcium ions act as seeds or nuclei for the growth of calcite crystals forming the shell. Some years ago my colleagues and I found that the isthmus contains extremely high concentrations of both calcium and citric acid, the former reaching a maximum of about 90 milligrams per 100 grams of fresh tissue and the latter about 360 milligrams. We concluded that the high level of calcium in this region may be of significance in the calcification of the mammillary knobs.

The main part of the shell was once known as the spongy layer but has more recently come to be called the palisade layer. It is composed of columns of tightly packed calcite crystals; the columns extend from the mammillary knobs to the cuticle. Occasional pores run up between the crystals from spaces formed where groups of knobs come together. The pores reach the surface in small depressions that are just visible to the unaided eye on the outside of the shell. It is through these pores that the embryo takes in oxygen and gives out carbon dioxide during the incubation of the egg.

The raw materials for the formation of the calcite crystals, namely the ions of calcium and carbonate, come from the blood plasma. The shell gland is provided with an extremely rich supply of blood. Careful measurements have shown that the level of plasma calcium

falls as the blood passes through the gland when the calcification of a shell is in progress but does not fall when there is no egg in the gland.

Changes in the level of calcium in the blood of female birds during the breeding season have engaged the attention of many workers since 1926, when Oscar Riddle and Warren H. Reinhart of the Carnegie Institution of Washington discovered that breeding hen doves and pigeons had blood calcium levels more than twice as high as those found in cocks or nonbreeding hens. Adult males, nonbreeding females and immature birds of both sexes have plasma calcium levels of about 10 milligrams per 100 milliliters, whereas the level in females during the reproductive period is usually between 20 and 30 milligrams per 100 milliliters. For many years it was assumed that the high level of plasma calcium found in laying females was related to the trait of producing eggs with calcified shells, but it is generally recognized now that it is related to the production of large, yolky eggs. The extra calcium in the blood of laying birds (as compared with nonlaying ones) is almost entirely bound to protein. In contrast, the level of ionic calcium, which is the form of calcium mainly used in the formation of the eggshell, is about the same in laying and nonlaying hens.

The particular protein concerned in the binding of the increased plasma calcium is the phosphorus-containing protein phosvitin. It is the characteristic protein of the egg yolk. Phosvitin has a great affinity for calcium: the greater the amount of this phosphoprotein in the blood, the higher the level of plasma calcium. Phosvitin is synthesized in the liver under the influence of estrogen and is carried in the blood (in combination with lipid material) to the follicles developing in the ovary. Similar proteins are found in the blood of all animals that lay yolky eggs, including fishes, amphibians and reptiles, and yet neither fishes nor amphibians lay eggs with calcified shells, and among the reptiles only the Chelonia (turtles and tortoises) and the Crocodilia do so.

In the passage of blood through the shell gland there is a fall in both the protein-bound calcium (also termed nondiffusible calcium because the molecules of the protein to which it is bound are too large to diffuse through a semipermeable membrane) and in the diffusible calcium, the latter being mainly in the form of calcium ions. The two forms of calcium appear to be in equilibrium with each other. It seems likely that calcium

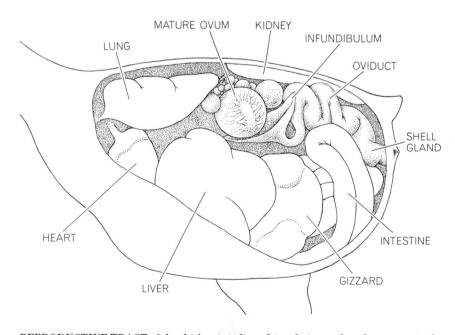

MATURE OVUM   KIDNEY
LUNG   INFUNDIBULUM
    OVIDUCT
    SHELL GLAND
HEART   INTESTINE
LIVER   GIZZARD

REPRODUCTIVE TRACT of the chicken is indicated in relation to the other organs in the body cavity. The single ovary and oviduct are on the hen's left side; an undeveloped ovary and an oviduct are sometimes found on the right side, having degenerated in the embryo.

in the form of ions is taken up from the plasma by the shell gland and that the level of ionic calcium is partly restored by the dissociation of a portion of the protein-bound calcium.

So much for the calcium ions. The origin of the carbonate ions is much harder to explain. At the slightly alkaline level of normal blood (pH 7.4) their concentration is extremely low, and it is the bicarbonate ion that predominates.

Theories to explain the formation of carbonate ions center on the enzyme carbonic anhydrase, which is present in high concentration in the cells lining the shell gland. One theory assumes that two bicarbonate ions are in equilibrium with a molecule of carbonic acid and a carbonate ion, with the equilibrium strongly in favor of the bicarbonate ions. The hypothesis is that the carbonic acid is continuously being dehydrated to carbon dioxide gas under the influence of the carbonic anhydrase, and that carbonate ions continuously diffuse or are pumped across the cell membranes into the shell gland, where they join calcium ions to form the calcite lattice of the growing crystals in the eggshell. An alternative theory, proposed by Kenneth Simkiss of Queen Mary College in London, is that the carbonate arises directly in the shell gland by the hydration of metabolic carbon dioxide under the influence of carbonic anhydrase.

The main evidence in support of the intimate involvement of carbonic anhydrase in eggshell formation is that certain sulfonamide drugs, which are powerful inhibitors of the enzyme, inhibit the calcification of shells. By feeding laying hens graded amounts of sulfanilamide, for example, it is possible to bring about a progressive thinning of the shells. Eventually, at the highest levels of treatment, completely shell-less eggs are laid.

On the average the shell of a chicken's egg weighs about five grams. Some 40 percent of the weight, or two grams, is calcium. Most of the calcium is laid down in the final 16 hours of the calcification process, which means that it is deposited at a mean rate of 125 milligrams per hour.

The total amount of calcium circulating in the blood of an average hen at any one time is about 25 milligrams. Hence an amount of calcium equal to the weight of calcium present in the circulation is removed from the blood every 12 minutes during the main period of shell calcification. Where does this calcium come from? The immediate source is the

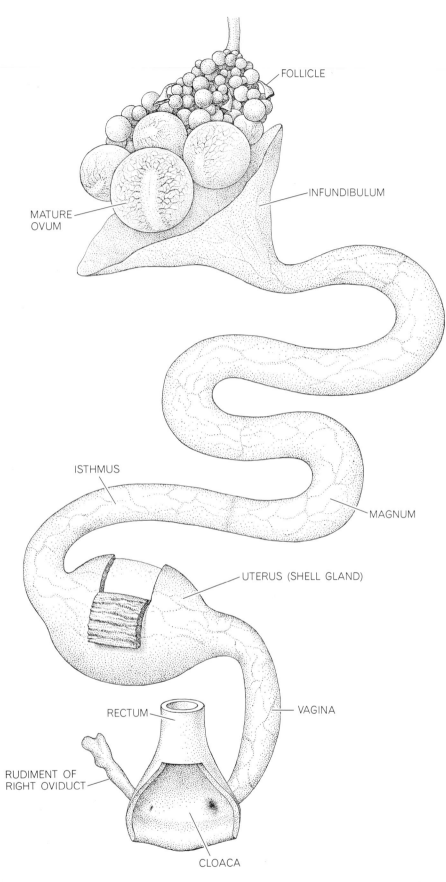

OVARY AND OVIDUCT of the chicken are involved in the formation of the egg. The shell is formed in the uterus, which is also called the shell gland. The principal steps in the formation of a chicken's egg are shown in the illustration at the top of the next two pages.

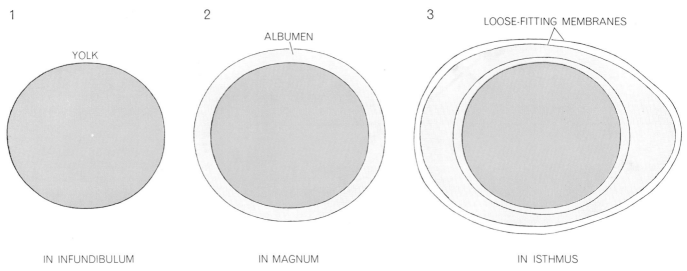

DEVELOPMENT OF EGG begins with the ovulation of a fully developed yolk from the ovary. It enters the infundibulum and begins moving along the oviduct. Layers of albumen are laid down in the magnum; the process takes about four hours. Two membranes

blood, but the ultimate source is the food. It has been demonstrated, however, that during the period of shell formation the hen is unable to absorb calcium from the intestines rapidly enough to meet the full requirement of the shell gland, no matter how much calcium is supplied in the food. When the rate of absorption from the gut falls short of the rate at which calcium is removed from the blood by the shell gland, the deficit is made good by the liberation of calcium from the skeleton.

This process has been demonstrated convincingly by the use of a radioactive isotope of calcium, calcium 45. Cyril Tyler of the University of Reading fed the isotope to laying hens daily and employed autoradiography to detect the amount of radioactive calcium deposited in the eggshells. (Beta particles given off by the calcium 45 of dietary origin blackened the X-ray film that was in contact with sections of shell, and the distribution of the isotope was thus visualized.) After the hens had been fed the radioactive calcium for a week the skeleton became intensely labeled, so that it was no longer possible to distinguish food calcium from bone calcium deposited in the shell. Accordingly the labeled calcium was withdrawn from the food, so that any calcium 45 deposited in the shells from then on must have come from the skeleton. Radioactive calcium appeared in abundance in the shells.

The mobilization of skeletal calcium for the formation of eggshell increases as the dietary supply of calcium decreases. When food completely devoid of calcium is fed, all the shell calcium comes from the bones. If a hen is fed a low-calcium diet, she will mobilize something like two grams of skeletal calcium in 15 to 16 hours. That is 8 to 10 percent of the total amount of calcium in her bones. Clearly hens cannot continue depleting their skeleton at this rate for long. When the food is continuously low in calcium, the shells become progressively thinner.

The hen's ability to mobilize 10 percent of her total bone substance in less than a day is quite fantastic but not unique: all birds that have been studied are able to call on their skeletal reserves of calcium for eggshell formation, and the rate of withdrawal is impressively high. This ability is associated with a system of secondary bone in the marrow cavities of most of the animal's bones. The secondary bone, which is called medullary bone, appears to have developed in birds during the course of evolution in direct relation to the laying of eggs with thick, calcified shells.

Strange to say, considering the fact that people had been killing birds for food for thousands of years and examining bones scientifically for at least a century, this unusual bone was not reported until 1916, when J. S. Foote of Creighton Medical College observed it in leg bones of the yellowhammer and the white pelican. The phenomenon was then forgotten until Preston Kyes and Truman S. Potter of the University of Chicago discovered it in the pigeon in 1934.

Medullary bone is quite similar in structure to the cancellous, or spongy, bone commonly found in the epiphyses (the growing ends) of bones. It occurs in the form of trabeculae, or fine spicules, which grow out into the marrow cavity from the inner surface of the structural bone. In males and nonbreeding females the marrow cavities of most bones are filled with red marrow tissue, which is involved in the production of blood cells. The spicules of medullary bone ramify through the marrow without interfering with the blood supply.

Medullary bone is found only in female birds during the reproductive period, which in the domestic fowl lasts many months. (In wild birds it lasts only a few weeks.) Medullary bone is never found in male birds under normal conditions, but it can be induced in males by injections of female sex hormones (estrogens). In hen birds medullary bone is produced under the combined influence of both estrogens and male sex hormones (androgens). It is thought that the developing ovary produces both kinds of hormone.

The formation and breakdown of medullary bone have been studied more closely in the pigeon than in any other bird. Pigeons lay only two eggs in a clutch; the second egg is laid two days after the first one. A pigeon normally lays the first egg about seven days after mating. The medullary bone is formed during this prelaying period. By the time the first egg is due to be provided with its shell, the marrow cavities of many bones of the skeleton are almost filled with bone spicules, which have grown steadily since the follicles developing in the ovary first started to secrete sex hormones.

About four hours after the egg enters the shell gland marked changes begin in

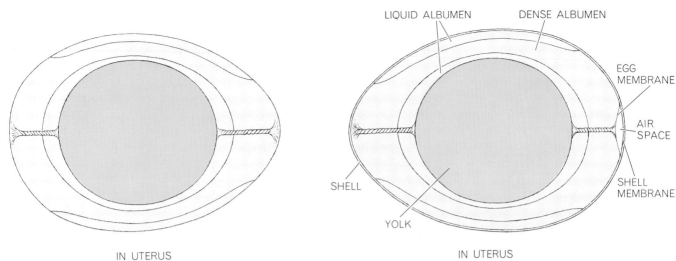

LIQUID ALBUMEN   DENSE ALBUMEN

EGG MEMBRANE

AIR SPACE

SHELL MEMBRANE

SHELL

YOLK

IN UTERUS   IN UTERUS

are added in the isthmus. At first they fit tightly, but by the time the egg enters the shell gland they have stretched so that the egg

can undergo a five-hour process called plumping. The formation of the shell occupies the 15 to 16 hours needed to complete the egg.

the medullary bone. Within a few hours its cellular population has been transferred from one dominated by osteoblasts, or bone-forming cells, to one dominated by osteoclasts, or bone-destroying cells. The phase of bone destruction continues throughout the period of shell calcification. The calcium released from the bone mineral is deposited on the shell as calcium carbonate, and the phosphate liberated simultaneously is excreted in the urine.

The breakdown of the medullary bone persists for a few hours after the egg is laid. Then, quite suddenly, another

phase of intense bone formation begins. This phase lasts until the calcification of the shell of the second egg starts; at that time another phase of bone destruction begins. No more bone is formed in this cycle. Resorption of the medullary bone continues after the second egg of the clutch is laid until, a week or so later, all traces of the special bone structure have disappeared and the marrow cavity regains its original appearance.

What mechanism might account for the rapid change from bone formation to bone destruction and vice versa?

One suggestion is that variations in the level of estrogen control the cyclic changes in the medullary bone. There can be little doubt that the high level of estrogen plus androgen in the blood plasma is primarily responsible for the induction of medullary bone during the prelaying period; the drop in the level of estrogen or androgen or both after the second egg of the clutch is laid might well give rise to the bone destruction. It is difficult to see, however, how the fine degree of control necessary to induce bone destruction when calcification of the first eggshell is due to start, and to

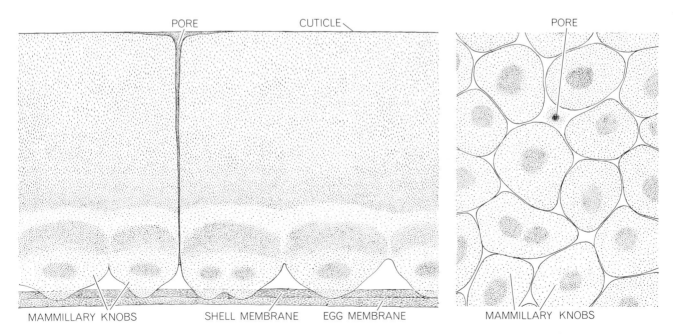

PORE   CUTICLE   PORE

MAMMILLARY KNOBS   SHELL MEMBRANE   EGG MEMBRANE   MAMMILLARY KNOBS

STRUCTURE OF EGGSHELL is portrayed in cross section (*left*) and in a tangential section (*right*) made through the layer containing the hemispherical mammillary knobs. The knobs, which attach

the shell to the outer membrane of the egg by fibers, have an organic core; the rest of the structure is made of oriented ions of calcium that apparently act as seeds for the shell's calcite crystals.

reverse the process soon after it is completed, can be exercised by changes in the secretion of sex hormones, presumably from the single follicle present in the ovary and possibly from the recently ruptured follicle.

The control mechanism that my colleagues and I consider more likely is one mediated by the parathyroid gland. The role of this gland is to regulate the level of calcium ions in the blood. A drop in the level of plasma calcium causes the release of parathyroid hormone from the gland, and the hormone brings about a resorption of bone tissue through the agency of the bone cells (osteoclasts and enlarged osteocytes). Both organic matrix and bone mineral are removed together, and the calcium and phosphate are released into the blood. The level of plasma calcium is thus restored; the phosphate is excreted.

Bone resorption under the influence of parathyroid hormone is largely due to an increase in the number and activity of osteoclasts. The histological picture

observed in the medullary bone of pigeons at the height of eggshell calcification bears a strong resemblance to the resorption of bone in rats and dogs following the administration of parathyroid hormone. Leonard F. Bélanger of the University of Ottawa and I have recently shown that the histological changes in the medullary bone of hens treated with parathyroid hormone were very similar to those occurring naturally during eggshell formation.

It has been shown that the level of diffusible calcium in the blood drops during eggshell calcification in the hen; the stimulus for the release of parathyroid hormone is therefore present. The hypothesis that the parathyroid hormone is responsible for the induction of bone resorption associated with shell formation is also consistent with the time lag between the end of the calcification of the eggshell of the first egg in the pigeon's clutch and the resumption of medullary bone formation.

When hens are fed a diet deficient

in calcium, they normally stop laying in 10 to 14 days, having laid some six to eight eggs. During this period they may deplete their skeleton of calcium to the extent of almost 40 percent. It is interesting to inquire why they should stop laying instead of continuing to lay but producing eggs without shells. Failure to lay is a result of failure to ovulate; once ovulation takes place and the ovum enters the oviduct, an egg will be laid, with or without a hard shell.

The question therefore becomes: Why do hens cease to ovulate when calcium is withheld from their diet? The most probable answer seemed to us to be that the release of gonadotrophic hormones from the anterior pituitary gland is reduced under these conditions. To test this hypothesis we placed six pullets, which had been laying for about a month, on a diet containing only .2 percent calcium—less than a tenth of the amount normally supplied in laying rations.

After five days on the deficient diet, when each hen had laid three or four eggs, we administered daily injections of an extract of avian pituitary glands to three of the experimental birds. During the next five days each of these hens laid an egg a day, whereas two of the untreated hens laid one egg each during the five days and the third untreated hen laid three eggs. We concluded that the failure to produce eggs on a diet deficient in calcium is indeed due to a reduction in the secretion of pituitary gonadotrophic hormones.

The mechanism of pituitary inhibition under these conditions has not been established. It is possible that the severe depression of the level of plasma calcium inhibits the part of the brain known as the hypothalamus, which is known to be sensitive to a number of chemical influences. The secretion of gonadotrophins in mammals is brought about by hormone-like factors released by the hypothalamus, but it is not known if the same mechanism operates in birds.

Plainly the laying of eggs with highly calcified shells has profound repercussions on the physiology of the bird. The success of birds in the struggle for existence indicates that they have been able to meet the challenge imposed on them by the evolution of shell making. Many facets of the intricate relations between eggshell formation, the skeletal mobilization of calcium, the ovary and the parathyroid and anterior pituitary glands await elucidation, but the general picture is now clear.

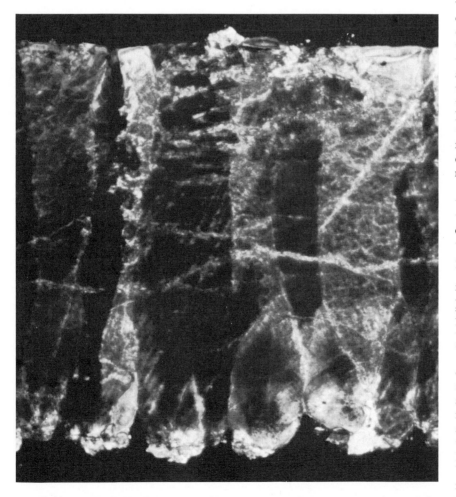

COLUMNAR STRUCTURE of an eggshell stands out in a photomicrograph also made by Terepka, using polarized light. The shell is seen in cross section at an enlargement of 325 diameters. The lumpy structures at bottom are the mammillary knobs of the eggshell.

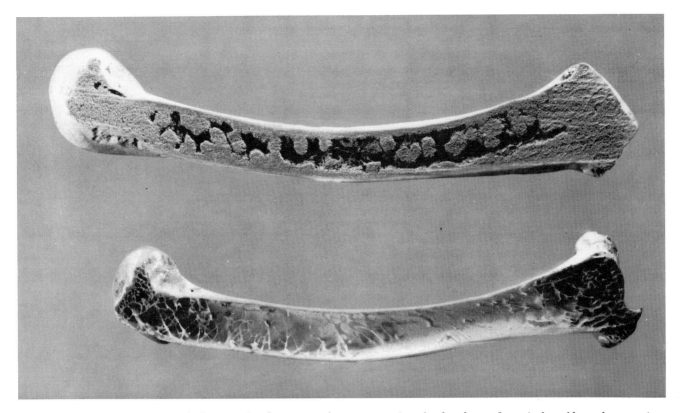

MEDULLARY BONE of a laying chicken contains the reserve of calcium that the hen draws on in forming the eggshell. Medullary bone in the femur of a laying chicken is shown at top; the struc- ture consists of trabeculae, or fine spicules, of bone that grow into the marrow cavity from the inside of the structural bone. The femur of a nonlaying bird (*bottom*) shows no medullary bone.

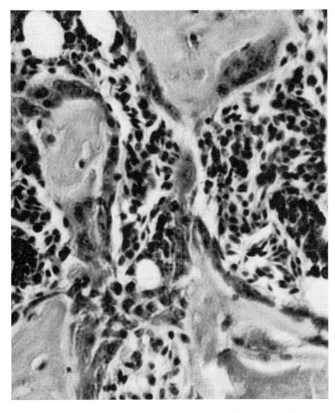 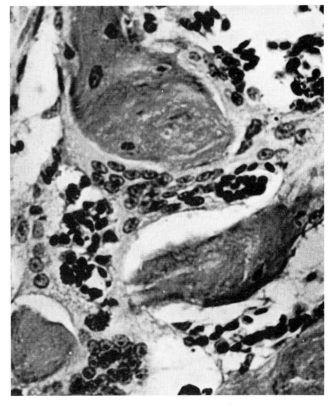

CELL POPULATION of medullary bone differs according to whether the bone is being built up or broken down. The bone is dominated by osteoblast cells (*left*) when the hen is accumulating a reserve of calcium and by osteoclast cells (*right*) when she is draw- ing on the reserve. The cells are the small, dark objects; the larger, gray objects are trabeculae. The bone is femur of a chicken; enlarge- ment is 600 diameters. These micrographs were made by Werner J. Mueller and A. Zambonin of Pennsylvania State University.

VIII
COMMUNICATION
BETWEEN VERTEBRATES

Nightingales pour out a ceaseless gush of song for fifteen days and nights on end when the buds of the leaves are swelling—a bird not in the lowest rank remarkable. In the first place there is so loud a voice and so persistent a supply of breath in such a tiny body; then there is consummate knowledge of music in a single bird: the sound is given out with modulations, and now is drawn out into a long note with one continuous breath, now varied by managing the breath, now made staccato by checking it, or linked together by prolonging it, or carried on by holding it back; or it is suddenly lowered, and at times sinks into a mere murmur, loud, low, bass, treble, with trills, with long notes, modulated when this seems good—soprano, mezzo, baritone; and briefly all the devices in that tiny throat which human science has devised with all the elaborate mechanisms of the flute, so that there can be no doubt that this sweetness was foretold by a convincing omen when it made music on the lips of the infant Stesichorus. And that no one may doubt its being a matter of science, the birds have several songs each, and not all the same but every bird songs of its own. They compete with one another, and there is clearly an animated rivalry between them; the loser often ends her life by dying, her breath giving out before her song. Often younger birds practise their music, and are given verses to imitate; the pupil listens with close attention and repeats the phrase, and the two keep silence by turns: we notice improvement in the one under instruction and a sort of criticism on the part of the instructress.

Pliny
NATURAL HISTORY, X, xliii.

VIII

COMMUNICATION BETWEEN VERTEBRATES

INTRODUCTION

In this short section we will consider three means of communication by vertebrates: fish and rabbit odors and bird songs. Communication within a species or between species depends, of course, on sensory cues, and it may elicit a variety of behavioral responses. Danger signals, feeding calls, and territorial warning devices are but a few of the many kinds of communication that animals employ. The complexity of the signal, the response, and the mode of communication (visual, odorous, vocal, or other) are highly variable and are appropriate for the circumstances of the situation and the environment. Thus the signals of night monkeys perched high in the branches of a tree in the dark are stereotyped relatively simple vocal calls, whereas equivalent signals among diurnal monkeys or apes in the daylight might include visual clues and more complex sound patterns. For these animals, visual signals and sound are appropriate vehicles of information exchange, whereas odors, electrical signals, and so on, which might function well for other animals in other types of environments, are less efficient and so are not used.

Territoriality provides good examples of simple communication. Territories may be feeding ranges, mating grounds, nest sites, and the like. It may be the exclusive area of one male who excludes all other males of the same species, it may be an area more loosely defended so that occasional intruders are tolerated, or it may be a region shared among one group of the species from which all other individuals of the same species are repelled. Nesting territories of birds are maintained by singing: the male repeats the song over and over as he flies from one part of the territory to another. As a result, other members of the species are made aware of the claim to the site. Birds, then, employ primarily sound and sometimes visual display for conveying their territorial information. In contrast, rabbits produce and distribute odorous chemicals that warn other rabbits to stay away from a territory. Because the fecal pellets retain their odor for a long time, "communication" can continue while the animal that left the pellets is active elsewhere. Unlike the odorous rabbit pellets which, of course, are fixed in space, fish such as bullheads are marked by mucus and probably other substances which diffuse into the surrounding water. As long as the fish remains in its nest, its territory can be marked.

The response of a rabbit coming upon pellets that carry the odor of a foreign territory is a fine example of complex behavior resulting from an apparently simple signal. Under such circumstances a rabbit dominant in its home range looks "nervous" and becomes subordinate even to juvenile rabbits belonging to the foreign territory (see "Territorial Marking by Rabbits," by Roman Mykytowycz). Analogous behavioral responses are described in "The Chemical Languages of Fishes" by John Todd, where dominance of individual fish is marked by odors. One can argue that in many vertebrates relatively simple stimuli elicit stereotyped behavioral responses of varying complexity. Primates in particular provide abundant examples: for instance, vervet monkeys produce three distinctive warning calls to identify a hawk, a snake, or a leopard. Each pattern of sound evokes very different evasive tactics. It is easy to imagine how natural selection has led to the association of a stimulus (such as the sighting of a hawk) with a response (emitting a specific sound) so that appropriate behavior by other individuals that can recognize the sound takes place.

The study of rabbit territoriality has raised at least two unanswered questions. First, it is not clear how each local population acquires a characteristic odor. Presumably each individual rabbit produces unique odors; the combination scattered around one warren may be the territorial smell. It is also possible that local inbreeding within each small territory might allow distinctive, group-specific odors to appear; this alternative does not seem likely, but it has not been ruled out.

Perhaps more intriguing is the question of what controls dominance and hypertrophied odor-producing glands. Mykytowycz points out that a dominant rabbit is not necessarily the largest rabbit in the territory, but its anal and chin glands are of extraordinarily large size and may be bigger than the glands of larger subordinate rabbits. One might guess that hormones cause this differential growth and the resultant increased scent-producing capacity, as well as aggressiveness and social dominance.

The article "How Birds Sing" by Crawford Greenewalt describes much of the physics and the biology of sound production by birds. Perhaps the most striking anatomical and neurological differences between bird song and mammalian communication sounds stem from the differences in the voice boxes themselves. The two sides of the avian syrinx function independently so that certain notes of a song are always produced by the right side, others by the left side. Severing the nerve to one side in an adult bird eliminates the notes from that side of the syrinx; the bird cannot "learn" that its song is deficient and cannot perform the whole repertoire on the intact side. However, if such an operation is performed on a young bird that has not yet perfected its full song, the intact side can be used for the whole song. Obviously, the avian voice box adds a versatility in sound production that cannot be matched by mammals, including man; it allows the fantastic diversity described so eloquently by Pliny and innumerable poets.

One of the main conclusions of recent work on bird song is that the slight variations observed between different birds of one species may be individual recognition markers. It is hard to generalize about as diverse a topic as bird song, but it does seem probable that the characteristics of the notes, or the syllables of the song are varied by individuals, while the basic time spans between the notes are kept relatively constant in order to provide species-recognition information. Since pitch, frequency modulation, and amplitude can all be varied, it is not difficult to imagine how even minute differences may identify each bird, and so allow pair maintenance, recognition, and so on to operate.

As a result of a number of recent experiments, we are able to determine quite precisely the stage at which young birds "learn" or acquire their songs. The zoologist Peter Marler and his collaborators have investigated populations of white-crowned sparrows living north, south, and east of San Francisco Bay. Each population sings its own dialect of the basic white-crowned sparrow song. A subsong, or a full song unlike that of any dialect, will develop in cage-reared, isolated white-crowned sparrows. If a mixed group of young from the three populations is never allowed to hear parental songs, the birds will develop their own unique dialect, which can in turn be taught to their offspring.

Normally a newly hatched bird hears its own dialect being sung for the first two or three months of life. Then in the fall and winter the older birds of the population stop singing. Several months later when spring time approaches, the maturing bird begins singing, and soon perfects the parental dialect. After a series of experiments in which captured young birds were raised in isolation, it was found that the period of learning the dialect occurs between about two weeks and two months of age. From then on, the bird "remembers" in some way what it heard early in life.

As might be expected, a nestling from one population placed with another group acquires the new dialect and sings it the following spring. This is an extraordinary feat. The young nestling acquires the song early, by listening. It does not practice or receive any reward for learning the song. Then months elapse before singing starts, and the bird practices and varies its song until it finally duplicates what it heard months before.

It is of interest, therefore, to investigate the relationship of hearing to dialect. A bird deafened at birth can at best sing only the species subsong. If a bird is deafened at two months of age, we might guess that it had heard the dialect for a sufficient length of time and that no interference with singing would occur. However, this is not true, because the dialect song has not been perfected. In fact, the experiments show that a bird must be able to hear during the weeks that it is perfecting its dialect song. Thus it appears that the bird must compare its own song noises with those that were heard months before, matching those it produces to the original sounds to produce the correct dialect. After the song has been perfected, deafening has little effect except over several months when the song gradually drifts away from the original melody; apparently, some feedback from hearing acts to maintain the normal dialect song.

Perhaps the most surprising example of this type of learning has been observed in female white-crowned sparrows. Normally, they never sing, but like the males, they are exposed to dialect singing during youth. A female injected with testosterone (a male gonadal hormone) soon breaks into song and before long is singing the dialect! Thus she too has retained the memory of early sounds, although as a rule she would never emit them. The utility of such retention is obvious if she can use it also to identify males of her regional population when pair formation and breeding take place.

One of the most fascinating variations on bird song is antiphonal singing. Some shrikes, warblers, and barbets use antiphony for pair identification and location under conditions of limited visibility, particularly in tropical habitats. As pointed out by Thorpe in "Duet Singing Birds," one bird begins by singing notes or groups of notes; the mate breaks in with its song; then the first may repeat, and so on.

Figure 1. The duet pattern on a pair of shrikes recorded on the Mozambique Coast. This is a duet of intermediate complexity; the A and B birds alternate as shown. After repeating this duet a number of times the birds may suddenly change to another song in their repertoire. On rare occasions W. H. Thorpe has heard three birds participating in a trio, but the biological function of such behavior is a mystery. The scientific pitch shown in this illustration is 256 hertz; the bar length equals 1.5 seconds. [From Thorpe, W. H., and North, Myles W., *Nature,* **208**, 1965.]

It is probable that the use of antiphonal singing as a form of personal identification and location has arisen in species in which territoriality is not rigidly enforced. Mixing, both between and within species, is much freer in the tropical habitats than in more strict territorial regions of the temperate zones. Consequently, accurate pair identification signals may have become so important that antiphony evolved. Examples of an extreme form of the imitative abilities of antiphonal singers are seen in the parrots and mynah birds, which have an amazing imitative repertoire. They too may have acquired such capabilities in order to have distinctive nuances in song for individual identification.

The final example of communication between vertebrates demonstrates that as basic a property as an individual's sex can be controlled by social interactions. There are certain "cleaner" fish (so called because they remove ectoparasites from other fishes) that live near the Great Barrier Reef and normally comprise small groups. A single dominant male and a harem of from three to six females will occupy a given territory. The females are hierarchical and have a dominant member that is usually the oldest (fittest?) of the sex. A fantastic event occurs if the dominant male dies. If foreign males do not invade the territory when this happens, the dominant female *changes sex*—in fact, within a few hours of the male's death, the dominant female may be showing "male" aggressive displays. Within a day or two, male courtship and spawning behaviors are evident! The transition to the point of releasing functional sperm takes longer—14 days or so—but that is surely a surprise in itself. D. R. Robertson, the Australian zoologist who has studied these fishes, suggests many advantages for this bizarre arrangement. For instance, the "male" genotype at any one time might be thought of as being the most highly adapted for local environmental conditions and survival, since that genotype is derived from the female genotype that survived longest in that locale. In anthropomorphic terms, however, some might question the desirability of a society in which "males are produced only when needed"!

THREE BULLHEAD CATFISH occupy a section of pipe in a laboratory tank. The largest of the three fish, at the bottom of the pipe, is recognized by the other two fish as being the dominant one in the tank, and it is usually the exclusive occupant of the shelter. In this instance, however, the two subordinate fish have sought refuge in the pipe after detecting the odor of an unfamiliar bullhead.

APPROACH OF STRANGE BULLHEAD induces an aggressive response from the dominant fish, as is shown by its gaping mouth display. The dominant fish often attacks the intruder, but its tankmates, with their mouths pacifically shut, usually stay in shelter.

37

THE CHEMICAL LANGUAGES OF FISHES

JOHN H. TODD
May 1971

*Many fishes have exquisitely sensitive organs of smell.
Experiments with catfishes demonstrate that they use this
sense for such social purposes as labeling the winners and
losers of hierarchical fights*

The fishes, of which there are some 25,000 living species, are unrivaled among the vertebrates in the ability to adapt to unpromising living conditions. They have managed to invade and occupy some of the earth's most extreme and inhospitable environments. There are catfishes in South America and in the Himalayas that spend their lives clinging to sheer rock faces in torrential mountain streams. *Anabas,* the "climbing perch" of the Orient, on occasion leaves its water habitat to walk about on land breathing air and even climbs low branches of trees in search of prey. The "annual" fish of Africa and South America, so called because its lifetime is only a single season, regularly shows up each year in tiny temporary pools that are limited to the wet season in semiarid areas. Perhaps the strangest of all the fishes are the bizarre creatures that live out their entire existence in the perpetual darkness of the ocean depths or murky waters. Most surprising, some of these animals, although sightless or nearly so, have developed sophisticated social behavior rivaling that of the higher animals in complexity.

The explanation is that these fishes have established communication, the prime necessity for social life, through their chemical senses. In many species the receptors and associated brain centers for taste and smell dominate the sensory side of the nervous system, overshadowing the optic lobes. Some fishes have been found to possess almost incredible chemosensory acuity. Harold Teichmann of the University of Giessen was able to condition eels to respond to concentrations of alcohol so dilute that he estimated the animals' olfactory receptors could not have received more than a few molecules.

I was led into the fascinating investigation of chemical communication among fishes by a study of the behavior of bullheads, or catfish, that I began at the University of Michigan under the stimulating direction of John E. Bardach. The bullhead, a predominantly nocturnal animal, has poor vision but highly developed senses of smell and taste. Its nose is almost as keen as the eel's, and its entire body is covered with hundreds of thousands of taste buds. Many years ago George Howard Parker of Harvard University had concluded from studies of the feeding behavior of bullheads that they found their way to food in the water by smelling it at a distance. I repeated and enlarged on his investigation and found that it was not smell that led bullheads to food. Blinded bullheads that had been deprived of their olfactory receptors were able to swim directly to a food source in still water. The guiding factor proved to be detection of the chemical stimulus from the distant food by the taste buds on the bullhead's body; this was demonstrated by the fact that when the taste buds on one side of the body were destroyed by surgery, the fish had difficulty finding the food and was able to do so only by constantly looping to the side where the taste buds were intact.

I was confronted then with a puzzling question: If the bullhead's remarkable sense of smell was not needed for finding food, what functions did it perform? By accident I was soon given a clue. One busy day, having an extra fish on hand without an aquarium in which to house it separately, I put the fish in a 50-gallon tank with a blinded bullhead that had occupied it alone for about a month. Immediately the two fish broke into frenzied activity, thrusting their tails against each other, quivering, opening their mouths in fierce displays and tearing at each other. I had to remove the newly introduced fish to prevent their seriously injuring each other.

These two bullheads were of the same species. When I later put a fish of a different species (but about the same size) in the tank with the resident bullhead, the resident did not attack it. In further experiments it developed that the blinded bullhead became excited and aggressive when water from the aquarium of the fish with which it had fought was introduced into its tank, but it was not aroused by a similar introduction of water from the aquarium of the different species. Evidently the blinded bullhead was able to recognize its own species from some chemical cue in the water. The animal made a surprising response to the next experiment: when water was introduced from the aquarium of a bullhead of the same species but not the fish with which the resident had fought, the resident circled the water area for several minutes but did not attack it and indicated a gradually waning interest in it.

The results of these experiments suggested that the chemical signals to which the blinded bullhead responded carried more than one message: the recipient not only was able to identify an individual of its own species but also received a signal that might identify the intruder as an individual representing a threat of some kind. Plainly the chemically sensitive bullheads must use their gift for purposes other than finding food. This realization inspired a series of investigations of fishes' communication and social behavior that I have carried out with various collaborators at the University of Michigan, San Diego State College and most recently at the Woods Hole Oceanographic Institution.

I shall relate first what we have

learned about communication of the most obvious kind: that having to do with sexual behavior. At San Diego State, Jack Nelson and I made extensive laboratory studies of the subject with two species of blennies, small fishes that abound in the tidal waters along the coast of California.

There had been earlier pioneering investigations in the area. It was George MacGinitie of the California Institute of Technology who first discovered some 35 years ago that fish use pheromones—chemical signals—for communication, as insects and some other animals do [see "Pheromones," by Edward O. Wilson; SCIENTIFIC AMERICAN Offprint 157]. MacGinitie studied the blind goby *Typhlogobius californiensis,* a small intertidal fish that loses its eyesight when young and spends its life in a burrow built by the mud-dwelling ghost shrimp. This fish, male or female, is extremely aggressive toward members of its own sex, and it will tolerate only a member of the opposite sex as a burrow-mate. Apparently a male and a female goby pair up early, before they are sexually mature, and usually remain paired for life. If an intruder invades the burrow, the resident of the same sex engages it in vicious combat that usually ends with the death of one or the other; if the intruder defeats the resident, it is accepted by the other inhabitant as if there had been no change.

MacGinitie found by experiment that the blind goby recognized the sex of a member of its species by means of a chemical signal. His experiment consisted in placing a goby in a small bag, puncturing the bag with a pin so that water could flow out of it, and then inserting the bag in a burrow occupied by a blind goby pair. When the odor from the enclosed fish reached the pair, the member of the same sex as the enclosed fish would immediately attack the bag.

Our own work with blennies was prompted by recent findings of George Losey of the Scripps Institution of Oceanography. He had shown that a male blenny could be stimulated to courtship behavior by exposure to water from a tank in which a pair of blennies of the same species were engaged in intensive courting or mating. Losey's research raised several intriguing questions, and Nelson and I began an investigation of the sexual behavior of two species of the genus *Hypsoblennius*—the mussel blenny (*H. jenkinsi*) and the tidepool blenny (*H. gilberti*)—that live in the tidal pools along the seacoast at La Jolla, the site of the Scripps Institution.

Both species bred readily in laboratory tanks simulating their natural environment, and we analyzed the details of their sexual and parental behavior. This behavior proved to be almost identical in the two species: in the building or selection of nests, courtship displays, actions during breeding and parental care there was little difference between the *jenkinsi* and the *gilberti* blennies. Indeed, when we tried to pair one species with the other, the mussel blenny males even showed some interest in courting tide-pool blenny females. This courtship was not consummated, however, by actual breeding. Furthermore, one had to find some explanation for the fact that the two species do not interbreed in nature, although they inhabit the same tide pools. The small differences in their action patterns that we observed did not seem sufficient to account for the species' isolation from each other. Evidently there must be signals of some kind that distinguished the species and prevented interbreeding.

We looked first for indications of how important visual stimuli alone were in invoking sexual behavior. Males and females were placed individually in sealed glass jars to exclude the transmission of all stimuli except the visual, and we observed their responses to one another. In this situation the males still courted females and usually made attacking moves toward males of their own kind. In the main each species ignored the blennies of the other species, except that the mussel blenny males did often court tide-pool blenny females [see upper illustration on page 393]. Hence apart from this one exception the blennies seemed to be able to recognize sex and species visually. Since the females, unlike the males, are similar in size and color, it was perhaps not too surprising that the male mussel blennies indiscriminately courted females of both species. Other signals seem to be indicated.

We then proceeded to hunt for chemical signals. The problem was complicated by the fact that we wanted to detect signals from the male and from the females of both species. Other signals had suggested to him that pheromones were produced only by the male. After many months of work and a number of unsuccessful attempts Nelson eventually developed a method for collecting the chemical emissions from an individual of either sex and presenting this uncontaminated material to a test animal. As the test animals we used mussel blenny males, since this provided us with the opportunity to find out if they were guided by a chemical signal to make a discrimination between species that they were unable to make visually. The mussel blenny male was first shown a sexually responsive female of the same species to determine whether or not he would show courtship behavior. If he did, he was then tested for his reactions to measured (100 milliliter) samples of water from various sources: a recently courted mussel blenny female, an uncourted mussel female, a courting mussel male, a courted tide-pool female or just plain seawater. The samples were uncontaminated by water that had been in contact with fish of the opposite sex. This was achieved by having the two courting fish separated by a glass barrier inside the tank.

Fortunately it was possible to determine unambiguously if the fish's response was due to a chemical signal; there are certain courtship actions in the blenny ("head jerking" and "quivering") that are associated with strong sexual motivation and that could be triggered by pheromones alone.

It turned out that in a high proportion of the tests the mussel blenny males did show this courting behavior when they were exposed to water that had bathed a courted female of their own species (no visual cues being present). They also responded to males of their own species, although much less often, by approaching water taken from the jar housing such a male. Presented with water from the jar of a tide-pool blenny female or with plain seawater, however, the mussel blenny male in almost every case made no response [see lower illustration on page 393].

Thus the results indicate that a pheromone produced by the female during courtship is the signal that sets the two species apart; the species difference in

AGONISTIC BEHAVIOR of the bullhead ranges from mild to severe; some typical actions are illustrated on the opposite page. Mouth display varies considerably in intensity: a low-key form (*a*) contrasts with more extreme displays (*b, left, and j*). Both approaching (*c*) and circling (*d*) occur in interactions of various intensities. Intermediate in intensity are the tail thrust (*e, top*), lateral display (*b, right*) and nip (*f*), all common actions among equals. The head thrust (*g*), mouth fight (*h*) and bite (*i*) are actions of greater intensity. Encounters between territorial fish usually end in a chase (*j*), with the dominant fish pursuing.

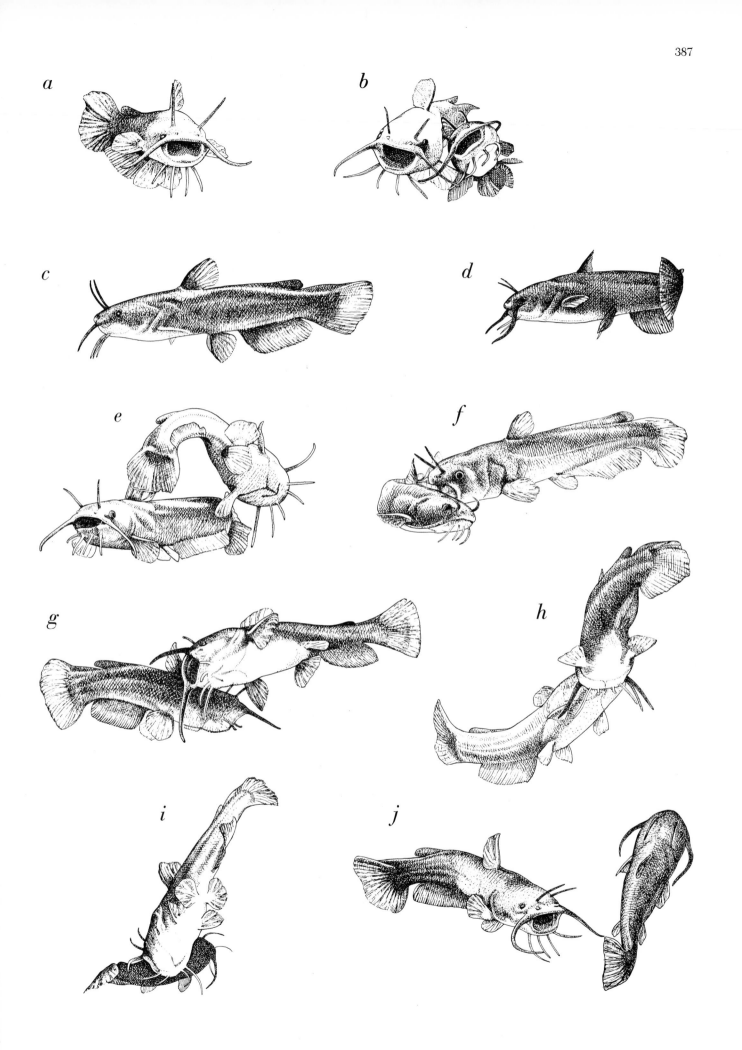

388

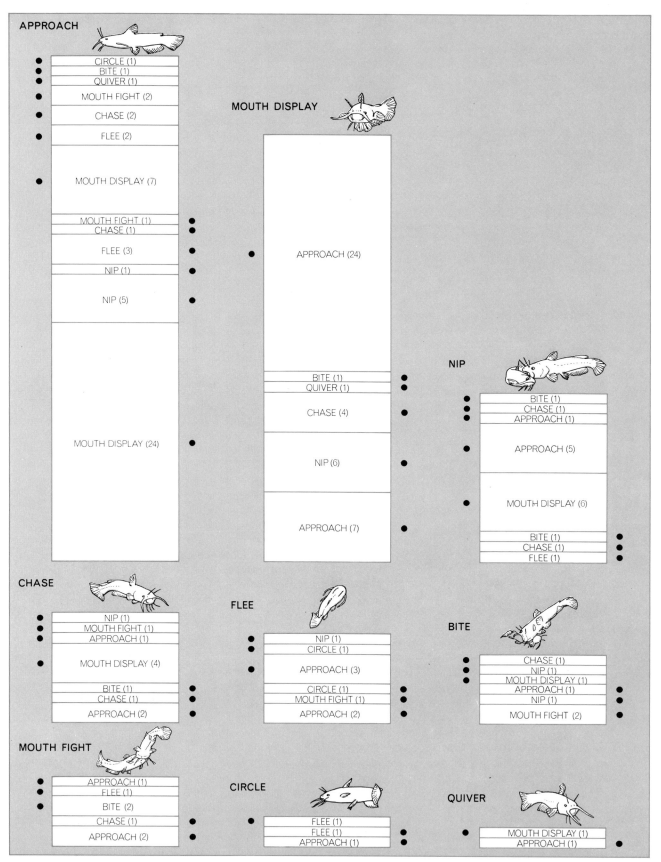

COMPLEX INTERACTIONS are characteristic of the encounters between bullhead strangers. Each of the columns in this diagram identifies one kind of action that one of the strangers frequently repeated during a prolonged encounter. The column subdivisions identify preceding or following actions by kind and frequency; the dots show whether the action preceded (*left*) or followed (*right*). Arrangement of the columns is in order of relative magnitude only; although "approach" is most frequently an initial action, temporal sequences are not represented. A simple encounter between dominant and subordinate is illustrated on the opposite page.

this chemical signal prevents interbreeding either in nature or in efforts to hybridize the two species in the laboratory.

Let us now return to the bullheads, because it is from these animals that we have gained a view of the wide extent of chemical communication among fishes and the sophistication of their chemical languages. Our helpful subjects in that inquiry have been the yellow catfish *Ictalurus natalis* and the brown catfish *Ictalurus nebulosus*, which are common residents in the small lakes and ponds of the eastern U.S. and are beloved by small boys who fish with hook and line. The bullheads may represent a pinnacle in the evolution of the chemical senses in fishes. Their organs of taste and smell are greatly extended and enlarged, and so are the regions of the brain involved in these senses, whereas their visual abilities are reduced or in some cases even absent. Our assumption that the bullheads would be ideal animals for the investigation of chemical communication has proved to be well founded.

At Michigan a neurologist who is a perceptive student of behavior, Jelle Atema, joined me in the investigation of bullhead behavior, and in preparation for controlled experiments on communications we spent months observing the social interactions of these fishes. One interesting feature of the behavior of bullheads is that they commonly establish a territory that they defend against intruders; in the natural habitat of a pond or lake the fish may occupy as its territory a hole in the bank or the space under a log or a burrow dug in the soft lake bed. In our laboratory tanks the fish similarly tended to form a territory, even if it was only a sharply defined region in the water. When a tank was occupied by two bullheads, each stayed in its own territory, turning back abruptly when it came to the border. If one of the pair was removed from the tank, the other still refrained for hours from invading the absent fish's territory. If a stranger was placed in the vacated territory, however, the resident soon crossed into that region and attacked the newcomer. On the other hand, when the stranger was removed and the former tankmate was returned, the resident again respected the border between the territories. Its performance suggested that a bullhead not only identifies but also remembers the identification of a particular individual, in this case an individual that, as experience had shown, would not contest the resident's territory

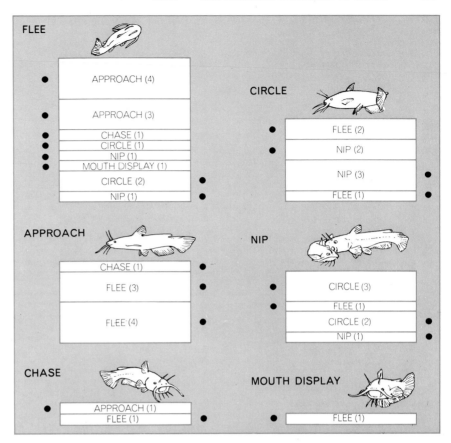

FLIGHT PREDOMINATES in this record of a subordinate bullhead's encounter with its dominant tankmate. Its most aggressive actions were a few nips and two chases. When two fish have interacted previously, their subsequent actions are limited in kind and number.

and would allow it to enjoy a peaceful coexistence in the tank.

We found that except during the mating season the females as well as the males occupied individual territories. In tanks containing a number of individuals, a hierarchy usually developed; one fish, dominating the rest, would establish the largest and most protected region as its territory, and those of lower status occupied smaller and more exposed areas. A detailed analysis showed that conflicts among the bullheads took various forms depending on the relative status of the combatants. In a confrontation between two community members of equal status they merely posture in a dancelike ritual, opening their mouths in an aggressive display, making nipping gestures and alternately approaching and fleeing. In the case of a dominant member and a subordinate member the action consists mainly in flight by the subordinate and pursuit by the dominant. Only in a rivalry between two strangers does the conflict become really violent; such a fight is characterized by a variety of aggressive actions, vicious biting and occasionally death for one of the contes-

tants. Curiously, we noted that even in nonfatal fights the loss of the fight sometimes resulted in serious consequences, apart from bodily wounds and the loss of territory, for the loser. After the battle the loser often failed to continue normal growth although it was adequately fed and housed in its own aquarium. The vanquished fish remained stunted, and after a few months it was hard to believe that the winner and the loser had been of comparable size.

Although much of the interaction between bullheads is aggressive, they exhibit cooperative behavior on occasion. When a stranger is inserted into a small community of a few fish, the dominant member allows its subordinate tankmates to swim into its territory and climb on its back for protection; after this the "boss" issues forth and engages the intruder in combat. On one occasion when a stranger swam into the boss's territory along with a subordinate tankmate of the boss, the boss evicted the stranger but allowed the tankmate to remain for a while. After the boss had defeated the stranger, however, it eventually drove the intruding tankmate out of its shelter.

The social behavior and recognition of bullheads apparently depend very little on vision; their behavior did not change significantly when fish in our laboratory were blinded (a condition that prevails in some bullhead populations in nature). For the bullheads, chemical cues are dominant, and their memory in the recognition of such cues is phenomenal. A single example will suffice to illustrate this point. Shortly after several small bullheads had been placed in a tank it happened that a large bullhead leaped in from an adjoining tank and began to maul the little fish severely. All but two jumped out to the floor and died, so that when the intruder was removed only two survivors were left in the tank. They established separate territories at the opposite ends of the tank. To test their response to a chemical cue, we introduced some water

from the tank of the bullhead that had attacked them. The two fish, disregarding territoriality, fled and hid together, returning to their own territories and shelters only after the danger signal had disappeared. Although the threatening bullhead itself did not enter their tank, for four months the two fish fled and hid every time they were exposed to water from the onetime attacker's tank. The memory of the chemical signal identifying their attacker stayed with them. They made no comparable response to water from other bullheads' tanks.

At this point in our research we began efforts to verify the existence of individual chemical "labels" in bullheads and to find the organic source of the identifying pheromones. Our first experiment was designed to test the fish's ability to discriminate between the signals from two bullhead individuals. One of

the individuals represented a positive stimulus offering a food reward for the correct response; the other represented a negative stimulus, and the wrong response to it would result in an electric shock. The fish to be tested was placed in a small tank in which there was a shelter made of half a flowerpot lying on its side. A small amount of water from the aquarium of one of the two individuals to be discriminated was poured into this tank. When the water came from the "positive" subject, the test fish would receive the food reward if it rose to the tank surface, stuck its head out of the water and opened its mouth within five seconds after sensing the chemical signal. When the water was from the "negative" individual, the test fish had to retreat to its shelter within five seconds after sensing the chemical signal in order to avoid the electric shock, delivered in the tank by means of silver electrodes.

PAIR OF AQUARIUMS provided the setting for an effort to see whether or not the placid behavior of bullheads when associating in groups might depend on the secretion of an antiaggression pheromone. A communal group, 10 in number, was established in one tank (*left*); two normally aggressive bullheads, separated by a barrier, occupied the other (*right*). Water could be transferred from

The tested bullheads readily distinguished one chemical signal from the other, and after about 25 training trials they were able to respond correctly in 95 percent of the trials (totaling 937 for the 10 tested fish). All the fish retained the learned discrimination for at least three weeks without retraining. Several of the animals developed an interesting reaction to the negative stimulus. They used their five seconds of grace to attack the stimulus with hostile mouth display and quivering before darting to the shelter!

In order to determine whether the chemical signal was detected by the sense of smell or by taste we destroyed the smelling receptors in some of the animals. These animals proved to be unable to tell one signal from the other, no matter how much training they were given. Clearly chemical communication in the bullheads is delivered through the smelling sense. This supports our hypothesis that the bullhead uses two channels: the taste buds to find its way to food and the nose to receive social communications.

For further study of the importance of the sense of smell in controlling the social behavior of bullheads we set up a community consisting entirely of fish that had been deprived of their nose tissues. These fish behaved as if they were all strangers to one another. They did not form mutually respected territories, engaged in vicious mouth fights lasting up to 20 minutes, did not flee when defeated but turned and attacked the opponent again and carried on their aggressive behavior for weeks. Only after their olfactory tissues had regenerated did the fish in this tank begin to act like members of a nonbelligerent community. We found also that when a bullhead deprived of its nose was introduced into a normal community, it failed to adjust in spite of repeated defeats and showed no sign of developing recognition of the individuals in the tank or awareness of the community's social structure. It became quite evident that the communication necessary for the formation and maintenance of a stable community depends entirely on the members' sense of smell.

In a natural environment bullheads often form a dense community, composed of hundreds of individuals, that is based not on a hierarchy or a collection of territories but on close togetherness, with the members swimming freely and peacefully throughout the pond that houses the community. We were able to reproduce such a community in the laboratory simply by placing a large number of newly trapped bullheads in one tank. The members then spent most of their time resting in close contact with one another in a kind of "love-in." These

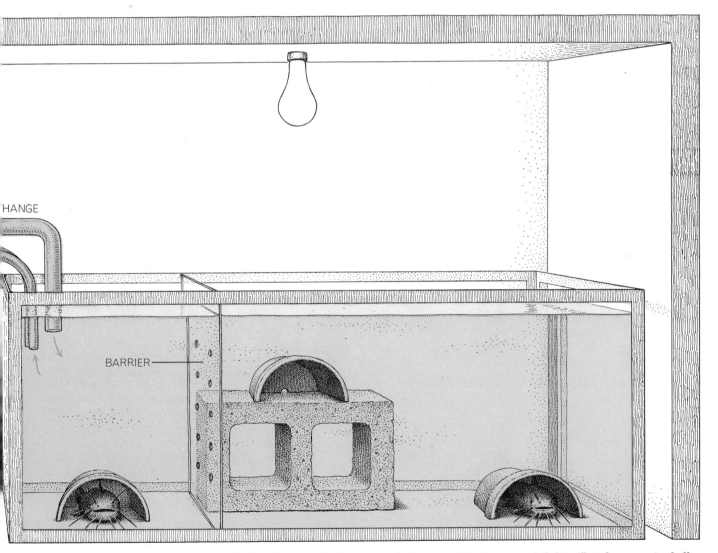

HANGE
BARRIER

one tank to the other (*arrows*). Removal of the barrier incited the isolated bullheads to aggressive action; the experimenters then pumped "love-in" water from the communal tank to the aggressors' tank. Short periods of exposure failed to affect the aggressive bullheads, but after seven consecutive days agonistic behavior was substantially diminished. Brief reexposures had the same effect.

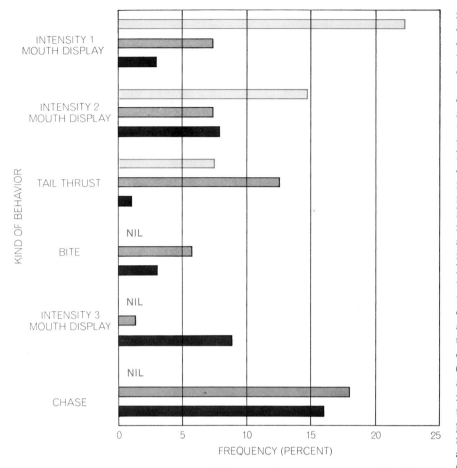

SIX AGONISTIC ACTIONS occur among bullheads with varying frequency under different living conditions. Two relatively unaggressive actions, for example, are mouth displays of the first and second level of intensity. In these categories fish in placid communities (*color*) outscore the dominant (*gray*) and subordinate (*black*) members of a territorial pair. Territorial bullheads are active, however, in other agonistic categories, such as biting and chasing; this behavior is not observed in communal groups. Territorial bullheads engage in about three times as many kinds of interaction as bullheads living in communal groups do.

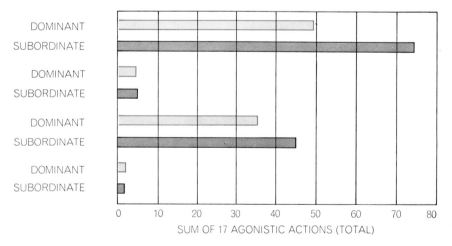

EFFECT OF "LOVE-IN" PHEROMONE on two isolated bullheads is shown in this graph. Length of each bar indicates the number of agonistic incidents in any of 17 categories of behavior recorded during eight hour-long observation periods. The gray bars present the behavior of the subordinate fish and the colored bars the behavior of the dominant. At first (*top*) the fish received no water from an adjacent communal tank (*see illustration on preceding two pages*). The bars second from the top record the bullheads' actions after a seven-day exposure to the pheromone. The next bars show the result of a 24-hour respite from the pheromone and the last (*bottom*) the result of resupplying it for 24 hours more.

individuals would become aggressive only if they were removed from the dense community, isolated for several weeks and then introduced into a low-density community.

With William McLarney, a behavioral ecologist, and Mark Park, a graduate student, I looked into the possibility that the establishment of the dense, peaceful type of community might be controlled by a specific chemical factor. We first compared the typical pattern of behavior in communal groups with that characteristic of bullheads in communities of the low-density, territorial type. The communal fish undertook very few aggressive acts, indulging only in mild manifestations such as head thrusts and low-level mouth displays. We then placed two identical tanks side by side, one containing 10 communal bullheads, the other a pair of territorial bullheads, with an arrangement for pumping water from the communal tank into the territorial one. The two fish in the territorial tank (a dominant and a subordinate) would be allowed to interact aggressively (by raising a glass partition that usually kept them apart) and then water would be pumped into their tank from the communal one. If the communal fish exuded an antiaggression, "love-in" substance, the receipt of this signal might be expected to calm the aggressive fish.

Our first trial, in which the territorial fish were exposed to a suffusion of the communal water for one hour, produced no such inhibitions; in fact the fish slightly increased the intensity of most of their aggressive acts. When we gave the fish continuous repetitions of this treatment, however, at the end of the seventh day they abruptly desisted almost completely from their aggressive behavior. On coming together the two fish behaved like communal bullheads. The introduction of communal water was then suspended for a day, and within 24 hours the high level of aggression was restored. The communal water treatment was renewed, and this time not seven days but one day of exposure sufficed to suppress the aggressive behavior.

Thus it seems that the clustering of bullheads in dense aggregations does produce a concentration of pheromone that acts to inhibit aggression. The experimental results suggest that some part of the bullhead's brain may need to be primed with a certain amount of this substance before it can take effect. It is not clear what benefits are conferred on bullhead survival by this curious system of two kinds of society: territorial and communal. Perhaps the communal aggregations provide a reserve for replenishment of the territorial populations. Field studies may throw some light on the question.

What are the sources and nature of the chemical signals—the alphabet—of the bullhead's language? A few clues have been uncovered. It turns out that the odor of the mucus on the fish's skin is sufficient to identify an individual for another bullhead of the same species. This cannot be the whole story, however, because a more intense response is evoked by the water that has bathed the fish, which suggests that other substances are involved. Perhaps the odor of secretions from the gonads serves to identify the fish's sex. The urine also may carry identifying information, although our tests of this material show that it alone does not supply enough information for recognition of an idividual. Possibly products in the urine indicating the state of stress of the fish may signal whether it is a dominant or a subordinate individual in its community.

We have done a few experiments on this question of the recognition of a bullhead's status. When a dominant fish is removed from its community, subjected to defeat by a fish we use as a "hatchet man" and then returned to the company of fish it dominated, the other fish no longer flee. Apparently the defeated fish has undergone a chemical change. In order to determine whether this alters the fish's identity as an individual or merely informs the other fish of its change in status, we applied the test to a pair, one dominant and the other subordinate, that had occupied the same tank. When the erstwhile dominant member was returned to the tank after a defeat, the originally subordinate tankmate attacked it. In this attack, however, the acts did not include those employed against a stranger (such as mouth fighting and biting) but were limited to the displays made against a known individual. In short, the attacker apparently recognized the returnee and also detected its change in status.

We are a long way still from deciphering the chemical language of bullheads. It is probably complex, communicating information about the individual's species, status, sex, age or size, reproductive state, individual characteristics and perhaps even family identification. No doubt pheromones are a means of communication between parents and their young. We also have evidence that there is some chemical communication between one species and another.

David Boylan, who is working with us in a research program at the Woods Hole Oceanographic Institution, and other chemists are working to identify the chemical substances that form the languages of fishes. In the not too distant future these substances will be used in marine farming as selective attractants, artificial baits, growth stimulators, inhibitors of aggression and cannibalism and repellents to deter predators from attacking cultivated fishes. Thus they may be even more useful than the application of insect pheromones to the control of pests and diseases in agriculture.

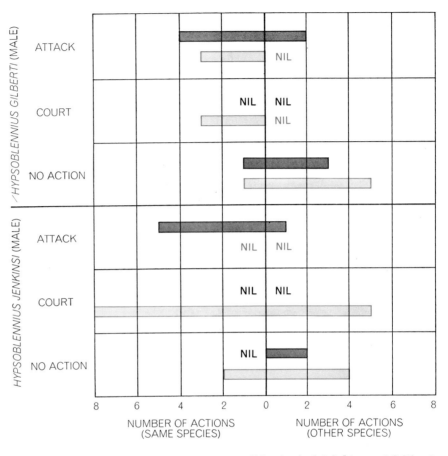

SEXUAL ISOLATION in the case of two species of blennies, both inhabitants of California coastal waters, is shown to depend on cues other than visual ones. The graph records the response of a male of each species to the sight of males (*gray*) and females (*color*) of the same or the opposite species, isolated in sealed jars and placed in a tank with the test subjects. *Hypsoblennius gilberti* males attacked males of both species but courted only females of their own kind. *H. jenkinsi* males, however, courted females of either species on sight.

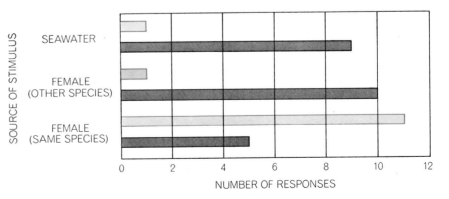

ISOLATING PHEROMONE proved to be a secretion that is produced by a female blenny during courtship ritual. The graph shows that the number of times a male *H. jenkinsi* reacts by courting (*color*) or fails to react (*gray*) is the same when either plain water or water from a jar containing a courted female of another species is added to its tank. Intense courting occurs only when the water comes from the jar of a female of the same species.

38

TERRITORIAL MARKING BY RABBITS

ROMAN MYKYTOWYCZ
May 1968

The wild rabbit of Australia lives in colonies from which the rabbits of other colonies are excluded. The rabbits mark their home territory by means of odorous substances that are secreted by specialized glands

Over the past few years it has become increasingly apparent that one of the principal ways animals communicate is by means of odorous glandular secretions that have been given the name pheromones. The best examples of such chemical communication are found among the ants, some species of which use pheromones to lead their fellows to food, to warn them of danger and to organize social behavior in the nest [see "Pheromones," by Edward O. Wilson; SCIENTIFIC AMERICAN Offprint 157]. It has also been shown that pheromones play an important role in the social behavior of mammals (perhaps including man), although exactly how they do so has been somewhat obscure. My colleagues and I have been investigating such mechanisms in the wild rabbit of Australia (*Oryctolagus cuniculus*), and we have been able to demonstrate a clear-cut relation between the animal's social behavior and pheromone-secreting glands.

As is well known, the rabbit was introduced into Australia about 100 years ago, and since that time it has become a serious pest. Over the past two decades the Division of Wildlife Research of the Australian Commonwealth Scientific and Industrial Research Organization has conducted a broad ecological study of the wild rabbit in order to gain a better understanding of the life of the species and to provide a basis for the development of more effective methods of controlling the rabbit population. One result of this study is that we have been able to detect a distinct social organization within the rabbit colony. The role individual animals play in such a society is reflected both in the size of certain glands and in the extent to which the glands secrete pheromones. These substances probably serve a number of functions. They may communicate in-

formation about age, sex, reproductive stage and group membership. They may also warn of danger, and they definitely serve to define the "territory" of the rabbit.

The area within which an animal confines its activities is not necessarily the same as its territory; in a strict sense the term territory refers to that part of an animal's home range which it protects, sometimes by fighting. The size, form and character of a territory of course vary according to the animal's way of life. Some animals occupy a territory permanently; others hold it only during a particular season when, for example, breeding or nesting takes place. The possession of a territory, and therefore available shelter and food, makes breeding possible for some individuals of a species and prevents breeding by other individuals. Hence territoriality is an important factor regulating the density of a population.

Many mammals are known to mark out a territory with substances manufactured by specialized glands. Indeed, nearly all species of mammals possess scent-producing glands, often in several places on the body. In the dromedary the glands are at the back of the head, in the elephant at the temples, in a number of deer species below the eyes (the American deer and some other species have such glands near the hoof and on the leg). The chamois's glands are around the horns, the golden hamster's between the ribs, the peccary's in the loin. The pika (mouse hare) and marmot (woodchuck) employ glands behind and under the eyes and on the cheek for territorial marking. We have identified odor-producing glands on the chest of the kangaroo, and other workers in Australia have found similar glands on the chest of the koala and the brush-tailed

possum. The rabbit has scent glands in the anal region, in the groin and under the chin.

In order to study the territorial behavior of the rabbit it was necessary to follow the activities of individual free-living animals. As in the study of other wild animals, there were certain difficulties. It is not easy to identify individual rabbits, even in daylight, and the animal is largely nocturnal. Live rabbits are hard to catch, and it was necessary to examine them regularly. To minimize these problems we confined our rabbit colonies within fenced areas. Each enclosure was small enough to allow close observation (from an elevated hiding place nearby) but not so small as to interfere with the rabbit's normal behavior. To facilitate the recognition of individual animals each one was marked with black dye in a distinctive pattern; tags of different shapes were also affixed to the rabbits' ears. We used a colored light-reflecting tape to mark the tags in a variety of patterns; at night under spotlights the rabbits could easily be identified. To keep track of the animals' movements in their burrows we drilled a few holes down to the burrow tunnels. When we were not using the holes to check the location of rabbits underground or to catch them, they were plugged with earth enclosed in wire netting.

When a rabbit was born in an enclosure, its ears were permanently marked by tattooing; in this way the behavior of the animal could be followed for its lifetime. The rabbits were observed every day and on certain occasions continuously for a 24-hour period. Some studies were continued for three years, which is twice the average lifespan of a free-living wild rabbit in Australia.

It was established by these observa-

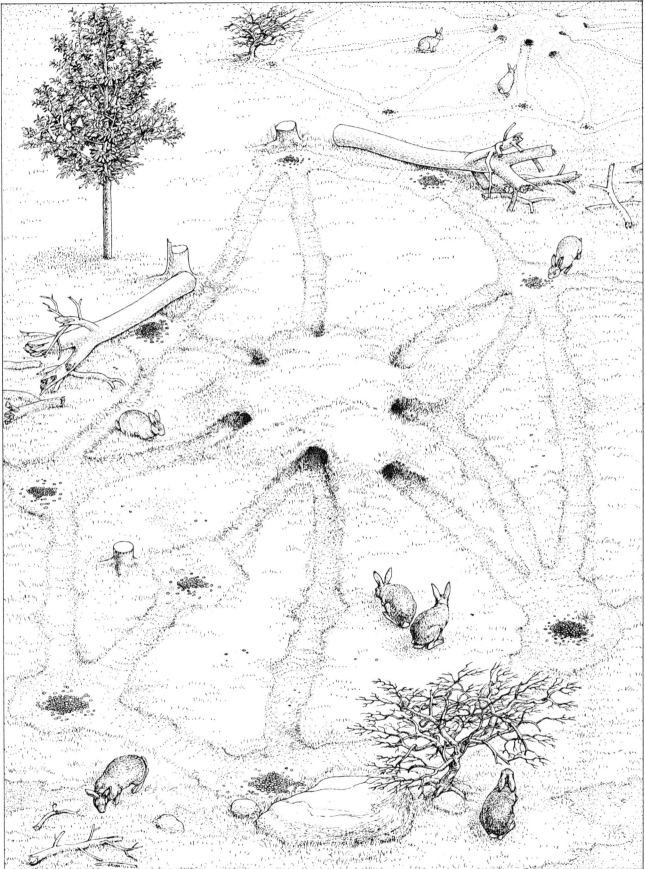

RABBIT TERRITORY, somewhat idealized in this drawing, lies around a central warren. Radiating from the entrances are intersecting runs. On the runs are small mounds of fecal pellets deposited by the rabbits of the colony. At top right a rabbit from another colony sniffs at a mound of pellets. It is warned by odor of an anal-gland secretion on the pellets that the territory is occupied.

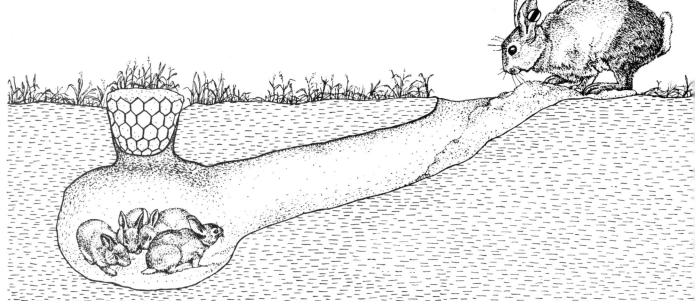

RABBIT EXPERIMENT followed the behavior of individual wild rabbits (*Oryctolagus cuniculus*) aboveground and underground. Holes dug down to the burrow tunnel gave access to the animals; when not in use, the holes were plugged with earth enclosed in wire

tions that the rabbit lives in small groups, each consisting of eight or 10 animals, depending on the density of the overall rabbit population. A group of rabbits occupies a distinct territory, with a warren—a burrow with a number of entrances—as its center. At the onset of the breeding season the males establish a hierarchy of descending dominance by fighting. The same is done by the females. The most dominant male and female rule over the group and its territory, which is defended by its occupants and recognized by rabbits of other groups. The rabbit hierarchy is constantly reinforced by chasing and submission.

The advantages of having a territory were evident in the different breeding patterns of female rabbits. High-ranking females give birth frequently and regularly, housing the young in a special breeding chamber dug as an extension to the burrow. Their young have a much higher survival rate and growth rate than the young of low-ranking females. Some of the subordinate females are chased away from a warren and forced to drop their litters in isolated breeding "stops"—short, shallow burrows dug at a distance of 10 to 50 yards from the warren. Both in the stop and later, when they come out of it, the young of such litters are vulnerable to predators, particularly foxes and crows. When they try to enter the central warren, they are attacked by the resident females and their young. Some subordinate females, although they are physi-

ologically fit, never give birth; the embryos they conceive do not develop to full term but are absorbed in the uterus.

It could also be seen that not all rabbit territories are of equal quality: some offer more protection and a better food supply than others. The most dominant animals tend to occupy the better territories. The breeding stops of subordinate females are often in places that are susceptible to flooding, another factor that contributes to the lower survival rate of their young.

A rabbit that crosses the border between its own territory and a neighboring one is immediately aware that it has done so. No matter how frequent and regular the visits, the animal's behavior changes. In its own territory it moves freely; it feeds and examines objects confidently. In a foreign territory the animal seems always on the alert. Its neck is stretched, the movement of its nostrils indicates that it is sniffing continuously and it does not feed. Although the interloper may be dominant in its own territory, when it is challenged outside by a rabbit permanently attached to the foreign territory, it will offer no resistance. It will not resist even if the challenger is half-grown.

It is not difficult to believe that a rabbit territory becomes saturated with the characteristic smell of the group and that within the area in which this group smell prevails the animals feel at home, much as a man may be able to perceive the odor of another man's home as be-

ing strange in contrast to the familiar smell of his own. Specific odors associated with an individual group or colony have been demonstrated in many social insects and also in some fishes. Apart from the effect on the odor of the food eaten by the animals, genetic factors are undoubtedly involved as well.

One component of a rabbit territory's odor is the smell of urine. During amatory behavior the males can be seen urinating on the females, and the animals also urinate on each other during aggressive displays. Young rabbits so marked by adult members of their group are identified with it. The smell of foreign rabbit urine releases aggression; females may even attack their own young when they have been smeared with foreign urine.

Another component of the territory odor comes from feces. The marking out of territory with feces seems to be common among animals, as has been pointed out by Heini Hediger of the University of Zurich. Among certain animals the odor of the feces comes not from the excrement itself but from the pheromone of anal glands; such glands have been identified in more than 100 species of mammals. Some animals whose anal glands are highly developed—for example the gray squirrel, the marten, the dormouse and the hyena—use the anal pheromone alone, rather than in combination with feces, for territorial marking.

The anal glands of the rabbit are well developed. They consist of two clusters

netting. Each adult animal was marked with dye in a distinctive pattern and wore an identification tag in its ear. The ears of rabbits born during the experiment were tattooed. To house their young, rabbits dig special breeding chambers as extensions to the burrow.

of brownish tissue forming a saddle-like mass around the end of the rectum. The secretions of the glands flow through a few ducts into the rectum, where they coat the pellets of feces passing out of the anus. Until recently it was generally accepted that the secretion facilitated the passage of the hard fecal pellets. We have found, however, that when the rabbit's anal glands are removed, defecation is not affected.

Our studies indicate that the fecal pellets serve to distribute the rabbit's anal pheromone.

When one examines a rabbit territory, one can see a number of places in which the animals have repeatedly deposited feces. Around a typical warren it is usual to find about 30 of these small dunghills. They are interconnected by paths that all the rabbits use in moving around the area. When a strange rabbit

enters the territory, it inevitably encounters a dunghill—a warning signal that the area is occupied. It then displays the wary behavior I have described.

A similar warning is deposited at the entrances to breeding stops. When a stop contains young, it is sealed by the female with soil. Once a day the female reopens the burrow, enters it and suckles the young. When the entrance

CHINNING rabbit enclosed in an experimental pen was photographed at night, when the animal is active. As the rabbit presses against a wooden peg, droplets from its submandibular (underchin) glands are forced through pores of the skin, "marking" the object.

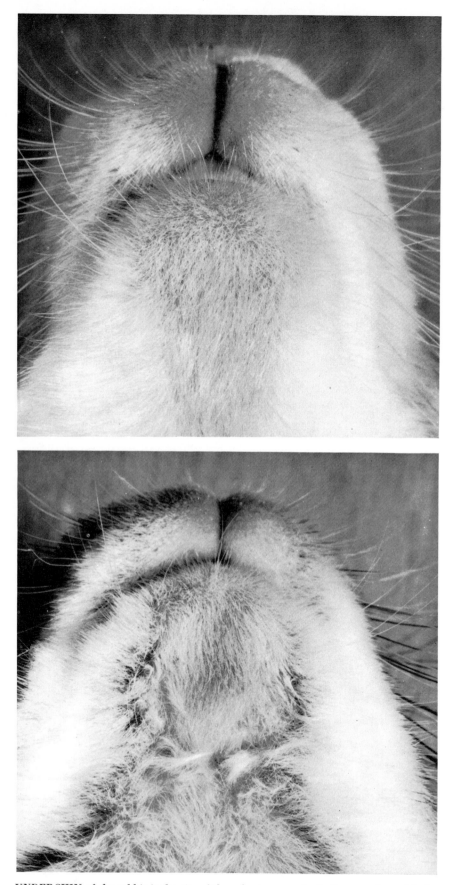

UNDERCHIN of the rabbit is the site of the subcutaneous glands that, together with the anal glands, function in territorial marking. The fur under the chin of the female rabbit (*top*) displays no trace of a glandular secretion. Under the male rabbit's chin (*bottom*) the fur is matted from the secretion, which the male produces more copiously than the female.

is resealed, the female is likely to deposit a few fecal pellets and some urine on top of the seal. It is remarkable that these earth seals are almost never disturbed by other rabbits. (The exceptions occur under abnormal circumstances, such as when the density of a confined population becomes exceptionally high and competition for breeding sites is intensive.)

Certain of our experiments support the idea that the rabbit's feces serve a communicative function, and indicate that the pellets of a dunghill differ from those that are randomly deposited around a territory. We presented a group of rabbits with artificial dunghills consisting of turf sprinkled with some of the pellets found at random locations in the territory of other rabbits. Around these artificial dunghills (and also on pieces of turf used as a control) the animals engaged mainly in digging and eating. When they were confronted with an actual dunghill from foreign territory, however, they did not eat, they sniffed intensely and they produced marking feces of their own. These pellets were similar to dunghill pellets in that their "rabbity" odor was (according to 30 human judges) decidedly stronger than the odor of the randomly distributed pellets. Presumably the difference between the two odors is perceived even more sharply by rabbits.

There is a highly significant relation between the size of a rabbit's anal gland and the place of the animal in the social hierarchy of a warren. The largest glands belong to dominant individuals, the smallest to subordinate animals. (Body weight is not the main factor determining glandular size.) The secretory activity of the anal gland also is higher in the dominant animals. Indeed, we found that the social rank of an individual rabbit could be guessed with a fair degree of accuracy merely from the appearance of the gland in section. It is also significant that the size and secretory activity of the gland were greatest during the breeding season—the period when territorial activity is most intense.

Male rabbits are mainly responsible for establishing dunghills (in Australia about 80 percent of the animals caught in traps set on dunghills are males), and the anal gland of the male rabbit is larger than the anal gland of the female. Our experiments indicate that the activity of the gland is under the control of sex hormones. When male rabbits were castrated before puberty, the growth and activity of the anal gland were inhibited; when the ovaries were removed from female rabbits, the size and activity of the gland were somewhat enhanced. When male sex hormones were given to both male and female rabbits, the gland grew larger and produced more secretion.

The rabbit's anal pheromone and its urine thus serve to establish its overall territory. For marking localized features such as logs, branches and blades of grass the animal employs a pheromone secreted by its chin gland. If one looks under the chin of a female rabbit, one cannot find any conspicuous marks, although the fur may be slightly moist or matted. Under the male rabbit's chin, however, there is a distinct yellowish encrustation and matted fur. One can feel the glands under the skin, and if pressure is applied, droplets of secretion can be forced out through a semicircular row of external pores. With this secretion (which is odorless to humans) the male rabbit marks not only objects that would be difficult to mark with feces or urine but also the entrances to its burrows, the fecal pellets of other rabbits, its own weathered pellets and its females and young. To describe marking with this gland we use the term "chinning."

Within its own territory a rabbit chins freely and frequently, particularly if it is a male. When it is in a foreign territory, it does not chin; when it is confronted on its own ground with foreign feces, it chins intensely. We have found that the individuals in a rabbit hierarchy that are most dominant chin more often than the subordinate animals. In fact, the frequency of chinning can

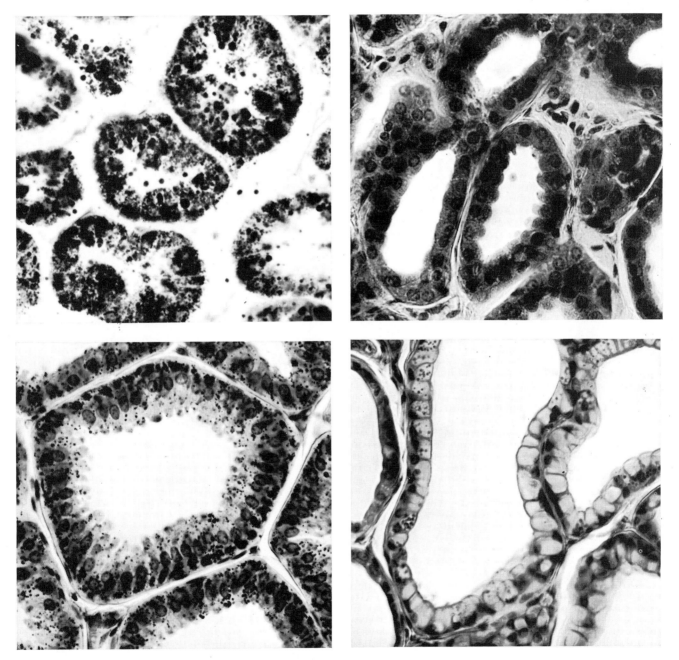

ANAL-GLAND ACTIVITY can be determined from the appearance of the gland in section. In the less actively secreting gland shown at top the tubules (*dark areas with light centers*) are smaller than those in the intensely secreting gland that appears below it.

CHIN-GLAND ACTIVITY is also reflected in contrasting size of tubules in less active gland (*top*) and more active one (*bottom*). These four photomicrographs were made by E. C. Slater of the Commonwealth Scientific and Industrial Research Organization.

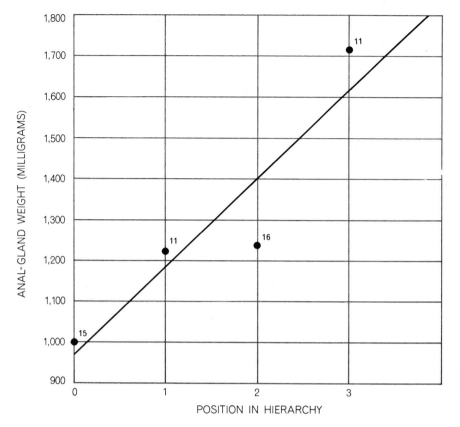

RELATION BETWEEN ANAL GLAND AND SOCIAL BEHAVIOR was established from observations of the activities of rabbits. The most dominant animals of a warren hierarchy (*here given the rating "3"*) were found to possess the heaviest anal glands. The number above a dot indicates the number of rabbits in the sample; the trend is indicated by a line.

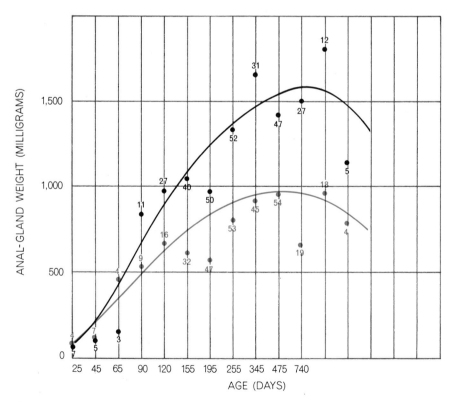

SEXUAL DIFFERENTIATION appears when the weight of the anal gland of male rabbits (*black*) is compared with that of female rabbits (*color*). The number of animals sampled is shown adjacent to a dot; the curves indicate the trend in the measurements. Age of the rabbits was estimated from the weight of the eye lens and is uncertain beyond 740 days.

be used as an indication of potential dominance: when two males completely strange to each other were brought together, the male with the highest chinning score always established his dominance over the other animal.

The rabbit's chin gland, like its anal gland, is larger and secretes more abundantly in dominant animals than in subordinate ones. This gland too, being larger in males than in females, appears to be under the control of sex hormones. Moreover, its activity, like that of the anal gland, fluctuates with the seasons; it secretes most freely at the time when the rabbit's territorial activity is at its most intense. Thus the animals that are dominant in the hierarchy are the ones most concerned with the demarcation and defense of territory. Indeed, their territorial behavior is rooted in their physiology.

Additional support for the territorial function of the anal and chin glands has come from the study of an animal that is closely related to the rabbit but that behaves differently with respect to territory. This animal is the European hare (*Lepus europaeus*), which like the rabbit has become widely distributed throughout Australia since being introduced there. Although a fully grown hare is four times as heavy as a rabbit and most of its glands (such as the thyroid and lachrymal glands) are larger, the anal and chin glands of the hare are only a tenth the size of those in rabbits.

This difference between the two animals reflects a difference in their territorial behavior. Unlike the rabbit, the hare does not live in a social group; it is a solitary animal and its home area is large. The hare protects only that part of its home range which is in its immediate vicinity, and it does not retain such a territory permanently. Although the animal makes dunghills (apparently at the sites it visits most frequently), the number of them is small compared with the number in a rabbit territory. There is no difference between the anal gland of the male hare and the gland of the female. In short, it appears to be unnecessary for the hare to mark its territory by odor.

Another difference in the glandular makeup of the hare and the rabbit also appears to have behavioral significance. The animals have similar glands in the groin, but in the hare these glands are larger than they are in the rabbit (and they are somewhat larger in the female hare than they are in the male). Observations of the hare's mating behavior suggest that the glands in the groin are the source of a powerful sexual attractant: when the solitary-living female hare is in estrus, she attracts males from some distance. The female of the gregarious rabbit species obviously has no need of a long-range attractant.

What we have learned about the wild rabbit in Australia is supported by studies of American species. The swamp rabbit (*Sylvilagus aquaticus*), a strongly territorial species, chins more frequently than the weakly territorial cottontail rabbit (*Sylvilagus floridanus*). Our examination of gland tissue from these species indicates that the anal and chin glands are larger in the swamp rabbit than in the cottontail. The social behavior of the two species was studied by Halsey M. Marsden and Nicholas R. Holler of the University of Missouri.

The pheromones of rabbits and other mammals have not yet been chemically separated and identified, as has been done with certain insect pheromones. With knowledge of the composition of specific odors it will be possible to establish their role in the life of a mammalian species. This will undoubtedly lead to a better understanding of the species' behavior. Some insect pheromones have even been synthesized; incorporated in traps, they have been used successfully in pest control. The synthesis of mammalian pheromones would no doubt be helpful in developing effective controls for mammals that are economically undesirable.

The existence of glands that function specifically and solely for territorial marking emphasizes the importance of territory and social organization in the life of an animal. In today's crowded world the question of space and territory is of interest not only to students of animal biology but also to all who are concerned with man's problems of overpopulation. The theory that in animals population control is achieved through social behavior of which territoriality is an integral part has been advanced by V. C. Wynne-Edwards of the University of Edinburgh [see "Population Control in Animals," by V. C. Wynne-Edwards; SCIENTIFIC AMERICAN Offprint 192]. More and more we are coming to realize that it is not only the availability of food that determines the size of a human population but also our own behavior and spacing, and that these factors must be considered in speculating on the fate of the human species. Every animal has, in addition to minimum requirements of things such as food, minimum requirements of living space and distance from others of its own species.

39

HOW BIRDS SING

CRAWFORD H. GREENEWALT
November 1969

The mechanism of bird song has traditionally been compared to either a wind musical instrument or the human vocal apparatus. The analysis of the bird songs themselves points toward an entirely different system

Some years ago I read a book in which, among other things, the author summarized the theories that had been advanced to explain the physiological processes employed by a singing bird. These theories, it seemed to me, were totally unacceptable, and I determined, perhaps with more rashness than wisdom, to see if I could find an explanation that did no violence to either the anatomical findings or the laws of physics.

The theories to which I took exception compared the singing bird either to a musical instrument—the clarinet, the oboe or the trumpet—or alternatively to the human voice (that is, a bird was believed to employ the same devices we do when we speak or sing). In a clarinet, for example, pitch and timbre are controlled by the effective length of the barrel of the instrument; the vibrating reed is in effect driven by and so forced to conform to the harmonic spectrum of the resonator. If a bird sang in this fashion, pitch could be varied only by extension and retraction of the neck. For a song sparrow, with its two-octave range, the neck would have to be extended by a factor of four—clearly a physical impossibility. Furthermore, birds sing many phrases in which the wave form is sinusoidal (a train of pure sine waves), without an associated harmonic spectrum. Resonators such as the barrel of a clarinet must by definition produce harmonics and cannot generate a sinusoidal wave form.

As for the human-voice theory, when we speak we produce at the glottis a series of puffs of air mathematically equivalent to a harmonic spectrum with an infinity of components, and with a fundamental frequency corresponding to the interval between puffs. The resulting acoustical disturbance is then modulated in our oral passages to produce the spoken word. This mechanism does not (and cannot) produce a purely sinusoidal wave form; every speech sound is associated with a harmonic spectrum of considerable complexity.

Negatives such as these are neither satisfying nor sporting, and to remedy both potential criticisms I undertook to develop a physiological and acoustical model that would describe a singing bird. The approach was to study in detail the bird songs themselves, and to develop from their constituent parts some notion of the associated acoustical proc-esses. It seemed useless to employ an anatomical approach because the many excellent anatomical studies in the scientific literature have produced no useful conclusions, and because the acoustical analysis of human speech has been so strikingly successful in elucidating our own vocal performance.

I might pause here to note that I have written a book—*Bird Song: Acoustics and Physiology* (Smithsonian Institution Press, 1968). In it much detail is given on the state of the literature, the instrumentation my associates and I have developed for the analysis of bird song, and

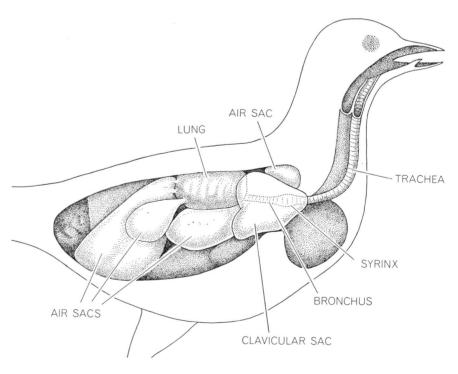

VOCAL APPARATUS OF A SONGBIRD is shown in schematic form. The vocal organ itself is the syrinx, which is located at the point where two bronchi join to form the trachea.

the proofs, mathematical and otherwise, for the findings I am about to describe. This article is in effect a summary of that book, and if the reader should wish more detail he will find it there.

The vocal organ of birds is called the syrinx. It is located deep in the thoracic cavity, at the point where the two bronchi join to form the trachea [see illustrations below]. It is difficult to decide from the detailed anatomy of the syrinx exactly what parts are important in vocalization, and precisely how sound is produced. At one time or another almost every anatomical element in the syrinx has been assigned a role in vocalization; unhappily there is an overabundance of possibilities.

I have proposed the relatively simple functional structure depicted in the top illustration on page 408. The elements are the tympanic membranes and their associated musculature, the external labia and the system of internal air sacs (not shown in the illustration) that force air through the bronchial passage and bulge the membrane into the bronchial lumen, or passageway. The illustration depicts the highly evolved syrinx of a songbird. At successively earlier stages of avian evolution the syrinx loses the external labia, then the syringeal muscles and ends up as a simple tube with a membrane on its periphery and contained within an air sac.

It should be kept in mind that the functional elements of the syrinx are doubled, that is, there is a set for each bronchus. Since each bronchial passage has its own membrane, musculature and nervous system, it is evident that birds can control each passage independently of the other; they can sing what might be called an internal duet.

The system operates as follows: When a bird undertakes to sing, it in effect closes a valve between the lung and the syrinx. Then it compresses (with its chest muscles) the air in a system of sacs. Pressure in the clavicular sac, which surrounds the syrinx, forces the exceedingly thin tympanic membrane into the bronchial passage, closing it momentarily. Tension is then applied to the syringeal muscles, which, acting in opposition to the sac pressure, withdraw the bulged membrane from the opposite bronchial wall, thus creating a passage through the bronchial tube. Air streaming through the passage past the tensed membrane stimulates it to vibrate, and song is produced. If only one of the two voices is to be used, no tension is applied to the other membrane, and the corresponding bronchial passage remains closed. When a duet is sung, both membranes are under tension; there are two airstreams and hence two vibrating membranes and two simultaneous sounds.

The illustrations on the next two pages show three complex songs analyzed in terms of amplitude and frequency. The songs embrace almost every vocal gymnastic of which birds are capable. Note, for example, the extremely rapid amplitude modulations, the wide range in the shape and extent of the amplitude envelopes and the relatively large frequency intervals: from just over two to just under seven kilocycles per second. The amplitude displays are precise and require no qualification. In the frequency displays there are a time delay and an integration of frequencies over small time intervals that introduce ambiguities for precise analysis.

The question of whether or not a bird can use its two acoustical sources independently is vital to the elucidation of the mechanics of bird song. If, for instance, the bird's trachea behaved like the barrel of a clarinet, both sources, that is, both vibrating membranes, would be forced to conform to the resonances of the trachea. The two sources would then merely reinforce each other, increasing the amplitude, or loudness, of the sound. If, on the other hand, the bird's trachea behaved like human oral cavities, and the vibrating membranes were analo-

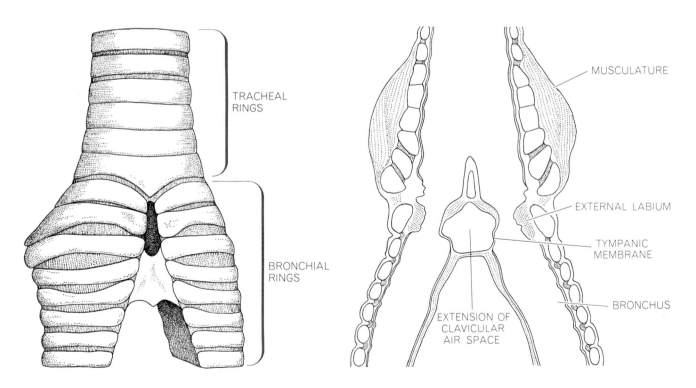

ANATOMY OF THE SYRINX sheds little light on its function. At left is the skeleton of the syrinx in the magpie; at right, a section through the syrinx of the European blackbird. Both drawings are based on studies published by the anatomist V. Häcker in 1900.

gous to the human glottis, the resulting wave form should be equivalent to a harmonic spectrum of great complexity involving three fundamental frequencies, one for each source and one for the first resonance of the trachea.

The fact that the two sources can produce two harmonically unrelated sinusoidal wave forms was first shown by Ralph K. Potter, George A. Kopp and H. C. Green of the Bell Telephone Lab-

oratories, for a fragment of the song of a brown thrasher. Since the phenomenon is an important one, I have searched for and found many similar examples in species representative of most families of birds. One example is provided by a relatively long phrase from the song of a mockingbird [see illustration on page 406]. In the displays the two voices were separated with sharply discriminating acoustical filters; frequency was determined by measuring time intervals for

10 successive sine waves. The two voices are harmonically unrelated and overlap on the time axis. We must conclude that neither the musical-instrument theory nor the human-voice theory is operative.

Other evidence may be found in phrases that cover a large frequency range. If such a "glissando" traverses a tracheal resonance or antiresonance, there should be a marked increase or decrease in amplitude at the appropriate frequencies. An example is a glissando

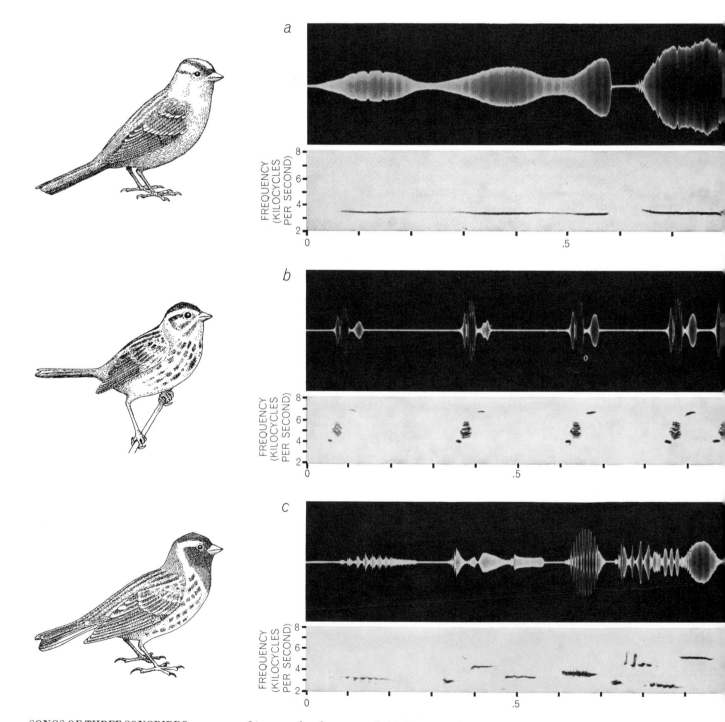

SONGS OF THREE SONGBIRDS are presented in traces that show amplitude and frequency as a function of time. The top record for each bird is an oscillogram that displays amplitude; the bottom record is a sound spectrogram that displays frequency. At a is the

sung by a yellow warbler that embraces a full octave [see illustration on page 407]. There are changes in amplitude in the glissando, but in an acoustical sense they are very small (less than three decibels), and it is most unlikely that they are associated with tracheal resonances. The two minima at 90 and 130 milliseconds appear at frequencies of 4.8 and 6.5 kilocycles per second. One of these could correspond to a resonance; both of them could not.

Although the evidence seems quite conclusive that the primary vibrations produced in the syrinx traverse the trachea without attenuation, amplification or change of any sort, we must offer a valid explanation for this rather surprising acoustical inertness of the trachea. The trachea is a tube. A tube, if its acoustical losses are small, exhibits resonances, and one would expect it to be an effective modulator, even with its relatively soft walls. A possible explana-

tion—indeed, the only likely explanation —is that the impedance of the source (the vibrating membrane in its constricted passage) closely matches the impedance of the trachea. Calculations based on reasonable assumptions indicate that this condition will obtain where the mean cross-sectional area of the trachea is about 10 times that of the syrinx. The curve relating this area ratio to attenuation is fairly flat and shows that the ratio can be substantially higher or lower than

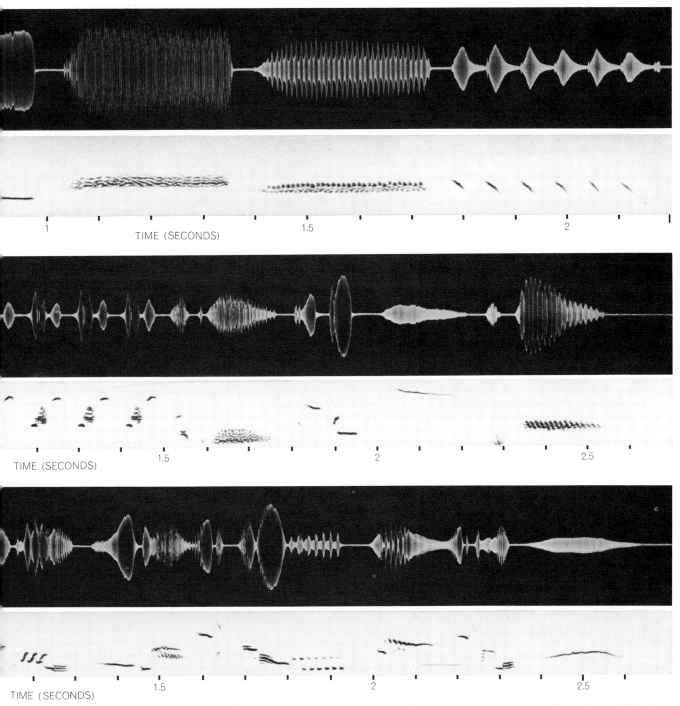

song of a white-crowned sparrow; at *b*, the song of a song sparrow; at *c*, the song of a Lapland longspur. These song recordings illus-

trate the extraordinary virtuosity of these singing birds. No human performance can match the complexity of the individual phrases.

10 before tracheal attenuation becomes important.

There is another way of expressing the effect of variation in the relative cross sections of source and resonator that may be easier to comprehend. If we take a tube closed at one end and open at the other, the resonances will occur at multiples 1, 3, 5, 7 and so on of the fundamental frequency. If the tube is open at both ends, the resonances will occur at multiples 0, 2, 4, 6 and so on of the fundamental frequency. If now we begin with a tube open at both ends and gradually close off one end, we will in due course arrive at a twilight zone within which the tube shows no resonant effects. This presumably is the zone in which birds sing.

Perhaps this is another example of Nature's ingenuity in dealing with the needs of her diverse creatures. The necks of birds (which contain and limit the trachea) have many functions, for instance feeding, preening and nest-building. It would be odd if such functions operated to restrict freedom in an activity as important as song.

So far I have undertaken to show that bird song has its origin in the syrinx,

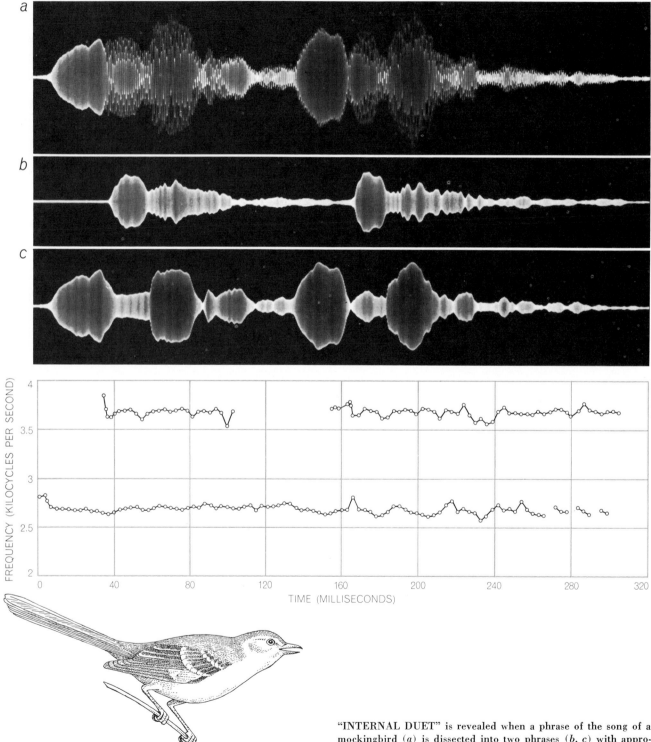

"INTERNAL DUET" is revealed when a phrase of the song of a mockingbird (a) is dissected into two phrases (b, c) with appropriate filters. The plots at bottom show the frequencies in b and c.

that the syrinx contains two acoustical sources which can be independently controlled to produce an internal duet, and that the sounds which originate in the syrinx pass through the trachea to the ear of the listener without further modulation. I must now show precisely how the sources in the syrinx operate and how the extraordinarily rapid and complex modulations so common in bird song are produced.

I have noted that pressure in the clavicular sac forces the internal tympanic membranes into the bronchial lumen against tension applied to the membrane by the muscles of the syrinx. Air streaming past the resulting constriction in the bronchial passage stimulates the membrane to vibrate. If this postulate is correct, it follows that in any rapid modulation amplitude and frequency must be coupled; an increase in frequency can be produced only by increasing the tension in the syringeal muscles. At a given pressure in the clavicular sac this tension will increase the cross section of the bronchial passage, which, as we shall see, must produce either an increase or a decrease in the acoustical amplitude.

The direct coupling of frequency and amplitude, that is, amplitude increasing with increasing frequency, appears in curves for Townsend's solitaire in which frequency and amplitude are simultaneously plotted against time [*see upper illustration on page 409*]. The reverse process, with amplitude falling as frequency rises, appears in similar curves for the red-winged blackbird [*see lower illustration on page 409*]. In curves for the song sparrow one can see both types of coupling within the modulating period: amplitude rises with increasing frequency up to 6.8 kilocycles and falls above that frequency.

These phenomena can best be explained by referring again to the simplified model of the syrinx shown on the next page. If we start with just sufficient tension (T) on the tympanic membranes to balance the pressure P, the distance across the passage (D) will be zero and no air will flow. A small increase in tension will open the passage and allow the membrane to vibrate at a frequency corresponding to the tension, but the amplitude of vibration will be restricted to low values by the small distance across the passage. As the tension increases, frequency and amplitude will increase together, as a larger distance across the passage permits greater vibrational amplitude in the membranes. With a further increase in membrane tension, and a correspondingly larger distance across

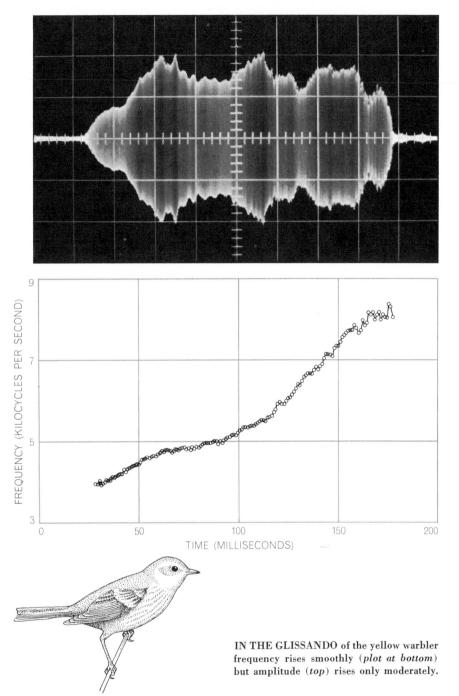

IN THE GLISSANDO of the yellow warbler frequency rises smoothly (*plot at bottom*) but amplitude (*top*) rises only moderately.

the passage, we arrive at a point where the airflow through the syringeal constriction can no longer stimulate the membrane to vibrate through the allowable distance D, and the amplitude will decrease. At the limit—a tension sufficiently large to open the passage fully—the amplitude will be zero, and the airflow from the bird's lungs will escape through the trachea without sound production. It should be understood that airflow is used here in an acoustical sense; it is a time-varying flow that is controlled by the vibration of the tympanic membrane. Air could well leak past the syringeal constriction, but this

airflow would be continuous and would produce no sound.

Amplitude-frequency coupling always accompanies rapid modulation, but it is not necessarily present when the period of the modulation is long. Consider a song sparrow phrase about 50 milliseconds long [*see top illustration on page 410*]. The modulating frequency remains constant at about 300 cycles per second, as does the frequency excursion. There is, however, a gradual rise and fall in amplitude from the beginning of the phrase to the end. Such comparatively long-term changes in amplitude without

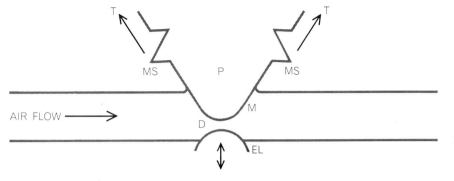

SIMPLIFIED MODEL OF THE SYRINX indicates how it functions. Air flows from the lungs at left into the trachea at right. The pressure of air in the clavicular sac (P) forces the tympanic membrane (M) into the bronchial lumen, or passage. Tension (T) is produced in the membrane by the syringeal muscles (MS). The resultant of the two forces P and T determines the distance across the lumen (D) and also the vibrating frequency. EL is external labium, which can regulate distance across bronchial lumen without affecting frequency.

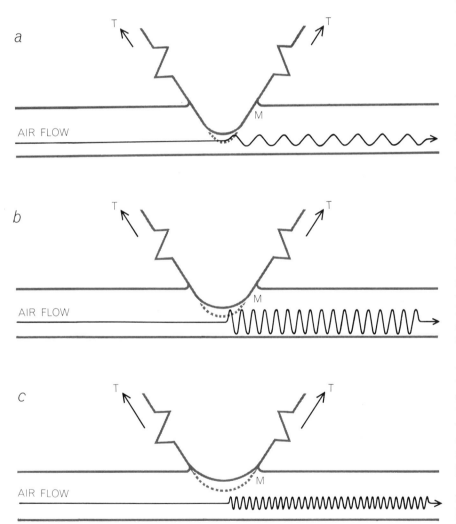

AMPLITUDE AND FREQUENCY ARE COUPLED when changes in either are rapid. In *a* a tone of a given amplitude is produced when the airflow causes the tensed tympanic membrane (M) to vibrate. In *b* the tension (T) is increased, causing the membrane to vibrate faster. Increased tension also retracts membrane, further opening the passage. The membrane can therefore vibrate through a larger distance, increasing amplitude of song. This is termed positive coupling. In *c* tension on the membrane has increased beyond a threshold; the tension is so great that the membrane can no longer vibrate across the entire passage. Beyond this threshold amplitude decreases with frequency, that is, the coupling is negative.

an accompanying change in frequency are relatively common. We find the necessary control mechanism in the external labium ["*EL*" *in the bottom illustration on page 410*], whose movement into and out of the bronchial lumen provides a simple device for changing amplitude without affecting muscle tension and hence without changing the frequencies produced in the tympanic membrane.

I must confess that this role for the external labium rests on a somewhat shaky anatomical base. Many anatomists report the presence of this member, particularly in the songbird syrinx; it is a pillow-like structure erected on the bronchial half-ring directly opposite the tympanic membrane. Only one anatomist, however, describes muscular attachments that could move the labium into and out of the bronchial lumen. I trust that some interested anatomist will note this opportunity for a definitive publication.

The bird could, of course, change the air pressure in its sac system, but this would have two effects whose resultant is at best ambiguous. Increasing the sac pressure would drive the membrane farther into the bronchial lumen, thereby decreasing its cross section; at the same time it would increase muscle tension. The increased pressure drop across the constricted bronchial passage would increase flow and hence would compensate for the reduction in cross section. I am not rash enough to predict the resultant of these phenomena, but it seems clear that they could not produce the overall rise and fall in amplitude exhibited in the song sparrow phrase on page 134. For that we need the external labium or some other anatomical feature that affects amplitude alone.

In any event, these rapid and highly variable modulations are so common as to be a prime characteristic in the songs of most birds. They can be repetitive, that is, there is a modulating frequency that continues for 50 to 500 milliseconds, or they can be nonrepetitive, that is, they may vary rapidly and randomly in amplitude, in frequency or in both.

Many phrases sung by the song sparrow comprise modulations that are repetitive. The modulating frequency averages 300 cycles per second, and amplitude ranges much more widely within a modulating period than frequency. In the songs of other species the modulating frequency ranges from about 100 cycles per second to something over 400. We are dealing here with two oscillating systems superimposed on each other. The first is the tympanic membrane vibrating

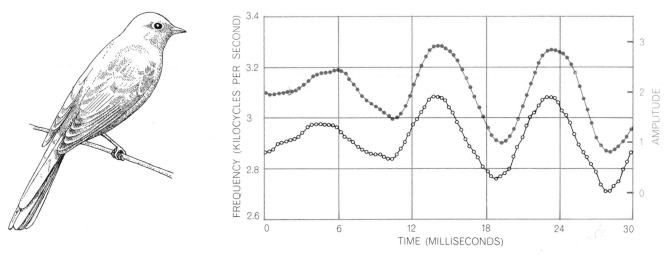

POSITIVE COUPLING of amplitude and frequency is shown in this analysis of a phrase sung by Townsend's solitaire. The colored curve traces the amplitude of the phrase; the black curve traces the frequency. The curves rise and fall together over the time period.

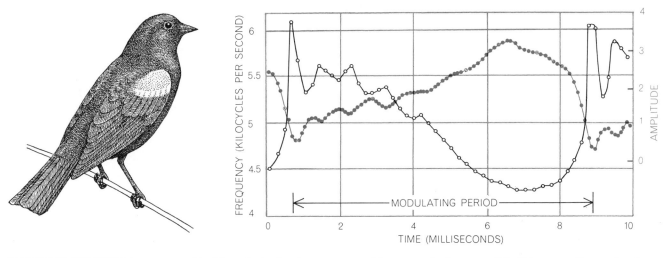

NEGATIVE COUPLING is represented in this analysis of a phrase sung by a red-winged blackbird. Here when the amplitude (*colored curve*) decreases, the frequency (*black*) increases. Curves also show the characteristic modulating period of this phrase in the song.

at its natural, or "carrier," frequency. The second includes the syringeal muscles, whose vibration would produce a periodic change in tension and hence a change in both frequency and amplitude. The mass and elasticity of the muscle system presumably limits its natural frequency of vibration to relatively low values—up to, say, 500 cycles per second. The much lower mass of the tympanic membrane allows vibration up to 10 kilocycles per second.

Nonrepetitive modulations can be seen in the courtship song of the brown-headed cowbird [*see illustration on page 411*]. This bird of unprepossessing appearance and habits is the undisputed winner in the decathlon of avian vocalization. Roger Tory Peterson characterizes the first phrase in the illustration as "glug" and the second phrase as "gleeee." Consider the following features: The frequency range in the two phrases is, by a large margin, wider than it is in any other bird song. It extends from .75 to 10.7 kilocycles per second—nearly four octaves! The maximum frequency at 10.7 kilocycles per second is higher than what we have found for any other bird, just nosing out the 10.5 kilocycles per second at the top of the blackpoll-warbler song. Both voices are used in the second subphrase of the "glug," and the frequency spread between the two voices (two full octaves) is exceeded only by that of the American bittern.

The first note in the "gleeee" is the shortest I have encountered. It lasts a bit less than two milliseconds and comprises a packet of 12 sine waves at 6.4 kilocycles per second. The glissando at 50 milliseconds in the "gleeee" is one of the most rapid, covering the range from five to eight kilocycles per second in four milliseconds and 23 sine waves, an average of 130 cycles per wave. The modulating frequency of the high voice in the second subphrase of the "glug" is about 700 cycles per second, higher by a large margin than any other. These performances are truly remarkable. What purpose is served by a "glug" comprising five widely different subphrases, together with a "gleeee" including a note of negligible duration, two rapid glissandi and a peak frequency of 10.7 kilocycles per second, only Madame Cowbird will know.

There is not much point in comparing the impressions bird songs make on human and on avian ears. The difference is moot; birds do not sing to us but to their own kind—to seduce a willing female, for example, or to warn off a potential male competitor. We are inclined to put everything we see or hear into our own standard of reference, and we wax lyrical over the song of a nightingale or a

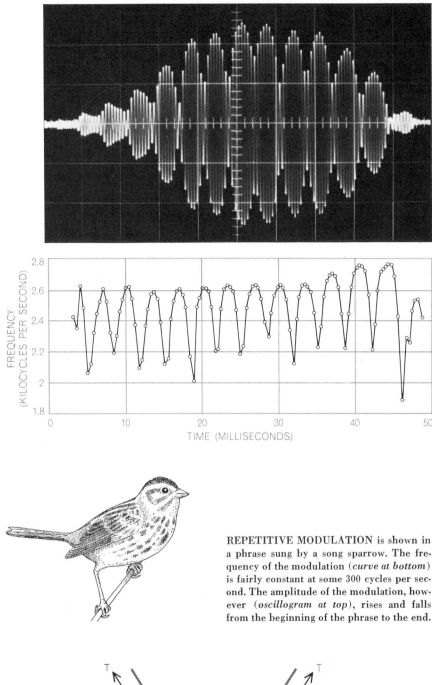

REPETITIVE MODULATION is shown in a phrase sung by a song sparrow. The frequency of the modulation (*curve at bottom*) is fairly constant at some 300 cycles per second. The amplitude of the modulation, however (*oscillogram at top*), rises and falls from the beginning of the phrase to the end.

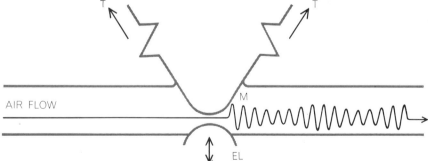

EXTERNAL LABIUM (*EL*), which is located in the bronchial wall opposite the tympanic membrane (*M*), may be the structure that enables the song sparrow and other birds to modulate amplitude without increasing or decreasing frequency. Normally a change in one alters the other, when both are controlled by tension on the tympanic membrane. If, however, the bird can diminish the distance across the bronchial lumen by extending the labium into it, it would be able to lower amplitude without affecting frequency, since this means of changing distance across the passage would not affect tension on the tympanic membrane.

thrush simply because they happen to sing within the ambit of our own musical experience. The fact is that the rapid modulations we have been discussing cannot be perceived as such by human ears. They are smeared, as it were, to produce what to us seems like a note of a different quality, or timbre, and one that on the whole we find unpleasant. It is easy for us to resolve a trill or a tremolo if its frequency is 30 cycles per second or less, but when the frequency rises to 100 cycles per second or more we hear a note of a rather unpleasant buzzy quality. Hence the beautiful complexity of the Lapland longspur song is completely lost in our ears, whereas it seems more than likely that Madame Longspur finds it delightful and enticing.

I have examined experimentally frequency perception and time perception for the avian ear. I conclude that its frequency discrimination, as expressed in the relation $\Delta f/f \times 10^{-3}$ (f is of course frequency), lies between 2 and 5. Time discrimination appears to be no greater than .5 millisecond. For human ears frequency discrimination is about the same, but time discrimination is perhaps 50 to 100 times worse for humans than it is for birds. There is then the strong presumption that birds hear *as such* the rapid modulations so characteristic of their songs, and that the information content even in relatively simple songs must be enormous. One readily understands how birds of the same species can recognize individuals from subtleties in their songs that are imperceptible to a human listener.

So far I have discussed only those phrases in bird song for which the basic wave form is sinusoidal without significant harmonic content. As we shall now see, harmonics do occur in bird song, but there is no broad generalization relating to the presence or absence of harmonics in the songs of the several bird families. One might say with some confidence that songs of the Passeriformes in which harmonics appear are relatively rare, whereas in the songs and calls of birds in other families phrases with substantial harmonic content are comparatively common. Each statement will, however, have numerous exceptions.

It can nonetheless be said with considerable assurance that for any given species there is a threshold frequency below which harmonics occur and above which one hears a phrase without significant harmonic content. This threshold varies widely for different species, from a value near 4,000 cycles per second for the blue-gray gnatcatcher to below 500 cycles per second for the barred owl. Such threshold frequencies can be determined only when a bird sings over a range of frequencies that embrace the threshold. This circumstance is uncommon, because the majority of the Passeriformes sing only in the frequency range giving rise to phrases free of harmonics, and birds of other families sing only in the range giving rise to harmonic phrases. Indeed, whether or not a bird passes through the harmonic threshold in its songs or calls appears to be a matter of choice; there is no physiological reason that would prevent a crow from singing a harmonic-free phrase, or a wood warbler, by way of contrast, from singing a phrase with substantial harmonic content.

As an example of the development of harmonic spectra as the fundamental descends below the threshold frequency, I offer a glissando sung by a smooth-billed ani that embraces nearly three octaves (from 485 to 3,500 cycles per second). As the frequency rises dominance shifts from the fourth harmonic through the third and second harmonics to the fundamental frequency. The transition frequencies (the frequency at which the relative amplitudes of adjacent harmonics are equal) are 1,600 cycles per second for the fundamental and the second harmonic, 950 cycles per second for the second and third harmonics and 560 cycles per second for the third and fourth harmonics. In all the many cases I have examined in which the fundamental passes through the threshold, curves of relative amplitude plotted against frequency show a small frequency interval embracing each transition between

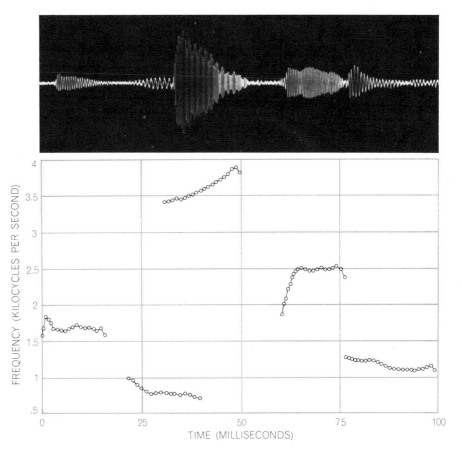

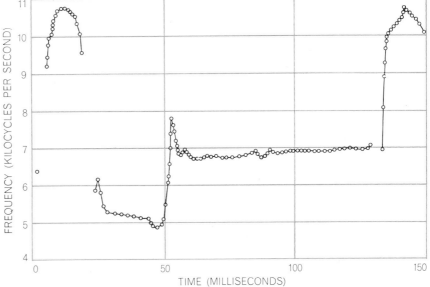

NONREPETITIVE MODULATION is seen in the courtship song of the brown-headed cowbird. In the oscillogram and frequency analysis at top is the phrase described as "glug"; in the corresponding records at bottom is the phrase "gleeee." Both of the phrases cover a large range of frequencies.

harmonics and a larger interval during which a particular harmonic contains a relatively large fraction of the acoustical energy. I have found no glissandi for which the dominant harmonic is higher than the fourth, but there are many calls in which the fundamental is constant, and in which the associated harmonic

spectra show similar characteristics, that is, a dominant harmonic with adjacent harmonics falling off rapidly in relative amplitude.

To understand how harmonic spectra with these characteristics can be generated in the syrinx, let us return once again to my simplified model of the organ. Imagine the tension in the tympanic membrane gradually being reduced, with the membrane at its vibrational peak approaching the opposing bronchial wall (or the external labium) more and more closely. The point will come when the bronchial wall will *constrain* the membrane, forcing it to depart from a pure sinusoidal vibration. At this point the second harmonic will become evident, increasing in amplitude as membrane tension falls and the constraint of the opposing bronchial wall influences

an increasing percentage of the period of vibration (the period of the fundamental). As the process continues, the amplitude of the fundamental will fall as the amplitude of the second harmonic rises. As membrane tension decreases still further, the second harmonic will become constrained and the third harmonic will become dominant. At this point the membrane can be visualized as being in a state of *rippling* vibration, with a fundamental fixed by membrane tension and the associated harmonic spectrum dictated by the constraints imposed by the passage within which the membrane is vibrating.

Among the ducks and geese harmonic spectra are found with many terms and with amplitudes showing no particular pattern. Such spectra must be associated with a form of membrane vibration re-

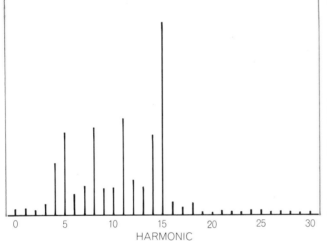

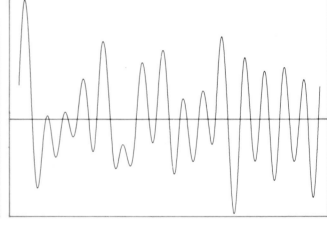

INDIAN HILL MYNA SAYS "AH" as in "Charlie." At left is the harmonic spectrum of the sound. At right is the wave form of the sound, which does not decay in amplitude as it would if it were generated by a resonant system such as the human vocal apparatus.

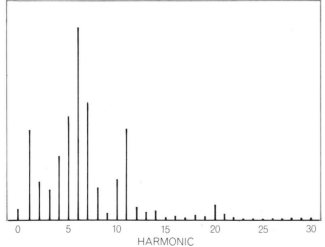

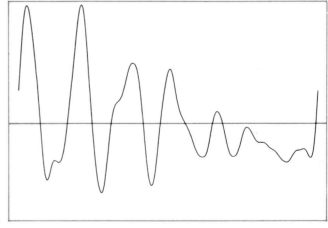

AUTHOR SAYS "AH" as in "Charlie." The wave form shows a decay in amplitude. The two "ahs" sound alike, but one is generated by a nonresonant mechanism and the other by a resonant one. This is a fundamental difference between the vocal apparatus of birds and the vocal apparatus of men. The vertical lines in the harmonic spectra represent the relative intensity of sound at each harmonic.

sembling the pulses produced in the human glottis, that is, a high-amplitude pulse followed by a series of less violent ripples of variable period.

Throughout the harmonic domain, whatever the species and whatever the wave form or harmonic spectrum may be, the evidence shows quite clearly that there is no tracheal modulation; the call is produced in the syrinx and passes unchanged through the trachea.

We come now to those birds that produce a more or less convincing imitation of the human voice. The question is whether such birds bring into play some completely new physiological or acoustical mechanism, or whether they achieve their imitations using the same mechanisms that produce their normal songs. I have selected the Indian hill myna as the standard of reference because its imitation of the human voice is excellent; it might well deceive the listener into thinking a person is speaking. A phrase was selected from the myna repertoire and was examined in detail in comparison with the same phrase spoken by me.

The reader will recall that the human voice originates as a series of puffs of air emanating from the glottis, the period between puffs corresponding to the fundamental frequency (80 to 180 puffs per second for the male voice). This acoustical disturbance is modulated in passing through our oral cavities so as to reinforce certain frequencies and attenuate others. Areas of reinforcement are called "formants" and are characteristic of particular vowel sounds. The wave form and the corresponding harmonic spectrum for the vowel "a" as enunciated by the myna and by me are shown in the illus-

trations below. The sound was unmistakably "ah" for both of us, but the associated wave forms are entirely different.

When the acoustical disturbance originating at the glottis stimulates a resonator (the oral cavities) to vibrate at a natural frequency, the generated wave form will have the frequency of the resonator, and the amplitude of the wave will decay exponentially over a period corresponding to that of the fundamental. One can readily isolate the formants for the myna and me corresponding to the vowel "a" and see if the resulting wave form fits these criteria. We find that the amplitude for my formants does indeed decay exponentially within the fundamental period; for the myna there is no decay and only random change in amplitude. This is the most convincing evidence we can offer that resonators are not involved in the myna "imitation."

How, then, is the imitation produced? It is important first to realize that the myna need not reproduce the human wave form at all. Since the human ear is not sensitive to phase, the myna need produce only an approximation of the amplitudes of the several harmonics, and this can be done with literally an infinity of wave forms. The ability to produce a separate harmonic spectrum with each of the two acoustical sources should be adequate to produce a reasonably good imitation, particularly since it has been shown that much of the human voice spectrum is redundant even if the criterion is recognition of the speech of a particular individual.

Let me summarize. The physiology and acoustics of bird vocalization are unique in the animal kingdom. Sound is produced at the syrinx in an air stream

modulated by an elastic membrane vibrating in a restricted passage bounded by the walls of the bronchus. This source-generated acoustical disturbance appears not to be modified in its passage through the trachea. The syrinx contains two independently controllable sources, one in each bronchus, enabling the bird to produce two notes or phrases simultaneously. Harmonics arise below a threshold frequency by mechanical constraints on the vibrating membrane, forcing a departure from a purely sinusoidal wave form. The source-generated sounds can be modulated in frequency or in amplitude or (more usually) in both with extraordinary rapidity, so rapidly that human ears cannot perceive the modulators as such, receiving instead impressions of notes of varying quality or timbre.

I end this account by pointing out that it has been in effect a scientific detective story, with conclusions reached by analyzing the evidence in the bird songs themselves. The criminal did not confess in the last chapter, and the evidence must remain circumstantial, without direct proof. Had I the deductive powers of an Albert Campion, a Gideon Fell or any of the other erudite detectives of fiction, coupled with the persuasiveness of Perry Mason, I might have done better. In any event I have developed a model, highly convincing to me, and I shall patiently await experimental evidence that will raise my spirits if the answer is yes but will not be too devastating if it should be no.

40

DUET-SINGING BIRDS

W. H. THORPE
August 1973

The male and female of certain tropical species join each other in remarkably precise song. The primary function of this behavior is to maintain close communication between the birds in dense foliage

Newcomers to East Africa soon become familiar with a striking birdsong, heard sometimes in large gardens or parks but more frequently in open forest and bushy savanna country. The song is brief, about a couple of seconds in duration, but it is often repeated with great regularity over long periods. What makes the song particularly outstanding is that its few notes possess a flutelike or bell-like quality, exceedingly pleasant to the ear, that strikes almost every listener as being in some curious sense "musical." The singing bird is a shrike (*Laniarius aethiopicus*), known in some parts of its range as the bellbird or bell shrike. Obvious and conspicuous though its song is, many quite observant people have lived in East Africa for years without realizing that the performance comes from two singers rather than one. Not until the listener happens to get between the two birds does he realize that the first few notes of the song come from one direction and the rest from another, yet with an almost incredible precision of timing.

It is usual to speak of such a performance, with one member of a pair starting the song and the other completing it, as antiphonal song. Antiphonal song, however, is only one particular kind of what may in general be called duetting, a term that also includes polyphonic performances, when the two birds sing at the same time, each coordinating its individual song pattern with the other. Occasionally each bird may sing exactly the same pattern of notes at the same moment, that is, in unison. Alternatively, the two contributions may be different in pattern but overlap each other, again with precise coordination [*see bottom illustration on page 416*].

Shrikes of the genus *Laniarius,* which are found only in Africa, provide the most striking examples of duetting. It appears that all 15 species of the genus, ranging throughout tropical Africa, exhibit the behavior. This mode of mutual singing between a paired male and female is not, however, confined to shrikes. It is well developed in eight or nine other families of birds in different parts of the world, nearly all of them tropical in range. Such behavior seems in fact to be most characteristic of species that live in very dense tropical vegetation, where it must be difficult for the two members of the pair to keep in sight of each other. In such a setting the birds would find a vocal pair bond useful, if not essential. Duetting also appears to be characteristic of species in which the male and female are of identical appearance, remain paired for life and maintain their territories for the greater part of the year, perhaps for their entire lives.

My colleagues and I first studied duetting in the field in Kenya and Uganda during the years from 1962 to 1967. We were able to extend the work greatly with studies of captive birds in large tropical aviaries in England from 1964 to 1970. In our field studies we concentrated on three populations of bell shrikes in widely separated areas: on the shores of Lake Nakuru in Kenya, in southwestern Uganda about 400 miles to the west of the lake, and near Kapenguria in Kenya, 150 miles northwest of the lake. Our aim was to record as far as possible all the main duet patterns in the area, to plot territories where possible, and again where possible to identify individual birds by marking them with colored rings.

In the Lake Nakuru area we recorded 102 different duet patterns, in the Uganda area we had 22 examples and in the Kapenguria region 24. The tape recordings were analyzed by sound spectrograph, but because of the very pure tonal quality of the bell shrike's notes we found it much more satisfactory to represent the songs by simple musical notation rather than by sound spectrograms. The songs can be almost completely specified by the pitch of the notes, by the intervals (the difference in pitch between any two notes), by whether the sounds come in harmonic intervals (simultaneously) or in melodic intervals (successively) and by the duration of the notes, their timing and their overall pattern.

Both absolute and relative pitch are of great importance in the recognition of the duets, and these features can only be roughly assumed from sound-spectrographic analyses. Fortunately standard musical notation has been developed over the centuries for the specific purpose of communicating details of pitch and time. It is an elegant and foolproof method for the purpose. On the other hand, sound-spectrographic and other electronic methods of analysis give much information about acoustic structure (tonal quality), relative intensity (loudness) and the minutiae of timing (which are particularly needed in assessing response times). Hence for these purposes vocalizations are better portrayed by such methods.

In the Lake Nakuru and Uganda field areas the vegetation was so dense that it was often extremely difficult, if not impossible, to map the paired birds' territorial boundaries. Birds in the Kapenguria area were living under more open conditions, so that here the mapping of the territories was much easier. In one of the Kapenguria areas four pairs of birds were holding territories, and characteristic song patterns were recorded for each [*see illustration on page 418*]. It will be seen that certain series of notes and certain patterns are fairly general to the species. Indeed, one quickly repeated series of low notes (either G or G-sharp)

DUET SINGERS, a paired male and female bell shrike, are shown in an untypically open environment. Also known as the bellbird, the bell shrike is an African species belonging to the genus *Laniarius*. The paintings on this page were made by David Bygott.

POLYPHONIC SINGER that also maintains contact with its mate by singing duets, the white-browed robin chat (*Cossypha heuglini*) is found in the thick forest undergrowth and scrub of East Africa.

UNMUSICAL SINGER, the black-headed gonolek is another bird of the genus *Laniarius*. Of the 15 African shrike species in the genus few are as musical as the bell shrike, but they all sing duets.

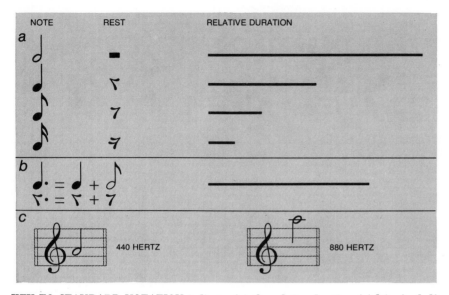

KEY TO STANDARD NOTATION indicates (*a*) the relative duration (*right*) of a half note and half rest compared with successively briefer intervals of sound or silence: quarter, eighth and 16th notes and rests respectively. The presence of a dot after either a note or a rest (*b*) increases the value of the symbol by one-half. Two staffs with G clefs are shown (*c*). The note on the staff at the left is a "tuning" A (440 hertz). The note on the staff at the right is one octave higher (880 hertz). The birdsongs are written one octave below true pitch.

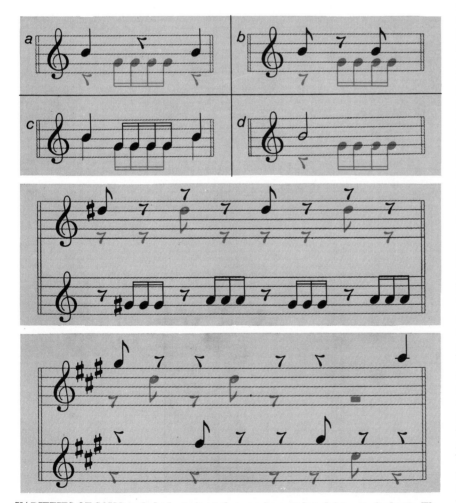

VARIETIES OF SONG include duets (*top*) that may be antiphonal (*a*) or polyphonic. The notes and rests of one singer appear in black and those of the other singer in color. Polyphonic duets (*b through d*) can include phrases sung in unison (*c*) or phrases that overlap (*d*). Not all bell shrike songs are duets. Shown here is a trio (*middle*) recorded at Lake Nakuru in Kenya and a quartet (*bottom*) by captive shrikes at the University of Cambridge.

has a mellow sound that has given rise to another common name for the bell shrike: the tropical boubou shrike. Another clearly characteristic feature of the species is its tendency to produce intervals between successive notes of about an octave, either an octave below or an octave above. The species can be recognized by these features wherever it is found throughout the thousands of square miles of its range. Yet the more individual pairs are studied, the clearer it becomes that each pair of bell shrikes works out its own particular repertoire of duet patterns, so that the repertoire tends to be unique to that pair. Once one knows it well enough, it provides a means for individual identification.

Quite early in our studies we came on examples of trio singing and sometimes even quartet and quintet singing. One way that trio singing arises is through the intervention of one bird of a pair in a neighboring territory. (We were seldom able to determine whether it was the male or the female.) The "outside" singer would interpose its notes between those of the pair in "home" territory in an ordered manner and with extremely precise timing. The home pair in turn would sometimes minutely adjust the pattern and timing of its own duet to allow the neighbor to participate [*see bottom illustration at left*]. On one or two occasions the mate of the outside bird was seen standing by but not taking part. Trios, and perhaps quartets, can also be formed when one or two grown offspring of a pair, still residing within the parental territory or on its margins, join in with their parents' songs.

Apart from certain snarling or buzzing sounds used as alarm notes, vocalizations by bell shrikes in the wild under normal conditions consist of antiphonal singing between the members of a mated pair. The male bird is usually, but not necessarily, the leader. When the female starts the duet, she usually does so with a particular snarling note that may then be incorporated into the overall pattern of musical notes. In other species of shrike (and in bell shrikes in captivity) it may be much more usual for the female to start the duet. We have good evidence that the vocal repertoire is worked out and developed between the two members of a mated pair. Indeed, we have noted that when a bird is isolated by some mischance in the wild (as in a Nairobi garden where one bell shrike was kept under observation for nearly six months), it appears unable to produce any complex pattern of vocalization.

In this connection we found that

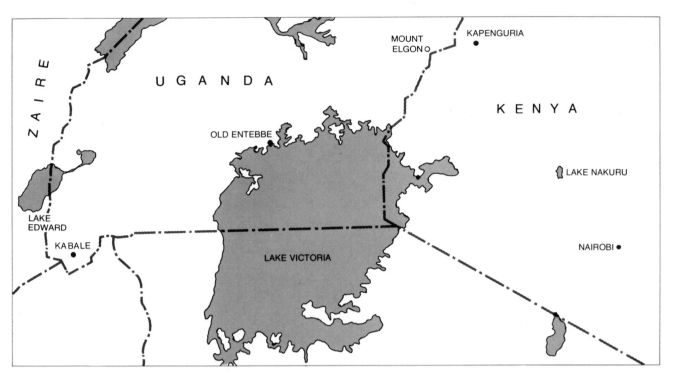

FIELD STUDIES of duetting birds were conducted at two areas in Kenya: Lake Nakuru, northwest of Nairobi, and Kapenguria, east of Mount Elgon. The third area was near Kabale, southeast of Lake Edward in Uganda. In all 148 different duet patterns were recorded.

where the birds seemed to be crowded in the wild the song patterns tend to be more elaborate than in areas where the birds have more space. This field observation was confirmed by studies of birds in captivity. When such birds, on arrival at our aviaries at the University of Cambridge, were all in cages in one room but were unable to see one another, trios and quartets of remarkable elaboration would quickly be built up, would be sung for a few hours or a day or two and then would be abandoned. When the birds were put out as pairs in the aviaries and therefore were no longer in close contact with other pairs, the duet patterns usually regressed to rather simple forms.

On the basis of both the field data and our aviary studies we can list the function of duetting under four headings in decreasing order of importance. The most important is the singer's recognition of, location of and maintenance of contact with its mate. The next is mutual stimulation between the two birds that form a pair, as a part of (or a substitute for) the ordinary methods of visual display. The third is an aggressive maintenance of the pair's territorial integrity. Last and least is mutual reassurance after some disturbance.

With respect to the first two functions it is important to realize that in the earth's temperate zones seasonal changes in the length of the day provide the most important of the cues that initiate the secretion of birds' sex hormones and so bring potential pairs into the breeding condition. In the Tropics the day-length cue is very slight, if not completely absent. Moreover, other possible "seasonal" cues (such as variations in humidity, rainfall, degree of cloud cover and the like) tend to be unpredictable and to give little advance warning. It follows that if birds in the Tropics are to take full advantage of the time when breeding conditions are at an optimum, paired males and females must be in constant contact so that their behavior and their reproductive cycles are fully coordinated.

Our aviary experiments yielded many interesting facts about duetting. To mention a few, studies in the aviaries confirmed our field observation that, if one member of a pair is absent, the bird remaining in the territory tends to sing the whole duet pattern by itself: both its own contribution and its mate's. When the mate returns, however, a period of unison singing is not uncommon. The duet in unison will last for a few seconds; thereafter the pair resumes its antiphonal song as before. These shrike studies evidently confirm an argument put forward by Konrad Lorenz to the effect that "whole duet" vocalization in a partner's absence may be intended to secure the partner's return. It is almost as if one bird were using the characteristic vocal performance of the other as a "name" that might serve as a recall.

A series of separation experiments in the aviaries yielded these conclusions. First, separation leads to an increase in the amount of vocalization by the deserted bird. In contrast, a bird that is moved to a new "territory" tends to decrease its vocalization. Second, if the bird remaining in the territory is a male, he employs all his usual vocalizations and in addition some of those of his missing partner. Third, if two members of a pair are separated for a long period, there is sooner or later a tendency to show a regression of vocalizations that can reduce the performance to what amounts to a juvenile condition.

If an isolated male hears a playback of his mate's voice, he will reply with the appropriate item in his own repertoire. The male is much less likely to respond to a playback of his own voice, however, and does so only by repeating the song that has just been played. This finding during our aviary studies confirmed still another conclusion based on fieldwork: even though notes of the bell shrike are very uniform in acoustic structure, they may be sufficiently different to be recognized on the basis of their vocal quality as well as by the pattern they form.

Our aviary experiments provided further evidence of this kind of recognizability. We found that a female bell shrike in the aviary will answer such notes of her own male as she may be able to hear but will not answer the notes of any neighboring pairs. In one set of experiments an extra male was kept in association with a duetting pair. The outside male was never heard to vocalize until the resident male was removed. When the removal took place, however, it became clear that the outside male had at least in part learned not only his rival male's repertoire but also that of the female. The resident female would respond to the educated outside male's song although she would not respond to the songs of strangers.

A remarkable result emerged from our difficulty in identifying the sex of captive birds. Because the male and female are exactly alike in appearance we made occasional mistakes in our caging, inadvertently constructing male-male and female-female "pairs." It sometimes happened that for weeks or months our mistakes had surprisingly little effect on the birds' behavior and vocalization. And so it became clear to us that two bell shrikes of the same sex, either male or female, can behave in a way that is indistinguishable from the behavior of a true pair. They may engage in mutual duetting, picking up and carrying nesting material, mutual preening, begging and even some forms of display.

The precision in timing in the bell shrike's duets is excellent. It was easier to study this, however, in another species of the same genus: the black-headed gonolek (*Laniarius erythrogaster*). The general build and pattern of the gonolek is very similar to the bell shrike's except that its underparts, instead of being white suffused with pink, are a brilliant crimson [*see bottom illustration at right on page 415*]. Throughout its range the gonolek is a bird of the thickest bush and undergrowth. We studied it both in Kenya and Uganda and in captivity. Its duet is unmusical and extremely simple. The initiating bird, in this species almost always a male, has a "yoick"-like note. This note is immediately followed by a tearing hiss, sometimes suggesting the ripping of cloth and sometimes being more like a sharp sneeze, from the female. Here, as with the bell shrike, the timing is so perfect as to make the duet sound like the song of one bird.

The timing of the gonolek is easy to investigate because the onset of the second bird's note is extremely sharp, giving the investigator an exact point of measurement. We recorded a consecutive series of eight duets of a pair of these birds at Old Entebbe in Uganda. The second bird, which was completely out of sight of the first, took its time cue

with extraordinary precision from the start of the first bird's note. The mean response time of the female in this series of duets was little more than 144 milliseconds, with a standard deviation of 12.6 milliseconds [*see illustration on page 420*]. In another series of seven consecutive duets the response time was much longer (425 milliseconds) but the standard deviation was even less (4.9 milliseconds). It is obvious that the species must have an extremely precise time sense; the accuracy does not decrease even if the response time is extended by a factor of four. I am not aware of any auditory reaction time in humans that has a standard deviation of less than 20 milliseconds.

Even more unusual is the performance of another African duetting species, a member of the genus *Cisticola* (the grass warblers) known as Chubb's cisticola (*Cisticola chubbi*). Chubb's cisticola is a small streaky-brown bird that inhabits long grass in bush-clad clearings at altitudes of between 5,000 and 8,000 feet. A series of six consecutive duets that I recorded, the two birds in this instance being in sight of each other, had a mean response time of 396 milliseconds and the remarkably small standard deviation of 2.9 milliseconds. That is about an eighth of the error a man would make under similar circumstances.

Is there any scientific basis for the impression of "musicality" the duets of the bell shrike so strongly suggest? The main investigator of this topic is Joan Hall-Craggs. As she has stated, it is useful to begin any such inquiry with the reasonable assumption that, since a bird's ear is similar to the human ear in its essential structure, it displays many of the same characteristics. At the same time we should remember that the perception of pitch probably begins at higher frequencies among most birds than among humans. The shortest time for the identification of pitch by man is approximately .05 second, assuming a signal in the middle range of frequency, but for birds it is quite likely that this time may be much reduced. It has also been calculated that the pitch of a tone is detectable when 70 percent of the energy in the spectrum lies within ±5 percent of the principal frequency. Such energy concentration is clearly discernible in the sound spectrograms of bell shrike songs, and most of the notes are of sufficient duration and concentration of sound energy to enable us to assess their pitch by ear.

Musical form, at its simplest, consists

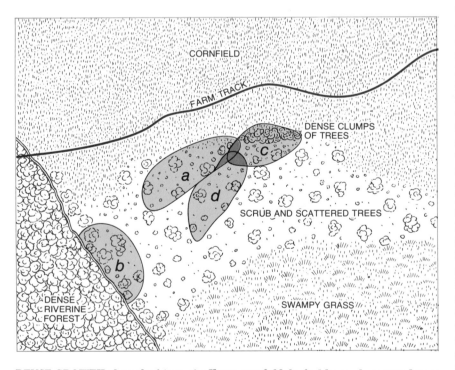

DENSE GROWTH along the fringe of a Kenya cornfield, backed by scrub, scattered trees and a thickly forested stream, contained four bell shrike breeding territories, three of them overlapping (*solid color*). Territories *a* and *b* contained paired adults only; the adult pairs in *c* and *d* shared the territory with juvenile birds. Both the territory overlap and the number of birds that were present led to frequent singing in trios, quartets and even quintets.

FOUR DUETS recorded in the Lake Nakuru area exhibit distinctly different patterns. In instances when the observer could not tell which notes were sung by which bird the notes and rests appear in black only. Otherwise the second singer's part is shown in color.

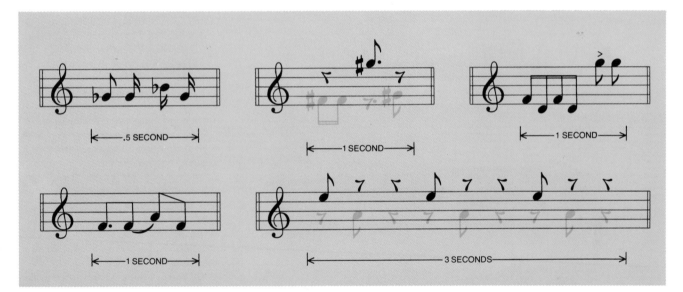

FIVE DUETS recorded in the Uganda study area provide a further example of differing bell shrike song patterns. In the third song, where the individual singers' contributions are unknown, an accent over the first G indicates that the singer put stress on the note.

TWO TRIOS were recorded in the Kapenguria study area shown on the facing page, where such singing was a frequent event. The notes of the first two singers appear in black and in color respectively; the notes of z, the third singer, are outlined in white.

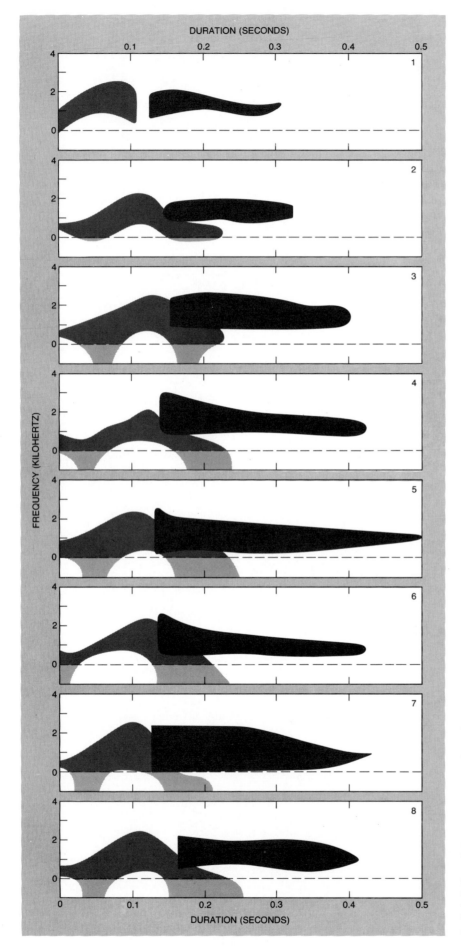

in the grouping together of units of sound energy in repeatable and consequently recognizable patterns of time and pitch. As Hermann von Helmholtz put it more than a century ago: "Melodic motion is change of pitch in time." It is obvious at a glance that the shrike songs epitomize melodic motion and therefore have musical form. To analyze that form one must first describe the intervals: the difference in pitch between any two tones.

We undertook such a study, using the duets of 10 pairs of one population of bell shrikes. The tones and intervals of the duets were subjected to physical analysis by taking single sounds from the field tapes and recording them on a sound spectrograph, which was then set to play the sound repeatedly at 2.5-second intervals. That signal was then matched, initially by ear, to a pure tone from a signal generator. The two tones were next matched precisely on a cathode ray oscilloscope by tuning the generated tone until a clear Lissajous figure with a ratio of 1:1 appeared repeatedly on the screen. (More recently we have employed a new instrument for this work. It is the "Melograph," designed and built at the University of Uppsala.) A frequency counter then gave a reading in hertz (cycles per second).

The absolute frequencies, as distinct from the intervals between them, were measured. It was found that the duets of all these birds fell within the range of 656 to 2,064 hertz. That is about one octave plus a sixth. The range of an individual pair varied from as little as a minor seventh (787 to 1,405 hertz) to a little less than a perfect 11th, that is, an octave plus a perfect fourth (798 to 2,064 hertz). The paired birds normally use only portions of the available frequency spectrum. In pairs where this uniformity is most marked a low band of sung frequencies is centered around 800 hertz, a middle band around 1,000 hertz and a smaller high band just below 1,200 to 1,220 hertz.

It was clear in almost every instance that the gaps of unused frequencies span

RAPID RESPONSE of female black-headed gonolek in eight successive duets is shown in the series of sound spectrograms at left. The song of the male is shown in gray; the female's response is in black. The series of duets was completed in some 30 seconds; the average response delay was .144 second. The apparent subzero frequencies are due to distortion and interference below 50 hertz.

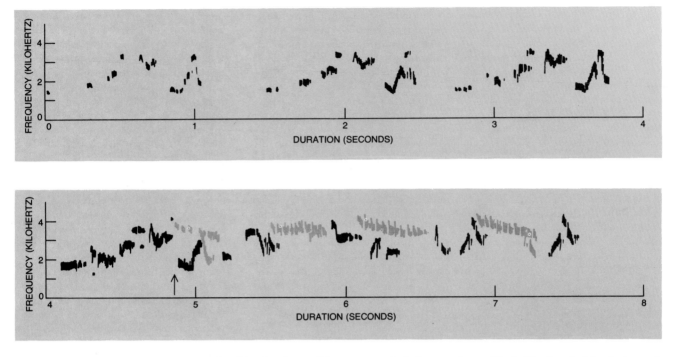

CRESCENDO DUET of the white-browed robin chat, some 7.5 seconds in duration, is seen in a sound spectrogram. During the first part of the duet the male utters four five-syllable phrases that progressively increase in amplitude. The female joins in (*arrow*), continuing the crescendo, whereupon the male mutes his higher notes but continues to accompany the female in a low-pitched song.

wider ranges in the available spectrum than the bands of sung notes do, giving the impression that the birds are using a series of notes resembling a gapped scale. There is not sufficient consistency, however, to postulate such a scale, nor is there good reason to suppose that the birds are unable to sing notes in the unused section of the spectrum. The distribution of the used frequencies suggests that the syrinx, the sound-producing organ of the bird, may be such that these notes are produced more easily and/or to greater effect than such notes as might have been expected to be present in the bands of unused frequency. The tendency in most pairs of birds for the notes to cluster around 800, 1,000 and 1,200 hertz, however, suggests that the birds are producing most readily the fourth, fifth and sixth harmonics of a fundamental at about 200 hertz. These tones, derived from the lower reaches of the harmonic series, give rise to the ordinary major triad approximately on G in octaves five to six. If indeed the birds are utilizing the harmonics in this manner, it goes part of the way toward explaining their decided proclivity for singing duets that incorporate, or are based on, a major or minor triad.

In music if two tones or frequencies are adjacent, then the interval is described as "melodic." When, as often happens in bell shrike duets, two sounds occur synchronously, or when one tone is sustained while another tone is sounded above or below it, the interval is termed harmonic. All 15 species of the genus *Laniarius* sing antiphonally, but the bell shrike and one other species I shall not discuss here appear to be unique among them in singing melodic lines that are sometimes antiphonal and sometimes polyphonic and sometimes even sung in unison.

According to musical theory, the sounding together of two notes may lead to a predominant use of consonant intervals in two-part or multipart singing. Helmholtz, who investigated the properties of consonance and dissonance from the physical point of view in the 1860's, showed that the "roughness" of dissonance that we (and presumably birds as well) experience is based on a physical phenomenon, namely "beats": periodic variations in the amplitude of the sound pressure due to the interference of two sound waves of different frequencies. According to Helmholtz, beats are maximally disturbing to man at about 33 per second. It might be expected, however, that with birds this critical figure would be substantially increased because of their faster identification of pitch.

Now, the original physiological explanation of beats was that two tones sounding together forced into vibration over-lapping regions of the basilar membrane of the ear. This explanation, which was once much in doubt, has now been largely rehabilitated, although no completely satisfactory physiological description of the phenomenon has yet been devised. This does not, however, affect the reality of the phenomena of dissonance and beats. Helmholtz drew a curve illustrating the degree of consonance and dissonance of intervals within the octave. If we draw this curve to the same scale as a curve that shows the incidence of these intervals as they are used by the bell shrikes, we find close agreement between them.

The shrikes' predisposition to sing consonant intervals is fully demonstrated by both aural and physical analyses. But since there are dissonant intervals as well, although in much smaller numbers, it cannot be argued that these birds are compelled by the structure of their syrinx to sing consonant intervals. It may be, however, that once such intervals are learned they are found to be functionally the best. It can also be argued that good, steady consonances might well assist in the effectiveness of the duets as a contact-maintenance system, particularly if it is important for the distance to be judged accurately.

Hence we can conclude the topic of bell shrike aesthetics by saying that the apparent musicality of the songs of these species depends primarily on the birds' having a hearing apparatus that responds in the same way to the roughness of dissonance as our own does. To quote Joan Hall-Craggs: "To the musical listener these songs may seem overharmonious; nevertheless, it is the kind of harmony to which man aspired and which probably reached its peak in Mozart. No musical listener could call the songs 'unmusical' or 'displeasing'; their only fault from our point of view rests in their brevity and simplicity."

Two other examples of antiphonal singing among East African birds are particularly interesting when they are compared with the songs of the shrikes

because they illustrate still other aspects of duetting. One of the birds is a member of the genus *Cossypha* (family Turdidae): the white-browed robin chat *Cossypha heuglini*. It is the sole known duettist among the 15 species of the genus.

Here again the sexes look alike. The birds are found in dense vegetation, in riverine forest and secondary scrub whether in farmland or garden, and from the East African coast up to an altitude of 6,000 feet. They are excessively shy, spending most of the time hidden in the undergrowth, feeding on the ground and usually singing at dawn and dusk from a perch in a low bush. The song is a long one, lasting five seconds or more, and starts with male solo phrases showing a

gradual crescendo. When the loudness reaches a critical level, it provides the signal for the female to join in. The birds then proceed together, continuing the crescendo and with the pitch steadily rising, although after the female takes part the male tends to cut out his higher notes, giving a lower-pitched accompaniment of the female's downward glissandos [see illustration on page 421].

The grass warblers (genus *Cisticola*) are a huge assemblage: 40 species and 153 races. Of these it seems likely that only four species are duettists, yet they are duettists of the highest precision. As I have mentioned, I have recorded the duets of Chubb's cisticola. One other species, *Cisticola nigriloris,* which is found in the highlands of northern Malawi and southern Tanzania, is a singer of special distinction. It is a persistent duettist, and the pair's usual theme is a very high-pitched four-note whistle: G (at 3,240 hertz), E-flat (near 2,568), E-flat again (but at 5,000) and B (at 4,064), all with a continuous squeaking or croaking accompaniment. In two separate recordings of this species, made in areas of Tanzania more than 100 miles apart, the entire four-note phrase is suddenly transposed to another key after the first half-dozen or so bouts. This suggests the interesting possibility that the musical transposition of the song is a species-specific character! If that is true, it is, as far as I am aware, unique.

To sum up, we can say that duetting clearly plays a very important part in the signal system between male and female in a large number of bird species, in particular species that inhabit tropical regions. These elaborate song patterns show many interesting features, of which only a few have been discussed here. Perhaps the most interesting result of our investigations of duetting is the light cast on the heretofore little appreciated precision and synthesizing power of avian aural perception, the great precision of response time and the equally great exactness of control of the vocal organs. The use of these vocal powers for individual recognition is in line with observations made over the past decade by Beat Tschanz of the University of Bern, by C. G. Beer of Rutgers University and by other investigators. Their work has shown that in many colonial nesting birds (for example auks, terns, gulls, gannets and penguins) brief calls of a half-second duration or less can have enough acoustic detail not only to serve as labels identifying the calling species but also to label the individual caller.

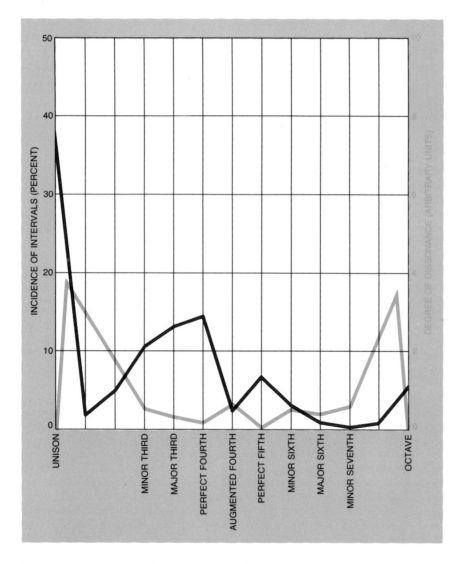

CONSONANT INTERVALS are characteristic of the bell shrike's songs. The black line indicates the intervals most frequently sung among selected semitones within the octave, expressed as percentages of the total number of intervals in the sample. The colored line is a modification of Hermann von Helmholtz' curve of consonance and dissonance between a note of fixed frequency and another note that changes smoothly from the same frequency to a frequency one octave higher. The closer the curve is to the base line, the greater the consonance is. Most of the intervals in the songs coincide with minimally dissonant frequencies.

BIBLIOGRAPHIES

These bibliographies have been revised by Norman K. Wessells.

INTRODUCTORY ESSAYS

An Essay on Vertebrates

Altman, S. A., SOCIAL COMMUNICATION AMONG PRIMATES. Chicago, University of Chicago Press, 1967.

Barrington, E. J. W., THE BIOLOGY OF HEMICHORDATA AND PROTOCHORDATA. London, Oliver and Boyd, 1965.

Campbell, B. G., HUMAN EVOLUTION, AN INTRODUCTION TO MAN'S ADAPTATIONS. Chicago, Aldine, 1966.

Diamond, I. T., and Hull, W. C., "Evolution of the neocortex." Science 164, 1969, pp. 251–62.

Dobzhansky, T., MANKIND EVOLVING. New Haven, Yale University Press, 1962.

Gordon, M. S., ANIMAL PHYSIOLOGY: PRINCIPLES AND ADAPTATIONS. Macmillan, New York, 1972.

Halstead, L. B., THE PATTERN OF VERTEBRATE EVOLUTION. W. H. Freeman, San Francisco, 1969.

Hoar, W. S., and Randall, D. J., FISH PHYSIOLOGY, Vols. I–VI. Academic Press, New York, 1969–1971.

Hochachka, P. W., and Somero, G. N., STRATEGIES OF BIOCHEMICAL ADAPTATIONS. Saunders, Philadelphia, 1973.

Howell, F. Clark, "Recent advances in human evolutionary studies." Quart. Rev. Biol. 42, 1967, pp. 471–513.

Le Gros Clark, W. E., THE ANTECEDENTS OF MAN. Quadrangle Books, Chicago, 1971.

Lewis, O. J., "Brachiation and early evolution of the hominoidea." Nature 230, 1971, pp. 577–78. But see also: Nature 237, 1972, pp. 103–104.

Marler, P. R., and Hamilton, W. J., MECHANISMS OF ANIMAL BEHAVIOR. Wiley, New York, 1966.

Romer, A. S., "Majors steps in vertebrate evolution." Science 158, 1967, pp. 1629–37.

Tobias, P. V., THE BRAIN IN HOMINID EVOLUTION. Columbia University Press, New York, 1971.

Young, J. Z., THE LIFE OF VERTEBRATES, 2nd ed. New York, Oxford University Press, 1962.

I. Structural Adaptations of Vertebrate Bodies

Budker, P., THE LIFE OF SHARKS. Columbia University Press, New York, 1971.

Carter, G. S., STRUCTURE AND HABIT IN VERTEBRATE EVOLUTION. Seattle, University of Washington Press, 1967.

Farner, D. S., and King, J. R., AVIAN BIOLOGY, Vols. I and II. Academic Press, New York, 1972.

Gray, J., ANIMAL LOCOMOTION. W. W. Norton, New York, 1968.

Hardisty, M. W., and Potter, I. C., THE BIOLOGY OF LAMPREYS, Vols. I and II. Academic Press, New York, 1972.

Tobias, P. V., "Early man in East Africa." Science 149, 1965, pp. 22–33.

Tutle, R. H., "Knuckle-walking and the problem of human origins." Science 166, 1969, pp. 953–61.

Voitkevich, A. A., THE FEATHERS AND PLUMAGE OF BIRDS. October House, New York, 1966.

Weihs, D., "Hydromechanics of fish schooling." Nature 241, 1973, pp. 290–91.

Welty, J. C., THE LIFE OF BIRDS. Philadelphia, Saunders, 1964.

II. Vascular System Biology

Antonini, E., and Brunori, M., "Hemoglobin." Ann. Rev. Biochem. 39, 1970, pp. 977–1043.

Guyton, A. C., Colman, T. G., and Granger, H. J., "Circulation: overall regulation." Ann. Rev. Physiol. 34, 1972, pp. 13–46.

Riggs, A., "Functional properties of hemoglobins." Physiol. Revs. 45, 1965, pp. 619–73.

Schmidt-Nielsen, K., and Taylor, C. R., "Red blood cells: why or why not?" Science 162, 1968, pp. 274–75. See also Science 162, 1968, pp. 275–77.

Zweifach, B. F., "Microcirculation." Ann. Rev. Physiol. 35, 1973, pp. 117–50.

III. Gas Exchange and the Lungs: and High Altitude. Adaptations for Diving

Andersen, H. T., THE BIOLOGY OF MARINE MAMMALS. Academic Press, New York, 1969.

Fänge, R., "Physiology of the swim bladder." Physiol. Revs. 46, 1966, pp. 299–322.

Pattle, R. E., "Surface lining of lung alveoli." Physiol. Revs. 45, 1965, pp. 48–79.

Slonim, N. B., and Hamilton, L. H., RESPIRATORY PHYSIOLOGY. C. V. Mosby, St. Louis, 1971.

West, J. B., "Respiration." Ann. Rev. Physiol. 34, 1972, pp. 91–116.

IV. Water Balance and Its Control

Fisher, J. W., KIDNEY HORMONES. Academic Press, New York, 1971.

Morel, F., and de Rouffignac, C., "Kidney." *Ann. Rev. Physiol.* **35**, 1973, pp. 17–54.

Norman, A. W., "The mode of action of vitamin D." *Biol. Revs.* **43**, 1968, pp. 97–137.

Potts, W. T. W., "Osmotic and ionic regulation." *Ann. Rev. Physiol.* **30**, 1968, pp. 73–104.

Poulson, T. L., "Countercurrent multipliers in avian kidneys." *Science,* **148**, 1965, pp. 389–391.

Salt, G. W., "Respiratory evaporation in birds." *Biol. Revs.* **39**, 1964, pp. 113–136.

Vander, A. J., "Control of renin release." *Physiol. Revs.* **47**, 1967, pp. 359–382.

V. Temperature Adaptations

Bartholomew, G. A., "Body Temperature and Energy Metabolism." *In* M. S. Gordon, Ed., ANIMAL FUNCTION: PRINCIPLES AND ADAPTATIONS. New York, Macmillan, 1968.

Gale, C. C., "Neuroendocrine aspects of thermoregulation." *Ann. Rev. Physiol.* **35**, 1973, pp. 391–430.

Hammel, H. T., "Regulation of internal body temperature." *Ann. Rev. Physiol.* **30**, 1968, pp. 641–710.

Hochachka, P. W., "Organization of metabolism during temperature compensation." *In* C. L. Prosser, Ed., MOLECULAR MECHANISMS OF TEMPERATURE ADAPTATION. Washington, D. C., American Association for the Advancement of Science, 1967, pp. 177–204.

Licht, P., "Thermal adaptation in the enzymes of lizards in relation to preferred body temperatures." *Ibid*, pp. 131–146.

Myers, R. D., and Tytell, M., "Fever: reciprocal shift in brain sodium to calcium ratios as the set point rises." *Science* **178**, 1972, pp. 765–67.

Smith, R. E., and Horwitz, B. A., "Brown fat and thermiogenesis." *Physiol. Rev.* **49**, 1969, pp. 330–425.

Waring, H., Moir, R. J., and Tyndale-Biscoe, C. H., "Comparative physiology of marsupials." *In* O. E. Lowenstein, Ed., ADVANCES IN COMPARATIVE PHYSIOLOGY AND BIOCHEMISTRY, Vol. 2, New York, Academic Press, 1966, pp. 237–376.

Whittow, G. C., *Comparative Physiology of Thermoregulation*, Vols. I and II. Academic Press, New York, 1971.

VI. Orientation and Navigation

Emlen, S. T., "Celestial rotation: its importance in the development of migratory orientation." *Science* **170**, 1970, pp. 1198–1202.

Griffin, D. R., LISTENING IN THE DARK. New Haven, Yale University Press, 1958.

Kalmus, H., "Comparative physiology: navigation by animals." *Ann. Rev. Physiol.* **26**, 1964, pp. 109–130.

Keeton, W. T., "Magnets interfere with pigeon homing." *Proc. Natl. Acad. Sci.* **68**, 1971, pp. 102–106.

Kellogg, W. N., PORPOISES AND SONAR. Chicago, University of Chicago Press, 1961.

Lang, T. G., and Smith, H. A. P., "Communication between dolphins in separate tanks by way of an electronic acoustic link." *Science* **150**, 1965, pp. 1839–1844.

Manley, G. A., "Some aspects of the evolution of hearing in vertebrates." *Nature* **230**, 1971, pp. 506–509.

Moncrieff, R. W., THE CHEMICAL SENSES. London, Leonard Hill, 1961.

Payne, R. S., and McVay, S., "Songs of humpback whales." *Science* **173**, 1971, pp. 585–97.

Pfaffman, C., OLFACTION AND TASTE, Vols. I–III. Rockefeller University Press, New York, 1969.

Pfeiffer, W., "The fright reaction of fish." *Biol. Revs.* **37**, 1962, pp. 495–511.

Rice, C. E., "Human echo perception." *Science* **155**, 1967, pp. 656–664.

Slijper, E. J., WHALES. New York, Basic Books, 1962, p. 475.

Storm, R. M., Ed., ANIMAL ORIENTATION AND NAVIGATION. Corvallis, Oregon State University Press, 1967.

Wimsatt, W. A., BIOLOGY OF BATS, Vols. I and II. Academic Press, New York, 1970.

VII. Hormones and Internal Regulation

Farrell, G., Fabre, L. F., and Rauschkolb, E. W. "The neurohypophysis." *Ann. Rev. Physiol.* **30**, 1968, pp. 557–588.

Gorbman, A., and Bern, H. A., A TEXTBOOK OF COMPARATIVE ENDOCRINOLOGY. New York, Wiley, 1962.

Jost, J. P., and Rickenberg, H. V., "Cyclic AMP." *Ann. Rev. Biochem.* **40**, 1971, pp. 741–74.

Loomis, W. F., "Skin-pigment regulation of vitamin D biosynthesis in man." *Science* **157**, 1967, pp. 501–506.

Lowry, P. J., and Chadwick, A., "Interelationships of some pituitary hormones." *Nature* **226**, 1970, pp. 219–22.

Proceedings of the Royal Society B. **170**, pp. 1–111. "A discussion of polypeptide hormones." May, 1968. (A series of papers on thyrocalcitonin, neurophysin, and neurohypophseal hormones. Included is the data of Archer on hormone evolution.)

Reiter, R. J., "Comparative physiology: pineal gland." *Ann. Rev. Physiol.* **35**, 1973, pp. 305–28.

Sutherland, E. W., "Studies on the mechanism of hormone action." *Science* **177**, 1972, pp. 401–408.

VIII. Communication between vertebrates

Bouhuys, A., "Physiology and musical instruments." *Nature* **221**, 1969, pp. 1199–1204.

Hinde, R. A., BIRD VOCALIZATIONS. Cambridge University Press, Cambridge, England, 1969.

Marler, P., "Animal communication signals." *Science* **157**, 1967, pp. 769–774.

Nottebohm, F., "Ontogeny of bird song." *Science* **167**, 1970, pp. 950–56.

Peters, M., and Ploog, D., "Communication among primates." *Ann. Rev. Physiol.* **35**, 1973, pp. 221–42.

Robertson, D. R., "Social control of sex reversal in a coral-reef fish." *Science* **177**, 1972, pp. 1007–1009.

Symposium of the Zoological Society of London 8, "Evolutionary aspects of animal communication. Imprinting and early learning." 1962.

Thorpe, W. H., "Antiphonal singing in birds as evidence for avian auditory reaction time." *Nature* **197**, 1963, pp. 774–776.

Thorpe, W. H., and North, M. E. W., "Origin and significance of the power of vocal imitation: with special reference to the antiphonal singing of birds." *Nature* **208**, 1965, pp. 219–222.

ARTICLES

1. How Fishes Swim

ASPECTS OF THE LOCOMOTION OF WHALES. R. W. L. Gawn in *Nature*, Vol. 161, No. 4,080, pages 44–46; January 10, 1948.

THE PROPULSIVE POWERS OF BLUE AND FIN WHALES. K. A. Kermack in *The Journal of Experimental Biology*, Vol. 25, No. 3, pages 237–240; September, 1948.

WHAT PRICE SPEED? G. Gabrielli and Th. von Kármán in *Mechanical Engineering*, Vol. 72, No. 10, pages 775–781; October, 1950.

2. How Animals Run

MOTIONS OF THE RUNNING CHEETAH AND HORSE. Milton Hildebrand in *Journal of Mammalogy*, Vol. 40, No. 4, pages 481–495; November, 1959.

QUADRUPEDAL AND BIPEDAL LOCOMOTION OF LIZARDS. Richard C. Snyder in *Copeia*, No. 2, pages 64–70; June, 1952.

SOME LOCOMOTORY ADAPTATIONS IN MAMMALS. J. Maynard Smith and R. J. G. Savage in *Journal of the Linean Society–Zoology*, Vol. 42, No. 288, pages 603–622; February, 1956.

SPEED IN ANIMALS. A. Brazier Howell. University of Chicago Press, 1944.

3. How Snakes Move

LOCOMOTION WITHOUT LIMBS. Carl Gans in *Natural History*, Vol. 75, No. 2, pages 10–17; February, 1966.

LOCOMOTION WITHOUT LIMBS: PART II. Carl Gans in *Natural History*, Vol. 75, No. 3, pages 36–41; March, 1966.

SIDEWINDING AND JUMPING PROGRESSION OF VIPERS. Carl Gans and H. Mendelssohn in *Proceedings of the Second International Symposium on Animals and Plant Toxins*, in press.

4. The Antiquity of Human Walking

THE FOOT AND THE SHOE. J. R. Napier in *Physiotherapy*, Vol. 43, No. 3, pages 65–74; March, 1957.

A HOMINID TOE BONE FROM BED 1, OLDUVAI GORGE, TANZANIA. M. H. Day and J. R. Napier in *Nature*, Vol. 211, No. 5052, pages 929–930; August 27, 1966.

5. The Evolution of the Hand

MAN THE TOOL-MAKER. Kenneth P. Oakley. British Museum of Natural History, 1950.

THE PREHENSILE MOVEMENTS OF THE HUMAN HAND. J. R. Napier in *The Journal of Bone and Joint Surgery*, Vol. 38-B, No. 4, pages 902–913; November, 1956.

PREHENSILITY AND OPPOSABILITY IN THE HANDS OF PRIMATES. J. R. Napier in *Symposia of the Zoological Society of London*, No. 5, pages 115–132; August, 1961.

6. Birds as Flying Machines

AVIAN BIOLOGY, Vols. I and II. D. S. Farner and J. R. King. Acedemic Press, New York, 1972.

7. Horns and Antlers

HISTOGENESIS OF BONE IN THE GROWING ANTLER OF THE CERVIDAE. Walter Modell and Charles V. Noback in *The American Journal of Anatomy*, Vol. 49, No. 1, pages 65–95; September 15, 1931.

THE POST-NATAL DEVELOPMENT OF THE HORN TUBULES AND FIBRES (INTERTUBULAR HORN) IN THE HORNS OF SHEEP. A. N. George in *The British Veterinary Journal*, Vol. 112, No. 1, pages 30–34; January, 1956.

STRUCTURE OF RHINOCEROS HORN, M. L. Ryder in *Nature*, Vol. 193, No. 4821, pages 1199–1201; March 24, 1962.

TROPHIC RESPONSES TO TRAUMA IN GROWING ANTLERS. Anthony B. Bubenik and R. Pavlansky in *The Journal of Experimental Zoology*, Vol. 159, No. 3, pages 289–302; August, 1965.

THERMOREGULATORY FUNCTION OF GROWING ANTLERS. Bernard Stonehouse in *Nature*, Vol. 218, No. 5144, pages 870–872; June 1, 1968.

8. The Physiology of Exercise

BEHAVIOR OF STROKE VOLUME AT REST AND DURING EXERCISE IN HUMAN BEINGS. Carleton B. Chapman, Joseph N. Fisher and Brian J. Sproule in *The Journal of Clinical Investigation*, Vol. 39, No. 8, pages 1208–1213; August, 1960.

THE PHYSIOLOGICAL MEANING OF THE MAXIMAL OXYGEN INTAKE TEST. Jere H. Mitchell, Brian J. Sproule and Carleton B. Chapman in *The Journal of Clinical Investigation*, Vol. 37, No. 4, pages 538–547; April, 1958.

9. The Heart

PHYSIOLOGY IN HEALTH AND DISEASE. Carl J. Wiggers. Lea & Febiger, 1949.

THE MOTION OF THE HEART: THE STORY OF CARDIOVASCULAR RESEARCH. Blake Cabot. Harper & Brothers, 1954.

10. The Microcirculation of the Blood

THE ANATOMY AND PHYSIOLOGY OF CAPILLARIES. August Krogh. Yale University Press, 1929.

GENERAL PRINCIPLES GOVERNING THE BEHAVIOR OF THE MICROCIRCULATION. B. W. Zweifach in *The American Journal of Medicine*, Vol. 23, No. 5, pages 684–696; November, 1957.

11. The Lymphatic System

THE LYMPHATIC SYSTEM WITH PARTICULAR REFERENCE TO THE KIDNEY. H. S. Mayerson in *Surgery, Gynecology & Obstetrics*, Vol. 116, No. 3, pages 259–272; March, 1963.

LYMPHATICS AND LYMPH CIRCULATION. István Rusznyák, Mihály Földi and György Szabó. Pergamon Press Ltd., 1960.

OBSERVATIONS AND REFLECTIONS ON THE LYMPHATIC SYSTEM. H. S. Mayerson in *Transactions & Studies of the College of Physicians of Philadelphia*, Fourth Series, Vol. 28, No. 3, pages 109–127; January, 1961.

12. "The Wonderful Net"

COUNTER-CURRENT VASCULAR HEAT EXCHANGE IN THE FINS OF WHALES. P. F. Scholander and William E. Schevill in *Journal of Applied Physiology*, Vol. 8, No. 3, pages 270–282; November, 1955.

THE RABBIT PLACENTA AND THE PROBLEM OF PLACENTAL TRANSMISSION. Harland W. Mossman in *The American Journal of Anatomy*, Vol. 37, No. 3, pages 433–497; July, 1926.

SECRETION OF GASES AGAINST HIGH PRESSURES IN THE SWIM-BLADDER OF DEEP SEA FISHES. II: THE RETE MIRABILE. P. F. Scholander in *The Biological Bulletin*, Vol. 107, No. 2, pages 260–277; October, 1954.

TEMPERATURE OF SKIN IN THE ARCTIC AS A REGULATOR OF HEAT. Laurence Irving and John Krog in *Journal of Applied Physiology*, Vol. 7, No. 4, pages 355–364; January, 1955.

13. Air-Breathing Fishes

AIR BREATHING IN THE TELEOST SYMBRANCHUS MARMORATUS. Kjell Johansen in *Comparative Biochemistry and Physiology*, Vol. 18, No. 2, pages 383–395; June, 1966.

CARDIOVASCULAR DYNAMICS IN THE LUNGFISHES. Kjell Johansen, Claude Lenfant and David Hanson in *Zeitschrift für vergleichende Physiologie*, Vol. 59, No. 2, pages 157–186; June 5, 1968.

GAS EXCHANGE AND CONTROL OF BREATHING IN THE ELECTRIC EEL, ELECTROPHORUS ELECTRICUS. Kjell Johansen, Claude Lenfant, K. Schmidt-Nielsen and J. A. Petersen in *Zeitschrift für vergleichende Physiologie* (in press).

OBSERVATIONS ON THE AFRICAN LUNGFISH PROTOPTERUS AETHIOPICUS, AND ON EVOLUTION FROM WATER TO LAND ENVIRONMENT. Homer W. Smith in *Ecology*, Vol. 12, No. 1, pages 164–181; January, 1931.

RESPIRATORY PROPERTIES OF BLOOD AND PATTERN OF GAS EXCHANGE IN THE LUNGFISH. Claude Lenfant, Kjell Johansen and Gordon C. Grigg in *Respiration Physiology*, Vol. 2, No. 1, pages 1–22; December, 1966–1967.

14. The Lung

THE MECHANISM OF BREATHING. Wallace O. Fenn in *Scientific American*, Vol. 202, No. 1, pages 138–148; January, 1960.

15. How Birds Breathe

BIRD RESPIRATION: FLOW PATTERN IN THE DUCK LUNG. William L. Bretz and Knut Schmidt-Nielsen in *The Journal of Experimental Biology*, Vol. 54, No. 1, pages 103–118; February, 1971.

A PRELIMINARY ALLOMETRIC ANALYSIS OF RESPIRATORY VARIABLES IN RESTING BIRDS. Robert C. Lasiewski and William A. Calder, Jr., in *Respiration Physiology*, Vol. 11, No. 2, pages 152–166; January, 1971.

RESPIRATORY PHYSIOLOGY OF HOUSE SPARROWS IN RELATION TO HIGH-ALTITUDE FLIGHT. Vance A. Tucker in *The Journal of Experimental Biology*, Vol. 48, No. 1, pages 55–66; February, 1968.

STRUCTURAL AND FUNCTIONAL ASPECTS OF THE AVIAN LUNGS AND AIR SACS. A. S. King in *International Review of General and Experimental Zoology: Vol. II*, edited by William J. L. Felts and Richard J. Harrison. Academic Press, Inc., 1964.

16. Surface Tension in the Lungs

MECHANICAL PROPERTIES OF LUNGS. Jere Mead in *Physiological Reviews*, Vol. 41, No. 2, pages 281–330; April, 1961.

THE PHYSICS AND CHEMISTRY OF SURFACES. Neil Kensington Adam. Oxford University Press, 1941,

SOAP-BUBBLES: THEIR COLOURS AND THE FORCES WHICH MOLD THEM. C. V. Boys. Dover Publications, Inc., 1959.

17. The Master Switch of Life

CIRCULATORY ADJUSTMENT IN PEARL DIVERS. P. F. Scholander, H. T. Hammel, H. LeMessurier, E. Hemmingsen and W. Garey in *Journal of Applied Physiology*, Vol. 17, No. 2, pages 184–190; March, 1962.

RESPIRATION IN DIVING MAMMALS. Laurence Irving in *Physiological Reviews*, Vol. 19, No. 1, pages 112–134; January, 1939.

SELECTIVE ISCHEMIA IN DIVING MAN. R. W. Elsner, W. F. Garey and P. F. Scholander in *American Heart Journal*, Vol. 65, No. 4, pages 571–572; April, 1963.

18. The Diving Women of Korea and Japan

THE ISLAND OF THE FISHERWOMEN. Fosco Maraini. Harcourt, Brace & World, Inc., 1962.

KOREAN SEA WOMEN: A STUDY OF THEIR PHYSIOLOGY. The departments of physiology, Yonsei University College of Medicine, Seoul, and the State University of New York at Buffalo.

THE PHYSIOLOGICAL STRESSES OF THE AMA. Hermann Rahn in *Physiology of Breath-Hold Diving and the Ama of Japan*. Publication 1341, National Academy of Sciences—National Research Council, 1965.

19. The Physiology of High Altitude

EFFECTS OF ALTITUDE ON BROWN FAT AND METABOLISM OF THE DEER MOUSE, PEROMYSCUS. Jane C. Roberts, Raymond J. Hock and Robert E. Smith in *Federation Proceedings*, Vol. 28, No. 3, pages 1065–1072; May-June, 1969.

HUMAN ADAPTATIONS TO HIGH ALTITUDE. Paul T. Baker in *Science*, Vol. 163, No. 3872, pages 1149–1156; March 14, 1969.

PHYSIOLOGICAL RESPONSES OF DEER MICE TO VARIOUS NATIVE ALTITUDES. R. J. Hock in *The Physiological Effects of High Altitude: Proceedings of a Symposium Held at Interlaken, September 18–22, 1962*, edited by W. H. Weihe. Pergamon Press, 1964.

20. The Eland and the Oryx

DESERT ANIMALS: PHYSIOLOGICAL PROBLEMS OF HEAT AND WATER. Knut Schmidt-Nielsen. Oxford University Press, 1964.

THE FIRE OF LIFE: AN INTRODUCTION TO ANIMAL ENERGETICS. Max Kleiber. John Wiley & Sons, Inc., 1961.

TERRESTRIAL ANIMALS IN DRY HEAT: UNGULATES. W. V. Macfarlane in *Handbook of Physiology, Section 4: Adaptation to the Environment*. The American Physiological Society, 1964.

21. Salt Glands

THE SALT GLANDS OF THE HERRING GULL. Ragnar Fänge, Knut Schmidt-Nielsen and Humio Osaki in *Biological Bulletin*, Vol. 115, pages 162–171; October, 1958.

SALT GLANDS IN MARINE REPTILES. Knut Schmidt-Nielsen and Ragnar Fänge in *Nature*, Vol. 182, No. 783–785; September 20, 1958.

22. Fishes with Warm Bodies

WARM-BODIED FISH. Francis G. Carey, John M. Teal, John W. Kanwisher, Kenneth D. Lawson and James S. Beckett in *American Zoologist*, Vol. 11, pages 137–145; 1971.

23. How Reptiles Regulate Their Body Temperature

A PRELIMINARY STUDY OF THE THERMAL REQUIREMENTS OF DESERT REPTILES. Raymond Bridgman Cowles and Charles Mitchill Bogert in *Bulletin of the American Museum of Natural History*, Vol. 83, Article 5, pages 265–296; 1944.

RATTLESNAKES: THEIR HABITS, LIFE HISTORIES, AND INFLUENCE ON MANKIND. Laurence M. Klauber. University of California Press, 1956.

TEMPERATURE TOLERANCES IN THE AMERICAN ALLIGATOR AND THEIR BEARING ON THE HABITS, EVOLUTION AND EXTINCTION OF THE DINOSAURS. Edwin H. Colbert, Raymond B. Cowles and Charles M. Bogert in *Bulletin of the American Museum of Natural History*, Vol. 86, Article 7, pages 331–373; 1946.

THERMOREGULATION IN REPTILES—A FACTOR IN EVOLUTION. Charles M. Bogert in *Evolution*, Vol. 3, No. 3, pages 195–211; September, 1949.

24. Adaptations to Cold

BODY INSULATION OF SOME ARCTIC AND TROPICAL MAMMALS AND BIRDS. P. F. Scholander, Vladimir Walters, Raymond Hock and Laurence Irving in *The Biological Bulletin*, Vol. 99, No. 2, pages 225–236; October, 1950.

BODY TEMPERATURES OF ARCTIC AND SUBARCTIC BIRDS AND MAMMALS. Laurence Irving and John Krog in *Journal of Applied Physiology*, Vol. 6, No. 11, pages 667–680; May, 1954.

EFFECT OF TEMPERATURE ON SENSITIVITY OF THE FINGER. Laurence Irving in *Journal of Applied Physiology*, Vol. 18, No. 6, pages 1201–1205; November, 1963.

TERRESTRIAL ANIMALS IN COLD: INTRODUCTION. Laurence Irving in *Handbook of Physiology, Section 4: Adaptation to the Environment*. American Physiological Society, 1964.

25. The Human Thermostat

ACTIVATION OF HEAT LOSS MECHANISMS BY LOCAL HEATING OF THE BRAIN. H. W. Magoun, F. Harrison, J. R. Brobeck and S. W. Ranson in *Journal of Neurophysiology*, Vol. I, No. 2, pages 101–114; March, 1938.

THE RELATION OF THE NERVOUS SYSTEM TO THE TEMPERATURE OF THE BODY. Isaac Ott in *The Journal of Nervous and Mental Disease*, Vol. XI, No. 2, pages 141–152; April, 1884.

THE ROLE OF THE ANTERIOR HYPOTHALAMUS IN TEMPERATURE REGULATION. R. S. Teague and S. W. Ranson in *The American Journal of Physiology*, Vol. 117, No. 3, pages 562–570; November 1, 1936.

26. The Production of Heat by Fat

BROWN ADIPOSE TISSUE AND THE RESPONSE OF NEW-BORN RABBITS TO COLD. M. J. R. Dawkins and D. Hull in *The Journal of Physiology*, Vol. 172, No. 2, pages 216–238; August, 1964.

BROWN FAT: A REVIEW. Bengt Johansson in *Metabolism: Clinical and Experimental*, Vol. 8, No. 3, pages 221–240; May, 1959.

ON THE ACTION OF HORMONES WHICH ACCELERATE THE RATE OF OXYGEN CONSUMPTION AND FATTY ACID RELEASE IN RAT ADIPOSE TISSUE IN VITRO. Eric G. Ball and Robert L. Jungas in *Proceedings of the National Academy of Sciences*, Vol. 47, No. 7, pages 932–941; July, 1961.

THERMOGENESIS OF BROWN ADIPOSE TISSUE IN COLD-ACCLIMATED RATS. Robert E. Smith and Jane C. Roberts in *American Journal of Physiology*, Vol. 206, No. 1, pages 143–148; January, 1964.

27. Desert Ground Squirrels

THE COMPETITIVE EXCLUSION PRINCIPLE. Garrett Hardin in *Science*, Vol. 131, No. 3409, pages 1292–1297; April 29, 1960.

EFFECTS OF SODIUM CHLORIDE ON WEIGHT AND DRINKING IN THE ANTELOPE GROUND SQUIRREL. George A. Bartholomew and Jack W. Hudson in *Journal of Mammalogy*, Vol. 40, No. 3, pages 354–360; August 20, 1959.

HEAT REGULATION IN SOME ARCTIC AND TROPICAL MAMMALS AND BIRDS. P. F. Scholander, Raymond Hock, Vladimir Walters, Fred Johnson and Laurence Irving in *The Biological Bulletin*, Vol. 99, No. 2, pages 237–258; October, 1950.

HIBERNATION. Charles P. Lyman and Paul O. Chatfield in *Scientific American*, Vol. 183, No. 6, pages 18–21; December, 1950.

WATER METABOLISM OF DESERT MAMMALS. Knut Schmidt-Nielsen and Bodil Schmidt-Nielsen in *Physiological Reviews*, Vol. 32, No. 2, pages 135–166; April, 1952.

28. The Homing Salmon

HOMING INSTINCT IN SALMON. Bradley T. Scheer in *The Quarterly Review of Biology*, Vol. 14, No. 4, pages 408–430; December, 1939.

SENSORY PHYSIOLOGY AND THE ORIENTATION OF ANIMALS. Donald R. Griffin in *American Scientist*, Vol. 41, No. 2, pages 209–244; April, 1953.

29. Electric Location by Fishes

ECOLOGICAL STUDIES ON GYMNOTIDS. H. W. Lissmann in *Bioelectrogenesis: A Comparative Survey of its Mechanisms with Particular Emphasis on Electric Fishes*. American Elsevier Publishing Co., Inc., 1961.

THE MECHANISM OF OBJECT LOCATION IN GYMNARCHUS NILOTICUS AND SIMILAR FISH. H. W. Lissmann and K. E. Machin in *Journal of Experimental Biology*, Vol. 35, No. 2, pages 451–486; June, 1958.

THE MODE OF OPERATION OF THE ELECTRIC RECEPTORS IN GYMNARCHUS NILOTICUS. K. E. Machin and H. W. Lissmann in *Journal of Experimental Biology*, Vol. 37, No. 4, pages 801–811; December, 1960.

30. The Infrared Receptors of Snakes

MEN AND SNAKES. Ramona and Desmond Morris. Hutchinson of London, 1965.

THE PIT ORGANS OF SNAKES. Robert Barrett in *Biology of the Reptilia-Morphology B: Vol. II*. Academic Press, 1970.

PROPERTIES OF AN INFRA-RED RECEPTOR. T. H. Bullock and F. P. J. Diecke in *The Journal of Physiology*, Vol. 134, No. 1, pages 47–87; October 29, 1956.

RADIANT HEAT RECEPTION IN SNAKES. T. H. Bullock and R. Barrett in *Communication in Behavioral Biology*, Part A, Vol. 1, pages 19–29; January, 1968.

SNAKE INFRARED RECEPTORS: THERMAL OR PHOTOCHEMICAL MECHANISM? John F. Harris and R. Igor Gamow in *Science*, Vol. 172, No. 3989, pages 1252–1253; June 18, 1971.

31. More about Bat "Radar"

BATS. Glover Morrill Allen. Harvard University Press, 1939.

BIRD SONAR. Donald R. Griffin in *Scientific American*, Vol. 190, No. 3, pages 78–83; March, 1954.

LISTENING IN THE DARK: THE ACOUSTIC ORIENTATION OF BATS AND MEN. Donald R. Griffin. Yale University Press, 1958.

32. The Navigation of Penguins

CURRENT PROBLEMS IN BIRD ORIENTATION. Klaus Schmidt-Koenig in *Advances in the Study of Behavior: Vol. I*, edited by Daniel S. Lehrman, Robert A. Hinde and Evelyn Shaw. Academic Press, 1965.

LONG-DISTANCE ORIENTATION. G. Kramer in *Biology and Comparative Physiology of Birds: Vol. II*, edited by A. J. Marshall. Academic Press, 1961.

33. The Hormones of the Hypothalamus

CHARACTERIZATION OF OVINE HYPOTHALAMIC HYPOPHYSIOTROPIC TSH-RELEASING FACTOR. Roger Burgus, Thomas F. Dunn, Dominic Desiderio, Darrell N. Ward, Wylie Vale and Roger Guillemin in *Nature*, Vol. 226, No. 5243, pages 321–325; April 25, 1970.

THE HYPOTHALAMUS: PROCEEDINGS OF THE WORKSHOP CONFERENCE ON INTEGRATION OF ENDOCRINE AND NON ENDOCRINE MECHANISMS IN THE HYPOTHALAMUS. Edited by L. Martini, M. Motta and F. Fraschini. Academic Press, 1970.

STRUCTURE OF THE PORCINE LH- AND FSH-RELEASING HORMONE, I: THE PROPOSED AMINO ACID SEQUENCE. H. Matsuo, Y. Baba, R. M. G. Nair, A. Arimura and A. V. Schally in *Biochemical and Biophysical Research Communications*, Vol. 43, No. 6, pages 1334–1339; June 18, 1971.

SYNTHETIC POLTPEPTIDE ANTAGONISTS OF THE HYPOTHALAMIC LUTEINIZING HORMONE RELEASING FACTOR. Wylie Vale, Geoffrey Grant, Jean Rivier, Michael Monahan, Max Amoss, Richard Blackwell, Roger Burgus and Roger Guillemin in *Science*, Vol. 176, No. 4037, pages 933–934; May 26, 1972.

SYPTHETIC LUTEINIZING HORMONE-RE-LEASING FACTOR: A POTENT STIMULATOR OF GONADOTROPIN RELEASE IN MAN. S. S. C. Yen, R. Rebar, G. VandenBerg, F. Naftolin, Y. Ehara, S. Engblom, K. J. Ryan, K. Benirschke, J. Rivier, M. Amoss and R. Guillemin in *The Journal of Clinical Endocrinology and Metabolism*, Vol. 34, No. 6, pages 1108–1111; June, 1972.

34. The Pineal Gland

MELATONIN SYNTHESIS IN THE PINEAL GLAND: EFFECT OF LIGHT MEDIATED BY THE SYMPATHETIC NERVOUS SYSTEM. Richard J. Wurtman, Julius Axelrod and Josef E. Fischer in *Science*, Vol. 143, No. 3612, pages 1328–1329; March 20, 1964.

STRUCTURE AND FUNCTION OF THE EPIPHYSIS CEREBRI. Edited by J. Ariëns Kappers and J. P. Schadé in *Progress in Brain Research*, Vol. X. Elsevier Publishing Company, 1965.

35. Calcitonin

THE AMINO ACID SEQUENCE OF PORCINE THYROCALCITONIN. J. T. Potts, Jr., H. D. Niall, H. T. Keutmann, H. B. Brewer, Jr., and L. J. Deftos in *Proceedings of the National Academy of Sciences*, Vol. 59, No. 4, pages 1321–1328; April 15, 1968.

CALCITONIN FROM ULTIMOBRANCHIAL GLANDS OF DOGFISH AND CHICKENS. D. H. Copp, D. W. Cockcroft and Yankoon Kueh in *Science*, Vol. 158, No. 3803, pages 924–925; November 17, 1967.

SYMPOSIUM ON THYROCALCITONIN. Edited by Maurice M. Pechet in *The American Journal of Medicine*, Vol. 43, No. 5, pages 645–726; November, 1967.

36. How an Eggshell is Made

CALCIFICATION AND OSSIFICATION. MEDULLARY BONE CHANGES IN THE REPRODUCTIVE CYCLE OF FEMALE PIGEONS. William Bloom, Margaret A. Bloom and Franklin C. McLean in *The Anatomical Record*, Vol. 81, No. 4, pages 443–475; December 26, 1941.

CALCIUM METABOLISM AND AVIAN REPRODUCTION. K. Simkiss in *Biological Reviews*, Vol. 36, No. 3, pages 321–367; August, 1961.

THE EFFECT OF PITUITARY HORMONES ON OVULATION IN CALCIUM-DEFICIENT PULLETS, T. G. Taylor, T. R. Morris and F. Hertelendy in *The Veterinary Record*, Vol. 74, No. 4, pages 123–125; January 27, 1962.

EGGSHELL FORMATION AND SKELETAL METABOLISM. T. G. Taylor and D. A. Stringer in *Avian Physiology*, edited by Paul D. Sturkie. Comstock Publishing Associates, 1965.

37. The Chemical Languages of Fish

CHEMICAL COMMUNICATION. John Ebling and Kenneth C. Highnam. Edward Arnold (Publishers) Ltd., 1969.

CHEMICAL COMMUNICATION IN FISH. John E. Bardach and John H. Todd in *Advances in Chemoreception, Vol. I: Communication by Chemical Signals*, edited by James W. Johnston, David G. Moulton and Amos Turk. Appleton-Century-Crofts, 1970.

OLFACTION AND BEHAVIORAL SOPHISTICATION IN FISH. J. Atema, J. H. Todd and J. E. Bardach in *Olfaction and Taste: Proceedings of the Third International Symposium*, edited by Carl Pfaffmann. The Rockefeller University Press, 1969.

38. Territorial Marking by Rabbits

ANIMAL DISPERSION IN RELATION TO SOCIAL BEHAVIOR. V. C. Wynne-Edwards. Hafner Publishing Company, 1962.

FURTHER OBSERVATIONS ON THE TERRITORIAL FUNCTION AND HISTOLOGY OF THE SUBMANDIBULAR CUTANEOUS (CHIN) GLANDS IN THE RABBIT, ORYCTOLAGUS CUNICULUS (L.). R. Mykytowycz in *Animal Behavior*, Vol. 13, No. 4, pages 400–412; October, 1965.

OBSERVATIONS ON ODORIFEROUS AND OTHER GLANDS IN THE AUSTRALIAN WILD RABBIT, ORYCTOLAGUS CUNICULUS (L.), AND THE HARE, LEPUS EUROPAEUS P, I: THE ANAL GLAND. R. Mykytowycz in *CSIRO Wildlife Research*, Vol. 11, No. 1, pages 11–29; October, 1966.

PHEROMONES. Edward O. Wilson in *Scientific American*, Vol. 208, No. 5, pages 100–114; May, 1963.

39. How Birds Sing

BIRD SONG: ACOUSTICS AND PHYSIOLOGY. Crawford H. Greenewalt. Smithsonian Institution Press, 1968.

INDEX